PREALGEBRA: MEDIA ENHANCED EDITION

edition **3**

PREALGEBRA: MEDIA ENHANCED EDITION

Stefan Baratto
Clackamas Community College

Barry Bergman
Clackamas Community College

Donald Hutchison
Clackamas Community College

 Higher Education

Boston   Burr Ridge, IL   Dubuque, IA   New York   San Francisco   St. Louis
Bangkok   Bogotá   Caracas   Kuala Lumpur   Lisbon   London   Madrid   Mexico City
Milan   Montreal   New Delhi   Santiago   Seoul   Singapore   Sydney   Taipei   Toronto

Higher Education

PREALGEBRA: MEDIA ENHANCED EDITION, THIRD EDITION

Published by McGraw-Hill, a business unit of The McGraw-Hill Companies, Inc., 1221 Avenue of the Americas, New York, NY 10020.

ISBN 978–0–07–340623–7
MHID 0–07–340623–6

ISBN 978–0–07–335782–9 (Annotated Instructor's Edition)
MHID 0–07–335782–0

Editorial Director: *Stewart K. Mattson*
Senior Sponsoring Editor: *Richard Kolasa*
Director of Development: *Kristine Tibbetts*
Senior Developmental Editor: *Michelle L. Flomenhoft*
Marketing Manager: *Victoria Anderson*
Lead Project Manager: *Peggy J. Selle*
Senior Production Supervisor: *Kara Kudronowicz*
Senior Media Project Manager: *Sandra M. Schnee*
Senior Designer: *David W. Hash*
Cover Designer: *Ellen Pettergell*
(USE) Cover Image: *Tate Modern and Millennium Bridge,* © *Jon Arnold Images/Superstock*
Senior Photo Research Coordinator: *Lori Hancock*
Supplement Producer: *Mary Jane Lampe*
Compositor: *ICC Macmillan Inc.*
Typeface: *10/12 Times Roman*
Printer: *R. R. Donnelley, Jefferson City, MO*

Design Element (man and woman at computer): © Getty RF; Chapter One Opener: © Corbis RF; p. 64: © Getty RF; Chapter Two, Three Openers: © Corbis RF; p. 234: © Getty RF; Chapter Four Opener: © Brand X/Jupiter RF; p. 365: © Getty RF; Chapter Five Opener: © Corbis RF; p. 477, 480: © Getty RF; Chapter Six Opener: © Getty RF; p. 505: © Getty RF; p. 560: © Stockbyte/Punchstock; Chapter Seven Opener: © BananaStck/PictureQuest RF; p. 612: © Getty RF; Chapter Eight Opener: © Corbis RF; p. 636: © Corbis RF; p. 673: © Getty RF; Chapter Nine Opener: © Corbis RF; p. 724, 766, 778: © Getty RF; Chapter Ten Opener: © Pixtal/age Fotostock; p. 823: © National Trust Photographic Library/Terry Cotterill/The Image Works.

Library of Congress Cataloging-in-Publication Data

Hutchison, Donald, 1948–
  Prealgebra : media enhanced / Donald Hutchison, Barry Bergman, Stefan Baratto. — 3rd ed.
    p. cm.
  Includes index.
  ISBN 978–0–07–340623–7 — ISBN 0–07–340623–6 (hard copy : alk. paper) 1. Algebra—Textbooks. 2. Equations—Textbooks. I. Baratto, Stefan. II. Bergman, Barry. III. Title.
  QA152.3.H877 2010
  512—dc22

                                                                                    2008040443

www.mhhe.com

About the Authors

Stefan Baratto   Stefan began teaching math and science in New York City middle schools in 1989. He went on to teach math at the University of Oregon, Southeast Missouri State University, and York County Technical College (in Maine). Currently, Stefan is the faculty chair of the Clackamas Community College math department and has found his niche there, delighting in the CCC faculty, staff, and student body.

Stefan's own education includes the University of Michigan (BGS, 1988), Brooklyn College (CUNY), and the University of Oregon (MS, 1996).

Stefan is currently serving on the AMATYC Executive Board as the organization's Northwest Vice President. He has also been involved with ORMATYC, NEMATYC, NCTM, and the State of Oregon Math Chairs group, as well as other local organizations.

On a more personal note, Stefan and his wife, Peggy, try to spend their leisure time enjoying the many wonders to be found in Oregon and the Pacific Northwest.

Barry Bergman   Barry has enjoyed teaching mathematics to a wide variety of students over the years. He began in the field of adult basic education, and moved into the teaching of high school mathematics in 1977. He taught at that level for 11 years, at which point he served as a K-12 mathematics specialist for his county. This work allowed him the opportunity to help promote the emerging NCTM Standards in his region.

In 1990 Barry began the next part of his career, having been hired to teach at Clackamas Community College. He maintains a strong interest in the appropriate use of technology and visual models in the learning of mathematics.

Throughout the past 30 years, Barry has played an active role in professional organizations. As a member of OCTM, he contributed several articles and activities to the group's journal. He has made presentations at AMATYC, ORMATYC, ICTCM, NCTM, and OCTM conferences. Barry also served as an officer of ORMATYC for four years and participated on an AMATYC committee to provide feedback to revisions of NCTM's Standards.

Don Hutchison   Don began teaching in a pre-school while he was an undergraduate. He subsequently taught children with disabilities, adults with disabilities, high school mathematics, and college mathematics. Although each of these positions offered different challenges and a different kind of satisfaction, it was always the process of breaking a lesson into teachable components that he most enjoyed.

It was at Clackamas Community College that he found his professional niche. The community college allowed him to focus on teaching within a department that constantly challenged faculty and students to expect more. Under the guidance of Jim Streeter, Don learned to present his approach to teaching in the form of a textbook.

Don has been an active member of many professional organizations. He has been President of ORMATYC, AMATYC committee chair, and ACM curriculum committee member. He has presented at AMATYC, ORMATYC, AACC, MAA, ICTCM, and numerous other conferences.

Above all, he encourages you to be involved, whether as a teacher or as a learner. Whether discussing curricula at a professional meeting or homework in the cafeteria, it is the process of communicating an idea that helps one to clarify it.

Dedications

Stefan Baratto
To my beautiful wife, Peggy, as we embark on yet another wonderful adventure.

Barry Bergman
To my beautiful wife, Marcia, and to my two sons, Joel and Adam, who have developed into fine young men.

Don Hutchison
Thank you to every teacher and every student I have ever encountered. You are each part of this text.

To the Student

You are about to begin a course in mathematics. We made every attempt to provide a text that will help you understand what mathematics is about and how to use it effectively. We made no assumptions about your previous experience with mathematics. Your progress through the course will depend on the amount of time and effort you devote to the course and your previous background in math. There are some specific features in this book that will aid you in your studies. Here are some suggestions about how to use this book. (Keep in mind that a review of *all* the chapter and summary material will further enhance your ability to grasp later topics and to move more effectively through the text.)

1. If you are in a lecture class, make sure that you take the time to read the appropriate text section *before* your instructor's lecture on the subject. Then take careful notes on the examples that your instructor presents during class.

2. After class, work through similar examples in the text, making sure that you understand each of the steps shown. Examples are followed in the text by *Check Yourself* exercises. You can best learn mathematics by being involved in the process, and that is the purpose of these exercises. Always have a pencil and paper at hand, work out the problems presented, and check your results immediately. If you have difficulty, go back and carefully review the previous exercises. Make sure you understand what you are doing and why. The best test of whether you do understand a concept lies in your ability to explain that concept to one of your classmates. Try working together.

3. At the end of each chapter section you will find a set of exercises. Work these carefully to check your progress on the section you have just finished. You will find the solutions for the odd-numbered exercises following the problem set. If you have difficulties with any of the exercises, review the appropriate parts of the chapter section. If your questions are not completely cleared up, by all means do not become discouraged. Ask your instructor or an available tutor for further assistance. A word of caution: Work the exercises on a regular (preferably daily) basis. Again, learning mathematics requires becoming involved. As is the case with learning any skill, the main ingredient is practice.

4. When you complete a chapter, review by using the *Summary.* You will find all the important terms and definitions in this section, along with examples illustrating all the techniques developed in the chapter. Following the *Summary* are *Summary Exercises* for further practice. The exercises are keyed to chapter sections, so you will know where to turn if you are still having problems.

5. When you finish with the *Summary Exercises,* try the *Self-Test* that appears at the end of each chapter. It is an actual practice test you can work on as you review for in-class testing. Again, answers with section references are provided.

6. Finally, an important element of success in studying mathematics is the process of regular review. We provide a series of *Cumulative Reviews* throughout the textbook, beginning at the end of Chapter 2. These tests will help you review not only the concepts of the chapter that you have just completed but also those of previous chapters. Use these tests in preparation for any midterm or final exams. If it appears that you have forgotten some concepts that are being tested, don't worry. Go back and review the sections where the idea was initially explained or the appropriate chapter *Summary*. That is the purpose of the *Cumulative Review.*

We hope that you will find our suggestions helpful as you work through this material, and we wish you the best of luck in the course.

Stefan Baratto
Barry Bergman
Donald Hutchison

Contents

PREFACE   x

Chapter 1

WHOLE NUMBERS   1

Pretest Chapter 1   2
1.1 Introduction to Whole Numbers and Place Value   3
1.2 Addition of Whole Numbers   11
1.3 Subtraction of Whole Numbers   28
1.4 Rounding, Estimation, and Ordering of Whole Numbers   41
1.5 Multiplication of Whole Numbers   50
1.6 Division of Whole Numbers   71
1.7 Exponents and Whole Numbers   88
1.8 Grouping Symbols and the Order of Operations   94
1.9 An Introduction to Equations   102

Activity 1: Wrapping Gifts   107
Summary   109
Summary Exercises   113
Self-Test for Chapter 1   117

Chapter 2

INTEGERS AND INTRODUCTION TO ALGEBRA   119

Pretest Chapter 2   120
2.1 Introduction to Integers   121
2.2 Addition of Integers   132
2.3 Subtraction of Integers   139
2.4 Multiplication of Integers   145
2.5 Division of Integers   151
2.6 Introduction to Algebra: Variables and Expressions   158
2.7 Evaluating Algebraic Expressions   167
2.8 Simplifying Algebraic Expressions   177
2.9 Introduction to Linear Equations   186
2.10 The Addition Property of Equality   192

Activity 2: Charting Temperatures   201
Summary   203
Summary Exercises   205
Self-Test for Chapter 2   209
Cumulative Review for Chapters 1 and 2   211

Chapter 3

FRACTIONS AND EQUATIONS   213

Pretest Chapter 3   214
3.1 Introduction to Fractions   215
3.2 Prime Numbers and Factorization   223
3.3 Equivalent Fractions   235
3.4 Multiplication and Division of Fractions   250
3.5 The Multiplication Property of Equality   264
3.6 Linear Equations in One Variable   272

Activity 3: Fractions in Diet and Exercise   281
Summary   283
Summary Exercises   285
Self-Test for Chapter 3   287
Cumulative Review for Chapters 1 to 3   289

Chapter 4

APPLICATIONS OF FRACTIONS AND EQUATIONS   291

Pretest Chapter 4   292
4.1 Addition and Subtraction of Fractions   293
4.2 Operations on Mixed Numbers   313
4.3 Applications Involving Fractions   330
4.4 Equations Containing Fractions   351
4.5 Applications of Linear Equations in One Variable   360
4.6 Complex Fractions   373

Activity 4: Home Remodeling   378
Summary   379

Summary Exercises   381
Self-Test for Chapter 4   385
Cumulative Review for
Chapters 1 to 4   387

Chapter 5

DECIMALS   **389**

Pretest Chapter 5   390
5.1 Introduction to Decimals,
    Place Value, and Rounding   391
5.2 Addition and Subtraction
    of Decimals   403
5.3 Multiplication of Decimals   414
5.4 Division of Decimals   423
5.5 Fractions and Decimals   434
5.6 Equations Containing
    Decimals   446
5.7 Square Roots and the
    Pythagorean Theorem   451
5.8 Applications   463

Activity 5: Statistics in Sports   483
Summary   485
Summary Exercises   487
Self-Test for Chapter 5   491
Cumulative Review for
Chapters 1 to 5   495

Chapter 6

RATIO, RATE, AND PROPORTION   **497**

Pretest Chapter 6   498
6.1 Ratios   499
6.2 Rates   507
6.3 Proportions   517
6.4 Similar Triangles and
    Proportions   537
6.5 Linear Measurement and
    Conversion   545

Activity 6: Measurements   560
Summary   561
Summary Exercises   563
Self-Test for Chapter 6   567
Cumulative Review for
Chapters 1 to 6   569

Chapter 7

PERCENT   **571**

Pretest Chapter 7   572
7.1 Percents, Decimals,
    and Fractions   573
7.2 Solving Percent Problems
    Using Proportions   588

7.3 Solving Percent Applications
    Using Equations   597
7.4 Applications: Simple
    and Compound Interest   607
7.5 More Applications of
    Percent   612

Activity 7: Population Changes   622
Summary   623
Summary Exercises   625
Self-Test for Chapter 7   629
Cumulative Review for
Chapters 1 to 7   631

Chapter 8

GEOMETRY   **633**

Pretest Chapter 8   634
8.1 Lines and Angles   635
8.2 Perimeter and Circumference   652
8.3 Area and Volume   662

Activity 8: Norman Windows   678
Summary   679
Summary Exercises   681
Self-Test for Chapter 8   683
Cumulative Review for
Chapters 1 to 8   685

Chapter 9

**GRAPHING AND INTRODUCTION
TO STATISTICS**   **689**

Pretest Chapter 9   690
9.1 Tables and Graphs of Data   693
9.2 The Rectangular Coordinate
    System   715
9.3 Linear Equations in Two
    Variables   729
9.4 Mean, Median, and Mode   753

Activity 9: Graphs in the
Media   766
Summary   767
Summary Exercises   769
Self-Test for Chapter 9   775
Cumulative Review for
Chapters 1 to 9   779

Chapter 10

POLYNOMIALS   **781**

Pretest Chapter 10   782
10.1 Properties of Exponents   783
10.2 Introduction to Polynomials   790

10.3 Addition and Subtraction
of Polynomials   797
10.4 Multiplying Polynomials   807
10.5 Introduction to Factoring
Polynomials   817

Activity 10: The Gravity
Model   823
Summary   825
Summary Exercises   827
Self-Test for Chapter 10   829
Practice Final Exam   831
Practice Final Exam Answers   837

APPENDIX A **THE INTERNET**   **838**
APPENDIX B **CIRCLE GRAPHS**   **840**
APPENDIX C **READING YOUR TEXT**
ANSWERS   **847**

ANSWERS TO PRETESTS, SUMMARY
EXERCISES, SELF-TESTS, AND
CUMULATIVE REVIEWS   **849**
INDEX   **857**

Preface

Message from the Authors

We believe the key to learning mathematics, at any level, is active participation—**MASTERING MATH THROUGH ACTIVE LEARNING.** Students who are active participants in the learning process have the opportunity to construct their own mathematical ideas and make connections to previously studied material. Such participation leads to understanding, retention, success, and confidence. We developed this text with that philosophy in mind and integrated many features throughout the book to reflect that philosophy. Our goal is to provide *content in context.* The opening vignette for each chapter, the activity in that chapter, and several section exercises all relate to the same topic in order to engage students and allow them to see the relevance of mathematics.

The *Check Yourself* exercises are designed to keep the students active and involved with every page of exposition. The optional calculator references involve students actively in the development of mathematical ideas. Almost every exercise set has application problems, challenging exercises, writing exercises, and/or collaborative exercises. Answer blanks for these exercises appear in the margin allowing the student to actively use the text, making it more than just a reference tool. Many of these exercises are designed to awaken interest and insight in students: all are meant to provide continual practice and reinforcement of the topics being learned. Not all of the exercises will be appropriate for every student, but each one gives another opportunity for both the instructor and the student. Our hope is that every student who uses this text will be a better mathematical thinker as a result.

As we developed the third edition of our *Prealgebra: Media Enhanced Edition,* we recognized that the use of technology was no longer just an optional supplement to our text, but an essential element. We recognized that today's students are learning in different modes besides attending class lectures, and that their daily schedules are pulling them in more directions than ever before. To address these different learning styles, we have further integrated videos associated with each and every section of the text—an icon is included next to those exercises that have a video. The goal of these videos is to provide students with a better framework—showing them how to solve a particular mathematical topic, no matter if they were taking this course online, or needed a refresher to review their lecture. In addition, we are now providing the videos in several formats, including iPOD/MP3 format, to accommodate the different ways students access information.

Finally, *Prealgebra: Media Enhanced Edition* focuses on **mastering math through active learning** with the integration of ALEKS®. ALEKS® icons accompany exercises in the text where a similar problem is available in ALEKS helping students to master topics when they are in need of remediation. ALEKS provides students with a map (pictorial graph) of their progress to identify mathematical skills they have mastered and skills where remediation is required.

Thank you for using our text, and we look forward to hearing about your success!

Stefan Baratto
Clackamas Community College

Barry Bergman
Clackamas Community College

Donald Hutchison
Clackamas Community College

Mastering Mathematics through Active Learning

In July 1996, Foster and Partners/Sir Anthony Caro/Ove Aru & Partners, proposed their innovative design for the Millenium Bridge which now spans the River Thames in London, England, linking the city of London at St. Paul's Cathedral and the Tate Modern Museum. This innovative design pushed the bounds of structural bridge design and has made the Millenium Bridge one of England's largest tourist attractions. This award winning architectural structure could not have been built without the use of today's modern technology. Likewise, *Prealgebra: Media Enhanced Edition, 3e* constitutes not only decades of work of experienced authors, but also the careful integration of some of the most developed mathematical technology available. Our reviewers have helped us build *Prealgebra* into a text that not only meets instructors' needs, but also presents mathematical concepts in such a manner that every math student can understand and master those concepts. The hallmarks of *Prealgebra: Media Enhanced Edition* include: 1) a building-block approach in presenting mathematical concepts; 2) a step-by-step presentation on how to solve examples, exercises and applications; and 3) a true integration of state-of-the art technology using ALEKS and videos for every section of the text. These hallmark all help to reinforce the authors' philosophy of *mastering math through active learning.*

The development of *Prealgebra* began over 15 years ago when Jim Streeter, along with the help of Don Hutchison, began writing developmental math texts to meet the growing need of struggling students across the country. With feedback from teachers around the country, this series, now led by Stefan Baratto and Barry Bergman, has become one of the most recognizable series in math. With hundreds of reviews taking place every year, and numerous focus groups, the author team has developed what many consider to be one of the most comprehensive and student-friendly Prealgebra texts on the market. The book has been greatly received by faculty all over the country. We feel confident that with the integration of our new technology that instructors and students will experience even greater success.

Besides the technology integration, another factor that differentiates this book from others on the market is that it utilizes an integrated equations approach, which pairs arithmetic concepts alongside corresponding algebraic concepts. Beginning in Chapter 1, students are gradually exposed to key algebraic concepts, such as variables and equations. In this way, students gradually build their confidence in dealing with basic algebra concepts and are better prepared for an introductory algebra course. After signed numbers, fractions, and decimals are introduced, they are used frequently in examples, exercises, and applications. Having this material should allow students to become more comfortable seeing and using these items in algebra and in life. This approach was implemented directly due to key market feedback.

Prealgebra: Media Enhanced Edition, 3e was designed to not only be a textbook, but also a complete resource for your students. Our online content, from videos to algorithmic problems keyed to specific topics within the text, will provide you and your students with one of the best learning packages available to mathematics students in the market.

The Hallmarks of *Prealgebra*

Prealgebra: Media Enhanced Edition, 3e is built around the following attributes:

- Building-Block Approach to Learning: The author team's philosophy is that students need to learn and master the basics before moving on to more difficult concepts—*mastering math through active learning.* Without this foundation, students will never truly understand the "why" behind solving a mathematical equation. In *Prealgebra: Media Enhanced Edition,* the authors have made a conscious effort to build on these basic foundations before introducing more difficult algebraic concepts.

  In terms of pedagogical tools, the book seeks to provide carefully detailed explanations and accessible pedagogy to introduce Prealgebra concepts to the students. The authors use a three-pronged approach to present the material and encourage critical thinking skills. The areas used to create the framework are communication, pattern recognition, and problem solving.

- Step-by-Step Presentation: Because many students struggle with understanding mathematics, the authors have integrated a step-by-step process to solving problems. This problem-solving technique is used consistently throughout all of their books. Items such as *Math Anxiety* boxes, *Check Yourself* exercises, and activities represent this presentation, and the underlying philosophy of mastering math through active learning.

- Technology Integration: In this edition, much effort has gone into properly integrating our homework system, which allows students to perform assignments keyed to the text, and receive immediate feedback. Videos have also been better integrated to help reinforce what the student is learning. Several different icons, such as the video and ALEKS icons, are used in the end-of-section exercises to designate where students can receive additional help. References to the book's website are also included in each set of end-of section exercises to consistently remind the student of various tools available to them.

What Stands out in the Third Edition?

Media Enhanced Edition integrated with ALEKS

We have integrated ALEKS throughout the text to identify topics where students can use ALEKS to remediate. Students have two modes of studying prealgebra using ALEKS: 1) students can use ALEKS in its current model by taking an assessment and learning mathematics at their own pace; and 2) students can use ALEKS as they progress through their course by chapter, using the artificial intelligence offered by ALEKS to master topics in the text.

New Media Integration with Videos

The videos that accompany the text present professors who work through selected exercises, and follow the solution methodology employed in the text. These are designated with marginal icons throughout the text for easy reference by the student. The videos can be viewed freely via the text website, downloaded to a computer, or viewed on an iPod/MP3 player.

New Applications

Developed out of information provided from focus groups and reviews, new applications were written in response to instructor feedback about the demographic of their student body, and the need for applications to relate to students' majors. Many of these applications relate to allied health, construction, and business fields.

What Keeps *Prealgebra* Users Coming Back?

Chapter-Opening Vignettes provide students interesting, relevant scenarios that will capture their attention and engage them in the upcoming material. Furthermore, exercises and chapter *Activities* related to the *Chapter-Opening Vignette* are included in each chapter.

Chapter Activities are included in each chapter. The *Activities* promote active learning by requiring students to find, interpret, and manipulate real-world data. Each *Activity* appearing before the chapter summary relates to the chapter-opening vignette, providing closure to the chapter. Students can complete the *Activities* on their own, but these are best solved in small groups.

---

chapter

2

INTEGERS AND INTRODUCTION TO ALGEBRA

CHAPTER 2 OUTLINE

Section 2.1   Introduction to Integers   page 121
Section 2.2   Addition of Integers   page 132
Section 2.3   Subtraction of Integers   page 139
Section 2.4   Multiplication of Integers   page 145
Section 2.5   Division of Integers   page 151
Section 2.6   Introduction to Algebra: Variables and Expressions   page 158
Section 2.7   Evaluating Algebraic Expressions   page 167
Section 2.8   Simplifying Algebraic Expressions   page 177
Section 2.9   Introduction to Linear Equations   page 186
Section 2.10   The Addition Property of Equality   page 192

INTRODUCTION

Of all the sciences, meteorology may be both the least precise and the most talked about. Meteorologists study the weather and climate. Together with geologists, they provide us with many predictions that play a part in our everyday decisions. Among the things we might look for are temperatures, rainfall, water level, and tide level.

On a day with unusual weather, we often become curious about the record for that day or even the all-time record. Here are a few of those records:

- The record high temperature for the United States is 134°F in 1913 in Death Valley, California.
- The record low temperature for the U.S. is –80°F in 1971 in Prospect Creek Camp, Alaska.
- The greatest one day temperature drop in the U.S. happened on Christmas Eve, 1924, in Montana. The temperature went from 63°F during the day to –21°F at night.

The use of negative numbers for temperatures below zero is common for such numbers, but there are many other applications. For example, the tide range in Delaware varies between +10 feet and –5 feet. We examine these and several other climate-related applications in Sections 2.1, 2.2, 2.3, and 2.7.

© 2010 McGraw-Hill Companies

119

ACTIVITY 7: POPULATION CHANGES

Each chapter in this text includes an activity. The activity is related to the vignette you encountered in the chapter opening. The activity provides you with the opportunity to apply the math you studied in the chapter to a relevant topic.

Your instructor will determine how best to use this activity in your instructional setting. You may find yourself working in class or outside of class; you may find yourself working alone or in small groups; or you may even be asked to perform research in a library or on the Internet.

The table below gives the population for the United States and each of the six largest states from both the 1990 and 2000 census. Use this table to answer the questions that follow. Round all computations of percents to the nearest tenth of a percent.

	1990 Population	2000 Population
United States	248,709,873	281,421,906
California	29,760,021	33,871,648
Texas	16,986,510	20,851,820

READING YOUR TEXT

The following fill-in-the-blank exercises are designed to assure that you understand the key vocabulary used in this section. Each sentence comes directly from the section. You will find the correct answers in Appendix C.

Section 5.2

(a) The first step in adding decimals is to write the numbers so that their decimal _____ align vertically.

(b) In adding decimals, the numbers may not have the same number of decimal _____.

(c) Because each place value is one-tenth of the value of the place to its left, _____, when subtracting decimals, works just as it did in subtracting whole numbers.

(d) We use the calculator sections in this text to show you how the calculator can be helpful as a _____.

The ***Reading Your Text*** feature is a set of quick exercises presented at the end of each section meant to quiz students' vocabulary knowledge. These exercises are designed to encourage careful reading of the text. If students do not understand the vocabulary, they cannot communicate effectively. The *Reading Your Text* exercises address the vocabulary issue which many students struggle with in learning and understanding mathematics. Answers to these exercises are provided at the end of the book.

Recall Notes are included to provide students with references to previously-learned material, when relevant. *Recall Notes* help students relate topics that they might otherwise feel are disconnected. They also allow students to see how the math builds upon itself.

Learning Objective References within each section are included with the examples when a new objective is covered. *Learning Objective References* are also included with the answers to Pre-Tests and the end-of-chapter exercises. These will help students quickly identify examples related to topics where they need more practice. In the Annotated Instructor's Edition, *Learning Objective References* appear in the exercise sets to facilitate the creation of homework assignments.

The ***conversational, friendly, non-threatening writing style*** seeks to make this the most accessible prealgebra book your students could own. Developmental math students who have reading difficulties will not be intimidated by the language, so they can focus on the math.

12.057 is read as "twelve and fifty-seven thousandths."

The rightmost digit, 7, is in the thousandths position.

Thousandths

NOTE An informal way of reading decimals is to simply read the digits in order and use the word *point* to indicate the decimal point. 2.58 can be read "two point five eight." 0.689 can be read "zero point six eight nine."

0.5321 is read as "five thousand three hundred twenty-one ten thousandths."

When the decimal has no whole-number part, we have chosen to write a 0 to the left of the decimal point. This simply makes sure that you don't miss the decimal point. However, both 0.5321 and .5321 are correct.

CHECK YOURSELF 4

Write 2.58 in words.

NOTE The number of digits to the right of the decimal point is called the number of **decimal places** in a decimal number. So, 0.35 has two decimal places.

One quick way to write a decimal as a common fraction is to remember that the number of decimal places must be the same as the number of zeros in the denominator of the common fraction.

OBJECTIVE 3

Example 5   Writing a Decimal Number as a Mixed Number

Write each decimal as a common fraction or mixed number.

$$0.35 = \frac{35}{100}$$

Two places   Two zeros

This fraction can then be simplified to $\frac{7}{20}$.

The same method can be used with decimals that are greater than 1. Here the result will be a mixed number.

NOTE The 0 to the right of the decimal point is a **placeholder** that is not needed in the common-fraction form.

$$2.058 = 2\frac{58}{1,000}$$

Three places   Three zeros

This mixed number can be simplified to $2\frac{29}{500}$.

CHECK YOURSELF 5

Write as common fractions or mixed numbers. Do not simplify.

(a) 0.528   **(b)** 5.08

RECALL By the Fundamental Principle of Fractions, multiplying the numerator and denominator of a fraction by the same nonzero number does not change the value of the fraction.

It is often useful to compare the sizes of two decimal fractions. One approach to comparing decimals uses this fact: Writing zeros to the right of the rightmost digit *does not change* the value of a decimal. 0.53 is the same as 0.530. Look at the fractional form:

$$\frac{53}{100} = \frac{530}{1,000}$$

The fractions are equivalent. We have multiplied the numerator and denominator by 10. We will see how this is used to compare decimals in Example 6.

Active learning is continuously promoted throughout, especially in the ***Check Yourself exercises*** that appear after every example. Students retain information and concepts better when they practice math immediately after learning it.

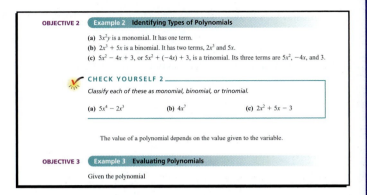

OBJECTIVE 2   **Example 2   Identifying Types of Polynomials**

(a) $3x^2y$ is a monomial. It has one term.
(b) $2x^3 + 5x$ is a binomial. It has two terms, $2x^3$ and $5x$.
(c) $5x^2 - 4x + 3$, or $5x^2 + (-4x) + 3$, is a trinomial. Its three terms are $5x^2$, $-4x$, and 3.

✓ CHECK YOURSELF 2

Classify each of these as monomial, binomial, or trinomial.

(a) $5x^4 - 2x^3$ (b) $4x^7$ (c) $2x^2 + 5x - 3$

The value of a polynomial depends on the value given to the variable.

OBJECTIVE 3   **Example 3   Evaluating Polynomials**

Given the polynomial

This edition of *Prealgebra* continues to provide ***extensive, graduated exercise sets*** including ***applications, challenge exercises, writing exercises,*** and ***collaborative exercises.*** The large number of exercises and exercise types give instructors flexibility in assigning homework, and accommodate multiple teaching/learning styles.

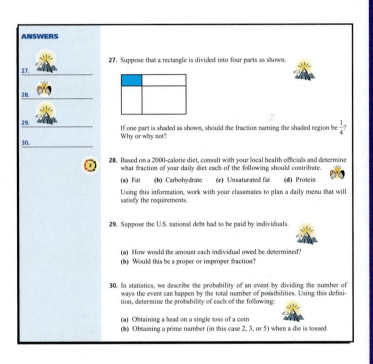

ANSWERS

27. _____
28. _____
29. _____
30. _____

27. Suppose that a rectangle is divided into four parts as shown.

If one part is shaded as shown, should the fraction naming the shaded region be $\frac{1}{4}$? Why or why not?

28. Based on a 2000-calorie diet, consult with your local health officials and determine what fraction of your daily diet each of the following should contribute.

(a) Fat   (b) Carbohydrate   (c) Unsaturated fat   (d) Protein

Using this information, work with your classmates to plan a daily menu that will satisfy the requirements.

29. Suppose the U.S. national debt had to be paid by individuals.

(a) How would the amount each individual owed be determined?
(b) Would this be a proper or improper fraction?

30. In statistics, we describe the probability of an event by dividing the number of ways the event can happen by the total number of possibilities. Using this definition, determine the probability of each of the following:

(a) Obtaining a head on a single toss of a coin
(b) Obtaining a prime number (in this case 2, 3, or 5) when a die is tossed

Problem relates to the Chapter Opener   **Check Yourself**   **Challenge Problem**   **Writing Exercise**   **Calculator Exercises**   **Caution**   **Collaborative Exercises**

Video   **ALEKS**

This edition continues to provide *Overcoming Math Anxiety boxes* giving students study hints and other tips to help them succeed in math. New methods for studying and thinking about math enable students who have fallen short in the past to approach the subject in a new way and succeed.

Overcoming Math Anxiety

Working Together

How many of your classmates do you know? Whether you are by nature outgoing or shy, you have much to gain by getting to know your classmates.

1. It is important to have someone to call when you have missed class or if you are unclear on an assignment.
2. Working with another person is almost always beneficial to both people. If you don't understand something, it helps to have someone to ask about it. If you do understand something, nothing will cement that understanding more than explaining the idea to another person.
3. Sometimes we need to share our feelings of distress. If an assignment is particularly frustrating, it is reassuring to find that it is also frustrating for other students.
4. Have you ever thought you had the right answer, but it doesn't match the answer in the text? Frequently the answers are equivalent, but that's not always easy to see. A different perspective can help you see that. Occasionally there is an error in a textbook (here, we are talking about *other* textbooks). In such cases, it is wonderfully reassuring to find that someone else has the same answer as you do.

Thorough coverage of geometry and measurement, including unit conversion were retained in this edition. The geometry chapter covers the topics that most courses need. Students who are involved in technical subject areas will especially benefit from studying unit conversions.

RECALL By the Fundamental Principle of Fractions, multiplying the numerator and denominator of a fraction by the same nonzero number does not change the value of the fraction.

NOTE The 0 to the right of the decimal point is a **placeholder** that is not needed in the common-fraction form.

Margin Notes help students focus on important concepts and techniques, while *Caution icons* help prevent mistakes.

Pretests, Summary Exercises, and Cumulative Reviews continue to appear in every chapter, with the exception of *Cumulative Reviews,* which naturally begin in chapter 2. These features allow students to test their knowledge of the chapter material from beginning to end.

9 Summary

DEFINITION/PROCEDURE	EXAMPLE	REFERENCE
Tables and Graphs of Data		Section 9.1
A **table** is a display of information in parallel columns or rows.		p. 693
A **graph** is a diagram that relates two different pieces of information. One of the most common graphs is the **bar graph.**		p. 694
In **line graphs,** one of the axes is usually related to time.		p. 698
The Rectangular Coordinate System		Section 9.2
The Rectangular Coordinate System The rectangular coordinate system is a system formed by two perpendicular axes that intersect at a point called the **origin.** The horizontal line is called the **x-axis.** The vertical line is called the **y-axis.**		p. 715
Graphing Points from Ordered Pairs The coordinates of an ordered pair allow you to associate a point in the plane with every ordered pair. To graph a point in the plane, **Step 1** Start at the origin. **Step 2** Move right or left according to the value of the x-coordinate: to the right if x is positive or to the left if x is negative. **Step 3** Then move up or down according to the value of the y-coordinate: up if y is positive and down if y is negative.	To graph the point corresponding to (2, 3):	p. 718
Linear Equations in Two Variables		Section 9.3
Solutions of Linear Equations A pair of values that satisfies the equation. Solutions for linear equations in two variables are written as *ordered pairs.* An ordered pair has the form (x, y)	If $2x - y = 10$, (6, 2) is a solution for the equation, because substituting 6 for x and 2 for y gives a true statement.	p. 730

x-coordinate   y-coordinate

Cumulative Review for Chapters 1 to 4

The following exercises are presented to help you review concepts from earlier chapters that you may have forgotten. This section is meant as review material and not as a comprehensive exam. The answers are presented in the back of the text. If you have difficulty with any of these exercises, be certain to at least read through the summary related to that section.

Name _____

Section _____ Date _____

ANSWERS

In exercises 1 to 14, perform the indicated operations.

1. $8 + (-4)$

2. $-7 + (-5)$

3. $\dfrac{7}{3} + \dfrac{11}{3}$

4. $\dfrac{4}{5} - \dfrac{2}{3}$

5. $(-6)(3)$

6. $\dfrac{1}{3} \cdot \dfrac{3}{5}$

7. $\dfrac{3}{7} \cdot \dfrac{-2}{3}$

8. $(-50) \div (-5)$

9. $\dfrac{-2}{5} \div \dfrac{3}{10}$

10. $15 \div 0$

11. $0 \div \dfrac{1}{2}$

12. $\dfrac{1}{5} + \dfrac{3}{4} - \dfrac{1}{3}$

13. $(-3)(-9)$

14. $\dfrac{3}{5} \cdot \dfrac{1}{3} \cdot \dfrac{5}{7}$

Convert to mixed numbers.

15. $\dfrac{16}{9}$

16. $\dfrac{36}{5}$

1._____
2._____
3._____
4._____
5._____
6._____
7._____
8._____
9._____
10._____
11._____
12._____
13._____
14._____
15._____

Supplements

How Does *Prealgebra* Help Students Improve Their Performance?

ALEKS (**A**ssessment and **LE**arning in **K**nowledge **S**paces) is an artificial intelligence-based system for individualized mathematics learning, available over the web 24/7.

ALEKS's unique adaptive questioning continually assesses each student's math knowledge. ALEKS then provides an individualized learning path, guiding the student in the selection of appropriate new study material. ALEKS 3.0 now links to text-specific video, multimedia tutorials, and text book pages in PDF format. The system records each student's progress toward mastery of curricular goals in a robust classroom management system.

ALEKS improves students' performance by assessing what they know, guiding them to what they are ready to learn, and helping them master key mathematical concepts. See www.aleks.com.

Video Lectures: In the videos, qualified teachers work through selected exercises from the text book, following the solution methodology employed in the text. The video series is available online at the book's website, and are downloadable for viewing on one's computer or iPod/MP3. The videos are closed-captioned for the hearing impaired, subtitled in Spanish, and meet the Americans with Disabilities Act Standards for Accessible Design. Instructors may use them as resources in a learning center, for online courses, and/or to provide extra help students who require extra practice.

MathZone—www.mhhe.com/baratto

McGraw-Hill's MathZone is a complete online homework system for mathematics and statistics. Instructors can assign textbook-specific content from over 40 McGraw-Hill titles as well as customize the level of feedback students receive, including the ability to have students show their work for any given exercise. Assignable content includes an array of videos and other multimedia along with algorithmic exercises, providing study tools for students with many different learning styles.

Within MathZone, a diagnostic assessment tool powered by ALEKS® is available to measure student preparedness and provide detailed reporting and personalized remediation. MathZone also helps ensure consistent assignment delivery across several sections through a course administration function and makes sharing courses with other instructors easy.

For additional study help students have access to NetTutor™, a robust online live tutoring service that incorporates whiteboard technology to communicate mathematics. The tutoring schedules are built around peak homework times to best accommodate student schedules. Instructors can also take advantage of this whiteboard by setting up a Live Classroom for online office hours or a review session with students.

For more information, visit the book's website (www.mhhe.com/baratto) or contact your local McGraw-Hill sales representative (www.mhhe.com/rep).

A Great Learning System Does Not Stop With the Book

Instructor's Supplements

Annotated Instructor's Edition: This ancillary contains answers to exercises in the text, including answers to all section exercises, all *Pretest, Summary and Review Exercises, Self-Tests,* and *Cumulative Reviews.* These answers are printed in a second color for ease of use by the instructor and are located on the appropriate pages throughout the text. The answers to the *Pretests, Self-Tests,* and *Cumulative Reviews* are annotated with section references to aid the instructor who may have omitted certain sections from study.

Computerized Test Bank (CTB) Online: Available through the book's website, this **computerized test bank,** utilizing Brownstone Diploma® algorithm-based testing software, enables users to create customized exams quickly. This user-friendly program enables instructors to search for questions by topic, format, or difficulty level; to edit existing questions or to add new ones; and to scramble questions and answer keys for multiple versions of the same test. Hundreds of text-specific open-ended and multiple-choice questions are included in the question bank. Sample chapter tests in Microsoft Word® and PDF formats are also provided.

Instructor's Solutions Manual: Available on the book's website, the Instructor's Solutions Manual provides comprehensive, **worked-out solutions** to all exercises in the text. The methods used to solve the problems in the manual are the same as those used to solve the examples in the textbook.

ALEKS–www.aleks.com

ALEKS (**A**ssessment and **LE**arning in **K**nowledge **S**paces) is a dynamic online learning system for mathematics education, available over the Web 24/7. ALEKS assesses students, accurately determines their knowledge, and then guides them to the material that they are most ready to learn. With a variety of reports, Textbook Integration Plus, quizzes, and homework assignment capabilities, ALEKS offers flexibility and ease of use for instructors.

- ALEKS uses artificial intelligence to determine exactly what each student knows and is ready to learn. ALEKS remediates student gaps and provides highly efficient learning and improved learning outcomes.

- ALEKS is a comprehensive curriculum that aligns with syllabi or specified textbooks. Used in conjunction with McGraw-Hill texts, students also receive links to text-specific videos, multimedia tutorials, and textbook pages.

- Textbook Integration Plus allows ALEKS to be automatically aligned with syllabi or specified McGraw-Hill textbooks with instructor chosen dates, chapter goals, homework, and quizzes.

- ALEKS with AI-2 gives instructors increased control over the scope and sequence of student learning. Students using ALEKS demonstrate a steadily increasing mastery of the content of the course.

- ALEKS offers a dynamic classroom management system that enables instructors to monitor and direct student progress towards mastery of course objectives.

This book can be customized with McGraw-Hill/Primis Online. A digital database offers the flexibility to customize a course including material from the largest online collection of textbooks, readings, and cases. Primis leads the way in customized eBooks with hundreds of titles available at prices that save your students over 20% off bookstore prices. Additional information is available at 800-228-0634.

Electronic Textbook: CourseSmart is a new way for faculty to find and review eTextbooks. It's also a great option for students who are interested in accessing their course materials digitally and saving money. CourseSmart offers thousands of the most commonly adopted textbooks across hundreds of courses from a wide variety of higher education publishers. It is the only place for faculty to review and compare the full text of a textbook online, providing immediate access without the environmental impact of requesting a print exam copy. At CourseSmart, students can save up to 50% off the cost of a print book, reduce their impact on the environment, and gain access to powerful web tools for learning including full text search, notes and highlighting, and email tools for sharing notes between classmates. **www.CourseSmart.com**

Student's Supplements

Student's Solutions Manual: The Student's Solutions Manual provides comprehensive, **worked-out solutions** to all of the odd-numbered exercises. The steps shown in the solutions match the style of solved examples in the textbook.

NetTutor: Available through the book's website, NetTutor is a revolutionary system that enables students to interact with a live tutor over the World Wide Web. NetTutor's web-based, graphical chat capabilities enable students and tutors to use mathematical notation and even to draw graphs as they work through a problem together. Students can also submit questions and receive answers, browse previously answered questions, and view previous live-chat sessions. Tutors are familiar with the textbook's objectives and problem-solving styles.

Acknowledgments

Those familiar with the publishing process will attest that change is inevitable. The same is true at MHHE. The amazing thing is that change invariably seems to lead to something positive. Throughout the development of this text, we have been most fortunate to work with Stewart Mattson, Rich Kolasa, Michelle Flomenhoft and Torie Anderson. All of them manage to be both demanding managers and supportive coworkers. We have appreciated their talents, energy, and time. As always, we encourage prospective authors to talk with the staff at MHHE. It will be a valuable use of your time.

We would like to thank the many reviewers who have taken the time to review and help improve this text through three editions:

Darla Aguilar, *Pima Community College*
Carla Ainsworth, *Salt Lake Community College*
Muhammad Akhtar, *El Paso Community College*
Chad Bemis, *Riverside Community College*
Monika Bender, *Central Texas College*
Norma Bisluca, *University of Maine–Augusta*
Debra Bryant, *Tennessee Technological University*
Susan Caldiero, *Consumnes River College*
Marc Campbell, *Daytona State College*
Pauline Chow, *Harrisburg Area Community College*
Camille Cochrane, *Shelton State Community College*
William Coe, *Montgomery College*
Pat Cook, *Weatherford College*
Nancy Desilet, *Carroll Community College*
Nerissa Felder, *Polk Community College*
Carol Flakus, *Lower Columbia College*
Robert Frye, *Polk Community College*
Lauryn Geritz, *Arizona Western College*

Nancy Graham, *Rose State College*
Jane Gringauz, *Minneapolis Community and Technical College*
Alberto Guerra, *St. Philip's College*
Celeste Hernandez, *Richland College*
Tammy Higson, *Hillsborough Community College*
Lori Holdren, *Manatee Community College*
Steven Howard, *Rose State College*
Paul Hrabovsky, *Indiana University of Pennsylvania*
Janice Hubbard, *Marshalltown Community College*
Mathew Hudock, *St. Philip's College*
Nicholas Huerta, *Fullerton College*
Victor Hughes III, *Shepard University*
John D. Jarvis, *Utah Valley State College*
Dr. Nancy Johnson, *Manatee Community College*
Rashunda Johnson, *Pulaski Technical College*
Judith Jones, *Valencia Community College*
Lisa Juliano, *El Paso Community College*
Judy Kasabian, *El Camino College*
Randa Kress, *Idaho State University-Pocatello*
Debra Laraway, *Polk Community College*
Paul Wayne Lee, *St. Philip's College*
Nancy Lehmann, *Austin Community College*
Pam Lipka, *University of Wisconsin-Whitewater*
Jean-Marie Magnier, *Springfield Technical Community College*
Robert Maxell, *Long Beach City College*
Robery Maynard, *Tidewater Community College*
Mikal McDowell, *Cedar Valley College*
Philip Meurer, *Palo Alto College*
Pam Miller, *Phoenix College*
Derek Milton, *Santa Barbara City College*
Peg Pankowski, *Community College of Allegheny County South*
Andrew Pitcher, *University of the Pacific*
Mary Romans, *Kent State University*
Linda Reist, *Macomb County Community College Center*
Patricia Rowe, *Columbus State Community College*
Liz Russell, *Glendale Community College*
Kelly Sanchez, *Columbus State Community College*
La Vache Scanlan, *Kapiolani Community College*
Sally Sestini, *Cerritos College*
Linda Spears, *Rock Valley College*
Jane St. Peter, *Mount Mary College*
Renée Starr, *Beaver College*
Daryl Stephens, *East Tennessee State University*
Bryan Stewart, *Tarrant County Community College*
Emily Sullivan, *Bates Technical College*
Abolhassan Taghavy, *Richard J. Daley College*
Patricia Taylor, *Thomas Nelson Community College*
Sharon Testone, *Onondaga Community College*
Tanya Townsend, *Pennsylvania Valley Community College*
Dr. Joseph Tripp, *Ferris State University*
Linda Tucker, *Rose State College*
Marjorie Whitmore, *Northwest Arkansas Community College*
Diane Williams, *Northern Kentucky University*
Alma Wlazlinski, *McLennan Community College*
Rebecca Wong, *West Valley College*
Kevin Yokoyama, *College of the Redwoods*

A COMMITMENT TO ACCURACY

You have a right to expect an accurate textbook, and McGraw-Hill invests considerable time and effort to make sure that we deliver one. Listed below are the many steps we take to make sure this happens.

OUR ACCURACY VERIFICATION PROCESS

First Round

Step 1: Numerous **college math instructors** review the manuscript and report on any errors that they may find, and the authors make these corrections in their final manuscript.

Second Round

Step 2: Once the manuscript has been typeset, the **authors** check their manuscript against the first page proofs to ensure that all illustrations, graphs, examples, exercises, solutions, and answers have been correctly laid out on the pages, and that all notation is correctly used.

Step 3: An outside, **professional mathematician** works through every example and exercise in the page proofs to verify the accuracy of the answers.

Step 4: A **proofreader** adds a triple layer of accuracy assurance in the first pages by hunting for errors, then a second, corrected round of page proofs is produced.

Third Round

Step 5: The **author team** reviews the second round of page proofs for two reasons: 1) to make certain that any previous corrections were properly made, and 2) to look for any errors they might have missed on the first round.

Step 6: A **second proofreader** is added to the project to examine the new round of page proofs to double check the author team's work and to lend a fresh, critical eye to the book before the third round of paging.

Fourth Round

Step 7: A **third proofreader** inspects the third round of page proofs to verify that all previous corrections have been properly made and that there are no new or remaining errors.

Step 8: Meanwhile, in partnership with **independent mathematicians,** the text accuracy is verified from a variety of fresh perspectives:
- The **test bank author** checks for consistency and accuracy as they prepare the computerized test item file.
- The **solutions manual author** works every single exercise and verifies their answers, reporting any errors to the publisher.
- A **consulting group of mathematicians,** who write material for the text's MathZone site, notifies the publisher of any errors they encounter in the page proofs.
- A video production company employing **expert math instructors** for the text's videos will alert the publisher of any errors they might find in the page proofs.

Final Round

Step 9: The **project manager,** who has overseen the book from the beginning, performs a **fourth proofread** of the textbook during the printing process, providing a final accuracy review.

⇒   What results is a mathematics textbook that is as accurate and error-free as is humanly possible, and our authors and publishing staff are confident that our many layers of quality assurance have produced textbooks that are the leaders of the industry for their integrity and correctness.

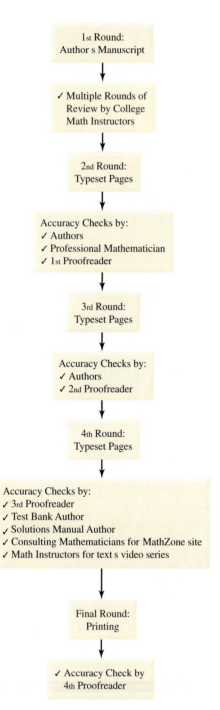

1st Round:
Author s Manuscript

↓

✓ Multiple Rounds of Review by College Math Instructors

↓

2nd Round:
Typeset Pages

↓

Accuracy Checks by:
✓ Authors
✓ Professional Mathematician
✓ 1st Proofreader

↓

3rd Round:
Typeset Pages

↓

Accuracy Checks by:
✓ Authors
✓ 2nd Proofreader

↓

4th Round:
Typeset Pages

↓

Accuracy Checks by:
✓ 3rd Proofreader
✓ Test Bank Author
✓ Solutions Manual Author
✓ Consulting Mathematicians for MathZone site
✓ Math Instructors for text s video series

↓

Final Round:
Printing

↓

✓ Accuracy Check by 4th Proofreader

Applications Index

Business and Finance

advertising, 2, 37, 56, 618, 619, 627, 712, 726
apples, cost of, 532
appliances, cost of, 369
assembly line, 532, 568
attendance, 117, 707–708, 771
bakery profits, 481
balance, 465–466, 476, 489, 493, 569
bankruptcy filings, 775–776
battery packs, 271, 533
bike purchase, 247
boat purchase, 65
book costs, 85, 288, 475, 532, 773
bottling company, 481
brand comparison, 479, 713
budget, 35, 36, 340, 346, 347, 617, 845
business trip expenses, 474, 476, 493, 585
butcher, 342
cans, cost of, 498, 513, 532
car dealership, 65, 101, 221
car payments, 80, 478, 831
car production, 707, 769, 776
car rentals, 54, 833
car repair shop, 476
cars, transporting, 64
car sales, 498, 708, 842–843
car tune-up, 476
cash, 390, 465
cash rebate, 114
charity fundraising, 490
checking account, 36, 38, 129, 135, 137, 140,
    143, 481, 833
checks, 9, 475
commission, 572, 597, 598, 604, 605, 626,
    630, 834
community college costs, 528
computer sales, 770–771
concession stand, 535
converting currencies, 519, 530
converting quarter dollars to dollars,
    330, 339
copy machine, 64
corn harvest, 65
cost per hour, 563
cost per item, 469
credit card payments, 114
daily pay, 101, 340, 524
deductions from paycheck, 347, 616, 617,
    618, 627, 630
deposits, 463, 475
depreciation, 617
discounts, 150, 489, 572, 599–600, 604, 605,
    606, 627, 628, 630, 834
down payment, 616
dresser, cost of, 496
exports, 130, 619
fabric, 335–336, 342, 513
farmer's estate, 343
first-class stamps, 701–702
fractional parts, 246, 247, 344
frequent-flyer miles, 37
fuel used, 370, 477
gas prices, 706–707
gas purchase, 390, 463, 474
gas sales, 39

grocery purchase, 48
home prices, 769–770, 775
hourly pay, 157, 271, 390, 450, 478,
    481, 489
imports, 130, 619
interest rate on loan, 349, 490, 572, 607, 608,
    610, 630
investment, 157, 166, 572, 602, 608, 627
kerosene purchase, 467
lightbulbs, 600
lunch, 47, 221
manufacturing, 64, 67, 232, 565, 795
markup, 599, 601–602, 604, 605, 630, 834
medication costs, 504
miles driven by salesperson, 24, 760
mortgage, 711
newspaper delivery, 64
night shift, 233
number of employees, 614, 618, 632
order costs, 478
pens, cost of, 478, 494, 567
photocopying costs, 481
pizza, 673
pizza parlor, 48
price increase, 601, 618, 627
price per acre, 342
printing labels, 64
production costs, 750–751, 795, 806
promotion sale, 490
quality control, 479
ratio of men to women employees, 505
rental agreement, 10
restaurant tax, 595
revenue, 763, 795, 806, 816
salary, 367, 369, 370, 725, 744, 757, 831
salary increase, 572, 617, 618, 627
sales, increase in, 835
sales tax, 572, 598–599, 604, 627, 629,
    630, 834
savings account, 129, 627
shipping container, 676
shirts, cost of, 477, 533
shopping trip, 386
shop's expenses, 24
stationary purchase, 48
stereo equipment purchase, 32
stocks, 513, 534, 618, 780
street improvement project, 494
string beans, cost of, 528
tapes, cost of, 490
tea bags, cost of, 532
television purchase, 35, 468, 475, 490, 632
tickets, 67, 85, 565, 776
time-deposit savings plan, 610
tips at restaurant, 73, 595, 612, 619, 628
trust fund, 711
typing costs, 795
unit price, 509–510, 514–515, 525–527,
    535–536, 564
U.S. government's budget deficit, 93
U.S. national debt, 93
utility costs, 618, 710, 761, 762
vacation expenses, 370
value, increase in, 617, 687
video rentals, 24

vineyard, 24
weekly pay, 35, 36, 114, 469
weekly sales, 774
wheat, 706, 835
working hours, 331, 348, 364–365
work stoppage, 764
wrapping packages, 87
year-end bonuses, 85

Construction

blueprint, 534
board, 369
bolts, 328, 346, 348, 388, 569
bricklayer, 514
cabin, 673
carpeting floor, 67, 86, 115, 348, 383, 477,
    673, 780
castle wall, 460
deck, 556
doweling, 346
driveway, 339
fence, 24, 68, 85, 464, 475, 543, 568
floor molding, 566
guywire, 472, 480, 490
home lots, 341, 383, 386, 481, 490,
    494, 570
home remodeling, 65, 291, 378
house addition, 672
house construction, 86
indoor track, 674
kennel, 68
kitchen counter, 67, 328, 346
ladder, 480
landfill, 342
linoleum floor, 340, 383, 631, 666, 682
lot area, 261, 673
lumber costs, 535
measurement units, 555
metal fitting, 474
nails, 478, 510, 516
Norman window, 660, 676, 678
painting walls, 67, 498, 564
parking lot design, 84
paving roads, 348
plastic pipe, 556
plywood, 340, 342, 346, 383
power saws, 764
property tax, 533
road inspector, 345
roof, 682
Sears Tower, 35
shelves, 342, 383, 386, 556, 631
shopping center, 673
sod, 339
street lighting, 85
strips of wood, 336
subflooring, 311
terrace, 672
tubing, 565
two-by-four, 532
water pump, 513
window molding, 23
window replacement, 67
wire, 333, 342
wrought-iron gate, 656–657

Crafts
circular rug, 660
coffee table, 672
cord of wood, 676
fabric, 335–336, 342, 513, 556, 832
kite, 672
labels, 64, 478
meat market's cooler, 676
orange juice, 343
picture frame, 556
plant food, 568
posters, 337, 673
potter's clay, 342
recipes, 339, 383
roast, 342, 478, 534
roll of paper, 347
storage bin, 676
string, 343
strips of cloth, 342
weight of spices, 335
wrapping packages, 1, 40, 49, 70, 87, 107–108

Geometry
area of circle, 570
area of paper, 331–332, 466–467
area of rectangle, 478, 494, 559, 816, 822
area of square, 559
area of triangle, 816, 834
baseball diamond, 461
circumference, 341
dimension in figure, 348, 388, 489, 558
dimensions of rectangle, 365–366
height of pole, 543, 568
height of tree, 543
length of diagonal, 471–472, 480, 490
length of rectangle, 210, 341
perimeter of figures, 345, 347, 475, 480
perimeter of rectangle, 174, 175, 344, 489, 496, 780, 805, 834
perimeter of square, 479, 559, 780
perimeter of triangle, 383, 480, 493, 805
ratio of length of pencils, 506
ratio of width to length of rectangle, 499, 504, 505, 563
shadow of building, 498

Number problems, 10, 35, 38, 48, 72, 84, 175, 210, 360–363, 367–371, 504, 528

Science and Medicine
acid solutions, 616, 627
area of Douglas fir tree, 674
average speed, 333, 340, 342, 383, 631
blood alcohol content, 479
blood types, 246
coffee consumption, 386
converting measurement units, 339, 468, 470–471, 478, 479
crop value, 37
daily caloric intake, 36, 248
dieting, 157, 213, 712–713
distance of Earth from Sun, 90
distance of Mars from Sun, 92
distance of Pluto from Sun, 92
dosage of medication, 401
extinction of species, 166
eye colors, 762
fractional parts, 246, 247
hair colors, 778

heart rate, 262, 280, 282
height of ball, 175, 795, 806
height of object, 806, 822
ideal body weight, 184
kinetic energy, 166
land areas of states, 39
landfill, 342, 349
lightning and thunder, 64
Mercury, Venus, and Earth lining up, 233
miles in laps, 421
minimum daily values, 583
odometer, 116
origin of universe, 8
programs for the disabled, 744
ratio problems, 500–502, 505
recommended daily allowance, 261, 281–282
state park, 332
temperature, average, 703–704, 724
temperature, charting, 201–202
temperature, lowest, 138
temperature changes, 129–130, 138, 143, 144, 150
temperature records, 119, 760
tidal changes, 144
topsoil, 129
weather, 762
yearly growth, 464

Social Science
active duty in military, 845
age and college education, 776
class enrollment, 20, 617, 627
college enrollment, 36, 288, 780
college scholarship, 841
college tuition, 690
course outline, 36
dropout rate, 617
education and income, 712
foreign-born residents, 709
fractional parts, 347
homework, 834, 835
length of Nile river, 8
Nobel Prize laureates, 845
number of doctors, 690
number of robberies, 710
petroleum consumption, 620
population, 8, 9, 38, 166, 606, 613, 618, 693, 695–697, 709
ratio of first-year students to second-year students, 565
ratio of men to women, 504, 532
scale on map, 69, 340, 383, 523–524, 534, 565, 632
school enrollment, 613–614, 617, 630, 761, 770
school lockers, 234
Social Security beneficiaries, 700–701
studying for exam, 724
unemployment, 617, 764
voters, 340
votes, 363–364, 367, 369, 504, 532, 568
women-owned firms, 25

Statistics
algebra class, 370
average age of students, 695
baseball, 243, 247, 345, 433, 483, 508, 513, 514, 528, 563, 725
basketball, 36, 220, 389, 563, 567, 568
bowling scores, 24

candy, 84, 505
census, 571, 586, 606, 621, 622, 689
distance between cities, 533–534, 661
driving with full tank, 525, 532, 568
field trip, 84
first prize, 8
football, 389, 504, 534, 699
golfing scores, 24
hockey, 726
lawn, 505
left-handed people, 617
length of run, 334, 349, 474, 489, 659
length of walk, 334, 347
marbles, 49
miles per gallon, 79, 85, 116, 469–470, 471, 478, 479, 490, 496, 512, 713, 760, 780
miles traveled, 24, 533, 760
nuclear energy, 582
number of cars registered, 698
parking spaces, 513
passengers, 114, 778
pet owners, 340
poker hands, 27
rainfall, 474
Senate members with military service, 708
slot machines, 150
sports, 389
survey, 617, 694, 840
test, 2, 35, 48, 220, 243, 247, 612, 616, 617, 630, 761–762, 765, 773, 774, 835
time for eating, 513
time for trip, 349, 388
Tour de France, 412, 450, 484
travel by car, 581
trip planning, 118
Yellowstone National Park visitors, 691

Technology
agriculture, 620
capacity of car tanks, 504
capacity of printer, 64, 513
capacity of refrigerators, 504
clearance of pin, 413
computer, 614–615, 618
computer-aided design, 413
crate shipment, 48
defective parts, 522–523, 534, 535, 565, 617, 777
energy use, 761–762
Ethernet network, 606
faster speed, 514
fertilizer, 512, 528, 672
fuel efficiency, 620
grass seed, 534
green laws, 606
historical number line, 143
locating cities, 727
machine screws, 48, 232
mechanical engineering, 137, 138
movie screen size, 500
newspaper recycling, 750
photocopy machine, 570
photo enlargement, 532, 565
plastics recycling, 723–724
production of cans, 524
teeth on gear, 533
virus-checking program, 618, 628

WHOLE NUMBERS

CHAPTER 1 OUTLINE

Section 1.1   Introduction to Whole Numbers and Place Value   page 3

Section 1.2   Addition of Whole Numbers   page 11

Section 1.3   Subtraction of Whole Numbers   page 28

Section 1.4   Rounding, Estimation, and Ordering of Whole Numbers   page 41

Section 1.5   Multiplication of Whole Numbers   page 50

Section 1.6   Division of Whole Numbers   page 71

Section 1.7   Exponents and Whole Numbers   page 88

Section 1.8   Grouping Symbols and the Order of Operations   page 94

Section 1.9   An Introduction to Equations   page 102

INTRODUCTION

Hundreds of millions of packages are wrapped each year. In fact, industry sources estimate that if we include industrial packaging, 15 million tons of wrapping paper was used in North America in the year 2000. Packages could be wrapped for mailing, for business delivery, or for gifts.

Wrapping paper sometimes comes in rolls that measure 1 yard (yd) by 25 feet (ft). Paper can also be bought in sheets that measure 30 inches (in.) by 18 in. In order to prevent a closet full of leftover wrapping paper, it is useful to estimate the amount of wrapping paper you will use during the holiday season. It can also be helpful to compare the cost of wrapping a present in papers of different quality.

In this chapter, we will encounter some ideas in geometry, such as area, that will help us to examine situations like these. We will also work on basic number skills and estimation.

In Sections 1.3 to 1.6, you will find exercises that refer to the topic of package wrapping. At the end of the chapter, but before the chapter summary, you will find a more in-depth activity on gift wrapping. We hope that by providing these elements we help you to see some of the ways that mathematics plays a part in your life.

Pretest Chapter 1

This pretest will provide a preview of the types of exercises you will encounter in each section of this chapter. The answers for these exercises can be found in the back of the text. If you are working on your own or are ahead of the class, this pretest can help you identify the sections in which you should focus more of your time.

[1.1] **1.** Write 107,945 in words.

[1.2] **2.** The statement $2 + (3 + 5) = (2 + 3) + 5$ illustrates which property of addition?

[1.5] **3.** The statement $5 \times 7 = 7 \times 5$ is an illustration of which property of multiplication?

[1.3] **4.** $35,147 + 2,873 - 7,305 - 3,101 = ?$

[1.5] **5.** $392 \times 51 = ?$

6. $187 \times 300 = ?$

7. (a) $5 + 3 \times 7 = ?$ **(b)** $7 \times (3 + 5) = ?$

[1.6] **8.** $3,234 \div 7 = ?$

Divide by long division.

9. $7\overline{)8,431}$ **10.** $267\overline{)21,758}$

11. Statistics Suppose that you need a total of 360 points on four tests during the semester to receive an A for the course. Your scores on the first three tests were 84, 91, and 92. What is the lowest score you can get on the fourth test and still receive the A?

12. Business and Finance A refrigerator is advertised as follows: "Pay $50 down and $30 a month for 24 months." If the cash price of the refrigerator is $619, how much extra will you pay if you buy on the installment plan?

[1.8] **13.** Evaluate $2^3 \div 2 \times 3 - (5 - 2 + 3)$.

14. Find the perimeter (P) and area (A) of the figure.

5 yd

2 yd

[1.4] **15.** Estimate the following sum by first rounding each value to the nearest hundred. Also, find the actual sum.

```
  921
  377
  855
+ 649
```

ANSWERS

1. _____
2. _____
3. _____
4. _____
5. _____
6. _____
7. _____
8. _____
9. _____
10. _____
11. _____
12. _____
13. _____
14. _____
15. _____

Introduction to Whole Numbers and Place Value

1.1 OBJECTIVES

1. Write numbers in expanded form
2. Determine the place value of a digit
3. Write a number in words
4. Write a number, given its word name

Overcoming Math Anxiety

Throughout this text, we will present you with a series of class-tested techniques that are designed to improve your performance in this math class.

Become familiar with your textbook.

Perform each of the following tasks.

1. Use the Table of Contents to find the title of Section 5.1.

2. Use the Index to find the earliest reference to the term *mean*. (By the way, this term has nothing to do with the personality of either your instructor or the textbook authors!)

3. Find the answer to the first Check Yourself exercise in Section 1.1.

4. Find the answers to the pretest for Chapter 1.

5. Find the answers to the odd-numbered exercises in Section 1.1.

6. In the margin notes for Section 1.1, find the origin for the term *digit*.

Now you know where some of the most important features of the text are. When you have a moment of confusion, think about using one of these features to help you clear up that confusion.

Number systems have been developed throughout human history. Starting with simple tally systems used to count and keep track of possessions, more and more complex systems were developed. The Egyptians used a set of picturelike symbols called **hieroglyphics** to represent numbers. The Romans and Greeks had their own systems of numeration. We see the Roman system today in the form of Roman numerals. Some examples of these systems are shown in the table.

Numerals	Egyptian	Greek	Roman
1	I	I	I
10	∩	Δ	X
100	ϙ	H	C

NOTE The prefix *deci* means 10. Our word *digit* comes from the Latin word *digitus,* which means finger.

Any number system provides a way of naming numbers. The system we use is described as a **decimal place-value system.** This system is based on the number 10 and uses symbols called **digits.** (Other numbers have also been used as bases. The Mayans used 20, and the Babylonians used 60.)

3

The basic symbols of our system are the digits 0, 1, 2, 3, 4, 5, 6, 7, 8, 9. These basic symbols, or digits, were first used in India and then adopted by the Arabs. For this reason, our system is called the **Hindu-Arabic numeration system.**

NOTE Any number, no matter how large, can be represented using the 10 digits of our system.

Numbers may consist of one or more *digits*. 3, 45, 567, and 2,359 are examples of the **standard form** for numbers. We say that 45 is a two-digit number, 567 is a three-digit number, and so on.

As we said, our decimal system uses a *place-value* concept based on the number 10. Understanding how this system works will help you see the reasons for the rules and methods of arithmetic that we will be introducing.

OBJECTIVE 1

| Example 1 | Writing a Number in Expanded Form |

NOTE Each digit in a number has its own place value.

Look at the number 438. We call 8 the **ones digit.** Moving to the left, the digit 3 is the **tens digit.** Again moving to the left, 4 is the **hundreds digit.**

438
4 hundreds | 8 ones
3 tens

NOTE Here the parentheses are used for emphasis. (4 × 100) means 4 is multiplied by 100. (3 × 10) means 3 is multiplied by 10. (8 × 1) means 8 is multiplied by 1.

If we rewrite a number such that each digit is written with its units, we have used the **expanded form** for the number. First think $400 + 30 + 8$; then we write 438 in expanded form as

$$(4 \times 100) + (3 \times 10) + (8 \times 1)$$

 CHECK YOURSELF 1 _____

Write 593 in expanded form.

The place-value diagram shows the place value of digits as we write larger numbers. For the number 3,156,024,798, we have

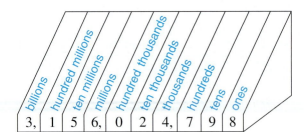

Of course, the naming of place values continues for larger and larger numbers beyond the chart.

For the number 3,156,024,798, the place value of 4 is thousands. As we move to the left, each place value is 10 times the value of the previous place value. Here the place value of 2 is ten thousands, the place value of 0 is hundred thousands, and so on.

OBJECTIVE 2   **Example 2   Identifying Place Value**

Identify the place value of each digit in the number 418,295.

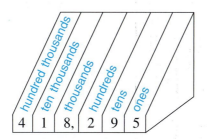

CHECK YOURSELF 2 _____

Use a place-value diagram to answer the following questions for the number 6,831,425,097.

(a) What is the place value of 2?   **(b)** What is the place value of 4?

(c) What is the place value of 3?   **(d)** What is the place value of 6?

Understanding place value will help you read or write numbers in word form. Look at the number

$$\underbrace{7\ 2,}_{\text{Millions}}\quad \underbrace{3\ 5\ 8,}_{\text{Thousands}}\quad \underbrace{6\ 9\ 4}_{\text{Ones}}$$

NOTE A four-digit number, such as 3,456, can be written with or without a comma. We have chosen to include the comma in these materials.

Commas are used to set off groups of three digits in the number. The name of each group—for example, millions, thousands, or ones—is then used as we write the number in words. To write a word name for a number, we work from left to right, writing the numbers in each group, followed by the group name. The chart summarizes the group names.

Billions Group			Millions Group			Thousands Group			Ones Group		
Hundreds	Tens	Ones	Hundreds	Tens	Ones	Hundreds	Tens	Ones	Hundreds	Tens	Ones

OBJECTIVE 3   **Example 3   Writing Numbers in Words**

NOTE The commas in the word statements are in the same place as the commas in the number.

27,345 is written in words as twenty-seven *thousand,* three hundred forty-five.

2,305,273 is two *million,* three hundred five *thousand,* two hundred seventy-three.

Note: We do *not* write the name of the ones group. Also, "and" is not used when a whole number is written in words. It will have a special meaning later.

CHECK YOURSELF 3 ⎯⎯⎯⎯⎯⎯⎯⎯⎯⎯⎯⎯⎯⎯⎯⎯⎯⎯⎯⎯⎯

Write each of the following numbers in words.

(a) 658,942   **(b)** 2,305

We now will reverse the process and write the standard form for numbers given in word form. Consider Example 4.

OBJECTIVE 4   **Example 4**   **Translating Words into Numbers**

Forty-eight thousand, five hundred seventy-nine in standard form is

48,579

Five hundred three thousand, two hundred thirty-eight in standard form is

503,238

Note the use of 0 as a placeholder
in writing the number.

CHECK YOURSELF 4 ⎯⎯⎯⎯⎯⎯⎯⎯⎯⎯⎯⎯⎯⎯⎯⎯⎯⎯⎯⎯⎯

Write twenty-three thousand, seven hundred nine in standard form.

READING YOUR TEXT ⎯⎯⎯⎯⎯⎯⎯⎯⎯⎯⎯⎯⎯⎯⎯⎯⎯⎯⎯⎯

The following fill-in-the-blank exercises are designed to assure that you understand the key vocabulary used in this section. Each sentence comes directly from the section. You will find the correct answers in Appendix C.

Section 1.1

(a) ⎯⎯⎯⎯⎯ are used to set off groups of three digits in a number.

(b) Our system of numbers is called the Hindu ⎯⎯⎯⎯⎯ numeration system.

(c) If we write a number such that each digit is written with its units, we have used the ⎯⎯⎯⎯⎯ form for that number.

(d) For the number 438, we call 8 the ⎯⎯⎯⎯⎯ digit.

CHECK YOURSELF ANSWERS ⎯⎯⎯⎯⎯⎯⎯⎯⎯⎯⎯⎯⎯⎯⎯⎯

1. $(5 \times 100) + (9 \times 10) + (3 \times 1)$ **2. (a)** Ten thousands; **(b)** hundred thousands; **(c)** ten millions; **(d)** billions   **3. (a)** Six hundred fifty-eight thousand, nine hundred forty-two; **(b)** two thousand three hundred five   **4.** 23,709

1.1 Exercises

Write each number in expanded form.

1. 456 [ALEKS]   **2.** 637 [ALEKS]   **3.** 5,073 [ALEKS]   **4.** 20,721 [ALEKS]

Give the place values for the indicated digits.

5. 4 in the number 416

6. 3 in the number 38,615

7. 6 in the number 56,489

8. 4 in the number 427,083

9. In the number 43,729,
   (a) What digit tells the number of thousands?
   (b) What digit tells the number of tens?

10. In the number 456,719,
   (a) What digit tells the number of ten thousands?
   (b) What digit tells the number of hundreds?

11. In the number 1,403,602,
   (a) What digit tells the number of hundred thousands?
   (b) What digit tells the number of ones?

12. In the number 324,678,903,
   (a) What digit tells the number of millions?
   (b) What digit tells the number of ten thousands?

Write each number in words.

13. 5,618   **14.** 21,812   **15.** 200,304   **16.** 103,900

Give the standard (numerical) form for each of the written numbers.

17. Two hundred fifty-three thousand, four hundred eighty-three [ALEKS]

18. Three hundred fifty thousand, three hundred fifty-nine [ALEKS]

19. Five hundred two million, seventy-eight thousand [ALEKS]

20. Four billion, two hundred thirty million [ALEKS]

Boost *your* GRADE at ALEKS.com!

ALEKS®

- Practice Problems
- Self-Tests
- NetTutor
- e-Professors
- Videos

Name _____

Section _____ Date _____

ANSWERS

1. _____
2. _____
3. _____
4. _____
5. _____
6. _____
7. _____
8. _____
9. _____
10. _____
11. _____
12. _____
13. _____
14. _____
15. _____
16. _____
17. _____
18. _____
19. _____
20. _____

Write the whole number in each sentence in standard form.

21. **Statistics**   The first place finisher in the 2002 U.S. Senior Open won four hundred fifteen thousand dollars.

22. **Science and Medicine**   Some scientists speculate that the universe originated in the explosion of a primordial fireball approximately fourteen billion years ago.

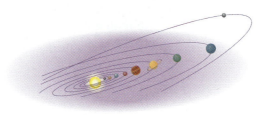

23. **Social Science**   The population of Kansas City, Missouri, in 2000 was approximately four hundred forty-five thousand, six hundred.

24. **Social Science**   The Nile river in Egypt is about four thousand, one hundred forty-five miles long.

Sometimes numbers found in charts and tables are abbreviated. The given table represents the population of the 10 largest cities in the 2000 U.S. census. Note that the numbers represent thousands. Thus, Detroit had a population of 952 thousand or 952,000.

City Name	Rank	Population (thousands)
New York, NY	1	8,008
Los Angeles, CA	2	3,695
Chicago, IL	3	2,896
Houston, TX	4	1,954
Philadelphia, PA	5	1,518
Phoenix, AZ	6	1,321
San Diego, CA	7	1,223
Dallas, TX	8	1,119
San Antonio, TX	9	1,145
Detroit, MI	10	952

Social Science   In exercises 25 to 28, write your answers in standard form using the given table.

25. What was the population of Phoenix in 2000?

26. What was the population of Chicago in 2000?

27. What was the population of Philadelphia in 2000?

28. What was the population of Dallas in 2000?

Assume that you have alphabetized the word names for every number from one to one thousand.

29. Which number would appear first in the list?

30. Which number would appear last?

Determine the number represented by the scrambled place values.

31. 4 thousands
1 ten
3 ten thousands
5 ones
2 hundreds

32. 7 hundreds
4 ten thousands
9 ones
8 tens
6 thousands

33. Business and Finance   Inci has to write a check for $2,565. There is a space on the check to write out the amount of the check in words. What should she write in this space?

ANSWERS

25. _____

26. _____

27. _____

28. _____

29. _____

30. _____

31. _____

32. _____

33. _____

34.

35. _____

36. _____

37. _____

38.

39.

40.

41. _____

34. In addition to personal checks, name two other places where writing amounts in words is necessary.

35. **Business and Finance** In a rental agreement, the initial deposit required is two thousand, five hundred forty-five dollars. Write this amount as a number.

36. **Number Problem** How many zeros are in the number for one billion?

37. **Number Problem** Write the largest five-digit number that can be made using the digits 6, 3, and 9 if each digit is to be used at least once.

38. Several early numeration systems did not use place values. Do some research and determine at least two of these systems. Describe the system that they used. What were the disadvantages?

39. What are the advantages of a place-value system of numeration?

40. The number 0 was not used initially by the Hindus in our number system (about 250 B.C.). Go to your library (or "surf the net"), and determine when a symbol for zero was introduced. What do you think is the importance of the role of 0 in a numeration system?

41. A *googol* is a very large number. Do some research to find out how big it is. Also try to find out where the name of this number comes from.

Answers

We will provide the answers (with some worked out in detail) for the odd-numbered exercises at the end of each exercise set. The answers for the even-numbered exercises are provided in the instructor's resource manual.

1. $(4 \times 100) + (5 \times 10) + (6 \times 1)$ **3.** $(5 \times 1,000) + (7 \times 10) + (3 \times 1)$
5. Hundreds **7.** Thousands **9. (a)** 3; **(b)** 2 **11. (a)** 4; **(b)** 2
13. Five thousand, six hundred eighteen
15. Two hundred thousand, three hundred four **17.** 253,483
19. 502,078,000 **21.** $415,000 **23.** 445,600 **25.** 1,321,000
27. 1,518,000 **29.** Eight **31.** 34,215
33. Two thousand, five hundred sixty-five **35.** $2,545
37. 99,963 **39.** **41.**

1.2 OBJECTIVES
1. Add single-digit numbers
2. Identify the properties of addition
3. Add groups of numbers
4. Solve applications involving a perimeter
5. Set up and solve addition applications

Overcoming Math Anxiety

Become familiar with your syllabus.

In the first class meeting, your instructor probably handed out a class syllabus. If you haven't done so already, you need to incorporate important information into your calendar and address book.

1. Write all important dates in your calendar. This includes homework due dates, quiz dates, test dates, and the date and time of the final exam. Never allow yourself to be surprised by any deadline!

2. Write your instructor's name, contact number, and office number in your address book. Also include the office hours. Make it a point to see your instructor early in the term. Although this is not the only person who can help clear up your confusion, he or she is the most important person.

3. Make note of other resources that are made available to you. These include CDs, videotapes, Web pages, and tutoring.

Given all of these resources, it is important that you never let confusion or frustration mount. If you can't "get it" from the text, try another resource. All of the resources are there specifically for you, so take advantage of them!

The *natural* or *counting numbers* are the numbers we use to count objects.

NOTE The three dots (. . .) are called an **ellipsis;** they mean that the list continues according to the pattern.

The natural numbers are 1, 2, 3, . . .

When we include the number 0, we then have the set of *whole numbers.*

The whole numbers are 0, 1, 2, 3, . . .

We will now look at the operation of *addition* on the whole numbers.

Definition: Addition

Addition is the combining of two or more groups of the same kind of objects.

This concept is extremely important, as we will see in our later work with fractions. We can only combine or add numbers that represent the same kind of objects.

From your first encounter with arithmetic, you were taught to add "3 apples plus 2 apples."

On the other hand, you have probably encountered a phrase such as, "like combining apples and oranges." That is to say, what do you get when you add 3 apples and 2 oranges?

You could answer "5 fruits" or "5 objects," but you can't combine the apples and the oranges.

What if you walked 3 miles and then walked 2 more miles? Clearly, you have now walked 3 miles + 2 miles = 5 miles. The addition was possible because you add groups of the same kind.

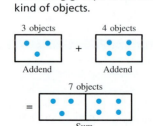

3 miles   +   2 miles

NOTE The first printed use of the symbol + dates back to 1500.

Each operation of arithmetic has its own special terms and symbols. The addition symbol + is read **plus.** When we write 3 + 4, 3 and 4 are called the **addends.**

We can use a number line to illustrate the addition process. To construct a number line, we pick a point on the line and label it 0. We then mark off evenly spaced units to the right, naming each point marked off with a successively larger whole number.

NOTE The point labeled 0 is called the **origin** of the number line.

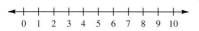

We use an arrowhead to show the number line continues.

OBJECTIVE 1   **Example 1   Representing Addition on a Number Line**

Represent 3 + 4 on the number line.

To represent an addition such as 3 + 4 on the number line, start by moving 3 spaces to the right of the origin. Then move 4 more spaces to the right to arrive at 7. The number 7 is called the **sum** of the addends.

NOTE Addition corresponds to combining groups of the same kind of objects.

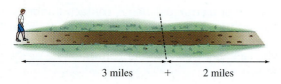

We can write 3 + 4 = 7

Addend   Addend   Sum

CHECK YOURSELF 1

Represent 5 + 6 on the number line.

A statement such as $3 + 4 = 7$ is one of the **basic addition facts.** These facts include the sum of every possible pair of digits. Before you can add larger numbers correctly and quickly, you must memorize these basic facts.

Basic Addition Facts

+	0	1	2	3	4	5	6	7	8	9
0	0	1	2	3	4	5	6	7	8	9
1	1	2	3	4	5	6	7	8	9	10
2	2	3	4	5	6	7	8	9	10	11
3	3	4	5	6	7	8	9	10	11	12
4	4	5	6	7	8	9	10	11	12	13
5	5	6	7	8	9	10	11	12	13	14
6	6	7	8	9	10	11	12	13	14	15
7	7	8	9	10	11	12	13	14	15	16
8	8	9	10	11	12	13	14	15	16	17
9	9	10	11	12	13	14	15	16	17	18

NOTE To find the sum 5 + 8, start with the row labeled 5. Move along that row to the column headed 8 to find the sum, 13.

Examining the basic addition facts leads us to several important properties of addition on the whole numbers. For instance, we know that the sum $3 + 4$ is 7. What about the sum $4 + 3$? It is also 7. This is an illustration of the fact that addition is a **commutative** operation.

NOTE *Commute* means to move back and forth, as in going to school or work.

> **Property:**   The Commutative Property of Addition
>
> The order of two numbers around an addition sign *does not* affect the sum.

OBJECTIVE 2

Example 2   Using the Commutative Property

NOTE The *order* does not affect the sum.

$8 + 5 = 5 + 8 = 13$

$6 + 9 = 9 + 6 = 15$

CHECK YOURSELF 2

Show that the sum on the left equals the sum on the right.

$7 + 8 = 8 + 7$

If we wish to add *more* than two numbers, we can group them and then add. In mathematics, this grouping is indicated by a set of parentheses ( ). This symbol tells us to perform the operation inside the parentheses first.

OBJECTIVE 3

| **Example 3** | **Using Grouping Symbols** |

NOTE We add 3 and 4 as the first step and then add 5.

$$(3 + 4) + 5 = 7 + 5 = 12$$ Here, the 4 is "associated" with the 3.

We also have

NOTE Here we add 4 and 5 as the first step and then add 3. Again the final sum is 12.

$$3 + (4 + 5) = 3 + 9 = 12$$ The 4 is "associated" with the 5.

This example suggests the following property of whole numbers.

NOTE The two equations of Example 3 demonstrate that the 4 can be "associated" with the 3 or the 5.

> **Property: The Associative Property of Addition**
>
> The way in which several whole numbers are grouped *does not* affect the final sum when they are added.

CHECK YOURSELF 3

Show that the two expressions have the same value.

(4 + 8) + 3   and   4 + (8 + 3)

The number 0 has a special property in addition.

> **Property: The Additive Identity Property**
>
> The sum of 0 and any whole number is just that whole number.

Because of this property, we call 0 the **identity** for the addition operation.

Example 4   Adding Zero

Find the sum of **(a)** 3 + 0 and **(b)** 0 + 8.

(a) $3 + 0 = 3$

(b) $0 + 8 = 8$

CHECK YOURSELF 4

Find the sum.

(a) 4 + 0 =   **(b)** 0 + 7 =

When we are adding larger numbers, we will apply the following rule.

> **Property: Adding Digits of the Same Place Value**
>
> We can add the digits of the same place value because they represent the same quantities.

Adding two numbers such as $25 + 34$ can be done in expanded form. Here we write out the place value for each digit.

RECALL 25 means 2 tens and 5 ones; 34 means 3 tens and 4 ones.

$$
\begin{array}{rl}
25 = & 2 \text{ tens} + 5 \text{ ones} \\
+\ 34 = & 3 \text{ tens} + 4 \text{ ones} \\
\hline
= & 5 \text{ tens} + 9 \text{ ones} \\[4pt]
= & 59
\end{array}
$$

Add down.

In actual practice, we use a more convenient short form and our basic addition facts to perform the addition.

Example 5   Adding Two Numbers

Add $352 + 546$.

NOTE In using the short form, be very careful to line up the numbers correctly so that each column contains digits of the same place value.

Step 1   Add in the ones column.

$$
\begin{array}{r}
352 \\
+\ 546 \\
\hline
8
\end{array}
$$

Step 2   Add in the tens column.

$$
\begin{array}{r}
352 \\
+\ 546 \\
\hline
98
\end{array}
$$

Step 3   Add in the hundreds column.

$$
\begin{array}{r}
352 \\
+\ 546 \\
\hline
898
\end{array}
$$

 CHECK YOURSELF 5

Add.

$$
\begin{array}{r}
245 \\
+\ 632 \\
\hline
\end{array}
$$

You have already seen that the word *sum* indicates addition. There are other words that also tell you to use the addition operation.

The *total* of 12 and 5 is written as

$$12 + 5 \qquad \text{or} \qquad 17$$

8 *more than* 10 is written as

$$10 + 8 \qquad \text{or} \qquad 18$$

12 *increased by* 3 is written as

$$12 + 3 \qquad \text{or} \qquad 15$$

Example 6 Translating Words that Indicate Addition

Find each number.

(a) 36 increased by 12.

36 increased by 12 is written as $36 + 12 = 48$.

(b) The total of 18 and 31.

The total of 18 and 31 is written as $18 + 31 = 49$.

CHECK YOURSELF 6

Find each number.

(a) 43 increased by 25   **(b)** The total of 22 and 73

In the examples and exercises of this section, the sum of the digits in each column was 9 or less. What about the situation in which a column has a two-digit sum? This involves the process of **carrying.** In Example 7, we look at the process in expanded form.

Example 7 Adding in Expanded Form when Carrying Is Needed

NOTE Carrying in addition is also called **regrouping** or **renaming.** Of course, the name makes no difference as long as you understand the process.

$$\begin{array}{r} 67 = 60 + 7 \\ + 28 = 20 + 8 \\ \hline 80 + 15 \end{array}$$

We have written 15 ones as 1 ten and 5 ones.

or $\quad 80 + 10 + 5$ The 1 ten is then combined with the 8 tens.

or $\qquad 90 \quad + 5$

or $\qquad 95$

NOTE Of course, this is true for any size number. The place value thousands is 10 times the place value hundreds, and so on.

The more convenient short form carries the excess units from one column to the next column left. Recall that the place value of the next column left is 10 times the value of the previous column. It is this property of our decimal system that makes carrying work.

Look at the problem again, this time done in the short, or "carrying," form.

Step 1   **Step 2**

Carry 1 ten.

$$\begin{array}{r} \overset{1}{6}7 \\ + 28 \\ \hline 5 \end{array} \qquad\qquad \begin{array}{r} \overset{1}{6}7 \\ + 28 \\ \hline 95 \end{array}$$

Step 1: The sum of the digits in the ones column is 15, so write 5 and carry 1 to the tens column. **Step 2:** Now add in the tens column, being sure to include the carried 1.

CHECK YOURSELF 7

Add.

$$\begin{array}{r} 58 \\ + 36 \\ \hline \end{array}$$

The addition process often requires more than one carrying step, as is shown in Example 8.

Example 8   Adding in Short Form when Carrying Is Needed

Add 285 and 378.

$$\begin{array}{r} \overset{1}{}\longleftarrow \text{ Carry 1 ten.} \\ 285 \\ + 378 \\ \hline 3 \end{array}$$

The sum of the digits in the ones column is 13, so write 3 and carry 1 to the tens column.

Carry ⟶ 1 hundred.
$$\begin{array}{r} \overset{1\,1}{285} \\ + 378 \\ \hline 63 \end{array}$$

Now add in the tens column, being sure to include the carry. We have 16 tens, so write 6 in the tens place and carry 1 to the hundreds column.

$$\begin{array}{r} \overset{1\,1}{285} \\ + 378 \\ \hline 663 \end{array}$$

Finally, add in the hundreds column.

 CHECK YOURSELF 8

Add.

$$\begin{array}{r} 479 \\ + 287 \\ \hline \end{array}$$

The carrying process is the same if we want to add more than two numbers.

Example 9   Adding in Short Form with Multiple Carrying Steps

Add 53, 2,678, 587, and 27,009.

$$\begin{array}{r} \overset{1\,1\,2\,2}{}\longleftarrow \text{ Carries} \\ 53 \\ 2{,}678 \\ 587 \\ + 27{,}009 \\ \hline 30{,}327 \end{array}$$

Add in the ones column: $3 + 8 + 7 + 9 = 27$. Write 7 in the sum and carry 2 to the tens column.

Now add in the tens column, being sure to include the carry. The sum is 22. Write 2 tens and carry 2 to the hundreds column. Complete the addition by adding in the hundreds column, the thousands column, and the ten thousands column.

 CHECK YOURSELF 9

Add 46, 365, 7,254, and 24,006.

One application of addition is in finding the *perimeter* of a figure.

The **perimeter** is the distance around a closed figure.

If the figure has straight sides, the perimeter is the sum of the lengths of its sides.

OBJECTIVE 4   **Example 10   Finding the Perimeter**

We wish to fence in the field shown in the figure. How much fencing, in feet (ft), is needed?

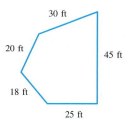

The fencing needed is the perimeter of (or the distance around) the field. We must add the lengths of the five sides.

NOTE Make sure to include the unit with each number.

20 ft + 30 ft + 45 ft + 25 ft + 18 ft = 138 ft

So the perimeter is 138 ft.

CHECK YOURSELF 10

What is the perimeter in inches (in.) of the region shown?

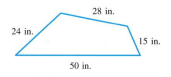

A **rectangle** is a figure, like a sheet of paper, with four sides and four equal corners. The perimeter of a rectangle is found by adding the lengths of the four sides.

Example 11   Finding the Perimeter of a Rectangle

Find the perimeter in inches of the rectangle pictured.

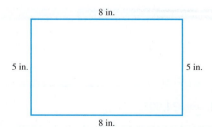

The perimeter is the sum of the lengths 8 in., 5 in., 8 in., and 5 in.

8 in. + 5 in. + 8 in. + 5 in. = 26 in.

The perimeter of the rectangle is 26 in.

 CHECK YOURSELF 11

Find the perimeter of the rectangle pictured.

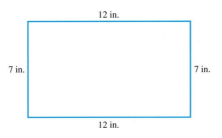

In general, we can find the perimeter of a rectangle by using a formula. A **formula** is a set of symbols that describe a general solution to a problem.

Look at a picture of a rectangle.

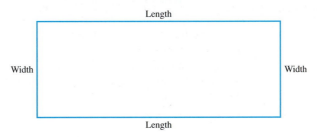

The perimeter can be found by adding the distances, so

Perimeter = length + width + length + width

To make this formula a little more readable, we abbreviate each of the words, using just the first letter.

Property: Formula for the Perimeter of a Rectangle

$P = L + W + L + W$

Example 12   **Finding the Perimeter of a Rectangle**

A rectangle has length 11 in. and width 8 in. What is its perimeter?

Start by drawing a picture of the problem.

NOTE We say the rectangle is 8 in. by 11 in.

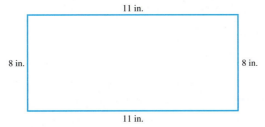

Use the formula $P = L + W + L + W$ with $L = 11$ in. and $W = 8$ in.

$P = 11$ in. $+ 8$ in. $+ 11$ in. $+ 8$ in.

$= 38$ in.

The perimeter is 38 in.

 CHECK YOURSELF 12

A bedroom is 9 ft by 12 ft. What is its perimeter?

NOTE You may very well be able to do some of these problems in your head. Get into the habit of writing down *all* your work, rather than just an answer.

Now we consider other applications that will use the operation of addition. An organized approach is the key to successful problem solving; we suggest this strategy.

Step by Step: Solving Addition Applications

Step 1 Read the problem carefully to determine the given information and what you are being asked to find.

Step 2 Decide upon the operation (in this case, addition) to be used.

Step 3 Write down the complete statement necessary to solve the problem and do the calculations.

Step 4 Write your answer as a complete sentence. Check to make sure you have answered the question of the problem and that your answer seems reasonable.

We use these steps in Example 13.

OBJECTIVE 5   **Example 13   Setting Up a Word Problem**

Four sections of algebra were offered in the fall quarter, with enrollments of 33, 24, 20, and 22 students. What was the total number of students taking algebra?

Step 1 The given information is the number of students in each section. We want the total number.

Step 2 Since we wish a total, we use addition.

NOTE Remember to attach the proper unit (here "students") to your answer.

Step 3 Write $33 + 24 + 20 + 22 = 99$ students.

Step 4 There were 99 students taking algebra.

 CHECK YOURSELF 13

Elva Ramos won an election for city council with 3,110 votes. Her two opponents had 1,022 and 1,211 votes. How many votes were cast in that election?

Using a Scientific Calculator to Add

Although part of this book is intended to help you review the basic skills of arithmetic, many of you will want to be able to use a handheld calculator for some of the problems that are presented. Ideally you should learn to do the basic operations *by hand*. So in each section of this book, start by learning to do the work *without* your calculator. We will then provide these special calculator sections to show how you can use the calculator.

Example 14   Using a Scientific Calculator to Add

Add 23 + 3,456 + 7 + 985.
Enter:

23 $\boxed{+}$ 3,456 $\boxed{+}$ 7 $\boxed{+}$ 985 $\boxed{=}$

> Enter each of the first three numbers followed by the plus key. Then enter the final number and press the equals key. The sum will be in the display.

Display   4,471

NOTE If you have a graphing calculator, follow the same steps, replacing $\boxed{=}$ with $\boxed{\text{ENTER}}$.

NOTE If you press the **clear** key ($\boxed{C}$) first, the calculator is ready to evaluate a new expression.

If you have a calculator, try the addition of this example now. If you have difficulty getting the answer, check the operating manual or ask your instructor for assistance.

CHECK YOURSELF 14 _____

Add 295 + 3 + 4,162 + 84.

READING YOUR TEXT _____

The following fill-in-the-blank exercises are designed to assure that you understand the key vocabulary used in this section. Each sentence comes directly from the section. You will find the correct answers in Appendix C.

Section 1.2

(a) The _____ numbers are 1, 2, 3, ...

(b) When we write 3 + 4, 3 and 4 are called the _____ .

(c) The sum of _____ and any whole number is just that number.

(d) The word _____ indicates addition.

CHECK YOURSELF ANSWERS _____

1.   5   +6   5 + 6 = 11

0 1 2 3 4 5 6 7 8 9 10 11

2. 7 + 8 = 15 and 8 + 7 = 15
3. (4 + 8) + 3 = 12 + 3 = 15; 4 + (8 + 3) = 4 + 11 = 15   **4. (a)** 4; **(b)** 7
5. 877   **6. (a)** 68; **(b)** 95   **7.** 94   **8.** 766   **9.** 31,671   **10.** 117 in.
11. 38 in.   **12.** 42 ft   **13.** 5,343 votes   **14.** 295 $\boxed{+}$ 3 $\boxed{+}$ 4,162 $\boxed{+}$ 84 $\boxed{=}$ 4,544

Name _____

Section _____ Date _____

ANSWERS

1. _____

2. _____

3. _____

4. _____

5. _____

6. _____

7. _____

8. _____

9. _____

10. _____

11. _____

12. _____

13. _____

14. _____

15. _____

16. _____

17. _____

18. _____

19. _____

20. _____

1.2   Exercises

1. Find each sum.
   (a) $5 + 9$
   (b) $8 + 3$
   (c) $0 + 7$

2. Find each sum.
   (a) $6 + 1$
   (b) $3 + 0$
   (c) $7 + 7$

Name the property of addition that is illustrated. Explain your choice of property.

3. $5 + 8 = 8 + 5$ **ALEKS®**

4. $(4 + 5) + 8 = 4 + (5 + 8)$ **ALEKS®**

5. $4 + (7 + 6) = 4 + (6 + 7)$ **ALEKS®**

6. $5 + (2 + 3) = (2 + 3) + 5$ **ALEKS®**

Perform the indicated addition.

7. $\begin{array}{r} 2{,}792 \\ +\ \ \ 205 \\ \hline \end{array}$ **ALEKS®**

8. $\begin{array}{r} 2{,}345 \\ +\ 6{,}053 \\ \hline \end{array}$ **ALEKS®**

9. $\begin{array}{r} 2{,}531 \\ +\ 5{,}354 \\ \hline \end{array}$ **ALEKS®**

10. $\begin{array}{r} 21{,}314 \\ +\ 43{,}042 \\ \hline \end{array}$ **ALEKS®**

11. $\begin{array}{r} 3{,}490 \\ 548 \\ +\ \ \ \ 25 \\ \hline \end{array}$ **ALEKS®**

12. $\begin{array}{r} 2{,}289 \\ 38 \\ 578 \\ +\ 3{,}489 \\ \hline \end{array}$ **ALEKS®**

13. $\begin{array}{r} 23{,}458 \\ +\ 32{,}623 \\ \hline \end{array}$

14. $\begin{array}{r} 26{,}735 \\ 259 \\ 3{,}056 \\ +\ 35{,}489 \\ \hline \end{array}$

15. Find the number that is 356 more than 1,213.

16. Find the number that is 567 more than 2,322.

17. Find the sum of 3,295, 9, 427, and 56.

18. Add 5,637, 78, 690, 28, and 35,589.

19. Find the total of 124 and 2,351.

20. Find the total of the three numbers 112, 24, and 532.

Find the perimeter of each figure.

21.

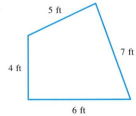

5 ft
7 ft
4 ft
6 ft

22.

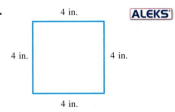

4 in. **ALEKS**
4 in.   4 in.
4 in.

23.

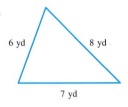

6 yd   8 yd
7 yd

24.

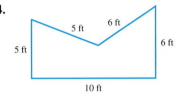

5 ft   6 ft
5 ft   6 ft
10 ft

25.

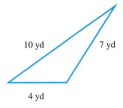

10 in.   **ALEKS**
3 in.   3 in.
10 in.

26.

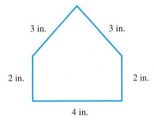

8 yd   10 yd
5 yd

27.

10 yd   7 yd
4 yd

28.

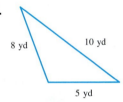

3 in.   3 in.
2 in.   2 in.
4 in.

Solve each of the addition applications.

29. Construction   A rectangular picture window is 4 ft by 5 ft. Meg wants to put a trim molding around the window. How many feet of molding should she buy?

ANSWERS

21. _____

22. _____

23. _____

24. _____

25. _____

26. _____

27. _____

28. _____

29. _____

ANSWERS

30. _____

31. _____

32. _____

33. _____

34. _____

35. _____

36. _____

30. Construction   You are fencing in a backyard that measures 30 ft by 20 ft. How much fencing should you buy?

31. Statistics   A golfer shot a score of 42 on the first nine holes and a score of 46 on the second nine holes. What was her total score for the round?

32. Statistics   A bowler scored 201, 153, and 215 in three games. What was the total score for those games?

33. Business and Finance   Angelo's vineyard shipped 4,200 pounds (lb) of grapes in August; 5,970 lb in September; and 4,850 lb in October. How many pounds were shipped?   ▼DEO

34. Business and Finance   A salesperson drove 68 miles (mi) on Tuesday, 114 mi on Thursday, and 79 mi on Friday. What was the mileage for those three days?

35. Business and Finance   The following chart shows Family Video's monthly rentals for the first three months of 2008 by category of film. Complete the totals.

Category of Film	Jan.	Feb.	Mar.	Category Totals
Comedy	4,568	3,269	2,189	_____
Drama	5,612	4,129	3,879	_____
Action/Adventure	2,654	3,178	1,984	_____
Musical	897	623	528	_____
Monthly Totals	_____	_____	_____	_____

36. Business and Finance   The chart shows Regina's Dress Shop's expenses by department for the last three months of the year. Complete the totals.

Department	Oct.	Nov.	Dec.	Department Totals
Office	$31,714	$32,512	$30,826	_____
Production	85,146	87,479	81,234	_____
Sales	34,568	37,612	33,455	_____
Warehouse	16,588	11,368	13,567	_____
Monthly Totals	_____	_____	_____	_____

ANSWERS

37. _____

38. _____

39. _____

40. _____

41. _____

Social Science For exercises 37 to 40, use the given table, which ranks the areas with the most women-owned firms in the United States.

Metro Area	Number of Firms	Employment	Sales
Los Angeles-Long Beach, Calif.	360,300	1,056,600	$181,455,900,000
New York	282,000	1,077,900	193,572,200,000
Chicago	260,200	1,108,800	161,200,900,000
Washington, D.C.	193,600	440,000	56,644,000,000
Philadelphia	144,600	695,900	90,231,000,000
Atlanta	138,700	331,800	50,206,800,000
Houston	136,400	560,100	78,180,300,000
Dallas	123,900	431,900	63,114,900,000
Detroit	123,600	371,400	50,060,700,000
Minneapolis-St. Paul, Minn.	119,600	337,400	51,063,400,000

37. How many firms in total are located in Washington, Philadelphia, and New York?

38. What is the total number of employees in all 10 of the areas listed?

39. What is the total sales for firms in Houston and Dallas?

40. How many firms in total are located in Chicago and Detroit?

41. A magic square is a square in which the sum along any row, column, or diagonal is the same. For example

35	10	15
0	20	40
25	30	5

Use the numbers 1 to 9 to form a magic square.

42. This puzzle gives you a chance to practice some of your addition skills.

Across

1. 23 + 22
3. 103 + 42
6. 29 + 58 + 19
8. 3 + 3 + 4
9. 1,480 + 1,624
11. 568 + 730
13. 25 + 25
14. 131 + 132
16. The total of 121, 146, 119, and 132
17. The perimeter of a 4 ft by 6 ft rug

Down

1. The sum of 224,000, 155, and 186,000
2. 20 + 30
4. 210 + 200
5. 500,000 + 4,730
7. 130 + 509
10. 90 + 92
12. 100 + 101
15. The perimeter of a 15 ft by 16 ft room

43. Adding of whole numbers is commutative. (The order in which you add does not affect the sum.) Can you think of two actions in your daily routine that are commutative? Explain. List two actions that are *not* commutative in your daily routine and explain.

44. Adding of whole numbers is associative. (The way you group whole numbers does not affect the final sum.) If you are following a recipe that lists 10 ingredients that need to be combined, do you think that adding these ingredients is associative? *Be daring!* Find a recipe and combine the ingredients in different groupings. Tell the class what happens in each case. (Better yet, bring in the completed product for all to sample.)

45. Complete each equation, using the given property.
    (a) Associative property of addition: $5 + (8 + 0) = $ _____
    (b) Commutative property of addition: $5 + (8 + 0) = $ _____
    (c) Additive identity property: $5 + (8 + 0) = $ _____

Calculator Exercises

46.
    458
    273
 + 568

47.
    2,743
     258
      35
 + 5,823

48.
    3,295,153
     573,128
      21,257
   2,586,241
 +   5,291

49. 23 + 5,638 + 385 + 27,345

50. Statistics The following table lists the number of possible types of poker hands. What is the total number possible?

Royal flush	4
Straight flush	36
Four of a kind	624
Full house	3,744
Flush	5,108
Straight	10,200
Three of a kind	54,912
Two pairs	123,552
One pair	1,098,240
Nothing	1,302,540

Answers

1. (a) 14; **(b)** 11; **(c)** 7 **3.** Commutative property of addition
5. Commutative property of addition **7.** 2,997 **9.** 7,885 **11.** 4,063
13. 56,081 **15.** 1,569 **17.** 3,787 **19.** 2,475 **21.** 22 ft
23. 21 yd **25.** 26 in. **27.** 21 yd **29.** 18 ft **31.** 88 **33.** 15,020 lb
35.

Category of Film	Jan.	Feb.	Mar.	Category Totals
Comedy	4,568	3,269	2,189	10,026
Drama	5,612	4,129	3,879	13,620
Action/Adventure	2,654	3,178	1,984	7,816
Musical	897	623	528	2,048
Monthly Totals	13,731	11,199	8,580	33,510

37. 620,200 **39.** $141,295,200,000
41.

8	3	4
1	5	9
6	7	2

43.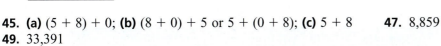

45. (a) (5 + 8) + 0; **(b)** (8 + 0) + 5 or 5 + (0 + 8); **(c)** 5 + 8 **47.** 8,859
49. 33,391

1.3 Subtraction of Whole Numbers

1.3 OBJECTIVES

1. Subtract whole numbers without borrowing
2. Use borrowing to subtract whole numbers
3. Solve subtraction applications

Overcoming Math Anxiety

Don't procrastinate!

1. Do your math homework while you're still fresh. If you wait until too late at night, you will have all that much more difficulty understanding the concepts.

2. Do your homework the day it is assigned. More recent explanations are easier to recall.

3. When your homework is finished, try reading the next section through once. This will give you a "sense of direction" when you next hear the material. This works whether you are in a lecture or lab setting.

Remember that, in a typical math class, you are expected to do two or three hours of homework for each weekly class hour. This means two or three hours per night. Schedule the time and stay to your schedule.

NOTE By *opposite* we mean that subtracting a number "undoes" an addition of that same number. Start with 1. Add 5 and then subtract 5. Where are you?

We are now ready to consider a second operation of arithmetic—subtraction. In Section 1.2, we described addition as the process of combining two or more groups of the same kinds of objects. Subtraction can be thought of as the *opposite operation* to addition. Every arithmetic operation has its own notation. The symbol for subtraction, $-$, is called a **minus sign.**

When we write $8 - 5$, we wish to subtract 5 from 8. We call 5 the **subtrahend.** This is the number being subtracted. And 8 is the **minuend.** This is the number we are subtracting from. The **difference** is the result of the subtraction.

To find the *difference* of two numbers, we will assume that we wish to subtract the smaller number from the larger. Then we look for a number that, when added to the smaller number, will give us the larger number. For example,

$$8 - 5 = 3 \qquad \text{because} \qquad 3 + 5 = 8$$

This special relationship between addition and subtraction provides a method of checking subtraction.

> **Property:** Relationship Between Addition and Subtraction
>
> The sum of the difference and the subtrahend must be equal to the minuend.

OBJECTIVE 1

Example 1 Subtracting a Single-Digit Number

$$12 - 5 = 7$$

Check:

$$7 + 5 = 12$$

Our check works because $12 - 5$ asks for the number that must be added to 5 to get 12.

Difference   Subtrahend   Minuend

CHECK YOURSELF 1

Subtract and check your work.

$13 - 9 =$

The procedure for subtracting larger whole numbers is similar to the procedure for addition. We subtract digits of the same place value.

Example 2   Subtracting a Larger Number

Step 1   **Step 2**   **Step 3**

$$\begin{array}{r} 789 \\ -\ 246 \\ \hline 3 \end{array} \qquad \begin{array}{r} 789 \\ -\ 246 \\ \hline 43 \end{array} \qquad \begin{array}{r} 789 \\ -\ 246 \\ \hline 543 \end{array}$$

We subtract in the ones column, then in the tens column, and finally in the hundreds column.

To check:
$$\left.\begin{array}{r} 789 \\ -\ 246 \\ \hline 543 \end{array}\right\} \text{Add} \quad 543 + 246 = 789$$

The sum of the difference and the subtrahend must be the minuend.

 CHECK YOURSELF 2

Subtract and check your work.

$$\begin{array}{r} 3{,}468 \\ -\ 2{,}248 \\ \hline \end{array}$$

You know that the word *difference* indicates subtraction. There are other words that also tell you to use the subtraction operation. For instance, 5 *less than* 12 is written as

$12 - 5$ or 7

20 *decreased* by 8 is written as

$20 - 8$ or 12

Example 3   Translating Words that Indicate Subtraction

Find each number.

(a) 4 less than 11

4 less than 11 is written $11 - 4$ and $11 - 4 = 7$.

(b) 27 decreased by 6

27 decreased by 6 is written $27 - 6$ and $27 - 6 = 21$.

CHECK YOURSELF 3 _____

Find each number.

(a) 6 less than 19   **(b)** 18 decreased by 3

Difficulties can arise in subtraction if one or more of the digits of the subtrahend are larger than the corresponding digits in the minuend. We solve this problem by using a process called **borrowing.**

First, we look at an example in expanded form.

OBJECTIVE 2   **Example 4   Subtracting when Borrowing Is Needed**

$$52 = 50 + 2$$
$$- \ 27 = 20 + 7$$

Do you see that we cannot subtract in the ones column?

Regrouping, we borrow 1 ten in the minuend and write that ten as 10 ones:

becomes $50 \quad + 2$

$40 + 10 + 2$

or $40 + \quad 12$

We now have

$$52 = 40 + 12$$
$$- \ 27 = 20 + \ 7$$
$$\overline{20 + \ 5}$$

We can now subtract as before.

or   25

In practice, we use a more convenient short form for the subtraction.

$$\begin{array}{r} 52 \\ -\ 27 \\ \hline \end{array} \qquad \begin{array}{r} 4\!\!1 \\ \cancel{5}2 \\ -\ 27 \\ \hline 25 \end{array}$$

We indicate the fact that we have borrowed 1 ten by putting a slash through the 5 and then writing 4 tens. Add 10 ones to the original 2 ones to get 12 ones. We can then subtract.

Check: $25 + 27 = 52$

CHECK YOURSELF 4 _____

Subtract and check your work.

$$\begin{array}{r} 64 \\ -\ 38 \\ \hline \end{array}$$

We will work through another subtraction example that requires a number of borrowing steps. Here, zero appears as a digit in the minuend.

Example 5   Subtracting when Borrowing Is Needed

Step 1

$$\begin{array}{r} \overset{4}{\cancel{~}}\overset{1}{~} \\ 4,0\cancel{5}3 \\ -\ 2,365 \\ \hline 8 \end{array}$$

In this first step, we borrow 1 ten. This is written as 10 ones and is combined with the original 3 ones. We can then subtract in the ones column.

NOTE Here we borrow 1 thousand; this is written as 10 hundreds.

Step 2

$$\begin{array}{r} \overset{3}{~}\overset{10}{~}\overset{4}{~}\overset{1}{~} \\ 4,\cancel{0}\cancel{5}3 \\ -\ 2,365 \\ \hline 8 \end{array}$$

We must borrow again to subtract in the tens column. There are no hundreds, and so we move to the thousands column.

NOTE Now we borrow 1 hundred; this is written as 10 tens and is combined with the remaining 4 tens.

Step 3

$$\begin{array}{r} 3\ 9\overset{14}{~}\overset{1}{~} \\ 4,\cancel{0}\cancel{5}3 \\ -\ 2,365 \\ \hline 8 \end{array}$$

The minuend is now renamed as 3 thousands, 9 hundreds, 14 tens, and 13 ones.

Step 4

$$\begin{array}{r} 9\ 14 \\ 3\cancel{10}\cancel{4}\cancel{1} \\ 4,\cancel{0}\cancel{5}3 \\ -\ 2,365 \\ \hline 1,688 \end{array}$$

The subtraction can now be completed.

To check our subtraction: $1{,}688 + 2{,}365 = 4{,}053$

CHECK YOURSELF 5

Subtract, and check your work.

$$\begin{array}{r} 5{,}024 \\ -\ 1{,}656 \\ \hline \end{array}$$

Units Analysis

This is the first in a series of essays that are designed to help you solve applications of mathematics. Questions in the exercise sets will require the skills that you build by reading these essays.

A number with a unit attached (like 7 *ft* or 26 *miles per gallon*) is called a **denominate** number. Any genuine application of mathematics involves denominate numbers.

When adding or subtracting denominate numbers, the units must be identical for both numbers. The sum or difference will have those same units.

Examples:

$4 + $9 = $13

Notice that, although we write the dollar sign first, we read it after the quantity, as in "four dollars."

7 ft + 9 ft = 16 ft
39 degrees − 12 degrees = 27 degrees
7 ft + 12 degrees yields no meaningful answer!

3 ft + 9 in. yields a meaningful result if the 3 ft is converted into 36 in. We will discuss conversion of units in later essays.

Now we consider subtraction word problems.

> **Step by Step:**   Solving Subtraction Applications
>
> **Step 1**   Read the problem carefully to determine the given information and what you are being asked to find.
> **Step 2**   Decide upon the operation (in this case, subtraction) to be used.
> **Step 3**   Write down the complete statement necessary to solve the problem and do the calculations.
> **Step 4**   Check to make sure you have answered the question of the problem and that your answer seems reasonable.

Often, you need to use both addition and subtraction to solve some problems.

OBJECTIVE 3   **Example 6   Solving a Subtraction Application**

Bernard wants to buy a new piece of stereo equipment. He has $142 and can trade in his old amplifier for $135. How much more does he need if the new equipment costs $449?

First, we must add to find out how much money Bernard has available. Then we subtract to find out how much more money he needs.

$142 + $135 = $277   The money available to Bernard
$449 − $277 = $172   The money Bernard still needs

 CHECK YOURSELF 6 _____

Martina spent on a business trip $239 in airfare, $174 for lodging, and $108 for food. Her company allowed her $375 for the expenses. How much of these expenses will she have to pay herself?

 # Using a Scientific Calculator to Subtract

Now that you have reviewed the process of subtracting by hand, we can look at the use of the calculator in performing that operation. Remember, the point is to brush up on your arithmetic skills by hand, *along with* learning to use the calculator in a variety of situations.

The calculator can be very helpful in a problem that involves both addition and subtraction operations.

Example 7   Using a Scientific Calculator to Subtract

NOTE As with addition, the same steps are used on a graphing calculator, except we press ENTER rather than =.

Find

$$23 - 13 + 56 - 29$$

Enter the numbers and the operation signs exactly as they appear in the expression.

23 ⊟ 13 ⊞ 56 ⊟ 29 ⊟

Display 37

CHECK YOURSELF 7

Find.

$58 - 12 + 93 - 67$

READING YOUR TEXT

The following fill-in-the-blank exercises are designed to assure that you understand the key vocabulary used in this section. Each sentence comes directly from the section. You will find the correct answers in Appendix C.

Section 1.3

(a) Subtraction can be thought of as the _____ operation to addition.

(b) The sum of the difference and the subtrahend must equal the _____.

(c) A number with a unit attached is called a _____ number.

(d) The last step in solving an application is to check to make sure you have answered the question and to make sure that your answer seems _____.

CHECK YOURSELF ANSWERS

1. $13 - 9 = 4$ **2.** 1,220   **3. (a)** 13; **(b)** 15

Check: $4 + 9 = 13$

4. $\overset{5}{\cancel{6}}\overset{1}{4}$ To check:
       $- 38$ $26 + 38 = 64$
       ——
        26

5. 3,368   Check: $3,368 + 1,656 = 5,024$

6. $239 $521 ←——— Total expenses
        174          $- \ 375$ ←——— Amount allowed
      $+ \ 108$ ——
        —— $146
       $521 ←——— Total expenses

7. 58 ⊟ 12 ⊞ 93 ⊟ 67 ⊟ 72

Name _____

Section _____ Date _____

ANSWERS

1. _____

2. _____

3. _____

4. _____

5. _____

6. _____

7. _____

8. _____

9. _____

10. _____

11. _____

12. _____

13. _____ 14. _____

15. _____ 16. _____

17. _____ 18. _____

19. _____ 20. _____

21. _____ 22. _____

23. _____ 24. _____

25. _____ 26. _____

27. _____ 28. _____

29. _____ 30. _____

31. _____ 32. _____

1.3 Exercises

1. In the statement $9 - 6 = 3$
9 is called the
6 is called the
3 is called the
Write the related addition statement.

2. In the statement $7 - 5 = 2$
5 is called the
2 is called the
7 is called the
Write the related addition statement.

In exercises 3 to 26, do the indicated subtraction and check your results by addition.

3. 347
      − 201

4. 575
      − 302

5. 689
      − 245

6. 598
      − 278

7. 3,446
      − 2,326

8. 5,896
      − 3,862

9. 64
      − 27

10. 73
       − 36

11. 627
       − 358

12. 642
       − 367

13. 6,423
       − 3,678

14. 5,352
       − 2,577

15. 6,034
       − 2,569

16. 5,206
       − 1,748

17. 4,000
       − 2,345

18. 6,000
       − 4,349

19. 33,486
       − 14,047

20. 53,487
       − 25,649

21. 29,400
       − 17,900

22. 53,500
       − 28,700

23. 59,000
       − 23,458

24. 41,000
       − 27,645

25. 3,537
       − 2,675

26. 4,693
       − 2,736

27. Find the number that is 25 less than 76.

28. Find the number that results when 58 is decreased by 23.

29. Find the number that is the difference between 97 and 43.

30. Find the number that is 125 less than 265.

31. Find the number that results when 298 is decreased by 47.

32. Find the number that is the difference between 167 and 57.

Based on units, determine whether each of the operations in exercises 33 to 38 produces a meaningful result. (°F = degrees Fahrenheit; °C = degrees Celsius; yd = yard; mi/h = miles per hour; ft/s = feet per second.)

33. 8 mi − 4 mi   **34.** $560 + $314   **35.** 7 ft + 11 in.

36. 18°F − 6°C   **37.** 17 yd − 10 yd   **38.** 4 mi/h + 6 ft/s

In exercises 39 to 42, for various treks by a hiker in a mountainous region, the starting elevations and various changes are given. Determine the final elevation of the hiker in each case.

39. Starting elevation of 1,053 ft, increase of 123 ft, decrease of 98 ft, increase of 63 ft.

40. Starting elevation of 1,231 ft, increase of 213 ft, decrease of 112 ft, increase of 78 ft.

41. Starting elevation of 7,302 ft, decrease of 623 ft, decrease of 123 ft, increase of 307 ft. VIDEO

42. Starting elevation of 6,907 ft, decrease of 511 ft, decrease of 203 ft, increase of 419 ft.

Solve the applications in exercises 43 to 59.

43. Statistics   Shaka's score on a math test was 87 and Tony's score was 23 points less than Shaka's. What was Tony's score on the test?   ALEKS

44. Business and Finance   Duardo's weekly pay of $879 was decreased by $175 for withholding. What amount of pay did he receive?   ALEKS

45. Number Problem   The difference between two numbers is 134. If the larger number is 655, what is the smaller number?

46. Business and Finance   In Jason's monthly budget, he set aside $875 for housing and $665 less than that for food. How much did he budget for food?

47. Business and Finance   Inez has $228 in cash and wants to buy a television set that costs $449. How much more money does she need?   VIDEO   ALEKS

48. Construction   The Sears Tower in Chicago is 1,454 ft tall. The Empire State Building is 1,250 ft tall. How much taller is the Sears Tower than the Empire State Building?   ALEKS

33. _____

34. _____

35. _____

36. _____

37. _____

38. _____

39. _____

40. _____

41. _____

42. _____

43. _____

44. _____

45. _____

46. _____

47. _____

48. _____

49. _____

50. _____

51. _____

52. _____

53. _____

54. _____

55. _____

56. _____

49. Social Science   A college's enrollment was 2,479 students in the fall of 2007 and 2,653 students in the fall of 2008. What was the increase in enrollment?

50. Business and Finance   In one week, Margaret earned $278 in regular pay and $53 for overtime work; $49 was deducted from her paycheck for income taxes and $18 for social security. What was her take-home pay?

51. Business and Finance   Rafael opened a checking account and made deposits of $85 and $272. He wrote checks during the month for $35, $27, $89, and $178. What was his balance at the end of the month?   VIDEO

52. Science and Medicine   Dalila is trying to limit herself to 1,500 calories per day (cal/day). Her breakfast was 270 cal, her lunch was 450 cal, and her dinner was 820 cal. By how much was she *under* or *over* her diet?

53. Statistics   A professional basketball team scored 98, 136, and 113 points in three games. If its opponents scored 102, 109, and 93 points, by how much did the team outscore its opponents?

54. Business and Finance   To keep track of a checking account, you must subtract the amount of each check from the current balance. Complete the given statement.

Beginning balance	$351
Check #1	29
Balance	
Check #2	139
Balance	
Check #3	75
Ending balance	

55. Business and Finance   Complete the given record of a monthly expense account.

Monthly income	$1,620
House payment	343
Balance	
Car payment	183
Balance	
Food	312
Balance	
Clothing	89
Amount remaining	

56. Social Science   A course outline states that you must have 540 points on five tests during the term to receive an A for the course. Your scores on the first four tests have been 95, 84, 82, and 89. How many points must you score on the 200-point final to receive an A?

ANSWERS

57. _____

58. _____

59. _____

60. _____

61. _____

62. _____

63. _____

57. Business and Finance Carmen's frequent-flyer program requires 30,000 mi for a free flight. During 2007, she accumulated 13,850 mi. In 2008, she took three more flights of 2,800, 1,475, and 4,280 mi. How much farther must she fly for her free trip?

58. Business and Finance Peter, Paul, and Mary all submitted advertising budgets for a student government dance.

Ad Medium	Peter	Paul	Mary
Radio ads	$500	$600	$300
Newspaper ads	$150	$200	$150
Posters	$225	$250	$275
Handbills	$175	$150	$250

If $900 is available for advertising, how much over budget would each student be?

59. Science and Medicine The value of all crops in the Salinas Valley in 2001 was about $2 billion. The top four crops are listed below. **(a)** How much greater is the combined value of both types of lettuce than broccoli? **(b)** How much greater is the value of the lettuce and broccoli combined than the strawberries?

Crop	Crop value, in millions
Head lettuce	$360
Broccoli	$246
Leaf lettuce	$210
Strawberries	$198

Complete the magic squares.

60.

	7	2
	5	
8		

61.

4	3	
	5	
		6

62.

16	3		13
	10	11	
9	6	7	
			1

63.

7			14
2	13	8	11
16			
	6	15	

64. **Business and Finance** Efrain has lost track of his checking account transactions. He knows he started with $50 and has deposited $120, $85, and $120. He also knows he has withdrawn $200 and $55. He just can't remember the order in which he did all this.

    (a) What is Efrain's balance after all these transactions?

    (b) Does the order of the transactions make any difference from the math point of view?

    (c) Does the order of transactions make any difference from the banking point of view?

    Explain your answers.

65. **Social Science** Using the World Wide Web, determine the population of Arizona, California, Oregon, and Pennsylvania in each of the last three censuses.

    (a) Find the total change in each state's population over this period.

    (b) Which state shows the most change over the past three censuses?

    (c) Write a brief essay describing the changes and any trends you see in this data. List any implications that they might have for future planning.

66. Describe in words each of the given equations. (Make sure you use a complete sentence.) Then exchange your sentence with other students and see if their interpretations result in the same equation you used.

    (a) $69 - 23 = 46$ **(b)** $17 + 13 = 30$

67. Evaluate the given expressions:

    (a) $8 - (4 - 2)$ **(b)** $(8 - 4) - 2$

    Do you obtain the same answer? What conclusion can you draw about subtraction and an associative property?

68. **Number Problem** Think of any whole number.

    Add 5.

    Subtract 3.

    Subtract two less than the original number.

    What number do you end up with?

    Check with other people. Does everyone have the same answer? Can you explain the results?

Calculator Exercises

Perform the indicated operations.

69.
$$\begin{array}{r} 89 \\ - 48 \\ \hline \end{array}$$

70.
$$\begin{array}{r} 576 \\ - 389 \\ \hline \end{array}$$

71.
$$\begin{array}{r} 5{,}830 \\ - 3{,}987 \\ \hline \end{array}$$

72.
$$\begin{array}{r} 15{,}280 \\ - 7{,}595 \\ \hline \end{array}$$

73. $193{,}243 - 49{,}285$

74. $257{,}500 - 78{,}750$

75. Subtract 235 from the sum of 534 and 678.

76. Subtract 476 from the sum of 306 and 572.

Solve the applications.

77. Business and Finance Readings from the Fast Service Station's storage tanks were taken at the beginning and the end of a month. How many gallons of each type of gas was sold? What was the total sold?

	Regular	Unleaded	Super Unleaded	Total
Beginning reading	73,255	82,349	81,258	
End reading	28,387	19,653	8,654	
Gallons sold	_____	_____	_____	_____

Science and Medicine The land areas, in square miles (mi^2), of three Pacific coast states are California, 158,693 mi^2; Oregon, 96,981 mi^2; and Washington, 68,192 mi^2.

78. How much larger is California than Oregon?

79. How much larger is California than Washington?

Determine whether each statement is true or false.

80. Subtraction is commutative.

81. Subtraction is associative.

82. If we add two numbers and then subtract a third, our result will be the same as if we subtracted the third number from the second and then added the result to the first number.

69. _____

70. _____

71. _____

72. _____

73. _____

74. _____

75. _____

76. _____

77. _____

78. _____

79. _____

80. _____

81. _____

82. _____

Solve each chapter-activity application.

83. Crafts A package is 28 inches (in.) long; a roll of wrapping paper is 36 in. wide (1 yd). How much wider is the wrapping paper than the package is long?

Note: The length of the package is being compared to the width of the wrapping paper.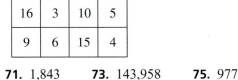

84. Crafts You have 36-in.-wide wrapping paper and need to wrap a 28-in.-long package. To account for the height of the package (3 in.), you need an additional 2 in. wrapping on each side in order to provide 1 in. of overlap per side. How wide a piece of scrap will you be left with?

Answers

1. 9 is the minuend, 6 is the subtrahend, and 3 is the difference. 3 + 6 = 9   **3.** 146
5. 444   **7.** 1,120   **9.** 37   **11.** 269   **13.** 2,745   **15.** 3,465
17. 1,655   **19.** 19,439   **21.** 11,500   **23.** 35,542   **25.** 862
27. 51   **29.** 54   **31.** 251   **33.** Yes   **35.** No   **37.** Yes
39. 1,141 ft   **41.** 6,863 ft   **43.** 64   **45.** 521   **47.** $221
49. 174 students   **51.** $28   **53.** 43 points
55. Balance: 1,277; balance: 1,094; balance: 782; amount remaining: 693
57. 7,595 mi   **59. (a)** $324,000,000; **(b)** $618,000,000

61.

4	3	8
9	5	1
2	7	6

63.

7	12	1	14
2	13	8	11
16	3	10	5
9	6	15	4

65.

67.   **69.** 41   **71.** 1,843   **73.** 143,958   **75.** 977

77. Regular: 44,868 gal; unleaded: 62,696 gal; super unleaded: 72,604 gal; total: 180,168 gal
79. 90,501 mi^2 **81.** False   **83.** 8 in.

1.4 Rounding, Estimation, and Ordering of Whole Numbers

1.4 OBJECTIVES

1. Round a whole number at any place value
2. Estimate sums by rounding
3. Use the symbols $<$ and $>$

It is a common practice to express numbers to the nearest hundred, thousand, and so on. For instance, the distance from Los Angeles to New York along one route is 2,833 mi. We might say that the distance is 2,800 mi. This is called **rounding** because we have rounded the distance to the nearest hundred miles.

One way to picture this rounding process is with the use of a number line.

OBJECTIVE 1

Example 1   Rounding to the Nearest Hundred

To round 2,833 to the nearest hundred:

Because 2,833 is closer to 2,800, we round *down* to 2,800.

 CHECK YOURSELF 1 _____

Round 587 to the nearest hundred.

Example 2   Rounding to the Nearest Thousand

To round 28,734 to the nearest thousand:

Because 28,734 is closer to 29,000, we round *up* to 29,000.

 CHECK YOURSELF 2 _____

Locate 1,375 and round to the nearest hundred.

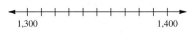

Instead of using a number line, we can apply a rule.

Step by Step: Rounding Whole Numbers

NOTE By a certain *place,* we mean tens, hundreds, thousands, and so on.

Step 1   Identify the place of the digit to be rounded.

Step 2   Look at the digit to the right of that place.

NOTE This is called **rounding up.**

Step 3   **a.** If that digit is 5 or more, that digit and all other digits to the right become 0. The digit in the place you are rounding to is increased by 1.

NOTE This is called **rounding down.**

   b. If that digit is less than 5, that digit and all other digits to the right become 0. The digit in the place you are rounding to remains the same.

Example 3   Rounding to the Nearest Ten

Round 587 to the nearest ten:

Tens
↓
5 **8** 7
   ↑
The digit to the right of the tens place
↓
5 **8** 7   is rounded to 590   We identify the tens digit. The digit to the right of the tens place, 7, is 5 or more. So round up.

 CHECK YOURSELF 3 _____

Round 847 to the nearest ten.

We look at some further examples of using the rounding rule.

Example 4   Rounding Whole Numbers

(a) Round 2,378 to the nearest hundred:

   ↓ ↓
2, **3** 78 is rounded to 2,400   We have identified the hundreds digit, 3. The digit to the right is 7. Because this is 5 or more, the 7 and all digits to the right become 0. The hundreds digit is increased by 1.

(b) Round 53,258 to the nearest thousand:

   ↓ ↓
5 **3** ,258 is rounded to 53,000   We have identified the thousands digit, 3. Because the digit to the right is less than 5, it and all digits to the right become 0, and the thousands digit remains the same.

(c) Round 685 to the nearest ten:

   ↓ ↓
6 **8** 5 is rounded to 690   The digit to the right of the tens place is 5 or more. Round up by our rule.

(d) Round 52,813,212 to the nearest million:

   ↓ ↓
5 **2** ,813,212 is rounded to 53,000,000

CHECK YOURSELF 4 _____

(a) Round 568 to the nearest ten.

(b) Round 5,446 to the nearest hundred.

In Example 5, we look at a case in which we round up a 9.

Example 5   Rounding to the Nearest Ten

Suppose we want to round 397 to the nearest ten. We identify the tens digit, which is 9, and look at the next digit to the right.

NOTE Which number is 397 closer to?

390   397 400

3 **9** 7   The digit to the right is 5 or more.
If this digit is 9 and it must be increased by 1, replace the 9 with 0 and increase the next digit to the *left* by 1.

So 397 is rounded to 400.

 CHECK YOURSELF 5 _____

Round 4,961 to the nearest hundred.

NOTE An estimate is basically a good guess. If your answer is close to your estimate, then your answer is reasonable.

Whether you are doing an addition problem by hand or using a calculator, rounding numbers gives you a handy way of deciding if an answer seems reasonable. The process is called **estimating.** We illustrate this with an example.

OBJECTIVE 2   ### Example 6   Estimating a Sum

Begin by rounding to the nearest hundred

NOTE Placing an arrow above the column to be rounded can be helpful.

456	500
235	200
976	1,000
+ 344	+ 300
2,011	2,000 ← Estimate

By rounding to the nearest hundred and adding quickly, we get an estimate or guess of 2000. Because this is close to the actual sum calculated, 2011, our estimate seems reasonable.

CHECK YOURSELF 6 _____

Round each addend to the nearest hundred and estimate the sum. Then find the actual sum.

287 + 526 + 311 + 378

Estimation is a wonderful tool to use while you're shopping. Every time you go to the store, you should try to estimate the total bill by rounding the price of each item. If you do this regularly, both your addition skills and your rounding skills will improve. The same holds true when you eat in a restaurant. It is always a good idea to know approximately how much you are spending.

Example 7   Estimating a Sum in a Word Problem

Samantha has taken the family out to dinner, and she's now ready to pay the bill. The dinner check has no total, only the individual entries, as below:

Soup	$2.95
Soup	2.95
Salad	1.95
Salad	1.95
Salad	1.95
Lasagna	7.25
Spaghetti	4.95
Ravioli	5.95

What is the approximate cost of the dinner?

Rounding each entry to the nearest whole dollar, we can estimate the total by finding the sum

$$3 + 3 + 2 + 2 + 2 + 7 + 5 + 6 = \$30$$

 CHECK YOURSELF 7

Jason is doing the weekly food shopping at FoodWay. So far, his basket has items that cost $3.99, $7.98, $2.95, $1.15, $2.99, and $1.95. Approximate the total cost of these items.

In the first three examples of this section, we used the number line to illustrate the idea of rounding numbers. The number line also gives us an excellent way to picture the concept of **order** for whole numbers, which means that numbers become larger as we move from left to right on the line.

For instance, we know that 3 is less than 5. On the number line

NOTE 3 is less than or smaller than 5.

we see that 3 lies *to the left* of 5.

We also know that 4 is greater than 2. On the number line

NOTE 4 is greater than or larger than 2.

we see that 4 lies *to the right* of 2.

Two symbols, < for "less than" and > for "greater than," are used to indicate these relationships.

Definition: Inequalities

NOTE The inequality always "points at" the smaller number.

For whole numbers, we can write

1. $2 < 5$ (read "2 is less than 5") because 2 is *to the left* of 5 on the number line.

2. $8 > 3$ (read "8 is greater than 3") because 8 is *to the right* of 3 on the number line.

Example 8 illustrates the use of this notation.

OBJECTIVE 3

Example 8   Indicating Order with < or >

Use the symbols $<$ or $>$ to complete each statement.

(a) 7 _____ 10

(b) 25 _____ 20

(c) 200 _____ 300

(d) 8 _____ 0

(a) $7 < 10$

(b) $25 > 20$

(c) $200 < 300$

(d) $8 > 0$

7 lies to the left of 10 on the number line.
25 lies to the right of 20 on the number line.

 CHECK YOURSELF 8

Use one of the symbols $<$ or $>$ to complete each of the statements.

(a) 35 _____ 25

(b) 0 _____ 4

(c) 12 _____ 18

(d) 1,000 _____ 100

READING YOUR TEXT

The following fill-in-the-blank exercises are designed to assure that you understand the key vocabulary used in this section. Each sentence comes directly from the section. You will find the correct answers in Appendix C.

Section 1.4

(a) The first step in rounding is to identify the _____ of the digit to be rounded.

(b) When you go to a store, you should try to estimate the total bill by _____ the price of each item.

(c) $2 < 5$ is read, "2 is _____ than 5."

(d) 7 lies to the _____ of 10 on the number line.

CHECK YOURSELF ANSWERS

1. 600   **2.** ◄┼┼┼┼┼┼┼┼┼┼┼┼► Round 1,375 *up* to 1,400
         1,300   1,375   1,400

3. 850   **4. (a)** 570; **(b)** 5,400   **5.** 5,000   **6.** 1,500; 1,502

7. $21   **8. (a)** $35 > 25$; **(b)** $0 < 4$; **(c)** $12 < 18$; **(d)** $1,000 > 100$

Name _____

Section _____ Date _____

ANSWERS

1. _____	2. _____
3. _____	4. _____
5. _____	6. _____
7. _____	8. _____
9. _____	
10. _____	
11. _____	
12. _____	
13. _____	
14. _____	
15. _____	
16. _____	
17. _____	
18. _____	
19. _____	
20. _____	
21. _____	
22. _____	
23. _____	
24. _____	
25. _____	
26. _____	

1.4   Exercises

Round each of the numbers to the indicated place.

1. 38, the nearest ten ALEKS   **2.** 72, the nearest ten   ALEKS

3. 253, the nearest ten   ALEKS   **4.** 578, the nearest ten   ALEKS

5. 696, the nearest ten   ALEKS   **6.** 683, the nearest hundred   ALEKS

7. 3,482, the nearest hundred   ALEKS   **8.** 6,741, the nearest hundred   ALEKS

9. 5,962, the nearest hundred   **10.** 4,352, the nearest thousand

11. 4,927, the nearest thousand   **12.** 39,621, the nearest thousand

13. 23,429, the nearest thousand   ALEKS   **14.** 38,589, the nearest thousand   ALEKS

15. 787,000, the nearest ten thousand   ALEKS   **16.** 582,000, the nearest hundred thousand   ALEKS

17. 21,800,000, the nearest million   ALEKS   **18.** 931,000, the nearest ten thousand   ALEKS

In exercises 19 to 40, estimate each of the sums or differences by rounding to the indicated place. Then do the addition or subtraction and use your estimate to see if your actual sum or difference seems reasonable.

Round to the nearest ten.

19.   58   ALEKS   **20.**   92   ALEKS
    27   37
  + 33   85
             + 64

21.   87   **22.**   78
    53   67
    41   53
    93   42
  + 62   + 86

23.   83   **24.**   97
  − 27   − 31

25.   33   **26.**   47
  − 21   − 36

Round to the nearest hundred.

27. 379
      1,215
    + 528   ALEKS

28. 967
      2,365
      544
    + 738 ALEKS

29. 1,378
      519
      792
    + 2,041

30. 3,145
      889
      259
      692
    + 2,518

31. 679
    − 231 VIDEO

32. 824
    − 358

33. 915
    − 411

34. 697
    − 539

Round to the nearest thousand.

35. 2,238
      3,925
    + 5,217 ALEKS

36. 3,678
      4,215
    + 2,032 ALEKS

37. 9,137
      2,315
      7,643
    + 3,092

38. 11,548
      3,874
      14,435
    + 5,398

39. 4,822
    − 2,134 ALEKS

40. 6,120
    − 4,890 ALEKS

Use the symbol < or > to complete each statement.

41. 4 _____ 8   ALEKS

42. 0 _____ 5   ALEKS

43. 500 _____ 400   ALEKS

44. 20 _____ 15   ALEKS

45. 100 _____ 1,000   VIDEO ALEKS

46. 3,000 _____ 2,000   ALEKS

Solve the applications.

47. Business and Finance   Ed and Sharon go to lunch. The lunch check VIDEO
has no total but only lists individual items:

Soup $1.95   Soup $1.95
Salad $1.80   Salad $1.80
Salmon $8.95   Flounder $6.95
Pecan pie $3.25   Vanilla ice cream $2.25

Estimate the total amount of the lunch check.

ANSWERS

27. _____

28. _____

29. _____

30. _____

31. _____

32. _____

33. _____

34. _____

35. _____

36. _____

37. _____

38. _____

39. _____

40. _____

41. _____

42. _____

43. _____

44. _____

45. _____

46. _____

47. _____

48. _____

49. _____

50. _____

51. _____

52. _____

53. _____

54. _____

55. _____

56. _____

48. Business and Finance Olivia will purchase several items at the stationery store. Thus far, the items she has collected cost $2.99, $6.97, $3.90, $2.15, $9.95, and $1.10. Approximate the total cost of these items.

49. Statistics Oscar scored 78, 91, 79, 67, and 100 on his arithmetic tests. Round each score to the nearest ten to estimate his total score.

50. Business and Finance Luigi's pizza parlor makes 293 pizzas on an average day. Estimate (to the nearest hundred) how many pizzas were made on a 3-day holiday weekend.

51. Number Problem A whole number rounded to the nearest ten is 60. **(a)** What is the smallest possible corresponding number? **(b)** What is the largest possible corresponding number? VIDEO

52. Number Problem A whole number rounded to the nearest hundred is 7,700. **(a)** What is the smallest possible corresponding number? **(b)** What is the largest possible corresponding number?

53. Number Problem A whole number rounded to the nearest thousand is 5,000. **(a)** What is the smallest possible corresponding number? **(b)** What is the largest possible corresponding number?

54. Business and Finance Maritza went to the local supermarket and purchased the following items: Milk, $3.89; butter, $2.75; bread, $2.10; orange juice, $3.25; cereal, $3.95; and coffee, $5.80. Approximate her total cost.

55. Manufacturing Technology An inventory of machine screws shows that Bin One contains 378 screws, Bin Two contains 192 screws, and Bin Three contains 267 screws. Estimate the total number of screws in the bins.

56. Manufacturing Technology A delivery truck must be loaded so that the heaviest crates are in the front (loaded first) and the lightest crates are in the back. One morning, crates weighing 378 pounds (lb), 221 lb, 413 lb, 231 lb, 208 lb, 911 lb, 97 lb, 188 lb, and 109 lb needed to be shipped. In what order should the crates be loaded?

57. Statistics A bag contains 60 marbles. The number of blue marbles, rounded to the nearest 10, is 40, and the number of green marbles in the bag, rounded to the nearest 10, is 20. How many blue marbles are in the bag? (List all the answers that satisfy the conditions of the problem.)

58. Describe some situations in which estimating and rounding would not produce a result that would be suitable or acceptable. Review the instructions for filing federal income tax. What rounding rules are used in the preparation of the tax returns? Do the same rules apply to the filing of the state tax returns? If not, what are these rules?

Solve each chapter-activity application.

59. Crafts To the nearest inch, the widths of a set of four packages are 51 in., 25 in., 31 in., and 6 in. Estimate the sum of the widths of your packages by rounding the width of each package to the nearest 10 inches.

60. Crafts To wrap your packages, you need enough paper to cover the width of each package twice as well as enough to cover the height of each package twice. You also need an extra inch to account for overlap. For each of the four packages in exercise 59, determine the length of the piece of wrapping paper required to wrap the packages if the packages are all 3 in. high.

61. Crafts Consider the four packages described in exercises 59 and 60.

(a) Estimate the total length of wrapping paper required to wrap the four packages.

(b) Give the actual length of wrapping paper required.

(c) If you have a 300-in. (25 ft) long roll of wrapping paper, how long a roll of wrapping paper will you be left with?

Answers

1. 40   **3.** 250   **5.** 700   **7.** 3,500   **9.** 6,000   **11.** 5,000
13. 23,000   **15.** 790,000   **17.** 22,000,000
19. Estimate: 120, actual sum: 118   **21.** Estimate: 330, actual sum: 336
23. Estimate: 50, actual difference: 56   **25.** Estimate: 10, actual difference: 12
27. Estimate: 2,100, actual sum: 2,122   **29.** Estimate: 4,700, actual sum: 4,730
31. Estimate: 500, actual difference: 448   **33.** Estimate: 500; actual difference: 504
35. Estimate: 11,000, actual sum: 11,380   **37.** Estimate: 22,000,
actual sum: 22,187   **39.** Estimate: 3,000, actual difference: 2,688   **41.** <
43. >   **45.** <   **47.** $29   **49.** 420   **51.** (a) 55; (b) 64
53. (a) 4,500; (b) 5,499   **55.** 900 screws   **57.** 36, 37, 38, 39, 40, 41, 42,
43, 44   **59.** 120 in.   **61.** (a) 260 in.; (b) 254 in.; (c) 46 in.

1.5 Multiplication of Whole Numbers

1.5 OBJECTIVES

1. Multiply single-digit numbers
2. Identify and use the properties of multiplication
3. Solve applications involving multiplication
4. Estimate products by rounding
5. Multiply whole numbers
6. Solve applications involving area

NOTE The use of the symbol × dates back to the 1600s.

Our work in this section deals with multiplication, another of the basic operations of arithmetic. Multiplication is closely related to addition. In fact, we can think of multiplication as a shorthand method for repeated addition. The symbol × is used to indicate multiplication.

3×4 can be interpreted as 3 rows of 4 objects. By counting we see that $3 \times 4 = 12$. Similarly, 4 rows of 3 means $4 \times 3 = 12$.

The fact that $3 \times 4 = 4 \times 3$ is an example of the **commutative property of multiplication.**

Property: The Commutative Property of Multiplication

Given any two numbers, we can multiply them in either order and get the same result.

In symbols, we say $a \cdot b = b \cdot a$.

NOTE The centered dot is the same as the times sign (×). We use the centered dot so the times sign will not be confused with the letter *x*.

OBJECTIVE 1

Example 1   Multiplying Single-Digit Numbers

$3 \cdot 5$ means 5 multiplied by 3. It is read 3 *times* 5. To find $3 \cdot 5$, we can add 5 three times.

$3 \cdot 5 = 5 + 5 + 5 = 15$

In a multiplication problem such as $3 \cdot 5 = 15$, we call 3 and 5 the **factors.** The answer, 15, is the **product** of the factors, 3 and 5.

$$3 \cdot 5 = 15$$

Factor   Factor   Product

CHECK YOURSELF 1 _____

Name the factors and the product in the statement.

$2 \cdot 9 = 18$

Statements such as $3 \cdot 4 = 12$ and $3 \cdot 5 = 15$ are called the **basic multiplication facts.** If you have difficulty with multiplication, it may be that you do not know some of these facts. This table will help you review them before you go on. Notice that, because of the commutative property, you need memorize only half of these facts!

NOTE To use the table to find the product of 7 · 6: Find the row labeled 7, then move to the right in this row until you are in the column labeled 6 at the top. We see that 7 · 6 is 42.

Basic Multiplication Facts Table

·	0	1	2	3	4	5	6	7	8	9
0	0	0	0	0	0	0	0	0	0	0
1	0	1	2	3	4	5	6	7	8	9
2	0	2	4	6	8	10	12	14	16	18
3	0	3	6	9	12	15	18	21	24	27
4	0	4	8	12	16	20	24	28	32	36
5	0	5	10	15	20	25	30	35	40	45
6	0	6	12	18	24	30	36	42	48	54
7	0	7	14	21	28	35	42	49	56	63
8	0	8	16	24	32	40	48	56	64	72
9	0	9	18	27	36	45	54	63	72	81

Armed with these facts, you can become a better, and faster, problem solver. Take a look at Example 2.

Example 2   Multiplying Instead of Counting

Find the total number of squares on the following checkerboard.

NOTE This checkerboard is an example of a rectangular array, a series of rows and columns that form a rectangle. When you see such an arrangement, be prepared to multiply to find the total number of units.

You could find the number of squares by counting them. If you counted one per second, it would take you just over a minute. You could make the job a little easier by simply counting the squares in one row (8), then adding 8 + 8 + 8 + 8 + 8 + 8 + 8 + 8. Multiplication, which is simply repeated addition, allows you to find the total number of squares by multiplying 8 · 8. How long that takes depends on how well you know the basic multiplication facts! By now, you know that there are 64 squares on the checkerboard.

 CHECK YOURSELF 2

Find the number of windows on the displayed side of the building.

The next property involves *both* multiplication and addition.

> **Property:**   **The Distributive Property of Multiplication over Addition**
>
> To distribute a factor over a sum of numbers, multiply the factor by each number inside the parentheses. Then add the products. (The result will be the same if we find the sum first and then multiply.)

OBJECTIVE 2   **Example 3   Using the Distributive Property**

$2 \cdot (3 + 4) = 2 \cdot 7 = 14$ We have added 3 + 4 and then multiplied.

Also,

$2 \cdot (3 + 4) = (2 \cdot 3) + (2 \cdot 4)$ We have multiplied 2 · 3 and 2 · 4 as the first step.

$ = 6 + 8$

$ = 14$ The result is the same.

We see that $2 \cdot (3 + 4) = (2 \cdot 3) + (2 \cdot 4)$. This is an example of the **distributive property of multiplication over addition** because we distributed the multiplication (in this case by 2) over the "plus" sign.

   CHECK YOURSELF 3 _____

Show that

$3 \cdot (5 + 2) = (3 \cdot 5) + (3 \cdot 2)$

Carrying must often be used to multiply larger numbers. We can see how carrying works in multiplication by looking at an example.

Example 4   Multiplying by a Single-Digit Number

$3 \cdot 25 = 3 \cdot (20 + 5)$ We use the distributive property again.

$ = 3 \cdot 20 + 3 \cdot 5$

$ = 60 \quad + 15$ Write the 15 as 10 + 5.

$ = 60 + 10 + 5$ Carry 10 ones or 1 ten to the tens place.

$ = \quad 70 \quad + 5$

$ = 75$

Here is the same multiplication problem using the short form.

          1 ⟵——— Carry

Step 1   25
           × 3
           ———
            5

Multiplying 3 · 5 gives us 15 ones. Write 5 ones and carry 1 ten.

Step 2
$$
\begin{array}{r}
1 \\
25 \\
\times\ 3 \\
\hline
75
\end{array}
$$
Now multiply $3 \cdot 2$ tens and add the carry to get 7, the tens digit of the product.

CHECK YOURSELF 4

Multiply.

$$
\begin{array}{r}
34 \\
\times\ 6 \\
\hline
\end{array}
$$

Two numbers, 0 and 1, have special properties in multiplication.

NOTE The number 1 is called the **multiplicative identity** for this reason.

Property: Multiplicative Identity Property

The product of 1 and any number is that number. In symbols,

$a \cdot 1 = 1 \cdot a = a$

Property: Multiplicative Property of Zero

The product of 0 and any number is 0. In symbols,

$a \cdot 0 = 0 \cdot a = 0$

Example 5   Multiplying by 1 or 0

Examine each product. What can you conclude about the use of the parentheses?

(a) $(7)(1) = 7$ **(b)** $(1)(5) = 5$ **(c)** $(0)(7) = 0$ **(d)** $(3)(9)(0) = 0$

CHECK YOURSELF 5

Find each product.

(a) $(19)(1)$ **(b)** $(1)(9)$ **(c)** $(0)(8)$ **(d)** $(5)(6)(0)$

Units Analysis

When multiplying a denominate number, like 6 ft, by an abstract number, like 5, the result has the same units as the denominate number. Some examples are

$5 \cdot 6 \text{ ft} = 30 \text{ ft}$
$3 \cdot \$7 = \$21$
$9 \cdot 4 \text{ A's} = 36 \text{ A's}$

When multiplying two different denominate numbers, the units must also be multiplied. We will discuss this when we look at the area of a geometric figure.

© 2010 McGraw-Hill Companies

RECALL It is best to write down the complete statement necessary for the solution of any application.

We now review our discussion of applications, or word problems.

The process of solving applications is the same no matter which operation is required for the solution. In fact, the four-step procedure we suggested in Section 1.2 can be effectively applied here.

Step by Step: Solving Applications

Step 1 Read the problem carefully to determine the given information and what you are being asked to find.

Step 2 Decide upon the operation or operations to be used.

Step 3 Write down the complete statement necessary to solve the problem and do the calculations.

Step 4 Check to make sure you have answered the question of the problem and that your answer seems reasonable.

OBJECTIVE 3

Example 6 Solving an Application Involving Multiplication

A car rental agency orders a fleet of seven new subcompact cars at a cost of $9,258 per automobile. What will the company pay for the entire order?

Step 1 We know the number of cars and the price per car. We want to find the total cost.

Step 2 Multiplication is the best approach to the solution.

Step 3 Write

$7 \cdot \$9,258 = \$64,806$ We could, of course, *add* $9,258, the cost, 7 times, but multiplication is certainly more efficient.

Step 4 The total cost of the order is $64,806.

 CHECK YOURSELF 6 _____

Tires sell for $47 apiece. What is the total cost for five tires?

There are some shortcuts that allow you to simplify your work when you are multiplying by a number that ends in 0. See what you can discover by looking at some examples.

Example 7 Multiplying by Ten

First we multiply by 10.

$$\begin{array}{r} 67 \\ \times\ 10 \\ \hline 670 \end{array}$$ $10 \cdot 67 = 670$

Next we multiply by 100.

$$\begin{array}{r} 537 \\ \times\ 100 \\ \hline 53,700 \end{array}$$ $100 \cdot 537 = 53,700$

Finally, we multiply by 1,000.

$$\begin{array}{r} 489 \\ \times\ 1{,}000 \\ \hline 489{,}000 \end{array}$$ 1,000 · 489 = 489,000

CHECK YOURSELF 7 _____

Multiply.

$$\begin{array}{r} 257 \\ \times\ 100 \\ \hline \end{array}$$

NOTE We talk about powers of 10 in more detail in Section 1.7.

Do you see a pattern? Rather than writing out the multiplication, there is an easier way! We call the numbers 10, 100, 1000, and so on, **powers of 10.**

Property: Multiplying by Powers of 10

When a whole number is multiplied by a power of 10, the product is just that number followed by as many zeros as there are in the power of 10.

Example 8   Multiplying by Numbers that End in Zero

Multiply 400 · 678.

Write

$$\begin{array}{r} 678 \\ \times\ \ \ 400 \\ \hline \end{array}$$ Shift 400 so that the two zeros are *to the right* of the digits above.

$$\begin{array}{r} 33\ \ \ \ \\ 678 \\ \times\ \ \ 4\,00 \\ \hline 271{,}2\,00 \end{array}$$ Bring down the two zeros, then multiply 4 · 678 to find the product.

There is no mystery about why this works. We know that 400 is 4 · 100. In this method, we are multiplying 678 by 4 and then by 100, adding two zeros to the product by our earlier rule.

CHECK YOURSELF 8 _____

Multiply.

300 · 574

Your work in this section, together with our earlier rounding techniques, provides a convenient means of using estimation to check the reasonableness of our results in multiplication, as Example 9 illustrates.

OBJECTIVE 4   **Example 9   Estimating a Product by Rounding**

Estimate the product below by rounding each factor to the nearest hundred.

Rounded

$$512 \longrightarrow 500$$
$$\times 289 \longrightarrow \times 300$$
$$\overline{150{,}000}$$

You might want to now find the *actual* product and use our estimate to see if your result seems reasonable.

 CHECK YOURSELF 9 _____

Estimate the product by rounding each factor to the nearest hundred.

689
× 425

Rounding the factors can be a very useful way of estimating the solution to an application problem.

Example 10   Estimating the Solution to a Multiplication Application

Bart is thinking of running an ad in the local newspaper for an entire year. The ad costs $19.95 per week. Approximate the annual cost of the ad.

Rounding the charge to $20 and rounding the number of weeks in a year to 50, we get

$$50 \cdot 20 = 1{,}000$$

The ad would cost approximately $1,000.

 CHECK YOURSELF 10 _____

Phyllis is debating whether to join the health club for $450 per year or just pay $7 per visit. If she goes about once a week, approximately how much would she spend at $7 per visit?

To multiply by numbers with more than one digit, we must multiply each digit of the first factor by each digit of the second. To do this, we form a series of partial products and then add them to arrive at the final product.

OBJECTIVE 5   **Example 11   Multiplying by a Two-Digit Number**

Multiply $56 \cdot 47$.

Step 1
$$\overset{4}{}56$$
$$\times 47$$
$$\overline{392}$$

The first partial product is $7 \cdot 56$, or 392. Note that we had to carry 4 to the tens column.

Step 2
$$\begin{array}{r} 2 \\ 56 \\ \times\ 47 \\ \hline 392 \\ 2240 \\ \end{array}$$

The second partial product is 40 · 56, or 2240.
We must carry 2 during the process.

Step 3
$$\begin{array}{r} 2 \\ 56 \\ \times\ 47 \\ \hline 392 \\ 2\ 240 \\ \hline 2{,}632 \\ \end{array}$$

We add the partial products for our final result.

 CHECK YOURSELF 11 —————————

Multiply.

$$\begin{array}{r} 38 \\ \times\ 76 \\ \end{array}$$

NOTE The three partial products are formed when we multiply by the ones, tens, and then the hundreds digits.

If multiplication involves two three-digit numbers, another step is necessary. In this case, we form three partial products. This will ensure that each digit of the first factor is multiplied by each digit of the second.

Example 12   Multiplying Two Three-Digit Numbers

Multiply.

$$\begin{array}{r} 2\ 2 \\ 3\ 3 \\ 2\ 2 \\ 278 \\ \times\ 343 \\ \hline 834 \\ 11\ 120 \\ 83\ 400 \\ \hline 95{,}354 \\ \end{array}$$

In forming the third partial product, we must multiply by 300. To indicate this, we shift that product *two* places left.

CHECK YOURSELF 12 —————————

Multiply.

$$\begin{array}{r} 352 \\ \times\ 249 \\ \end{array}$$

We will now look at an example of multiplying by a number involving 0 as a digit. There are several ways to arrange the work, as Example 13 shows.

> **Example 13   Multiplying Larger Numbers**

Multiply $573 \cdot 205$.

Method 1

$$
\begin{array}{r}
1 \\
31 \\
573 \\
\times\ 205 \\
\hline
2865 \\
0000 \\
114600 \\
\hline
117465
\end{array}
$$

⟵ We can write the second partial product as 0000 to indicate the multiplication by 0 in the tens place.

Next, we look at a second approach to the problem.

Method 2

$$
\begin{array}{r}
1 \\
31 \\
573 \\
\times\ 205 \\
\hline
2865 \\
114600 \\
\hline
117465
\end{array}
$$

⟵ We can write a double 0 as our second step. If we place the third partial product on the same line, that product will be shifted *two* places left, indicating that we are multiplying by 200.

Because this second method is more compact, it is usually used.

CHECK YOURSELF 13 _____

Multiply.

$$
\begin{array}{r}
489 \\
\times\ 304 \\
\hline
\end{array}
$$

Example 14 leads us to another property of multiplication.

> **Example 14   Using the Associative Property**

$(2 \cdot 3) \cdot 4 = 6 \cdot 4 = 24$ We do the multiplication in the parentheses first, $2 \cdot 3 = 6$. Then multiply $6 \cdot 4$.

or,

$2 \cdot (3 \cdot 4) = 2 \cdot 12 = 24$ Here we multiply $3 \cdot 4$ as the first step. Then multiply $2 \cdot 12$.

We see that

$$(2 \cdot 3) \cdot 4 = 2 \cdot (3 \cdot 4)$$

The product is the same no matter which way we *group* the factors. This is called the **associative property** of multiplication.

> **Property:** The Associative Property of Multiplication
>
> Multiplication is an *associative* operation. The way in which you group numbers in multiplication does not affect the final product.

 CHECK YOURSELF 14

Find the products.

(a) $(5 \cdot 3) \cdot 6$ **(b)** $5 \cdot (3 \cdot 6)$

Units Analysis

What happens when we multiply two denominate numbers? The units of the result turn out to be the product of the units. This makes sense when we look at an example from geometry.

We write the area, A, of a square with sides of length s as

NOTE s^2 is read as "s squared."

$$A = s \cdot s = s^2.$$

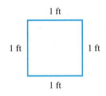

This tile is 1 ft by 1 ft.

$$A = s^2 = (1 \text{ ft})^2 = 1 \text{ ft} \cdot 1 \text{ ft} = 1 \, (\text{ft}) \cdot (\text{ft}) = 1 \text{ ft}^2$$

In other words, its area is one square foot.

If we want to find the area of a room, we are actually finding how many of these square feet can be placed in the room.

We can look now at the idea of **area.** Area is a measure that we give to a surface. It is measured in terms of **square units.** The area is the number of square units that are needed to cover the surface.

One standard unit of area measure is the **square inch,** written as in.^{2} and is the measure of the surface contained in a square with sides of 1 in.

One square inch

NOTE The unit inch (in.) can be treated as though it were a number. So in. · in. can be written as in.^{2}. It is read as "square inches."

Other units of area measure are the square foot (ft^2), the square yard (yd^2), the square centimeter (cm^2), and the square meter (m^2).

Finding the area of a figure means finding the number of square units it contains. One simple case is a rectangle.

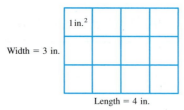

NOTE The length and width must be in terms of the same unit.

The length of the rectangle is 4 in., and the width is 3 in. The area of the rectangle is measured in terms of square inches. We can simply count to find the area, 12 in.². However, because each of the four vertical strips is 3 in. long, we can multiply:

Area = 4 in. · 3 in. = 12 in.²

> **Property:** Formula for the Area of a Rectangle
>
> In general, we can write the formula for the **area of a rectangle**: If the length of a rectangle is L units and the width is W units, then the formula for the area, A, of the rectangle can be written as
>
> $A = L \cdot W$ (square units)

OBJECTIVE 6

Example 15   Find the Area of a Rectangle

A room has dimensions 12 ft by 15 ft. Find its area.

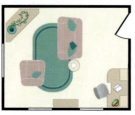

Use the formula $A = L \cdot W$, with $L = 15$ ft and $W = 12$ ft.

$A = L \cdot W$

$= 15 \text{ ft} \cdot 12 \text{ ft} = 180 \text{ ft}^2$

The area of the room is 180 ft².

 CHECK YOURSELF 15

A desktop has dimensions 50 in. by 25 in. What is the area of its surface?

We can also write a convenient formula for the area of a square. If the sides of the square have length s, we can write

> **Property:** Formula for the Area of a Square
>
> $A = s \cdot s = s^2$

Example 16   Finding the Area

You wish to cover a square table with a plastic laminate that costs 60¢ a square foot. If each side of the table measures 3 ft, what will it cost to cover the table?

We first must find the area of the table. Use the formula $A = s^2$, with $s = 3$ ft.

$$A = s^2$$
$$= (3 \text{ ft})^2 = 3 \text{ ft} \cdot 3 \text{ ft} = 9 \text{ ft}^2$$

Now, multiply by the cost per square foot.

$$\text{Cost} = 9 \cdot 60¢ = 540¢ = \$5.40$$

 CHECK YOURSELF 16

You wish to carpet a room that is a square, 4 yd by 4 yd, with carpet that costs $12 per square yard. What will be the total cost of the carpeting?

Sometimes the total area of an oddly shaped figure, frequently called a **composite figure,** is found by adding the smaller areas. Example 17 shows how this is done.

Example 17   Finding the Area of an Oddly Shaped Figure

Find the area of the figure.

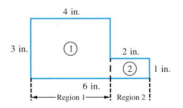

The area of the figure is found by adding the areas of regions 1 and 2. Region 1 is a 4 in. by 3 in. rectangle; the area of region 1 is 4 in. $\cdot$ 3 in. = 12 in.^{2} Region 2 is a 2 in. by 1 in. rectangle; the area of region 2 is 2 in. $\cdot$ 1 in. = 2 in.^{2}

The total area is the sum of the two areas:

Total area = 12 in.^{2} + 2 in.^{2} = 14 in.^{2}

 CHECK YOURSELF 17

Find the area of the figure.

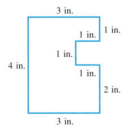

Hint: You can find the area by adding the areas of three rectangles, or by subtracting the area of the "missing" rectangle from the area of the "completed" larger rectangle.

READING YOUR TEXT

The following fill-in-the-blank exercises are designed to assure that you understand the key vocabulary used in this section. Each sentence comes directly from the section. You will find the correct answers in Appendix C.

Section 1.5

(a) Given any two numbers, we can multiply them in either order and get the same _____.

(b) When multiplying a denominate number, like 6 ft, by an abstract number like 5, the result has the same _____ as the denominate number.

(c) When a whole number is multiplied by a power of ten, the product is just that number followed by as many _____ as there are in that power of ten.

(d) The last step in solving an application is to check to make sure you have answered the question and to make sure that your answer seems _____.

CHECK YOURSELF ANSWERS

1. Factors 2, 9; product 18   **2.** 24   **3.** $3 \cdot 7 = 21$ and $15 + 6 = 21$
4. 204   **5.** **(a)** 19; **(b)** 9; **(c)** 0; **(d)** 0   **6.** $235   **7.** 25,700   **8.** 172,200
9. 280,000   **10.** $350   **11.** 2,888   **12.** 87,648   **13.** 148,656
14. **(a)** 90; **(b)** 90   **15.** 1,250 in.^{2} **16.** $192   **17.** 11 in.^{2}

1.5 Exercises

1. Find $3 \cdot 7$ and $7 \cdot 3$ by repeated addition.

2. Find $4 \cdot 5$ and $5 \cdot 4$ by repeated addition.

3. If $6 \cdot 7 = 42$, we call 6 and 7 _____ of 42. And 42 is the _____ of 6 and 7.

4. If $5 \cdot 8 = 40$, we call 5 and 8 _____ of 40. And 40 is the _____ of 5 and 8.

5. Find the number of 1-ft^2 tiles in a floor that is 9 ft long and 12 ft wide.

6. Find the number of squares in a quilt that has four squares in each of four rows.

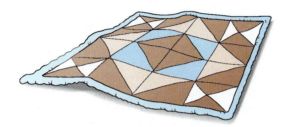

Multiply.

7. $8 \cdot 3$ yd

8. $9 \cdot \$15$

9. $6 \cdot 4°C$

10. $8 \cdot 11$ mi

11. $\begin{array}{r} 2 \\ \times\ 9 \\ \hline \end{array}$ **ALEKS**

12. $\begin{array}{r} 1 \\ \times\ 7 \\ \hline \end{array}$ **ALEKS**

13. $\begin{array}{r} 23 \\ \times\ 2 \\ \hline \end{array}$ **ALEKS**

14. $\begin{array}{r} 32 \\ \times\ 3 \\ \hline \end{array}$ **ALEKS**

15. $\begin{array}{r} 523 \\ \times\ 8 \\ \hline \end{array}$ **ALEKS**

16. $\begin{array}{r} 2,035 \\ \times\ 9 \\ \hline \end{array}$ **ALEKS**

17. $\begin{array}{r} 327 \\ \times\ 59 \\ \hline \end{array}$ **VIDEO**
ALEKS

18. $\begin{array}{r} 2,364 \\ \times\ 67 \\ \hline \end{array}$

19. $\begin{array}{r} 4,075 \\ \times\ 84 \\ \hline \end{array}$

20. $\begin{array}{r} 315 \\ \times\ 243 \\ \hline \end{array}$

21. $\begin{array}{r} 58 \\ \times\ 40 \\ \hline \end{array}$ **ALEKS**

22. $\begin{array}{r} 562 \\ \times\ 400 \\ \hline \end{array}$

23. $\begin{array}{r} 907 \\ \times\ 900 \\ \hline \end{array}$

24. $\begin{array}{r} 345 \\ \times\ 230 \\ \hline \end{array}$

Name _____

Section _____ Date _____

ANSWERS

1. _____
2. _____
3. _____
4. _____
5. _____
6. _____
7. _____
8. _____
9. _____
10. _____
11. _____
12. _____
13. _____
14. _____
15. _____
16. _____
17. _____
18. _____
19. _____
20. _____
21. _____
22. _____
23. _____
24. _____

25. _____

26. _____

27. _____

28. _____

29. _____

30. _____

31. _____

32. _____

33. _____

34. _____

35. _____

36. _____

37. _____

38. _____

39. _____

40. _____

25. Find the product of 304 and 7.   **26.** Find the product of 409 and 4.

27. What is the product of 21 and 551?   **28.** What is the product of 112 and 168?

Name the property of addition and/or multiplication that is illustrated.

29. $5 \cdot 8 = 8 \cdot 5$ **ALEKS**   **30.** $2 \cdot (3 \cdot 5) = (2 \cdot 3) \cdot 5$ **ALEKS**

31. $3 \cdot (2 + 8) = (3 \cdot 2) + (3 \cdot 8)$ **ALEKS**   **32.** $4 \cdot 1 = 4$ **ALEKS**

In exercises 33 and 34, complete the statement using the given property.

33. $3 \cdot (2 + 7) =$ Distributive property   **ALEKS**

34. $9 \cdot (8 \cdot 5) =$ Associative property of multiplication   **ALEKS**

In exercises 35 to 44, solve the applications.

35. Business and Finance A convoy company can transport eight new cars on one of its trucks. If 34 truck shipments were made in 1 week, how many cars were shipped?

36. Business and Finance A computer printer can print 40 mailing labels per minute. How many labels can be printed in 1 h?

37. Business and Finance A ream of paper is 500 sheets. If 29 reams of paper were used in a copy machine during 1 week, how many copies were made?

38. Science and Medicine If sound waves travel at a rate of 1,088 feet per second (ft/s) and you hear thunder 23 s after seeing a lightning flash, how far away did the lightning flash?

39. Business and Finance The manufacturer of woodburning stoves can make 15 stoves in 1 day. How many stoves can be made in 28 days?

40. Business and Finance Each bundle of newspapers contains 25 papers. If 43 bundles are delivered to Jose's house, how many papers are delivered?

41. Business and Finance Erin agrees to buy a boat by paying $2,500 down and $139 a month for 36 months. What is the total cost of the boat?

42. Business and Finance Celeste harvested 34 bushels of corn per acre from 32 acres in June and 43 bushels of corn per acre from 36 acres in July. How many bushels of corn did Celeste harvest?

43. Construction The table shows the hourly wages of four different types of jobs at a home remodeling company.

Job	Hourly Wage
Electrician	$27
Plumber	$22
Clerk	$12
Accountant	$18

Based on the architectural plans for an addition, it is estimated that the remodel will require four electricians each working 50 h and two plumbers each working 21 h. In addition, 4 h of clerical work and 6 h of accounting are needed. What is the total cost of the job?

44. Business and Finance The monthly sales at the Magic Carpet used car dealership are given as

Car Model	Average Price per Sale	No. of Cars Sold
Honda Accord	$29,000	23
Volkswagon Jetta	$26,900	38
Pontiac Grand AM	$31,700	18

What are the gross receipts for the month at the dealership?

45. _____

46. _____

47. _____

48. _____

49. _____

50. _____

51. _____

52. _____

53. _____

54. _____

55. _____

56. _____

57. _____

58. _____

59. _____

60. _____

Estimate each of the products by rounding each factor to the nearest hundred.

45. 212
× 278

46. 179
× 431

47. 391
× 531

48. 729
× 481

Multiply. Be sure to use the proper units in your answer.

49. 3 ft · 2 ft

50. 5 mi · 13 mi

51. 17 in. · 11 in.

52. 143 yd · 26 yd

Label the statements as true or false.

53. $(10 \text{ ft})^2 = 100 \text{ ft}$

54. $(5 \text{ mi})^2 = 25 \text{ mi}^2$

55. $(8 \text{ yd})^2 = 512 \text{ yd}^2$

56. $(9 \text{ in.})^2 = 9 \text{ in.}^2$

Find the area of each figure.

57.

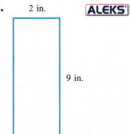

6 yd

6 yd

ALEKS

58.

2 in.

9 in.

ALEKS

59.

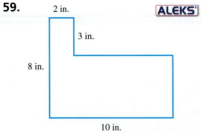

2 in.

3 in.

8 in.

10 in.

ALEKS

60.

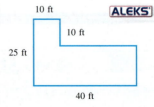

10 ft

10 ft

25 ft

40 ft

ALEKS

61.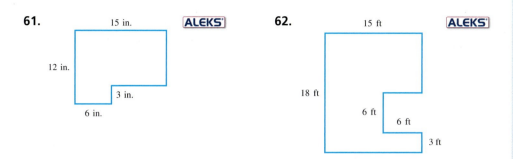

15 in.

12 in.

3 in.

6 in.

62.

15 ft

18 ft

6 ft

6 ft

3 ft

In exercises 63 to 68, solve the applications.

63. Business and Finance   A company can manufacture 45 sleds per day. Approximately how many can this company make in 128 days?

64. Business and Finance   The attendance at a basketball game was 2,845. The cost of admission was $12 per person. Estimate the total gate receipts for the game.

65. Construction   A plate glass window measures 5 ft by 7 ft. If glass costs $8 per square foot, how much will it cost to replace the window?

66. Construction   In a hallway, Bill is painting two walls that are 10 ft high by 22 ft long. The instructions on the paint can say that it will cover 400 ft^2 per gallon (gal). Will 1 gal be enough for the job?

67. Construction   Tile for a kitchen counter will cost $7 per square foot to install. If the counter measures 12 ft by 3 ft, what will the tile cost?

68. Construction   You wish to cover a floor 4 yd by 5 yd with a carpet costing $13 per square yard. What will the carpeting cost?

ANSWERS

61. _____

62. _____

63. _____

64. _____

65. _____

66. _____

67. _____

68. _____

69. **Construction** Suppose you wish to build a small, rectangular pen, and you have enough fencing for the pen's perimeter to be 36 ft. Assuming that the length and width are to be whole numbers, answer each question. **ALEKS**

   (a) What are the possible dimensions that the pen could have? (Note: a square is a type of rectangle.)
   (b) For each set of dimensions (length and width), what is the area that the pen would enclose?
   (c) Which dimensions give the greatest area?
   (d) What is the greatest area?

70. **Construction** Suppose you wish to build a rectangular kennel that encloses 100 ft². Assuming that the length and width are to be whole numbers, answer each question. **ALEKS**

   (a) What are the possible dimensions that the kennel could have? (Note: a square is a type of rectangle.)
   (b) For each set of dimensions (length and width), what is the perimeter that would surround the kennel?
   (c) Which dimensions give the least perimeter?
   (d) What is the least perimeter?

71. We have seen that addition and multiplication are commutative operations. Decide which of the listed activities are commutative.

   (a) Taking a shower and eating breakfast
   (b) Getting dressed and taking a shower
   (c) Putting on your shoes and your socks
   (d) Brushing your teeth and combing your hair
   (e) Putting your key in the ignition and starting your car

72. The associative properties of addition and multiplication indicate that the result of the operation is the same regardless of where the grouping symbol is placed. This is not always the case in the use of the English language. Many phrases can have different meanings based on how the words are grouped. For each phrase, explain why the associative property would not hold.

   (a) Cat fearing dog   (b) Hard test question
   (c) Defective parts department   (d) Man eating animal

   Write some phrases in which the associative property is satisfied.

73. Social Science Most maps contain legends that allow you to convert the distance between two points on the map to actual miles. For instance, if a map uses a legend that equates 1 in. to 5 mi and the distance between two towns is 4 in. on the map, then the towns are actually 20 mi apart.

(a) Obtain a map of your state and determine the shortest distance between any two major cities.

(b) Could you actually travel the route you measured in part **(a)**?

(c) Plan a trip between the two cities you selected in part **(a)** over established roads. Determine the distance that you actually travel using this route.

74. Calculate the product 378 · 215 in two ways.

Method 1: Round each factor to the nearest hundred and then multiply.

Method 2: Multiply first and then round the product to the nearest hundred.

(a) Compare your answers and comment on the difference between the two results.

(b) List the advantages and disadvantages of each method.

(c) Describe situations in which each method is the preferred approach.

75. There are many different ways of rounding. One way used in computer applications is called **truncating.**

(a) Determine what rules would be used in truncating and compare them to the rules used in rounding.

(b) Round 7,473 to the hundreds place using truncating and rounding.

(c) State some possible problems that could occur in truncating.

76. Maria has been asked to estimate the number of pieces of paper in five large piles. She does not want to count every piece. Devise a plan to help her estimate the total number of pieces of paper.

77. Complete the following number cross.

Across
1. 6 × 551
5. 7 × 8
6. 27 × 27
7. 19 × 50
10. 3 × 67
12. 6 × 25
13. 9 × 8
15. 16 × 303

Down
1. 5 × 7
2. 9 × 41
3. 67 × 100
4. 2 × (49 + 100)
8. 4 × 1,301
9. 100 + 10 + 1
11. 2 × 87
14. 25 + 3

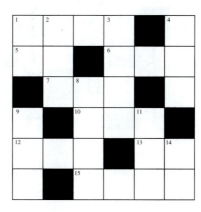

Solve each chapter-activity application.

78. Crafts   A package has a base that is 13 in. wide and 28 in. long. Its height is 3 in. Find the area of the base of the package. Find the area of each of the other two distinct sides of the package. Give the sum of the areas of the six sides of this package (this is called the **total surface area** of the box).

79. Crafts   The amount of wrapping paper needed to wrap the box in exercise 78 is subject to the following criteria. The width of the wrapping paper needs to be equal to the length of the package plus 2 in. per side (for a total of 4 in. added to the length of the package). The length of the wrapping paper needs to be 1 in. more than the girth of the package (twice the width of the package plus twice the height of the package). What are the dimensions of a piece of wrapping paper that will wrap the package? What is the area of this piece of wrapping paper?

80. Crafts   Referring to exercises 78 and 79, how much more wrapping paper do you need than the total surface area of your package?

Answers

1. 21   **3.** Factors, product   **5.** 108 tiles   **7.** 24 yd   **9.** 24°C
11. 18   **13.** 46   **15.** 4,184   **17.** 19,293   **19.** 342,300
21. 2,320   **23.** 816,300   **25.** 2,128   **27.** 11,571
29. Commutative property of multiplication
31. Distributive property of multiplication over addition   **33.** $(3 \cdot 2) + (3 \cdot 7)$
35. 272 cars   **37.** 14,500 copies   **39.** 420 stoves   **41.** $7,504
43. $6,480   **45.** 60,000   **47.** 200,000   **49.** 6 ft^2 **51.** 187 in.^{2}
53. False   **55.** False   **57.** 36 yd^2 **59.** 56 in.^{2} **61.** 153 in.^{2}
63. 6,000 sleds $(50 \cdot 120)$ **65.** $280   **67.** $252   **69.**

71.    **73.**   **75.**

77.

3	3	0	6	■	2
5	6	■	7	2	9
■	9	5	0	■	8
1	■	2	0	1	■
1	5	0	■	7	2
1	■	4	8	4	8

79. 32 in. by 33 in.; 1,056 in.^{2}

1.6 Division of Whole Numbers

1.6 OBJECTIVES

1. Use repeated subtraction to divide whole numbers
2. Check the results of a division problem
3. Divide whole numbers using long division
4. Estimate a quotient

Overcoming Math Anxiety

Learn to Take Useful Notes

Although some students find it easier to be organized than do other students, every student can become a better note taker. Below are some hints that can help you learn to take more useful notes. Good note taking begins with your preparation for class. Note that the first several items refer to your preparation.

1. **Read the assigned material before the lecture.** This helps you become familiar with both the vocabulary and the concepts.

2. **Review your notes from the previous class meeting.** If you already have a concept in your lecture notes and the idea is referred to again, you then simply jot a note to refer back to previous material in your notes.

3. **Get to class a few minutes early.** Have your materials out and ready to go when your instructor walks in the door.

4. **Be ready to listen** as soon as the instructor walks in the door.

5. Have more than one pencil sharpened and ready to use.

6. Have a highlighter or colored pen available to mark particularly important segments of your notes.

7. **Know how to spell.** This includes both plain old English words and technical words that have already been presented in class. If you can spell them, you won't have to waste time trying to figure out *How* to spell them.

8. **Be aggressive in notetaking.** Don't wait for an idea to strike you—it's better to have too much material than too little.

9. If the professor repeats something, **write it down**.

10. **Take notes, not dictation.** That means being able to develop your own form of shorthand.

11. **Develop abbreviations** for words that are used frequently in the course.
    - Real numbers = R
    - Natural numbers = N

12. Use symbols when you can.
    - & = and
    - B = but
    - ∀ = for each
    - ∴ = therefore
    - ∃ = there exists
    - ∋ = such that
    - ∈ = is an element of

13. **Skip lines.** Leave visual breaks between definitions, lists, or explanations.

14. If you miss something, **leave a blank in your notes.** You can fill it in later. If you try to copy it from your neighbor during the lecture, both of you will lose more material.

15. **Get together with your classmates.** Do so after lecture and pool your notes. That way, you can be sure you have everything down. It will also help make sure you understand what you have written down.

We now examine a fourth arithmetic operation, division. Just as multiplication was repeated addition, division is repeated subtraction. Division asks *how many times* one number is contained in another.

OBJECTIVE 1

> **Example 1** **Dividing by Using Subtraction**

Joel needs to set up 48 chairs in the student union for a concert. If there is room for 8 chairs per row, how many rows will it take to set up all 48 chairs?

This problem can be solved by subtraction. Each row subtracts another 8 chairs.

48	40	32	24	16	8
−8	−8	−8	−8	−8	−8
40	32	24	16	8	0

Because 8 can be subtracted from 48 six times, there will be 6 rows.

This can also be seen as a division problem

NOTE Each of these notations represent the same division problem, "48 divided by 8 is 6."

$$48 \div 8 = 6 \qquad \text{or} \qquad 8\overline{)48}^{\,6} \qquad \text{or} \qquad \frac{48}{8} = 6$$

No matter which notation we use, we call the 48 the **dividend,** the 8 the **divisor,** and the 6 the **quotient.**

✓ **CHECK YOURSELF 1** _____

Carlotta is creating a garden path made of bricks. She has 72 bricks. Each row will have 6 bricks in it. How many rows can she make?

Units Analysis

When dividing a denominate number by an abstract number, the result has the units of the denominate number. Here are a couple of examples.

76 trombones ÷ 4 = 19 trombones

$55 ÷ 11 = $5

When one denominate number is divided by another, the result has the units of the dividend over the units of the divisor.

144 mi ÷ 6 gal = 24 mi/gal (which we read as "miles per gallon")

$120 ÷ 8 h = 15 dollars/h ("dollars per hour")

To solve a problem that requires division, you first set up the problem as a division statement. Example 2 illustrates this.

Example 2   Writing a Division Statement

Write a division statement that corresponds to the following situation. You need not do the division.

The staff at the Wok Inn Restaurant splits all tips at the end of each shift. Yesterday's evening shift collected a total of $224. How much should each of the seven employees get in tips?

$224 ÷ 7 employees   Note that the units for the answer will be "dollars per employee."

 CHECK YOURSELF 2 _____

Write a division statement that corresponds to the following situation. You need not do the division.

All nine sections of basic math skills at SCC (Sum Community College) are full. There are a total of 315 students in the classes. How many students are in each class? What are the units for the answer?

In Section 1.5, we used a rectangular array of stars to represent multiplication. These same arrays can represent division. Just as $3 \cdot 4 = 12$ and $4 \cdot 3 = 12$, so is it true that $12 \div 3 = 4$ and $12 \div 4 = 3$.

```
★ ★ ★ ⎤
★ ★ ★ ⎥
★ ★ ★ ⎬ 4 • 3 = 12      ★ ★ ★ ★ ⎤
★ ★ ★ ⎦                  ★ ★ ★ ★ ⎬ 3 • 4 = 12
                         ★ ★ ★ ★ ⎦
     or                        or
12 ÷ 3 = 4                12 ÷ 4 = 3
```

This relationship allows us to check our division results by doing multiplication.

OBJECTIVE 2

NOTE For a division problem to check, the *product* of the divisor and the quotient *must equal the dividend.*

Example 3   Checking Division by Using Multiplication

(a) $7\overline{)21}$ with quotient 3   Check: $7 \cdot 3 = 21$

(b) $48 \div 6 = 8$ Check: $6 \cdot 8 = 48$

CHECK YOURSELF 3 _____

Complete the division statements and check your results.

(a) $9\overline{)45}$ **(b)** $28 \div 7 =$

NOTE Because $36 \div 9 = 4$, we say that 36 is *exactly divisible* by 9.

In our examples so far, the product of the divisor and the quotient has been equal to the dividend. This means that the dividend is *exactly divisible* by the divisor. That is not always the case. In Example 4, we are again using repeated subtraction.

| Example 4 | Dividing by Using Subtraction, Leaving a Remainder |

How many times is 5 contained in 23?

NOTE The remainder must be smaller than the divisor or we could subtract again.

$$
\begin{array}{cccc}
23 & 18 & 13 & 8 \\
-\ 5 & -\ 5 & -\ 5 & -\ 5 \\
\hline
18 & 13 & 8 & 3
\end{array}
$$

We see that 5 is contained 4 times in 23, but 3 is "left over."

23 is not exactly divisible by 5. The "leftover" 3 is called the **remainder** in the division.

CHECK YOURSELF 4 _____

How many times is 7 contained in 38?

Property: Remainder

Dividend = divisor · quotient + remainder

We can check our work in a division problem with a remainder as follows.

| Example 5 | Checking Division by a Single-Digit Number |

Using the work of Example 4, we can write

NOTE Another way to write the result is

$$
\begin{array}{r}
4\ \text{r3} \\
5\overline{)23}
\end{array}
$$

The "r" stands for remainder.

$$
\begin{array}{r}
4 \\
5\overline{)23}
\end{array}
$$ with remainder 3

To apply the remainder property, we have

NOTE The multiplication is done before the 3 is added.

Divisor   Quotient

Dividend $\longrightarrow$ $23 = 5 \cdot 4 + 3$ $\longleftarrow$ Remainder

$23 = \ 20 \ + 3$

$23 = \ 23$ The division checks.

CHECK YOURSELF 5 _____

Evaluate $7\overline{)38}$. Check your answer.

We must be careful when 0 is involved in a division problem. There are two special cases.

> **Property:** Division and Zero
>
> **1.** 0 divided by any whole number (except 0) is 0.
> **2.** Division by 0 is undefined.

The first case involving zero occurs when we are dividing into zero.

Example 6   Dividing into Zero

$0 \div 5 = 0$ because $0 = 5 \cdot 0$.

 CHECK YOURSELF 6

(a) $0 \div 7 =$ **(b)** $0 \div 12 =$

Our second case illustrates what happens when 0 is the *divisor*. Here we have a special problem.

Example 7   Dividing by Zero

$8 \div 0 = ?$ This means that $8 = 0 \cdot ?$

Can 0 times some number ever be 8? From our multiplication facts, the answer is *no!* There is no answer to this problem, so we say that $8 \div 0$ is undefined.

 CHECK YOURSELF 7

Decide whether each problem results in 0 or is undefined.

(a) $9 \div 0$ **(b)** $0 \div 9$ **(c)** $0 \div 15$ **(d)** $15 \div 0$

It is easy to divide when small whole numbers are involved, because much of the work can be done mentally. In working with larger numbers, we turn to a process called **long division.** This is a method for performing the steps of repeated subtraction.

To start, we can look at an example in which we subtract multiples of the divisor.

OBJECTIVE 3

Example 8   Dividing by a Single-Digit Number

NOTE With larger numbers, repeated subtraction is just too time-consuming to be practical.

Divide 176 by 8.

Because 20 eights are 160, we know that there are at least 20 eights in 176.

Step 1 Write

$$
\begin{array}{r}
20 \\
8)\overline{176} \\
160 \\
\hline
16
\end{array}
$$

20 eights ⟶ (160)

Subtracting 160 is just a shortcut for subtracting eight 20 times.

After subtracting the 20 eights, or 160, we are left with 16. There are 2 eights in 16, and so we continue.

Step 2
$$
\begin{array}{r}
2 \\
20 \\
\hline
8)\overline{176} \\
160 \\
\hline
16 \\
16 \\
\hline
0
\end{array}
$$
22   Adding 20 and 2 gives us the quotient, 22.

2 eights ⟶ 16

Subtracting the 2 eights, we have a 0 remainder. So $176 \div 8 = 22$.

CHECK YOURSELF 8

Verify the results of Example 8, using multiplication.

The next step is to simplify this repeated-subtraction process one step further. The result is the long-division method.

Example 9   Dividing by a Single-Digit Number

Divide 358 by 6.

The dividend is 358. We look at the first digit, 3. We cannot divide 6 into 3, so we look at the *first two digits,* 35. There are 5 sixes in 35, and so we write 5 above the tens digit of the dividend.

$$
\begin{array}{r}
5 \\
\hline
6)\overline{358}
\end{array}
$$
When we place 5 as the tens digit, we really mean 5 tens, or 50.

Now multiply $5 \cdot 6$, place the product below 35, and subtract.

$$
\begin{array}{r}
5 \\
\hline
6)\overline{358} \\
30 \\
\hline
5
\end{array}
$$
We have actually subtracted 50 sixes (300) from 358.

Because the remainder, 5, is smaller than the divisor, 6, we bring down 8, the ones digit of the dividend.

$$
\begin{array}{r}
5 \\
\hline
6)\overline{358} \\
30\downarrow \\
\hline
58
\end{array}
$$

Now divide 6 into 58. There are 9 sixes in 58, and so 9 is the ones digit of the quotient. Multiply $9 \cdot 6$ and subtract to complete the process.

NOTE Because the 4 is smaller than the divisor, we have a remainder of 4.

$$
\begin{array}{r}
59 \\
\hline
6)\overline{358} \\
30\downarrow \\
\hline
58 \\
54 \\
\hline
4
\end{array}
$$
We now have:
$358 \div 6 = 59\ \text{r}4$

NOTE Verify that this is true and that the division checks.

To check: $358 = 6 \cdot 59 + 4$.

✔ CHECK YOURSELF 9 _____

Divide $7\overline{)453}$

Long division becomes a bit more complicated when we have a two-digit divisor. It becomes, in part, a matter of trial and error. We round the divisor and dividend to form a *trial divisor* and a *trial dividend.* We then estimate the proper quotient and must determine whether our estimate is correct.

Example 10   Dividing by a Two-Digit Number

Divide

$38\overline{)293}$

Round the divisor and dividend to the nearest ten. So 38 is rounded to 40, and 293 is rounded to 290. The trial divisor is then 40, and the trial dividend is 290.

NOTE Think: $4\overline{)29}^{7}$

Now look at the nonzero digits in the trial divisor and dividend. They are 4 and 29. We know that there are 7 fours in 29, and so 7 is our first estimate of the quotient. Now we will see if 7 works.

$$
\begin{array}{r}
7 \;\longleftarrow\; \text{Your estimate} \\
38\overline{)293} \\
\underline{266} \\
27
\end{array}
$$

Multiply 7 · 38. The product, 266, is less than 293, and so we can subtract.

The remainder, 27, is less than the divisor, 38, and so the process is complete.

$293 \div 38 = 7 \text{ r}27$

Check: $293 = 38 \cdot 7 + 27.$ You should verify that this statement is true.

✔ CHECK YOURSELF 10 _____

Divide.

$57\overline{)482}$

Because this process is based on estimation, our first guess will sometimes be wrong.

Example 11   Dividing by a Two-Digit Number

Divide

$54\overline{)428}$

Rounding to the nearest ten, we have a trial divisor of 50 and a trial dividend of 430.
Looking at the nonzero digits, how many fives are in 43? There are 8. This is our first estimate.

NOTE Think: $5\overline{)43}^{8}$

$$
\begin{array}{r}
8 \\
54\overline{)428} \\
\underline{432} \;\longleftarrow\; \text{Too large}
\end{array}
$$

We multiply 8 · 54. Do you see what's wrong? The product, 432, is too large. We cannot subtract. Our estimate of the quotient must be adjusted *downward.*

© 2010 McGraw-Hill Companies

NOTE If we tried 6 as the quotient

$$
\begin{array}{r}
6 \\
54\overline{)428} \\
324 \\
\hline 104
\end{array}
$$

We have 104, which is too large to be a remainder.

We adjust the quotient downward to 7. We can now complete the division.

$$
\begin{array}{r}
7 \\
54\overline{)428} \\
378 \\
\hline 50
\end{array}
$$

We have

$428 \div 54 = 7 \text{ r}50$

Check: $428 = 54 \cdot 7 + 50$.

✔ **CHECK YOURSELF 11**

Divide.

$63\overline{)557}$

We have to be careful when a 0 appears as a digit in the quotient. Next, we look at an example in which this happens with a two-digit divisor.

Example 12   Dividing with Large Dividends

Divide

$32\overline{)9871}$

NOTE Our divisor, 32, divides into 98, the first two digits of the dividend.

Rounding to the nearest ten, we have a trial divisor of 30 and a trial dividend of 100. Think, "How many threes are in 10?" There are 3, and this is our first estimate of the quotient.

$$
\begin{array}{r}
3 \\
32\overline{)9871} \\
96 \\
\hline 2
\end{array}
$$ Everything seems fine so far!

Bring down 7, the next digit of the dividend.

$$
\begin{array}{r}
30 \\
32\overline{)9871} \\
96\downarrow \\
\hline 27
\end{array}
$$ Now do you see the difficulty? We cannot divide 32 into 27, and so we place 0 in the tens place of the quotient to indicate this fact.

We continue by multiplying by 0. After subtraction, we bring down 1, the last digit of the dividend.

$$
\begin{array}{r}
30 \\
32\overline{)9871} \\
96 \\
\hline 27 \\
00\downarrow \\
\hline 271
\end{array}
$$

Another problem develops here. We round 32 to 30 for our trial divisor, and we round 271 to 270, which is the trial dividend at this point. Our estimate of the last digit of the quotient must be 9.

```
      309
32)9871
     96
     ‾‾
     27
     00
     ‾‾
    271
    288  ←——— Too large
    ‾‾‾
```

We cannot subtract. The trial quotient must be adjusted downward to 8. We can now complete the division.

```
      308
32)9871
     96
     ‾‾
     27
     00
     ‾‾
    271
    256
    ‾‾‾
     15
```

$9{,}871 \div 32 = 308 \text{ r}15$

Check: $9{,}871 = 32 \cdot 308 + 15$.

CHECK YOURSELF 12

Divide.

43)8857

Because of the availability of the handheld calculator, it is rarely necessary that people find the exact answer when performing long division. On the other hand, it is frequently important that one be able to either estimate the result of long division or confirm that a given answer (particularly from a calculator) is reasonable. As a result, the emphasis in this section will be to improve your estimation skills in division.

In Example 13, we divide a four-digit number by a two-digit number. Generally, we round the divisor to the nearest ten and the dividend to the nearest hundred.

OBJECTIVE 4   **Example 13   Estimating the Result of a Division Application**

The Ramirez family took a trip of 2,394 mi in their new car, using 77 gal of gas. Estimate their gas mileage (mi/gal).

Our estimate will be based on dividing 2,400 by 80.

```
     30
80)2400
```

They got approximately 30 mi/gal.

CHECK YOURSELF 13 _____

Troy flew a light plane on a trip of 2,844 mi that took 21 h. What was his approximate speed in miles per hour?

As before, we may have to combine operations to solve an application of the mathematics you have learned.

> **Example 14** **Estimating the Result of a Division Application**

Charles purchases a used car for $8,574. He agrees to make payments for 4 years. Interest charges will be $978. Approximately what should his monthly payments be?

First, we find the amount that Charles owes:

$8,574 + $978 = $9,552

Now, to find the monthly payment, we divide that amount by 48 (months). To estimate the payment, we divide $9,600 by 50 months.

$$\begin{array}{r} 192 \\ 50\overline{)9600} \end{array}$$

The payments will be approximately $192 per month.

CHECK YOURSELF 14 _____

One $10 bag of fertilizer will cover 310 ft². Approximately what would it cost to cover 2,200 ft²?

Using a Scientific Calculator to Divide

Of course, division is easily done using your calculator. However, as we will see, some special things come up when we use a calculator to divide. First we outline the steps of division as it is done on a calculator.

Divide $35\overline{)2380}$.

Step 1 Enter the dividend.   2380

Step 2 Press the divide key. $\div$

Step 3 Enter the divisor.   35

RECALL A graphing calculator uses the [Enter] key rather than $=$.

Step 4 Press the equals key. $=$ The desired quotient is now in your display.

The display shows 68.

We have already mentioned some of the difficulties related to division with 0. We will experiment on the calculator.

Example 15   Using a Scientific Calculator to Divide

To find $0 \div 5$, we use this sequence:

$0 \boxed{\div} 5 \boxed{=}$

Display   0

There is no problem with this. Zero divided by any whole number other than 0 is just 0.

 CHECK YOURSELF 15 _____

What is the result when you use your calculator to perform the given operation?

$0 \div 17$

We see what happens when dividing zero by another number, but what happens when we try to divide by zero? More importantly to this section, how does the calculator handle division by zero? Example 16 illustrates this concept.

Example 16   Using a Scientific Calculator to Divide

To find $5 \div 0$, we use this sequence:

$5 \boxed{\div} 0 \boxed{=}$

Display   Error

NOTE You may find that you must "clear" your calculator after trying this.

If we try this sequence, the calculator gives us an error! Do you see why? Division by 0 is not allowed. Try this on your calculator to see how this error is indicated.

 CHECK YOURSELF 16 _____

What is the result when you use your calculator to perform the given operation?

$17 \div 0$

Another special problem comes up when a remainder is involved in a division problem.

Example 17   Using a Scientific Calculator to Divide

Dividing 293 by 38 gives 7 with remainder 27.

NOTE Be aware that the calculator will not give you a remainder in the form we have been using in this chapter.

$293 \boxed{\div} 38 \boxed{=} \boxed{7.7105263}$

Quotient   Remainder

7 is the *whole-number* part of the quotient as before.

0.7105263 is the *decimal form* of the remainder, 27, as a fraction of 38.

CHECK YOURSELF 17 _____

What is the result when you use your calculator to perform the given operation?

$458 \div 36$

The calculator can also help you combine division with other operations.

Example 18   Using a Scientific Calculator to Divide

To find $18 \div 2 + 3$, use this sequence:

18 ÷ 2 + 3 =

Display   12   Do you see that the calculator has
                        done the division as the first step?

CHECK YOURSELF 18 _____

Use your calculator to compute.

$15 \div 5 + 7$

Example 19   Using a Scientific Calculator to Divide

To find $6 \div 3 \cdot 2$, use this sequence:

6 ÷ 3 × 2 =

Display   4

CHECK YOURSELF 19 _____

Use your calculator to compute.

$18 \div 6 \cdot 5$

READING YOUR TEXT

The following fill-in-the-blank exercises are designed to assure that you understand the key vocabulary used in this section. Each sentence comes directly from the section. You will find the correct answers in Appendix C.

Section 1.6

(a) If you read the assigned material before the lecture, you will become familiar with both the _____ and the concepts.

(b) Just as multiplying was repeated addition, division is repeated _____.

(c) You can check division by using _____.

(d) Division by _____ is undefined.

CHECK YOURSELF ANSWERS

1. 12   **2.** 315 students ÷ 9 classes; students per class   **3.** **(a)** 5; $9 \cdot 5 = 45$;
(b) 4; $7 \cdot 4 = 28$ **4.** 5   **5.** 5 with remainder 3   **6.** **(a)** 0; **(b)** 0
7. **(a)** undefined; **(b)** 0; **(c)** 0; **(d)** undefined   **8.** $8 \cdot 22 = 176$
9. 64 with remainder 5   **10.** 8 with remainder 26   **11.** 8 with remainder 53
12. 205 with remainder 42   **13.** 140 mi/h   **14.** $70   **15.** 0
16. Error message   **17.** 12.72222   **18.** 10   **19.** 15

Name _____

Section _____ Date _____

ANSWERS

1. _____
2. _____
3. _____ 4. _____
5. _____ 6. _____
7. _____ 8. _____
9. _____ 10. _____
11. _____ 12. _____
13. _____
14. _____
15. _____ 16. _____
17. _____ 18. _____
19. _____ 20. _____
21. _____
22. _____
23. _____ 24. _____
25. _____ 26. _____
27. _____ 28. _____
29. _____ 30. _____
31. _____ 32. _____
33. _____ 34. _____
35. _____ 36. _____
37. _____
38. _____

1.6 Exercises

1. Given $48 \div 8 = 6$, 8 is the _____, 48 is the _____, and 6 is the _____.

2. In the statement $5)\overline{45}$ (with 9 above), 9 is the _____, 5 is the _____, and 45 is the _____.

3. Find $36 \div 9$ by repeated subtraction.

4. Find $40 \div 8$ by repeated subtraction.

5. **Problem Solving** Stefanie is planting rows of tomato plants. She wants to plant 63 plants with 9 plants per row. How many rows will she have? **ALEKS**

6. **Construction** Nick is designing a parking lot for a small office building. He must make room for 42 cars with 7 cars per row. How many rows should he plan for? **ALEKS**

Divide and identify the correct units for the quotient.

7. 36 pages $\div$ 4 **ALEKS**

8. $96 \div 8$ **ALEKS**

9. 4,900 km $\div$ 7 **ALEKS**

10. 360 gal $\div$ 18 **ALEKS**

11. 160 mi $\div$ 4 h **VIDEO** **ALEKS**

12. 264 ft $\div$ 3 s

13. 3,720 h $\div$ 5 months

14. 560 cal $\div$ 7 g **ALEKS**

Divide using long division and check your work.

15. $5)\overline{43}$ **ALEKS**

16. $40 \div 9$ **ALEKS**

17. $9)\overline{65}$ **VIDEO** **ALEKS**

18. $6)\overline{51}$ **ALEKS**

19. $57 \div 8$ **ALEKS**

20. $74 \div 8$ **ALEKS**

21. $0 \div 6$

22. $18 \div 0$

Divide.

23. $5)\overline{83}$ **ALEKS**

24. $9)\overline{78}$ **ALEKS**

25. $8)\overline{293}$ **ALEKS**

26. $7)\overline{346}$ **ALEKS**

27. $8)\overline{3136}$ **ALEKS**

28. $9)\overline{3527}$

29. $8)\overline{22,153}$ **VIDEO**

30. $5)\overline{43,287}$

31. $48)\overline{892}$ **ALEKS**

32. $54)\overline{372}$ **ALEKS**

33. $45)\overline{2367}$

34. $53)\overline{3480}$

35. $763)\overline{3071}$

36. $871)\overline{4321}$

Solve the applications.

37. **Statistics** Ramon bought 56 bags of candy. There were 8 bags in each box. How many boxes were there?

38. **Statistics** There are 32 students who are taking a field trip. If each car can hold 4 students, how many cars are needed for the field trip?

39. Business and Finance Ticket receipts for a play were $552. If the tickets were $4 each, how many tickets were purchased? **ALEKS**

40. Construction Construction of a fence section requires 8 boards. If you have 256 boards, how many sections can you build? **ALEKS**

41. Construction The homeowners along a street must share the $2,030 cost of new street lighting. If there are 14 homes, what amount will each owner pay? **VIDEO**

42. Business and Finance A bookstore ordered 325 copies of a textbook at a cost of $7,800. What was the cost to the store for an individual textbook?

43. Business and Finance A company distributes $16,488 in year-end bonuses. If each of the 36 employees receives the same amount, what bonus will each receive?

VIDEO

44. Complete the following number cross.

Across	**Down**
1. 48 ÷ 4	**1.** (12 + 16) ÷ 2
3. 1,296 ÷ 8	**2.** 67 × 3
6. 2,025 ÷ 5	**4.** 744 ÷ 12
8. 4 × 5	**5.** 2,600 ÷ 13
9. 11 × 11	**7.** 6,300 ÷ 12
12. 15 ÷ 3 × 111	**10.** 304 ÷ 2
14. 144 ÷ (2 × 6)	**11.** 5 × (161 ÷ 7)
16. 1,404 ÷ 6	**13.** 9,027 ÷ 17
18. 2,500 ÷ 5	**15.** 400 ÷ 20
19. 3 × 5	**17.** 9 × 5

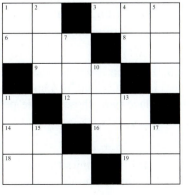

ANSWERS

39. _____

40. _____

41. _____

42. _____

43. _____

44. _____

45. _____

46. _____

47. _____

48. _____

49. _____

50. _____

51. _____

52. _____

53. _____

54. _____

55. _____

Estimate the result in the division problems. (Remember to round divisors to the nearest ten and dividends to the nearest hundred.)

45. 810 divided by 38 **ALEKS**   **46.** 458 divided by 18   **ALEKS**

47. 4,967 divided by 96   **ALEKS**   **48.** 3,971 divided by 39   **ALEKS**

49. 3,812 divided by 188   **ALEKS**   **50.** 5,245 divided by 255   **ALEKS**

In exercises 51 to 54, solve the applications.

51. Technology   Jose drove 279 mi on 18 gal of gas. Estimate his mileage. (*Hint:* Find the number of miles per gallon.)

52. Construction   A contractor can build a house in 27 days. Estimate how many houses can be built in 265 days.

53. Construction   You are going to recarpet your living room. You have budgeted $1,500 for the carpet and installation.

  (a) Determine how much carpet you will need to do the job. Draw a sketch to support your measurements.
  (b) What is the highest price per square yard you can pay and still stay within budget?
  (c) Go to a local store and determine the total cost of doing the job for three different grades of carpet. Be sure to include padding, labor costs, and any other expenses.
  (d) What considerations (other than cost) would affect your decision about what type of carpet to install?
  (e) Write a brief paragraph indicating your final decision and give supporting reasons.

54. Division is the inverse operation of multiplication. Many daily activities have inverses. For each of the following activities, state the inverse activity:

  (a) Spending money
  (b) Going to sleep
  (c) Turning down the volume on your CD player
  (d) Getting dressed

55. Division is not associative. For example, $8 \div 4 \div 2$ will produce different results if 8 is divided by 4 and then divided by 2 or if 8 is divided by the result of $4 \div 2$. Place parentheses in the proper place so that each expression is true.

  (a) $16 \div 8 \div 2 = 4$ **(b)** $16 \div 8 \div 2 = 1$
  (c) $125 \div 25 \div 5 = 1$ **(d)** $125 \div 25 \div 5 = 25$
  (e) Is there any situation in which the order of how the operation of division is performed produces the same result? Give an example.

56. Division is not commutative. For example, $15 \div 5 \neq 5 \div 15$. What must be true of the numbers a and b if $a \div b = b \div a$?

Solve each chapter-activity application.

57. Crafts A set of same-sized packages are each 13 in. wide and have a height of 3 in. Find the girth of each package (twice the width plus twice the height). You need a length of wrapping paper equal to one more inch than the girth of the package. What length of wrapping paper do you need for each package? How many packages can you wrap if your wrapping paper is 300 in. (25 ft) long? How long a piece of scrap will you be left with?

58. Business and Finance If a business needs to wrap 50 packages, each as described in exercise 57, how many rolls of 300-in.-long wrapping paper will they need? What if they needed to wrap 200 such packages?

🖩 Calculator Exercises

Use your calculator to perform the indicated operations.

59. $36{,}182 \div 79$ **60.** $464{,}184 \div 189$

61. $6 + 9 \div 3$ **62.** $18 - 6 \div 3$

63. $24 \div 6 \cdot 4$ **64.** $1{,}176 \div 42 - 1{,}572 \div 524$

65. $3 \cdot 8 \cdot 8 \cdot 8 \div 12$ **66.** $(89 - 14) \div 25$

Answers

1. Divisor, dividend, quotient   **3.** 4   **5.** 7   **7.** 9 pages   **9.** 700 km
11. 40 mi/h   **13.** 744 h/month   **15.** 8 r3   **17.** 7 r2   **19.** 7 r1
21. 0   **23.** 16 r3   **25.** 36 r5   **27.** 392   **29.** 2,769 r1   **31.** 18 r28
33. 52 r27   **35.** 4 r19   **37.** 7 boxes   **39.** 138 tickets   **41.** $145
43. $458   **45.** 20   **47.** 50   **49.** 20   **51.** 15 mi/gal   **53.**
55.   **57.** 32 in.; 33 in.; 9 packages; 3 in.

59. 458   **61.** 9   **63.** 16   **65.** 128

1.7 Exponents and Whole Numbers

1.7 OBJECTIVES

1. Write a repeated product using exponent notation
2. Evaluate a whole number raised to a power
3. Expand a whole number using powers of 10

RECALL In Section 1.5, we saw that we can write repeated addition as multiplication:

$3 + 3 + 3 + 3 = 4 \cdot 3$.

Earlier, we described multiplication as a shorthand for repeated addition. There is also a shorthand for repeated multiplication. It uses **powers of a whole number.**

OBJECTIVE 1

Example 1 Writing Repeated Multiplication as a Power

$3 \cdot 3 \cdot 3 \cdot 3$ can be written as 3^4.
This is read as "3 to the fourth power."

In this case, repeated multiplication is written as the power of a number.

In this example, 3 is the **base** of the expression, and the raised number, 4, is the **exponent,** or **power.**

NOTE René Descartes, a French philosopher and mathematician, is generally credited with first introducing our modern exponent notation in about 1637.

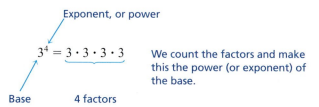

Exponent, or power

$3^4 = \underbrace{3 \cdot 3 \cdot 3 \cdot 3}_{\text{4 factors}}$

Base

We count the factors and make this the power (or exponent) of the base.

 CHECK YOURSELF 1 _____

Write $2 \cdot 2 \cdot 2 \cdot 2 \cdot 2 \cdot 2$ as a power of 2.

Definition: Exponents

The **exponent** tells us the number of times the base is to be used as a factor.

We actually encountered exponents in Section 1.5 when we discussed area. There, our units of area were quantities like 1 in.^{2} and 1 ft^2, which meant 1 in $\cdot$ in and 1 ft $\cdot$ ft, respectively.

We even used exponents to write a *formula* for the area of a square whose side has length s: $A = s^2$.

We can evaluate a number raised to a power by multiplying.

OBJECTIVE 2

Example 2 Evaluating a Number Raised to a Power

Read and evaluate each of the following.

(a) 2^5 is read "2 to the fifth power."

$2^5 = \underbrace{2 \cdot 2 \cdot 2 \cdot 2 \cdot 2}_{\text{5 times}} = 32$

Here 2 is the base, and 5 is the exponent.

2^5 tells us to use 2 as a factor 5 times. The result is 32.

CAUTION

Be careful: 5^3 is *entirely different* from $5 \cdot 3$.

NOTE $5^3 = 125$ whereas $5 \cdot 3 = 15$.

(b) $8^2 = 8 \cdot 8 = 64$ Use 2 factors of 8.

And 8^2 is read "8 to the second power" or "8 squared."

(c) $5^3 = 5 \cdot 5 \cdot 5 = 125$ Use 3 factors of 5.

5^3 is read "5 to the third power" or "5 cubed."

 CHECK YOURSELF 2

Evaluate each of the following.

(a) 3^5 **(b)** 6^2 **(c)** 2^6

We need two special definitions for powers of whole numbers.

Property: Raising a Number to the First Power

A whole number raised to the first power is just that number.

For example, $9^1 = 9$.

NOTE 0^0 is undefined.

Property: Raising a Number to the Zero Power

A whole number, other than 0, raised to the zero power is 1.

For example, $7^0 = 1$.

Example 3   Evaluating Numbers Raised to the Power of Zero or One

(a) $8^0 = 1$ **(b)** $4^0 = 1$ **(c)** $5^1 = 5$ **(d)** $3^1 = 3$

 CHECK YOURSELF 3

Evaluate.

(a) 9^0 **(b)** 7^1

We talked about *powers of 10* when we multiplied by numbers that end in 0. Because the powers of 10 have a special importance, we will list some of them.

$10^0 = 1$

$10^1 = 10$

$10^2 = 10 \cdot 10 = 100$

NOTE 10^3 is just a 1 followed by *three zeros.*

$10^3 = 10 \cdot 10 \cdot 10 = 1,000$

$10^4 = 10 \cdot 10 \cdot 10 \cdot 10 = 10,000$

NOTE 10^5 is a 1 followed by *five zeros.*

$10^5 = 10 \cdot 10 \cdot 10 \cdot 10 \cdot 10 = 100,000$

Do you see why the powers of 10 are so important?

NOTE Archimedes (about 250 B.C.) reportedly estimated the number of grains of sand in the universe to be 10^{63}. This would be a 1 followed by 63 zeros!

> **Property: Powers of 10**
>
> The powers of 10 correspond to the place values of our number system, ones, tens, hundreds, thousands, and so on.

This is what we meant earlier when we said that our number system was based on the number 10.

OBJECTIVE 3

> **Example 4   Expanding Whole Numbers Using Powers of 10**

Expand 2,607 using powers of 10 notation.

We recall from Section 1.1 that we can write

NOTE In Section 1.8, we find that this can be written without parentheses as
$2 \times 10^3 + 6 \times 10^2 + 7 \times 10^0$.

$$2{,}607 = (2 \cdot 1{,}000) + (6 \cdot 100) + (0 \cdot 10) + (7 \cdot 1)$$
$$= (2 \cdot 10^3) + (6 \cdot 10^2) + (7 \cdot 10^0) \qquad 10^0 = 1.$$

 CHECK YOURSELF 4

Expand 42,371 using powers of 10 notation.

NOTE Later, we will see that very small numbers can also be written in scientific notation.

One use of exponent notation is for rounding. Very large numbers are often rounded and written using **scientific notation.**

We can round very large numbers to *one significant digit,* that is, we round them so there is a single number followed by zeros. Then, we write them as that number times ten to the appropriate power.

> **Example 5   Writing Numbers in Scientific Notation**

NOTE The symbol ≈ can be read as "can be rounded as" or "is approximately."

Earth is approximately 89,760,000 miles from the Sun. Round this number to one significant digit and write it in scientific notation.

Begin by rounding to one significant digit. In this case, that means round to the nearest ten million.

$$89{,}760{,}000 \approx 90{,}000{,}000$$

Now write 90,000,000 using powers of 10 notation.

$$90{,}000{,}000 = 9 \cdot 10^7 \text{ miles}$$

 CHECK YOURSELF 5

The Moon is approximately 230,640 miles from Earth. Round this number to one significant digit and write it in scientific notation.

Using a Scientific Calculator to Evaluate Exponent Notation

Scientific and graphing calculators have either $\boxed{y^x}$ or $\boxed{\wedge}$ keys to indicate powers. Example 6 demonstrates how to evaluate an exponential expression on a calculator.

Example 6   Evaluating Exponential Expressions on a Calculator

Evaluate 13^7.

13 $\boxed{y^x}$ 7 $\boxed{=}$ or   13 $\boxed{\wedge}$ 7 $\boxed{\text{ENTER}}$

Your display should read 62,748,517.

CHECK YOURSELF 6 ───────────

Use your calculator to evaluate 6^{12}.

READING YOUR TEXT

The following fill-in-the-blank exercises are designed to assure that you understand the key vocabulary used in this section. Each sentence comes directly from the section. You will find the correct answers in Appendix C.

Section 1.7

(a) We first encountered exponents in Section 1.5 when we discussed _____ .

(b) A whole number, other than zero, raised to the _____ power is 1.

(c) Very large numbers are often rounded and written in _____ notation.

(d) Earth is approximately 89,760,000 _____ from the Sun.

CHECK YOURSELF ANSWERS

1. 2^6 **2. (a)** 243; **(b)** 36; **(c)** 64   **3. (a)** 1; **(b)** 7
4. $(4 \cdot 10^4) + (2 \cdot 10^3) + (3 \cdot 10^2) + (7 \cdot 10^1) + (1 \cdot 10^0)$
5. $200,000 = 2 \times 10^5$ miles   **6.** 2,176,782,336

Name _____

Section _____ Date _____

ANSWERS

1. _____	2. _____
3. _____	4. _____
5. _____	6. _____
7. _____	8. _____
9. _____	10. _____
11. _____	12. _____
13. _____	14. _____
15. _____	
16. _____	
17. _____	18. _____
19. _____	
20. _____	
21. _____	
22. _____	
23. _____	
24. _____	
25. _____	
26. _____	
27. _____	
28. _____	
29. _____	
30. _____	

1.7 Exercises

Write each product using exponent notation.

1. $7 \cdot 7 \cdot 7$

2. $12 \cdot 12 \cdot 12 \cdot 12 \cdot 12 \cdot 12 \cdot 12$

3. $6 \cdot 6 \cdot 6 \cdot 6 \cdot 6 \cdot 6 \cdot 6 \cdot 6$

4. $8 \cdot 8 \cdot 8 \cdot 8 \cdot 8 \cdot 8$

Evaluate.

5. 3^2

6. 2^3

7. 2^4

8. 5^2

9. 5^1

10. 6^0

11. 9^0

12. 7^1

13. 10^3

14. 10^2

15. 10^6

16. 10^7

17. 1^{10}

18. 0^9

19. 1^0

20. 0^0

21. 11^2

22. 12^2

23. 11^3

24. 12^3

Expand each number using powers of 10.

25. 367

26. 5,406

27. 14,750

28. 9,325,002

Round each quantity to one significant digit and write it in scientific notation.

29. Science and Medicine Mars is approximately 136,764,000 miles from the Sun.

30. Science and Medicine The average distance between Pluto and the Sun is approximately 5,913,520,000 miles.

31. Business and Finance Research the U.S. government's budget deficit or surplus for the current fiscal year. Round this number to one significant digit and write it in scientific notation. Write a brief paragraph explaining why it might be better to express this number with scientific notation.

32. Business and Finance Research the U.S. government's overall national debt. Round this number to one significant digit and write it in scientific notation. Write a brief paragraph explaining why it might be better to express this number with scientific notation.

ANSWERS

31. _____

32. _____

33. _____

34. _____

35. _____

36. _____

37. _____

38. _____

Calculator Exercises

Use your calculator to evaluate.

33. 2^{10} **34.** 2^{20}

35. 3^{6} **36.** 3^{8}

37. 11^{5} **38.** 12^{5}

Answers

1. 7^{3} **3.** 6^{8} **5.** 9   **7.** 16   **9.** 5   **11.** 1   **13.** 1,000
15. 1,000,000   **17.** 1   **19.** 1   **21.** 121   **23.** 1,331
25. $(3 \cdot 10^{2}) + (6 \cdot 10^{1}) + (7 \cdot 10^{0})$
27. $(1 \cdot 10^{4}) + (4 \cdot 10^{3}) + (7 \cdot 10^{2}) + (5 \cdot 10^{1}) + (0 \cdot 10^{0})$
29. $1 \cdot 10^{8}$ miles   **31.**   **33.** 1,024   **35.** 729   **37.** 161,051

Grouping Symbols and the Order of Operations

1.8 OBJECTIVES

1. Use grouping symbols to write an expression
2. Use the order of operations to evaluate an expression
3. Expand a whole number using powers of 10

Overcoming Math Anxiety

Preparing for a Test

Preparation for a test really begins on the first day of class. Everything you have done in class and at home has been part of that preparation. However, there are a few things that you should focus on in the last few days before a scheduled test.

1. Plan your test preparation to end at least 24 hours before the test. The last 24 hours is too late, and besides, you will need some rest before the test.

2. Go over your homework and class notes with pencil and paper in hand. Write down all of the problem types, formulas, and definitions that you think might give you trouble on the test. Rework the class examples from your notes.

3. The day before the test, take the page(s) of notes from step 2 and transfer the most important ideas to a 3 by 5 card.

4. Just before the test, review the information on the card. You will be surprised at how much you remember about each concept.

5. Understand that, if you have been successful at completing your homework assignments, you can be successful on the test. This is an obstacle for many students, but it is an obstacle that can be overcome. Truly anxious students are often surprised that they scored as well as they did on a test. They tend to attribute this to blind luck. It is not. It is the first sign that you really do "get it." Enjoy the success.

If multiplication is combined with addition or subtraction, you must know which operation to do first in finding an expression's value. For example, how should we evaluate the expression $3 + 4 \cdot 5$?

We can perform the addition first and then multiply the result as shown.

$$3 + 4 \cdot 5 \overset{?}{=} 7 \cdot 5 = 35$$

Alternatively, we can perform the multiplication first and then add as a final step.

$$3 + 4 \cdot 5 \overset{?}{=} 3 + 20 = 23$$

Only one of these can be the correct answer. In fact, the second method, giving 23 as an answer, is correct.

If our intention is to add before multiplying, we must include parentheses and write $(3 + 4) \cdot 5$. But, if we are simply given the expression $3 + 4 \cdot 5$, then $3 + 4 \cdot 5 = 3 + 20 = 23$ is the correct way to evaluate the expression.

NOTE An explanation involving the proper order of operations of an expression when parentheses are involved will follow shortly.

OBJECTIVE 1   **Example 1   Using Grouping Symbols**

(a) Write an algebraic expression to mean, "5 times the sum of 12 and 3."
Here, we want to add 12 and 3 first and then multiply the resulting sum by 5. To do this, we need grouping symbols.

$$5 \cdot (12 + 3)$$

(b) Write an algebraic expression to mean, "3 added to the product of 5 and 12."
We do not need grouping symbols here because we are asked to multiply first.

$$3 + 5 \cdot 12$$

CHECK YOURSELF 1

Write algebraic expressions to mean each phrase.

(a) The cube of 5 added to 12.
(b) The cube of the sum of 5 and 12.

Mathematicians have agreed upon an **order of operations** to evaluate expressions.

NOTE Students sometimes remember this order by relating each step to part of the phrase

Please	P
Excuse	E
My Dear	MD
Aunt Sally.	AS

Step by Step: The Order of Operations

Mixed operations in an expression should be done in this order:

Step 1   Do any operations inside the parentheses.
Step 2   Evaluate any exponents.
Step 3   Do all multiplication or division in order from left to right.
Step 4   Do all addition or subtraction in order from left to right.

Step 1, above, says to do operations inside the parentheses first. This is true for any grouping symbol. Parentheses ( ), brackets [ ], and absolute value bars | |, are all examples of grouping symbols. Expressions inside any set of grouping symbols should be evaluated before moving on to the next step.

OBJECTIVE 2   **Example 2   Evaluating an Expression**

Evaluate $4 \cdot 2^3$.

Step 1   There are no parentheses.

Step 2   Evaluate exponents.

$$4 \cdot 2^3 = 4 \cdot 8$$

Step 3   Multiply or divide.

$$4 \cdot 8 = 32$$

CHECK YOURSELF 2 _____

Evaluate.

$3 \cdot 3^2$

Example 3   Evaluating an Expression

Evaluate $(2 + 3)^2 + 4 \cdot 3$.

Step 1   Do operations inside the parentheses.

$(2 + 3)^2 + 4 \cdot 3 = 5^2 + 4 \cdot 3$

Step 2   Evaluate the exponents.

$5^2 + 4 \cdot 3 = 25 + 4 \cdot 3$

Step 3   Multiply or divide.

$25 + 4 \cdot 3 = 25 + 12$

Step 4   Add or subtract.

$25 + 12 = 37$

 CHECK YOURSELF 3 _____

Evaluate.

$4 + (8 - 5)^2$

Example 4   Using the Order of Operations

(a) Evaluate $20 \div 2 \cdot 5$.

$$20 \div 2 \cdot 5$$
$$= 10 \quad \cdot 5$$
$$= 50$$

Because the multiplication and division appear next to each other, work in order from left to right. Try it the other way and see what happens!

So $20 \div 2 \cdot 5 = 50$.

(b) Evaluate $(5 + 13) \div 6$.

$$(5 + 13) \div 6$$
$$= 18 \quad \div 6$$
$$= 3$$

Do the addition in the parentheses as the first step.

So $(5 + 13) \div 6 = 3$.

CHECK YOURSELF 4 _____

Evaluate.

(a) $36 \div 4 \cdot 2$

(b) $(8 + 22) \div 5$

For very complex algebraic expressions, you may need to follow the order of operations within a set of grouping symbols before moving to the rest of the expression.

OBJECTIVE 3 | **Example 5   Evaluating Expressions**

Evaluate $3 + (15 - 3 \cdot 2^2)^2 + 11$.

First, we follow the order of operations within the parentheses. Then, we square that result. Finally, we perform the additional operations.

$$3 + (15 - 3 \cdot 2^2)^2 + 11 = 3 + (15 - 3 \cdot 4)^2 + 11$$
$$= 3 + (15 - 12)^2 + 11$$
$$= 3 + (3)^2 + 11$$
$$= 3 + 9 + 11$$
$$= 23$$

 CHECK YOURSELF 5 _____

Evaluate $11 \cdot 8 - (3 \cdot 7 + 2 \cdot 5^2) + 3 \cdot 2^3$.

 **Using a Scientific Calculator to Evaluate an Expression**

Scientific calculators are designed to correctly apply the order of arithmetic operations. For each of the following two examples, be certain that you understand *why* the given answer is correct.

Example 6   Using a Scientific Calculator to Evaluate an Expression

To evaluate $6 + 3 \cdot 5$, use this sequence:

6 $\boxed{+}$ 3 $\boxed{\times}$ 5 $\boxed{=}$

Display   21

Again the calculator follows the "order of operations." Try this with your calculator.

 CHECK YOURSELF 6 _____

Evaluate $9 + 3 \cdot 7$ with your calculator.

Most calculators have parentheses keys that will allow you to evaluate more complicated expressions easily.

> ### Example 7 Using a Scientific Calculator to Evaluate an Expression
>
> To evaluate $3 \cdot (4 + 5)$, use this sequence:
>
> 3 ⨯ $($ 4 ⊞ 5 $)$ $=$
>
> **Display** 27
>
> Now the calculator does the addition in the parentheses as the first step. Then it does the multiplication.

CHECK YOURSELF 7

Evaluate $4 \cdot (2 + 9)$ *with your calculator.*

READING YOUR TEXT

The following fill-in-the-blank exercises are designed to assure that you understand the key vocabulary used in this section. Each sentence comes directly from the section. You will find the correct answers in Appendix C.

Section 1.8

(a) Mathematicians have agreed upon an order of _____ to evaluate expressions.

(b) When evaluating an expression, the first step is to do any _____ inside parentheses.

(c) Parentheses, brackets, and _____ value bars are all examples of grouping symbols.

(d) Most calculators have parentheses _____ that will help you to evaluate complicated expressions.

CHECK YOURSELF ANSWERS

1. (a) $5^3 + 12$; **(b)** $(5 + 12)^3$ **2.** 27   **3.** 13   **4. (a)** 18; **(b)** 6   **5.** 41
6. 30   **7.** 44

Name _____

Section _____ Date _____

ANSWERS

Evaluate.

1. $2 \cdot 4^3$

2. $(2 \cdot 4)^3$

3. $5 + 2^2$

4. $(5 + 2)^2$

5. $(3 \cdot 2)^4$

6. $3 \cdot 2^4$

7. $2 \cdot 6^2$

8. $(2 \cdot 6)^2$

9. $14 - 3^2$

10. $12 + 4^2$

11. $(3 + 2)^3 - 20$

12. $5 + (9 - 5)^2$

13. $(7 - 4)^4 - 30$

14. $(5 + 2)^2 + 20$

15. $8 \div 4 + 2$

16. $3 \cdot 5 + 2$

17. $24 - 6 \div 3$

18. $3 + 9 \div 3$

19. $(24 - 6) \div 3$ **ALEKS**

20. $(3 + 9) \div 3$ **ALEKS**

21. $12 + 3 \div 3$

22. $6 \cdot 12 \div 3$

23. $18 \div 6 \cdot 3$

24. $30 \div 5 \cdot 2$

25. $30 \div 6 - 12 \div 3$

26. $5 + 8 \div 4 - 3$ **ALEKS**

27. $4^2 \div 2$

28. $2 \cdot 4^2$

29. $5^2 \cdot 3$

30. $6^2 \div 3$

31. $3 \cdot 3^3$

32. $2^5 \cdot 3$

33. $(3^3 + 3) \div 10$

34. $(2^4 + 4) \div 5$

35. $15 \div (5 - 3 + 1)$

36. $20 \div (3 + 4 - 2)$

37. $27 \div (2^2 + 5)$

38. $48 \div (2^3 + 4)$

1. _____	2. _____
3. _____	4. _____
5. _____	6. _____
7. _____	8. _____
9. _____	10. _____
11. _____	12. _____
13. _____	14. _____
15. _____	16. _____
17. _____	18. _____
19. _____	20. _____
21. _____	22. _____
23. _____	24. _____
25. _____	26. _____

27. _____

28. _____

29. _____

30. _____

31. _____

32. _____

33. _____

34. _____

35. _____

36. _____

37. _____

38. _____

39. $13 + 2 \cdot (18 - 12 \div 2^2) - 12 \cdot 3$ **40.** $4 \cdot (1^4 + 18 - 2 \cdot 3^2)^6 - 2^2$

41. $2 \cdot [3 \cdot 2^3 - (2^4 + 2^3)]^6$ **42.** $3 \cdot (2 + 3^2) - 4 \cdot 5^0$

Numbers such as 3, 4, and 5 are called **Pythagorean triples,** after the Greek mathematician Pythagoras (sixth century B.C.), because

$3^2 + 4^2 = 5^2$

Which of the following sets of numbers are Pythagorean triples?

43. 6, 8, 10   **44.** 6, 11, 12

45. 5, 12, 13   **46.** 7, 24, 25

47. 8, 16, 18    **48.** 8, 15, 17

49. Is $(a + b)^P$ equal to $a^P + b^P$?

Try a few numbers and decide if you think this is true for all whole numbers, for some whole numbers, or never true. Write an explanation of your findings and give examples.

50. Does $(a \cdot b)^P = a^P \cdot b^P$?

Try a few numbers and decide if you think this is true for all whole numbers, for some whole numbers, or never true. Write an explanation of your findings and give examples.

Calculator Exercises

Multiply, using your calculator.

51. 256
         $\times\ 508$

52. $18,569$
         $\times\ 3,286$

53. $78 \cdot 145 \cdot 36$ **54.** $37 \cdot 15 \cdot 42 \cdot 29$

Use your calculator to evaluate each of the following expressions.

55. $4 \cdot 5 - 7$ **56.** $6 \cdot 0 + 3$

57. $4 + 5 \cdot 0$ **58.** $8 \cdot (6 + 5)$

59. $5 \cdot 4 + 5 \cdot 7$ **60.** $8 \cdot 6 + 8 \cdot 5$

Solve the applications in exercises 61 and 62, using a calculator.

61. **Business and Finance**   A car dealer kept a record of a month's sales. Complete the table.

Model	Number Sold	Profit per Sale	Monthly Profit
Subcompact	38	$528	_____
Compact	33	647	_____
Standard	19	912	_____
		Monthly Total Profit	_____

62. **Business and Finance**   You take a job paying $1 the first day. On each following day, your pay doubles. That is, on day 2 your pay is $2, on day 3 the pay is $4, and so on. Complete the table.

Day	Daily Pay	Total Pay
1	$1	$1
2	2	3
3	4	7
4	_____	_____
5	_____	_____
6	_____	_____
7	_____	_____
8	_____	_____
9	_____	_____
10	_____	_____

Answers

1. 128 **3.** 9 **5.** 1,296 **7.** 72 **9.** 5 **11.** 105 **13.** 51
15. 4 **17.** 22 **19.** 6 **21.** 13 **23.** 9 **25.** 1 **27.** 8
29. 75 **31.** 81 **33.** 3 **35.** 5 **37.** 3 **39.** 7 **41.** 0
43. Yes **45.** Yes **47.** No **49.** **51.** 130,048

53. 407,160 **55.** 13 **57.** 4 **59.** 55 **61.** Monthly profit:
$20,064
21,351
17,328
$58,743

1.9 An Introduction to Equations

1.9 OBJECTIVES

1. Identify expressions and equations
2. Determine whether an equation is true
3. Translate a sentence into an equation

Overcoming Math Anxiety

Taking a Test

Doing homework and asking questions are the best ways to prepare for a test. Once you are thoroughly prepared for the test, you must learn how to take it.

There is much to the psychology of anxiety that we can't readily address. There is, however, a physical aspect to anxiety that can be addressed rather easily. When people are in a stressful situation, they frequently start to panic. One symptom of the panic is shallow breathing. In a test situation, this starts a vicious cycle. If you breathe too shallowly, then not enough oxygen reaches the brain. When that happens, you are unable to think clearly. In a test situation, being unable to think clearly can cause you to panic. Hence, we have a vicious cycle.

How do you break that cycle? It's pretty simple. Take a few deep breaths. We have seen students whose performance on math tests improved markedly after they got in the habit of writing "remember to breathe!" at the bottom of every test page. Try breathing; it will almost certainly improve your math test scores!

An **expression** is a number or a meaningful collection of operations $(+, -, \cdot, \div)$ and numbers. Each of the following is an expression.

$$14 \qquad 4 + 5 \cdot 2 \qquad 3 \cdot 5 - 1 \qquad 2^3 + 3 \cdot 2$$

In Section 1.8, we used the order of operations to determine the value of an expression. Expressions are used to build equations.

An **equation** is two expressions connected by an equal sign. Each of the following is an equation.

$$14 = 4 + 5 \cdot 2 \qquad 14 = 3 \cdot 5 - 1 \qquad 4 + 5 \cdot 2 = 3 \cdot 5 - 1$$

OBJECTIVE 1

Example 1 Identifying Expressions and Equations

Label each as an expression or an equation.

(a) $2 + 8 - 5$ is an expression.

(b) $2 \cdot 4 - 5 = 3$ is an equation.

(c) $2^3 + 4 = 12$ is an equation.

(d) $3^3 - 5 \cdot 4 - 6$ is an expression.

CHECK YOURSELF 1 _____

Label each as an expression or an equation.

(a) $12 + 8 - 5 = 15$ (b) $3^3 - 5 = 22$

(c) $2^3 - 2 \cdot 4 + 5$

Did you notice that each of the equations in Example 1 was true? Unfortunately, that is not always the case. An important part of algebra is determining when an equation is true. Example 2 will help you to develop this skill.

OBJECTIVE 2   **Example 2   Determining Whether an Equation Is True**

An equation can also be called a **statement.** Label each equation as true or false.

(a) $17 + 8 - 9 = 16$ is a true statement.

(b) $2^3 - 5 = 1$ is a false statement.

(c) $3^2 - 2 = 2^2 + 3$ is a true statement because $3^2 - 2 = 9 - 2 = 7$ and $2^2 + 3 = 4 + 3 = 7$.

CHECK YOURSELF 2 _____

Label each equation as true or false.

(a) $5 \cdot 4 - 7 = 13$ (b) $2^3 \cdot 4 = 32$

(c) $12 - 3^2 = 3^2 - 12$

In our final example in this chapter, we will translate sentences into equations.

OBJECTIVE 3   **Example 3   Translating Sentences into Equations**

An equation can be read as a sentence. Given $2 \cdot 4 = 8$, we can read it as "two times four equals eight," or simply, "two times four is eight." For each sentence, write an equation.

(a) Three plus seven is two more than eight. This can be written as $3 + 7 = 8 + 2$. Notice that "two more than eight" is written as $8 + 2$.

(b) Four squared is the same as four times four. This can be written as $4^2 = 4 \cdot 4$.

(c) Three less than two squared is one. This is translated as $2^2 - 3 = 1$. Notice here that "three less than two squared" is written as $2^2 - 3$.

CHECK YOURSELF 3 _____

For each of the sentences, write an equation.

(a) Twenty-seven minus seven is two times ten.

(b) Four squared less than twenty is four.

(c) Thirteen more than two squared is seventeen.

READING YOUR TEXT

The following fill-in-the-blank exercises are designed to assure that you understand the key vocabulary used in this section. Each sentence comes directly from the section. You will find the correct answers in Appendix C.

Section 1.9

(a) Doing homework and asking _____ are the best ways to prepare for a test.

(b) An _____ is two expressions connected by an equal sign.

(c) An important part of algebra is determining when an equation is _____.

(d) An equation can be read as a _____ .

CHECK YOURSELF ANSWERS

1. **(a)** Equation; **(b)** equation; **(c)** expression   **2.** **(a)** True; **(b)** true; **(c)** false
3. **(a)** $27 - 7 = 2 \cdot 10$; **(b)** $20 - 4^2 = 4$; **(c)** $2^2 + 13 = 17$

1.9 Exercises

Name _____

Section _____ Date _____

Label each of the following as an expression or an equation.

1. $3 + 4 = 7$

2. $45 - 3 \cdot 5$

3. $36 \div 4 + 6 \cdot 3$

4. $2^2 + 15$

5. $2^2 + 1 = 5$

6. $27 - 13 + 6 = 2 \cdot 10$

7. $13 + 29 - 3$

8. $32 = 2^5$

Determine whether each equation is true or false.

9. $24 + 13 = 37$

10. $15 \cdot 3 = 35$

11. $2 \cdot 3 = 5$

12. $3 + 4 \cdot 2 = 11$

13. $12 - 3 \cdot 3 = 3$

14. $3^2 + 1 = 7$

15. $2 \cdot 3^2 = 18$

16. $2 \cdot 3^2 = 36$

17. $5 \cdot 2^2 = 100$

18. $5 \cdot 2^2 = 20$

Translate each phrase into an expression.

19. Three plus two minus five

20. Seven less than 12

21. Nine more than three

22. Six squared plus 5

Translate each expression into a phrase.

23. $5 + 3$

24. $2 \cdot 3$

25. $5 \cdot 2 - 1$

26. $3^2 + 1$

Translate each equation into a sentence.

27. $3 + 2 = 5$

28. $5 \cdot 2 = 10$

29. $15 - 3 = 12$

30. $3^2 + 2 = 11$

ANSWERS

1. _____
2. _____
3. _____
4. _____
5. _____
6. _____
7. _____
8. _____
9. _____ 10. _____
11. _____ 12. _____
13. _____ 14. _____
15. _____ 16. _____
17. _____ 18. _____
19. _____
20. _____
21. _____ 22. _____
23. _____
24. _____
25. _____
26. _____
27. _____
28. _____
29. _____
30. _____

Translate each sentence into an equation.

31. Four more than seven is eleven.

32. Two more than three times five is seventeen.

33. Thirteen is five more than eight.

34. Six less than five squared is nineteen.

Answers

1. Equation   **3.** Expression   **5.** Equation   **7.** Expression   **9.** True
11. False   **13.** True   **15.** True   **17.** False   **19.** $3 + 2 - 5$
21. $3 + 9$ **23.** Five plus three   **25.** One less than five times two
27. Two more than three is five   **29.** Three less than fifteen is twelve
31. $7 + 4 = 11$ **33.** $13 = 8 + 5$

ACTIVITY 1: WRAPPING GIFTS

Each chapter in this text includes an activity. The activity is related to the vignette you encountered in the chapter opening. The activity provides you with the opportunity to apply the math you studied in the chapter.

Your instructor will determine how best to use this activity in your class. You may find yourself working in class or outside of class; you may find yourself working alone or in small groups; or you may even be asked to perform research in a library or on the Internet.

We begin this activity with a package. Traditionally, we call the longest side the length, and the shortest side we call the height (unless the package has a "this side up" side, in which case, this determines which side is called the height).

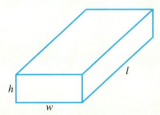

When wrapping a package, we try to align the length of the package with the width of the wrapping paper. In order to do this, the width of the wrapping paper must be longer than the length of the package. How much longer?

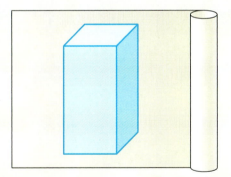

Each side of the wrapping paper must cover half the height (when folded down). In fact, we would like the wrapping paper folded down from the top to overlap the wrapping paper folded up from the bottom. So, we need a piece of wrapping paper to have a width equal to the length of the package plus its height plus an inch.

For the purposes of this activity, we would like all overlaps to be exactly 1 inch (in.). Assume you have a roll of wrapping paper that is 1 yard (yd) wide and 25 feet (ft) long.

1. If your package is 18 in. long and 2 in. high, how wide does your piece of wrapping paper need to be?
2. If your package is 28 in. long and 2 in. high, how wide does your piece of wrapping paper need to be?

3. Write an algebraic expression that gives the width of a piece of wrapping paper that should be used to wrap a package of length l and height h.

4. What is the longest package length that can be wrapped when aligned with the width of the paper?

The U.S. Post Office defines the **girth** of a package to be the distance around a cross-section, that is, the perimeter of the rectangle given by the width and height of the package.

If we align the length of the package with the width of the wrapping paper, then we need to use a piece of wrapping paper whose length is 1 in. larger than the girth of our package. The extra inch enables us to overlap the wrapping paper so that we may tape it closed.

5. Sketch a package and piece of wrapping paper, showing the extra wrapping paper necessary to wrap a package with a given length and girth, including the overlap.

6. If your package has a height of 4 in. and a width of 16 in., how long does the piece of wrapping paper need to be?

7. If a package has dimensions 24 in. by 18 in. by 5 in., how long does a piece of wrapping paper need to be in order to wrap the package?

8. Write an algebraic expression that gives the length of wrapping paper that should be used if a package has width w and height h.

9. If a package has dimensions 38 in. by 24 in. by 4 in., what dimensions will you need for a piece of wrapping paper if the wrapping paper has a width of 1 yd.

10. If you have six packages whose dimensions (in inches) are shown below, how much wrapping paper will you need?

Package no.	l	w	h
1	18	16	4
2	28	16	4
3	24	18	5
4	38	24	4
5	6	6	2
6	6	6	2

1 Summary

DEFINITION/PROCEDURE	EXAMPLE	REFERENCE
Introduction to Whole Numbers and Place Value		Section 1.1
Digits Digits are the basic symbols of the system.	0, 1, 2, 3, 4, 5, 6, 7, 8, and 9 are digits.	p. 3
Place Value The value of a digit in a number depends on its position or place.	7,352,589 — Ones, Tens, Hundreds, Thousands, Ten thousands, Hundred thousands, Millions	p. 3–4
The **value of a number** is the sum of each digit multiplied by its place value.	$2,345 = (2 \cdot 1,000) + (3 \cdot 100) + (4 \cdot 10) + (5 \cdot 1)$	p. 4
Addition of Whole Numbers		Section 1.2
Addends The numbers that are being added. **Sum** The result of the addition.	$\begin{array}{r} 5 \\ + 8 \end{array}$ Addends 13 Sum	p. 12
The Properties of Addition **The Commutative Property** The order in which you add two whole numbers does not affect the sum.	$5 + 4 = 4 + 5$	p. 13
The Associative Property The way in which you group whole numbers in addition does not affect the final sum.	$(2 + 7) + 8 = 2 + (7 + 8)$	p. 14
The Additive Identity The sum of 0 and any whole number is just that whole number.	$6 + 0 = 0 + 6 = 6$	p. 14
Measuring Perimeter The perimeter is the total distance around the outside edge of a shape. The perimeter of a rectangle is $P = L + W + L + W$ (which can be written as $P = 2 \cdot L + 2 \cdot W$).	6 ft, 2 ft, 2 ft, 6 ft $P = L + W + L + W$ $= 6\text{ ft} + 2\text{ ft} + 6\text{ ft} + 2\text{ ft}$ $= 16\text{ ft}$	p. 18

Continued

DEFINITION/PROCEDURE	EXAMPLE	REFERENCE
Subtraction of Whole Numbers		**Section 1.3**
Minuend The number we are subtracting from. **Subtrahend** The number that is being subtracted. **Difference** The result of the subtraction.	15 ←—Minuend − 9 ←—Subtrahend ——— 6 ←—Difference	*p. 28*
Rounding, Estimation, and Ordering of Whole Numbers		**Section 1.4**
Step 1 To round a whole number to a certain decimal place, look at the digit to the *right* of that place. **Step 2** **a.** If that digit is 5 or more, that digit and all digits to the right become 0. The digit in the place you are rounding to is increased by 1. **b.** If that digit is less than 5, that digit and all digits to the right become 0. The digit in the place you are rounding to remains the same.	To the nearest hundred, 43,578 is rounded to 43,600. To the nearest thousand, 273,212 is rounded to 273,000.	*p. 42*
Order on the Whole Numbers For the numbers a and b, we can write **1.** $a < b$ (read "a is less than b") when a is *to the left* of b on the number line.	$8 < 12$ 8 9 10 11 12	*p. 44–45*
2. $a > b$ (read "a is greater than b") when a is *to the right* of b on the number line.	$15 > 10$ 10 11 12 13 14 15	*p. 44–45*
Multiplication of Whole Numbers		**Section 1.5**
Factors The numbers being multiplied. **Product** The result of the multiplication.	$\underline{7 \cdot 9} = 63$ ←—Product ↑ Factors	*p. 50*
The Properties of Multiplication **The Commutative Property** Multiplication, like addition, is a *commutative* operation. The order in which you multiply two whole numbers does not affect the product.	$7 \cdot 9 = 9 \cdot 7$	*p. 50*
The Distributive Property To multiply a factor by a sum of numbers, multiply the factor by each number inside the parentheses. Then add the products.	$2 \cdot (3 + 7) = (2 \cdot 3) + (2 \cdot 7)$	*p. 52*
Multiplicative Property of Zero The product of zero and any number is zero.	$3 \cdot 0 = 0 \cdot 3 = 0$	*p. 53*
Multiplicative Identity Property The product of one and any number is that number.	$3 \cdot 1 = 1 \cdot 3 = 3$	*p. 53*

DEFINITION/PROCEDURE	EXAMPLE	REFERENCE
Multiplication of Whole Numbers		**Section 1.5**
The Associative Property Multiplication is an *associative* operation. The way in which you group numbers in multiplication does not affect the final product.	$(3 \cdot 5) \cdot 6 = 3 \cdot (5 \cdot 6)$	*p. 58*
Finding the Area of a Rectangle The area of a rectangle is found using the formula $A = L \cdot W$.	6 ft 2 ft $A = L \cdot W = 6\text{ ft} \cdot 2\text{ ft} = 12\text{ ft}^2$	*p. 60*
Division of Whole Numbers		**Section 1.6**
Divisor The number we are dividing by. **Dividend** The number being divided. **Quotient** The result of the division. **Remainder** The number "left over" after the division.	Divisor Quotient 5 $7\overline{)38}$ ← Dividend $\underline{35}$ 3 ← Remainder	*p. 72*
Dividend = divisor · quotient + remainder	$38 = 7 \cdot 5 + 3$	*p. 74*
The Role of 0 in Division Zero divided by any whole number (except 0) is 0.	$0 \div 7 = 0$	*p. 75*
Division by 0 is undefined.	$7 \div 0$ is undefined.	*p. 75*
Exponents and Whole Numbers		**Section 1.7**
Using Exponents **Base** The number that is raised to a power. **Exponent** The exponent is written to the right and above the base. The exponent tells the number of times the base is to be used as a factor.	Exponent $5^3 = 5 \cdot 5 \cdot 5 = 125$ Base Three factors This is read "5 to the third power" or "5 cubed."	*p. 88*
Grouping Symbols and the Order of Operations		**Section 1.8**
The Order of Operations *Mixed operations* in an expression should be done in the following order: **Step 1** Do any operations inside parentheses. **Step 2** Evaluate any exponents. **Step 3** Do all multiplication and division in order from left to right. **Step 4** Do all addition and subtraction in order from left to right. Remember *P*lease *E*xcuse *M*y *D*ear *A*unt *S*ally.	$4 \cdot (2 + 3)^2 - 7$ $= 4 \cdot 5^2 - 7$ $= 4 \cdot 25 - 7$ $= 100 - 7$ $= 93$	*p. 95*

Continued

DEFINITION/PROCEDURE	EXAMPLE	REFERENCE
An Introduction to Equations		**Section 1.9**
An **expression** is a number or a meaningful collection of operations $(+, -, \cdot, \div)$ and numbers.	$9 + 5 \cdot 2$ $2^3 - 5 \cdot 2$	*p. 102*
An **equation** is two expressions connected by an equal sign. An equation can be true or false.	$19 = 9 + 5 \cdot 2$ $14 - 3 \cdot 2 = 8$ is true $14 - 3 \cdot 2 = 22$ is false	

Summary Exercises

This summary exercise set is provided to give you practice with each of the objectives of this chapter. Each exercise is keyed to the appropriate chapter section. When you are finished, you can check your answers to the odd-numbered exercises against those presented in the back of the text. If you have difficulty with any of these questions, go back and reread the examples from that section. The answers to the even-numbered exercises appear in the *Instructor's Solutions Manual.* Your instructor will give you guidelines on how to best use these exercises in your instructional setting.

[1.1]   In exercises 1 and 2, give the place value of each of the indicated digits.

1. 6 in the number 5,674

2. 5 in the number 543,400

In exercises 3 and 4, write each number in words.

3. 27,428

4. 200,305

Write each of the following as a number.

5. Thirty-seven thousand, five hundred eighty-three

6. Three hundred thousand, four hundred

[1.2]   In exercises 7 and 8, name the property of addition that is illustrated.

7. $4 + 9 = 9 + 4$

8. $(4 + 5) + 9 = 4 + (5 + 9)$

In exercises 9 to 13, perform the indicated operations.

9.
```
   784
   385
+ 247
```

10.
```
    2,570
      498
   21,456
+     28
```

11.
```
     367
     289
   1,463
+ 2,682
```

12.
```
     6,389
     1,567
       315
+ 113,602
```

13. Find the value of 7 more than 4.

Find the perimeter of each figure.

14.

15.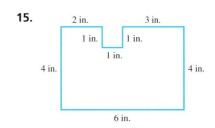

Solve the applications.

16. **Statistics** An airline had 173, 212, 185, 197, and 202 passengers on five morning flights between Washington, D.C., and New York. What was the total number of passengers?

17. **Business and Finance** The Future Stars summer camp employs five junior counselors. Their weekly salaries last week were $324, $405, $243, $405, and $243. What was the total salary for the junior counselors?

[1.3] In exercises 18 to 23, perform the indicated operations.

18. 5,325	19. 38,400	20. 86,000	21. 2,682
− 847	− 19,600	− 2,169	− 108

22. Find the difference of 7,342 and 5,579.

23. Find the value of 34 decreased by 7.

Solve the applications.

24. **Business and Finance** Chuck owes $795 on a credit card after a trip. He makes payments of $75, $125, and $90. Interest of $31 is charged. How much remains to be paid on the account?

25. **Business and Finance** Juan bought a new car for $16,785. The manufacturer offers a cash rebate of $987. What was the cost after the rebate?

[1.4] In exercises 26 to 28, round the numbers to the indicated place.

26. 6,975 to the nearest hundred

27. 15,897 to the nearest thousand

28. 548,239 to the nearest ten thousand

In exercises 29 and 30, complete the statements by using the symbol $<$ or $>$.

29. 60 _____ 70

30. 38 _____ 35

[1.5] In exercises 31 to 34, name the property that is illustrated.

31. $7 \cdot 8 = 8 \cdot 7$

32. $3 \cdot (4 + 7) = 3 \cdot 4 + 3 \cdot 7$

33. $(8 \cdot 9) \cdot 4 = 8 \cdot (9 \cdot 4)$

34. $6 \cdot 1 = 6$

In exercises 35 to 37, perform the indicated operations.

35. 58
     × 32

36. 25
     × 43

37. 378
     × 409

Solve the application.

38. Construction   You wish to carpet a room that is 5 yd by 7 yd. The carpet costs $18 per square yard. What will be the total cost of the materials?

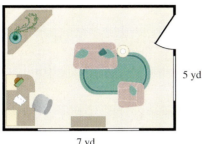

5 yd

7 yd

In exercise 39, perform the indicated operation.

39. 129
     × 240

Estimate the product by rounding each factor to the nearest hundred.

40. 1,217
     × 494

Find the area of the given figures.

41.

6 in.

3 in.   3 in.

6 in.

42.

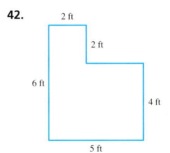

2 ft

2 ft

6 ft

4 ft

5 ft

[1.6]   In exercises 43 and 44, divide if possible.

43. $0 \div 8$

44. $5 \div 0$

In exercises 45 to 48, divide.

45. $8\overline{)2,469}$

46. $39\overline{)2,157}$

47. $64\overline{)31,809}$

48. $362\overline{)86,915}$

Solve the application.

49. **Science and Medicine** Hasina's odometer read 25,235 mi at the beginning of a trip and 26,215 mi at the end. If she used 35 gal of gas for the trip, what was her mileage (mi/gal)?

Estimate the following.

50. 356 divided by 37

51. 2,125 divided by 28

Perform the indicated operations.

52. Find the value for the product of 9 and 5, divided by 3.

[1.7] Evaluate each expression.

53. 3^4

54. 4^3

[1.8] In exercises 55 to 62, evaluate the expressions.

55. $4 + 8 \cdot 3$

56. $48 \div (2^3 + 4)$

57. $(4 + 8) \cdot 3$

58. $4 \cdot 3 + 8 \cdot 3$

59. $8 \div 4 \cdot 2 - 2 + 1$

60. $63 \cdot 2 \div 3 - 54 \div (12 \cdot 2 \div 4)$

61. $(3 \cdot 4)^2 - 100 \div 5 \cdot 6$

62. $(16 \cdot 2) \div 8 - (6 \div 3 \cdot 2)$

[1.9] Determine whether each is an expression or an equation.

63. $5 + 3 - 2 \cdot 2$

64. $7 - 2 \cdot 3 = 1$

65. $4^2 - 3 \cdot 2$

66. $2 \cdot 3^2 = 18$

Determine whether each equation is true or false.

67. $3 \cdot 2 + 1 = 7$

68. $5 + 3 \cdot 4 = 32$

69. $7 + 2 \cdot 2 = 11$

70. $3 \cdot 2^2 + 1 = 13$

Translate each of the sentences to an equation.

71. Five more than two squared is nine.

72. Eight less than four times three is four.

Self-Test for Chapter 1

Name _____

Section _____ Date _____

The purpose of this self-test is to help you check your progress so that you can find sections and concepts that you need to review before the next in-class exam. Allow yourself about an hour to take this test. At the end of that hour, check your answers against those given in the back of this text. If you missed a question, notice the section reference that accompanies the question. Go back to that section and reread the examples until you have mastered that particular concept.

ANSWERS

1. Give the place value of 7 in 3,738,500.

2. Give the word name for 302,525.

3. Write two million, four hundred thirty thousand as a number.

In exercises 4 and 5, name the property of addition that is illustrated.

4. $5 + 12 = 12 + 5$

5. $(7 + 3) + 8 = 7 + (3 + 8)$

In exercises 6 and 7, perform the indicated operations.

6.
$$\begin{array}{r} 489 \\ 562 \\ 613 \\ + 254 \\ \hline \end{array}$$

7.
$$\begin{array}{r} 13 \\ 2,543 \\ + 10,547 \\ \hline \end{array}$$

8. What is the total of 392, 95, 9,237, and 11,972?

Solve the application.

9. Statistics The attendance for the games of a playoff series in basketball was 12,438, 14,325, 14,581, and 14,634. What was the total attendance for the series?

In exercises 10 to 13, subtract.

10. $289 - 54$

11. $53,294 - 41,074$

12. $32,345 - 1,575$

13. $55,342 - 14,787$

14. The maximum load for a light plane with full gas tanks is 500 lb. Mr. Whitney weighs 215 lb, his wife 135 lb, and their daughter 78 lb. How much luggage can they take on a trip without exceeding the load limit?

In exercise 15, estimate the sum by rounding each addend to the nearest hundred.

15.
$$\begin{array}{r} 943 \\ 3,281 \\ 778 \\ 2,112 \\ + 570 \\ \hline \end{array}$$

1. _____

2. _____

3. _____

4. _____

5. _____

6. _____

7. _____

8. _____

9. _____

10. _____

11. _____

12. _____

13. _____

14. _____

15. _____

16. _____

17. _____

18. _____

19. _____

20. _____

21. _____

22. _____

23. _____

24. _____

25. _____

26. _____

27. _____

28. _____

29. _____

30. _____

In exercises 16 and 17, complete the statements by using the symbol $<$ or $>$.

16. 49 _____ 47

17. 80 _____ 90

In exercises 18 to 20, name the property that is illustrated.

18. $3 \cdot (2 \cdot 7) = (3 \cdot 2) \cdot 7$

19. $4 \cdot (3 + 6) = (4 \cdot 3) + (4 \cdot 6)$

20. $5 \cdot 1 = 5$

Find the products.

21. $\begin{array}{r} 89 \\ \times\ 56 \\ \hline \end{array}$

22. $\begin{array}{r} 538 \\ \times\ 103 \\ \hline \end{array}$

23. A truck rental firm has ordered 25 new vans at a cost of \$12,350 per van. What will be the total cost of the order?

Divide using long division.

24. $28\overline{)2135}$

25. Statistics Eight people estimate that the total expenses for a trip they are planning to take together will be \$1,784. If each person pays an equal amount, what will be each person's share?

26. Evaluate the expression $15 - 12 \div 2^2 \cdot 3 + (12 \div 4 \cdot 3)$.

Find the perimeter of the figure shown.

27.

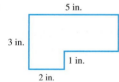

2 in.
2 in.   2 in.
2 in.   2 in.
2 in.

Find the area of the given figure.

28.

5 in.

3 in.

1 in.

2 in.

29. Is the equation $3 \cdot 5^2 = 75$ a true or false statement?

30. Translate into an equation: Two less than three squared is seven.

INTEGERS AND INTRODUCTION TO ALGEBRA

CHAPTER 2 OUTLINE

Section 2.1   Introduction to Integers   page 121

Section 2.2   Addition of Integers   page 132

Section 2.3   Subtraction of Integers   page 139

Section 2.4   Multiplication of Integers   page 145

Section 2.5   Division of Integers   page 151

Section 2.6   Introduction to Algebra: Variables and Expressions   page 158

Section 2.7   Evaluating Algebraic Expressions   page 167

Section 2.8   Simplifying Algebraic Expressions   page 177

Section 2.9   Introduction to Linear Equations   page 186

Section 2.10   The Addition Property of Equality   page 192

INTRODUCTION

Of all the sciences, meteorology may be both the least precise and the most talked about. Meteorologists study the weather and climate. Together with geologists, they provide us with many predictions that play a part in our everyday decisions. Among the things we might look for are temperatures, rainfall, water level, and tide level.

On a day with unusual weather, we often become curious about the record for that day or even the all-time record. Here are a few of those records:

- The record high temperature for the United States is 134°F in 1913 in Death Valley, California.
- The record low temperature for the U.S. is −80°F in 1971 in Prospect Creek Camp, Alaska.
- The greatest one day temperature drop in the U.S. happened on Christmas Eve, 1924, in Montana. The temperature went from 63°F during the day to −21°F at night.

The use of negative numbers for temperatures below zero is common for such numbers, but there are many other applications. For example, the tide range in Delaware varies between +10 feet and −5 feet. We examine these and several other climate-related applications in Sections 2.1, 2.2, 2.3, and 2.7.

ANSWERS

1. _____

2. _____

3. _____

4. _____ 5. _____

6. _____

7. _____

8. _____

9. _____

10. _____

11. _____

12. _____

13. _____

14. _____

15. _____

16. _____

17. _____

18. _____

19. _____

20. _____ 21. _____

22. _____ 23. _____

24. _____ 25. _____

26. _____ 27. _____

28. _____

Pretest Chapter 2

This pretest will provide a preview of the types of exercises you will encounter in each section of this chapter. The answers for these exercises can be found in the back of the text. If you are working on your own or are ahead of the class, this pretest can help you identify the sections in which you should focus more of your time.

[2.1] Represent the integers on the number line shown.

1. $6, -8, 4, -2, 10$

2. Place the following data set in ascending order: $5, -2, -4, 0, -1, 1$.

3. Determine the maximum and minimum of the following data set: $-4, 1, -5, 7, 3, 2$.

Evaluate:

4. $|-5|$ 　　　　　5. $|6|$ 　　　　　6. $|11 - 5|$

7. $|-11| - |5|$ 　　　8. $|4 + 5| - |6 - 3|$

Find the opposite of each integer.

9. -16 　　　　　　　　10. 23

[2.6] Write each of the phrases using symbols.

11. 8 less than x

12. the quotient when w is divided by the product of x and 17

Identify which are expressions and which are not.

13. $7x - 5 = 11$ 　　　　　14. $3x - 2(x + 1)$

[2.2 to 2.5] Perform the indicated operations.

15. $-7 + (-3)$ 　　16. $8 + (-9)$ 　　17. $(-3) + (-2)$

18. $8 - 11$ 　　　　19. $-8 - 11$ 　　　20. $9 - (-3)$

21. $6 + (-6)$ 　　　22. $(-7)(-3)$ 　　23. $\dfrac{-27 + 6}{-3}$

[2.7] Evaluate each expression.

24. $5 - 4^2 \cdot 3 \div 6$ 　　　　　25. $(45 - 3 \cdot 5) + 5^2$

26. If $x = -2$, $y = 7$, and $w = -4$, evaluate the expression $\dfrac{x^2 y}{w}$.

[2.8] Combine like terms.

27. $5w^2 t + 3w^2 t$ 　　　　　28. $4a^2 - 3a + 5 + 7a - 2 - 5a^2$

2.1 Introduction to Integers

2.1 OBJECTIVES

1. Represent integers on a number line
2. Place a set of integers in ascending order
3. Determine the extreme values of a data set
4. Find the opposite of a given integer
5. Evaluate expressions involving absolute value

When numbers are used to represent physical quantities (altitudes, temperatures, and amounts of money are examples), it may be necessary to distinguish between *positive* and *negative* quantities. The symbols + and − are used for this purpose. For instance, the altitude of Mount Whitney is 14,495 ft *above* sea level (+14,495 ft).

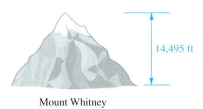

14,495 ft

Mount Whitney

The altitude of Death Valley is 282 ft *below* sea level (−282 ft).

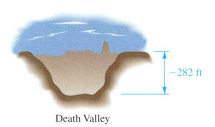

−282 ft

Death Valley

On a given day the temperature in Chicago might be 10°F *below* zero (−10°F).

An account could show a *gain* of $100 (+100) or a *loss* of $100 (−100).

These numbers suggest the need to extend the whole numbers to include both positive numbers (like +100) and negative numbers (like −282).

To represent the negative numbers, we extend the number line to the *left* of zero and name equally spaced points.

Numbers corresponding to points to the right of zero are positive numbers. They are written with a positive (+) sign or with no sign at all.

+6 and 9 are positive numbers

Numbers corresponding to points to the left of zero are negative numbers. They are always written with a negative (−) sign.

−3 and −20 are negative numbers

Read "negative 3."

The positive and negative numbers, as well as zero, are called the **real numbers.**

Here is the number line extended to include positive and negative numbers, and zero.

NOTE On the number line, we call zero the **origin.**

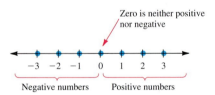

NOTE Braces { and } are used to hold a collection of numbers. We call the collection a **set.** The dots are called **ellipses** and indicate that the pattern continues.

The numbers used to name the points shown on the number line are called the **integers.** The integers consist of the natural numbers, their negatives, and the number zero. We can represent the set of integers by

$\{\ldots, -3, -2, -1, 0, 1, 2, 3, \ldots\}$

OBJECTIVE 1

Example 1   Representing Integers on the Number Line

Represent the integers on the number line shown.

−3, −12, 8, 15, −7

CHECK YOURSELF 1

Represent the integers on a number line.

$-1, -9, 4, -11, 8, 20$

The set of numbers on the number line is *ordered.* The numbers get smaller moving to the left on the number line and larger moving to the right.

When a group of numbers is written from smallest to largest, the numbers are said to be in *ascending order.*

OBJECTIVE 2   **Example 2**   **Ordering Integers**

Place each group of numbers in ascending order.

(a) $9, -5, -8, 3, 7$

From smallest to largest, the numbers are

$-8, -5, 3, 7, 9$ Note that this is the order in which the numbers appear on a number line as we move from left to right.

(b) $3, -2, 18, -20, -13$

From smallest to largest, the numbers are

$-20, -13, -2, 3, 18$

 CHECK YOURSELF 2

Place each group of numbers in ascending order.

(a) $12, -13, 15, 2, -8, -3$ **(b)** $3, 6, -9, -3, 8$

The least and greatest numbers in a group are called the **extreme values.** The least number is called the **minimum,** and the greatest number is called the **maximum.**

OBJECTIVE 3   **Example 3**   **Labeling Extreme Values**

For each group of numbers, determine the minimum and maximum values.

(a) $9, -5, -8, 3, 7$

From our previous ordering of these numbers, we see that -8, the least number, is the minimum, and 9, the greatest number, is the maximum.

(b) $3, -2, 18, -20, -13$

-20 is the minimum and 18 is the maximum.

CHECK YOURSELF 3 _____

For each group of numbers, determine the minimum and maximum values.

(a) $12, -13, 15, 2, -8, -3$ **(b)** $3, 6, -9, -3, 8$

Each point on the number line corresponds to a **real number.** There are more points on the number line than integers. The real numbers include decimals, fractions, and other numbers.

Example 4   **Identifying Real Numbers that Are Integers**

Which of the real numbers, (a) 145, (b) -28, (c) 0.35, and (d) $-\dfrac{2}{3}$, are also integers?

(a) 145 is an integer.

(b) -28 is an integer.

(c) 0.35 is not an integer.

(d) $-\dfrac{2}{3}$ is not an integer.

CHECK YOURSELF 4 _____

Which of the real numbers are also integers?

$$-23 \quad 1{,}054 \quad -0.23 \quad 0 \quad -500 \quad -\frac{4}{5}$$

Sometimes we refer to the negative of a number as its *opposite*. For a nonzero number, this corresponds to a point the same distance from the origin as the given number, but on the other side of zero. Example 5 illustrates this.

OBJECTIVE 4   **Example 5**   **Find the Opposite of Each Number**

(a) 5   The opposite of 5 is -5.

(b) -9 The opposite of -9 is 9.

CHECK YOURSELF 5 _____

Find the opposite of each number.

(a) 17   **(b)** -12

Definition: Absolute Value

The **absolute value** of a number represents the distance of the point named by the number from the origin on the number line.

Because we think of distance as a positive quantity (or as zero), the absolute value of a number is never negative.

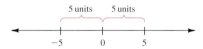

The absolute value of 5 is 5. The absolute value of -5 is also 5.

As a consequence of the definition, the absolute value of a positive number or zero is itself. The absolute value of a negative number is its opposite.

In symbols, we write

$$|5| = 5 \qquad \text{and} \qquad |-5| = 5$$

Read "the absolute value of 5."

Read "the absolute value of negative 5."

The absolute value of a number does *not* depend on whether the number is to the right or to the left of the origin, but on its *distance* from the origin.

OBJECTIVE 5

Example 6   Simplifying Absolute Value Expressions

(a) $|7| = 7$

(b) $|-7| = 7$

(c) $-|-7| = -7$ This is the *negative*, or opposite, of the absolute value of negative 7.

CHECK YOURSELF 6 _____

Evaluate.

(a) $|8|$ **(b)** $|-8|$ **(c)** $-|-8|$

To determine the order of operation for an expression that includes absolute values, note that the absolute value bars are treated as a grouping symbol.

Example 7   Adding or Subtracting Absolute Values

(a) $|-10| + |10| = 10 + 10 = 20$

(b) $|8 - 3| = |5| = 5$ Absolute value bars, like parentheses, serve as a set of grouping symbols, so do the operation *inside* first.

(c) $|8| - |3| = 8 - 3 = 5$ Evaluate the absolute values, then subtract.

CHECK YOURSELF 7 _____

Evaluate.

(a) $|-9| + |4|$ **(b)** $|9 - 4|$ **(c)** $|9| - |4|$

READING YOUR TEXT ───────

The following fill-in-the-blank exercises are designed to assure that you understand the key vocabulary used in this section. Each sentence comes directly from the section. You will find the correct answers in Appendix C.

Section 2.1

(a) When numbers are used to represent physical quantities, it may be necessary to distinguish between positive and _____ quantities.

(b) Numbers that correspond to points to the _____ of zero are negative numbers.

(c) When a set of numbers is written from smallest to largest, the numbers are said to be in _____ order.

(d) The absolute value of a number depends on its _____ from the origin.

CHECK YOURSELF ANSWERS ───────

1.

2. (a) $-13, -8, -3, 2, 12, 15$
   (b) $-9, -3, 3, 6, 8$

3. (a) minimum is -13; maximum is 15; **(b)** minimum is -9; maximum is 8

4. $-23, 1054, 0,$ and -500 **5. (a)** -17; **(b)** 12

6. (a) 8; **(b)** 8; **(c)** -8.   **7. (a)** 13; **(b)** 5; **(c)** 5

2.1 Exercises

Represent each quantity with an integer.

1. An altitude of 400 ft above sea level

2. An altitude of 80 ft below sea level

3. A loss of $200

4. A profit of $400

5. A decrease in population of 25,000

6. An increase in population of 12,500

Represent the integers on the number lines shown.

7. 5, −15, 18, −8, 3

8. −18, 4, −5, 13, 9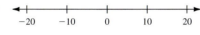

Which numbers in the sets are integers?

9. $\left\{ 5, -\dfrac{2}{9}, 175, -234, -0.64 \right\}$

10. $\left\{ -45, 0.35, \dfrac{3}{5}, 700, -26 \right\}$

Place each group of numbers in ascending order.

11. 3, −5, 2, 0, −7, −1, 8

12. −2, 7, 1, −8, 6, −1, 0

13. 9, −2, −11, 4, −6, 1, 5

14. 23, −18, −5, −11, −15, 14, 20

15. −6, 7, −7, 6, −3, 3

16. 12, −13, 14, −14, 15, −15

For each group of numbers, determine the maximum and minimum values.

17. 5, −6, 0, 10, −3, 15, 1, 8

18. 9, −1, 3, 11, −4, 2, 5, −2

19. 21, −15, 0, 7, −9, 16, −3, 11

20. −22, 0, 22, −31, 18, −5, 3

21. 3, 0, 1, −2, 5, 4, −1

22. 2, 7, −3, 5, −10, −5

23. _____

24. _____

25. _____

26. _____

27. _____

28. _____

29. _____

30. _____

31. _____

32. _____

33. _____

34. _____

35. _____

36. _____

37. _____

38. _____

39. _____

40. _____

41. _____

42. _____

43. _____

44. _____

45. _____

46. _____

47. _____   48. _____

49. _____   50. _____

51. _____   52. _____

53. _____   54. _____

Find the opposite of each number.

23. 15   **24.** 18

25. 11   **26.** 34

27. -19 **28.** -5

29. -7 **30.** -54

Evaluate.

31. $|17|$ **ALEKS**   **32.** $|28|$ **ALEKS**

33. $|-10|$ **ALEKS**   **34.** $|-7|$ **ALEKS**

35. $-|3|$ **36.** $-|5|$

37. $-|-8|$ VIDEO   **38.** $-|-13|$

39. $|-2|+|3|$ **ALEKS**   **40.** $|4|+|-3|$ **ALEKS**

41. $|-9|+|9|$ **ALEKS**   **42.** $|11|+|-11|$ **ALEKS**

43. $|4|-|-4|$ **ALEKS**   **44.** $|5|-|-5|$ **ALEKS**

45. $|15|-|8|$ **ALEKS**   **46.** $|11|-|3|$ **ALEKS**

47. $|15-8|$ **ALEKS**   **48.** $|11-3|$ **ALEKS**

49. $|-9|+|2|$ **ALEKS**   **50.** $|-7|+|4|$ **ALEKS**

51. $|-8|-|-7|$ VIDEO **ALEKS**   **52.** $|-9|-|-4|$ **ALEKS**

Label each statement as true or false.

53. All whole numbers are integers.

54. All nonzero integers are real numbers.

55. All integers are whole numbers.

56. All real numbers are integers.

57. All negative integers are whole numbers.

58. Zero is neither positive nor negative.

Place absolute value bars in the proper location on the left side of the expression so that the equation is true.

59. $(-6) + 2 = 8$ **60.** $(-8) + (-3) = 11$

61. $6 + (-2) = 8$ **62.** $8 + (-3) = 11$

Represent each quantity with a real number.

63. Science and Medicine The erosion of 5 centimeters (cm) of topsoil from an Iowa cornfield.

64. Science and Medicine The formation of 2.5 cm of new topsoil on the African savanna.

65. Business and Finance The withdrawal of $50 from a checking account.

66. Business and Finance The deposit of $200 in a savings account.

67. Science and Medicine The temperature change pictured.

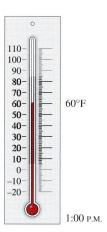

60°F

1:00 P.M.

50°F

2:00 P.M.

ANSWERS

55. _____

56. _____

57. _____

58. _____

59. _____

60. _____

61. _____

62. _____

63. _____

64. _____

65. _____

66. _____

67. _____

68. Science and Medicine   The temperature change indicated.

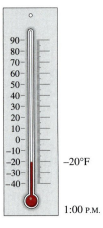

1:00 P.M.   2:00 P.M.

69. Science and Medicine   The temperature change indicated.

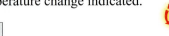

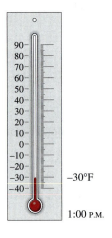   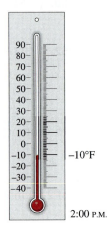

1:00 P.M.   2:00 P.M.

70. Science and Medicine   The temperature change indicated.

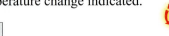

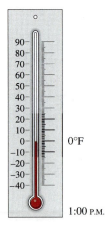   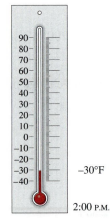

1:00 P.M.   2:00 P.M.

71. Business and Finance   A country exported \$90,000,000 more than it imported, creating a positive trade balance.

72. Business and Finance   A country exported \$60,000,000 less than it imported, creating a negative trade balance.

For each group of numbers given in exercises 73 to 76, answer questions **(a)** to **(d)**:

(a) Which number is smallest?

(b) Which number lies farthest from the origin?

(c) Which number has the largest absolute value?

(d) Which number has the smallest absolute value?

73. $-6, 3, 8, 7, -2$

74. $-8, 3, -5, 4, 9$

75. $-2, 6, -1, 0, 2, 5$

76. $-9, 0, -2, 3, 6$

77. Simplify each of the following:

$$-(-7) \qquad -(-(-7)) \qquad -(-(-(-7)))$$

Based on your answers, generalize your results.

Answers

1. 400 or $(+400)$ **3.** -200 **5.** $-25,000$

7.

	−15	−8		3 5		18	
−20		−10		0	10		20

9. $5, 175, -234$

11. $-7, -5, -1, 0, 2, 3, 8$ **13.** $-11, -6, -2, 1, 4, 5, 9$

15. $-7, -6, -3, 3, 6, 7$ **17.** Max: 15; Min: -6 **19.** Max: 21; Min: -15

21. Max: 5; Min: -2 **23.** -15 **25.** -11 **27.** 19   **29.** 7

31. 17   **33.** 10   **35.** -3 **37.** -8 **39.** 5   **41.** 18   **43.** 0

45. 7   **47.** 7   **49.** 11   **51.** 1   **53.** True   **55.** False   **57.** False

59. $\left|-6\right| + 2 = 8$ **61.** $6 + \left|-2\right| = 8$ **63.** -5 cm   **65.** -50 dollars

67. $-10°$ **69.** $20°$ **71.** $+90,000,000$ **73.** $-6; 8; 8; -2$

75. $-2; 6; 6; 0$ **77.**

ANSWERS

73. _____

74. _____

75. _____

76. _____

77. _____

2.2 Addition of Integers

2.2 OBJECTIVES

1. Add two integers with the same sign
2. Add two integers with opposite signs
3. Solve applications involving integers

In Section 2.1 we introduced the idea of negative numbers. Here we examine the four arithmetic operations (addition, subtraction, multiplication, and division) and see how those operations are performed when integers are involved. We start by considering addition.

An application may help. We will represent a gain of money as a positive number and a loss as a negative number.

If you gain $3 and then gain $4, the result is a gain of $7:

$$3 + 4 = 7$$

If you lose $3 and then lose $4, the result is a loss of $7:

$$-3 + (-4) = -7$$

If you gain $3 and then lose $4, the result is a loss of $1:

$$3 + (-4) = -1$$

If you lose $3 and then gain $4, the result is a gain of $1:

$$-3 + 4 = 1$$

The number line can be used to illustrate the addition of integers. Starting at the origin, we move to the *right* for positive integers and to the *left* for negative integers.

OBJECTIVE 1 **Example 1 Adding Integers**

(a) Add $3 + 4$.

Start at the origin and move 3 units to the right. Then move 4 more units to the right to find the sum. From the number line, we see that the sum is

$$3 + 4 = 7$$

(b) Add $(-3) + (-4)$.

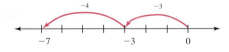

Start at the origin and move 3 units to the left. Then move 4 more units to the left to find the sum. From the number line, we see that the sum is

$$(-3) + (-4) = -7$$

CHECK YOURSELF 1

Add.

(a) $(-4) + (-5)$ **(b)** $(-3) + (-7)$
(c) $(-5) + (-15)$ **(d)** $(-5) + (-3)$

You have probably noticed a helpful pattern in the previous example. This pattern will allow you to do the work mentally without having to use the number line. Look at the following rule.

Property: Adding Integers Case 1: Same Sign

> **NOTE** This means that the sum of two positive integers is positive and the sum of two negative integers is negative.

If two integers have the same sign, add their absolute values. Give the result the sign of the original integers.

We can use the number line to illustrate the addition of two integers. This time the integers will have *different* signs.

OBJECTIVE 2   **Example 2   Adding Integers**

(a) Add $3 + (-6)$.

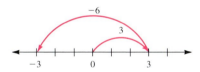

First move 3 units to the right of the origin. Then move 6 units to the left.

$3 + (-6) = -3$

(b) Add $-4 + 7$.

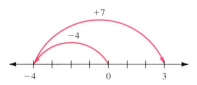

This time move 4 units to the left of the origin as the first step. Then move 7 units to the right.

$-4 + 7 = 3$

CHECK YOURSELF 2

Add.

(a) $7 + (-5)$ **(b)** $4 + (-8)$ **(c)** $-1 + 16$ **(d)** $-7 + 3$

You have no doubt noticed that, in adding a positive integer and a negative integer, sometimes the sum is positive and sometimes it is negative. This depends on which of the integers has the larger absolute value. This leads us to the second part of our addition rule.

RECALL We first encountered absolute values in Section 2.1.

> **Property: Adding Integers Case 2: Different Signs**
>
> If two integers have different signs, subtract their absolute values, the smaller from the larger. Give the result the sign of the integer with the larger absolute value.

> **Example 3   Adding Integers**

(a) $7 + (-19) = -12$

Because the two integers have different signs, subtract the absolute values ($19 - 7 = 12$). The sum of 7 and -19 has the sign ($-$) of the integer with the larger absolute value, -19.

(b) $-13 + 7 = -6$

Subtract the absolute values ($13 - 7 = 6$). The sum of -13 and 7 has the sign ($-$) of the integer with the larger absolute value, -13.

CHECK YOURSELF 3

Add mentally.

(a) $5 + (-14)$ **(b)** $-7 + (-8)$
(c) $-8 + 15$ **(d)** $7 + (-8)$

In Section 1.2, we discussed the commutative, associative, and additive identity properties. There is another property of addition that we should mention.

Recall that every number has an *opposite*. It corresponds to a point the same distance from the origin as the given number but in the opposite direction.

NOTE The opposite of a number is also called the **additive inverse** of that number.

NOTE 3 and -3 are opposites.

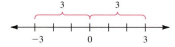

The opposite of 9 is -9.
The opposite of -15 is 15.

The additive inverse property states that the sum of any number and its opposite is 0.

NOTE Here $-a$ represents the opposite of the number a. If a is positive, $-a$ is negative. If a is negative, $-a$ is positive.

> **Property: Additive Inverse Property**
>
> For any number a, there exists a number $-a$ such that
>
> $a + (-a) = (-a) + a = 0$
>
> The sum of any number and its opposite, or additive inverse, is 0.

> Example 4   Adding Integers

NOTE Later, we will show that $-0 = 0$; therefore, the opposite of 0 is 0.

(a) $9 + (-9) = 0$

(b) $-15 + 15 = 0$

CHECK YOURSELF 4

Add.

(a) $(-17) + 17$

(b) $12 + (-12)$

When solving an application of integer arithmetic, the first step is to translate the phrase or statement using integers. Example 5 illustrates this step.

OBJECTIVE 3

> Example 5   An Application of the Addition of Integers

Shanique has $250 in her checking account. She writes a check for $120 and makes a deposit of $90. What is the resulting balance?

First, translate the phrase using integers. Such problems will usually include something that is represented by negative integers and something that is represented by positive integers. In this case, a check can be represented as a negative integer and a deposit as a positive integer. We have

$$250 + (-120) + 90$$

This expression can now be evaluated.

$$250 + (-120) + 90$$
$$= 130 + 90$$
$$= 220$$

The resulting balance is $220.

CHECK YOURSELF 5

Translate the problem into an integer expression and then answer the question.

When Kirin awoke, the temperature was twelve degrees below zero, Fahrenheit. Over the next six hours, the temperature increased by seventeen degrees. What was the temperature at that time?

READING YOUR TEXT

The following fill-in-the-blank exercises are designed to assure that you understand the key vocabulary used in this section. Each sentence comes directly from the section. You will find the correct answers in Appendix C.

Section 2.2

(a) In Section 2.1, we introduced the idea of _____ numbers.

(b) To add two numbers with different signs, _____ their absolute values.

(c) The sum of any number and its opposite, or additive _____, is 0.

(d) The opposite of zero is _____.

CHECK YOURSELF ANSWERS

1. **(a)** -9; **(b)** -10; **(c)** -20; **(d)** -8 **2.** **(a)** 2; **(b)** -4; **(c)** 15; **(d)** -4
3. **(a)** -9; **(b)** -15; **(c)** 7; **(d)** -1 **4.** **(a)** 0; **(b)** 0
5. $-12 + 17$; the temperature was 5°F.

2.2 Exercises

Name _____

Section _____ Date _____

ANSWERS

1. _____
2. _____
3. _____
4. _____
5. _____
6. _____
7. _____
8. _____
9. _____
10. _____
11. _____
12. _____
13. _____
14. _____
15. _____
16. _____
17. _____
18. _____
19. _____

Add.

1. $3 + 6$ **ALEKS**

2. $5 + 9$ **ALEKS**

3. $11 + 5$ **ALEKS**

4. $8 + 7$ **ALEKS**

5. $(-2) + (-3)$ **ALEKS**

6. $(-1) + (-9)$ **ALEKS**

7. $9 + (-3)$ **ALEKS**

8. $10 + (-4)$ **ALEKS**

9. $-9 + 0$ **ALEKS**

10. $-15 + 0$ **ALEKS**

11. $7 + (-7)$ **ALEKS**

12. $12 + (-12)$ **ALEKS**

13. $7 + (-9) + (-5) + 6$

14. $(-4) + 6 + (-3) + 0$

15. $7 + (-3) + 5 + (-11)$

16. $-6 + (-13) + 16$

In exercises 17 to 22, restate the problem using an expression involving integers and then answer the question.

17. Business and Finance   Amir has $100 in his checking account. He writes a check for $23 and makes a deposit of $51. What is his new balance?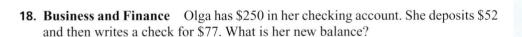

18. Business and Finance   Olga has $250 in her checking account. She deposits $52 and then writes a check for $77. What is her new balance?

19. Mechanical Engineering   A pneumatic actuator is operated by a pressurized air reservoir. At the beginning of an operator's shift, the pressure in the reservoir was 126 pounds per square inch ($lb/in.^2$). At the end of each hour, the operator recorded the reservoir's change in pressure. The values recorded (in $lb/in.^2$) were a drop of 12, a drop of 7, a rise of 32, a drop of 17, a drop of 15, a rise of 31, a drop of 4, and a drop of 14. What was the pressure in the tank at the end of the shift?

20. **Mechanical Engineering** A diesel engine for an industrial shredder has an 18-quart (qt) oil capacity. When a maintenance technician checked the oil, it was 7 qt low. Later that day, she added 4 qt to the engine. What was the oil level after the 4 qt were added?

21. **Science and Medicine** The lowest one-day temperature in Helena, Montana, was −21°F at night. The temperature increased by 25 degrees by noon. What was the temperature at noon?

22. **Science and Medicine** At 7 A.M., the temperature was −15°F. By 1 P.M., the temperature had increased by 18 degrees. What was the temperature at 1 P.M.?

23. In this chapter, it is stated that "every number has an opposite." The opposite of 9 is −9. This corresponds to the idea of an opposite in English. In English, an opposite is often expressed by a prefix, for example, *un-* or *ir-*.

   (a) Write the opposite of these words: unmentionable, uninteresting, irredeemable, irregular, uncomfortable.
   (b) What is the meaning of these expressions: not uninteresting, not irredeemable, not irregular, not unmentionable?
   (c) Think of other prefixes that *negate* or change the meaning of a word to its *opposite*. Make a list of words formed with these prefixes and write a sentence with three of the words you found. Make a sentence with two words and phrases from parts **(a)** and **(b)**.
   What is the value of $-[-(-5)]$? What is the value of $-(-6)$? How does this relate to the given examples? Write a short description about this relationship.

Answers

1. 9   **3.** 16   **5.** −5   **7.** 6   **9.** −9   **11.** 0   **13.** −1
15. −2   **17.** $128   **19.** 120 lb/in.²   **21.** 4°F   **23.**

Subtraction of Integers

2.3 OBJECTIVES

1. Find the difference of two integers
2. Solve applications involving the subtraction of integers

To begin our discussion of subtraction when integers are involved, we can look back at a problem using natural numbers. We know that

$$8 - 5 = 3$$

From our work in adding integers, we know that it is also true that

$$8 + (-5) = 3$$

Comparing these equations, we see that the results are the same. This leads us to an important pattern. Any subtraction problem can be written as a problem in addition. Subtracting 5 is the same as adding the opposite of 5, or -5. We can write this fact as follows:

$$8 - 5 = 8 + (-5) = 3$$

This leads us to the following rule for subtracting integers.

Property: Subtracting Integers

1. To rewrite the subtraction problem as an addition problem:
   a. Change the subtraction operation to addition.
   b. Replace the integer being subtracted with its opposite.
2. Add the resulting integers as before.
   In symbols,

   $$a - b = a + (-b)$$

Example 1 illustrates the use of this property while subtracting.

OBJECTIVE 1

Example 1 Subtracting Integers

Change the subtraction symbol $(-)$ to an addition symbol $(+)$.

(a) $15 - 7 = 15 + (-7)$

Replace 7 with its opposite, -7.

$$= 8$$

(b) $9 - 12 = 9 + (-12) = -3$

(c) $-6 - 7 = -6 + (-7) = -13$

(d) Subtract 5 from -2. We write the statement as $-2 - 5$ and proceed as before:

$$-2 - 5 = -2 + (-5) = -7$$

CHECK YOURSELF 1

Subtract.

(a) $18 - 7$ **(b)** $5 - 13$ **(c)** $-7 - 9$ **(d)** $-2 - 7$

The subtraction rule is used in the same way when the integer being subtracted is negative. Change the subtraction to addition. Replace the negative integer being subtracted with its opposite, which is positive. Example 2 illustrates this principle.

Example 2   Subtracting Integers

Change the subtraction to addition.

(a) $5 - (-2) = 5 + (+2) = 5 + 2 = 7$

Replace -2 with its opposite, $+2$ or 2.

(b) $7 - (-8) = 7 + (+8) = 7 + 8 = 15$

(c) $-9 - (-5) = -9 + 5 = -4$

(d) Subtract -4 from -5. We write

$-5 - (-4) = -5 + 4 = -1$

CHECK YOURSELF 2

Subtract.

(a) $8 - (-2)$ **(b)** $3 - (-10)$
(c) $-7 - (-2)$ **(d)** $7 - (-7)$

OBJECTIVE 2   **Example 3   An Application of the Subtraction of Integers**

Susanna's checking account shows a balance of $285. She has discovered that a deposit for $47 was accidently recorded as a check for $47. Write an integer expression that represents the correction on the balance. Then find the corrected balance.

$285 - (-47) + (47)$

Subtract the check and then add the deposit.

$285 - (-47) + (47) = 285 + 47 + 47 = 379$

The corrected balance is $379.

CHECK YOURSELF 3

It appears that Marshal, a running back, gained 97 yards in the last game. A closer inspection of the statistics revealed that a 9-yard gain had been recorded as a 9-yard loss. Write an integer expression that represents the corrected yards gained and then find that number.

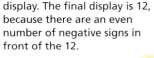

Using Your Calculator to Add and Subtract Integers

Your scientific (or graphing) calculator has a key that makes a number negative. This key is different from the "subtraction" key. The negative key is marked as either $\boxed{+/-}$ or $\boxed{(-)}$. With a scientific calculator, this key is pressed *after* the number you wish to make negative is entered. All of the instructions in this section assume that you have a scientific calculator.

Example 4   Entering a Negative Integer into the Calculator

Enter each of the following into your calculator.

(a) -24

24 $\boxed{+/-}$

NOTE The 12 changes between positive and negative in the display. The final display is 12, because there are an even number of negative signs in front of the 12.

(b) $-(-(-(-12)))$

12 $\boxed{+/-}$ $\boxed{+/-}$ $\boxed{+/-}$ $\boxed{+/-}$

CHECK YOURSELF 4 _____

Enter each number into your calculator.

(a) -36 **(b)** $-(-(-6))$

Example 5   Adding Integers

Find the sum for each pair of integers.

(a) $256 + (-297)$

256 $\boxed{+}$ 297 $\boxed{+/-}$ $\boxed{=}$

Your display should read -41.

(b) $-312 + (-569)$

312 $\boxed{+/-}$ $\boxed{+}$ 569 $\boxed{+/-}$ $\boxed{=}$

Your display should read -881.

CHECK YOURSELF 5 _____

Find the sum for each pair of integers.

(a) $-368 + 547$ **(b)** $-596 + (-834)$

Example 6   Subtracting Integers

Find the difference for $-356 - (-469)$.

356 $\boxed{+/-}$ $\boxed{-}$ 469 $\boxed{+/-}$ $\boxed{=}$

Your display should read 113.

CHECK YOURSELF 6

Find each difference.

(a) $349 - (-49)$

(b) $-294 - (-137)$

READING YOUR TEXT

The following fill-in-the-blank exercises are designed to assure that you understand the key vocabulary used in this section. Each sentence comes directly from the section. You will find the correct answers in Appendix C.

Section 2.3

(a) Any subtraction problem can be written as a problem in _____.

(b) To rewrite a subtraction problem as an addition problem, change the subtraction operation to addition and replace the integer being subtracted with its _____.

(c) The opposite of -2 is _____.

(d) The calculator key that makes a number negative is different from the _____ key.

CHECK YOURSELF ANSWERS

1. (a) 11; **(b)** -8; **(c)** -16; **(d)** -9 **2. (a)** 10; **(b)** 13; **(c)** -5; **(d)** 14
3. $97 - (-9) + 9 = 115$ yards   **4. (a)** -36; **(b)** -6 **5. (a)** 179; **(b)** $-1,430$
6. (a) 398; **(b)** -157

2.3 Exercises

Name _____

Section _____ Date _____

ANSWERS

Subtract.

1. $21 - 13$ ALEKS

2. $36 - 22$ ALEKS

3. $82 - 45$ ALEKS

4. $103 - 56$ ALEKS

5. $8 - 10$ ALEKS

6. $14 - 19$ ALEKS

7. $24 - 45$ ALEKS

8. $136 - 352$ ALEKS

9. $-5 - 3$ ALEKS

10. $-15 - 8$ ALEKS

11. $-9 - 14$ VIDEO ALEKS

12. $-8 - 12$ ALEKS

13. $5 - (-11)$ ALEKS

14. $7 - (-5)$ ALEKS

15. $7 - (-12)$ ALEKS

16. $3 - (-10)$ ALEKS

17. $-36 - (-24)$ ALEKS

18. $-28 - (-11)$ ALEKS

19. $-19 - (-27)$ VIDEO ALEKS

20. $-11 - (-16)$ ALEKS

For exercises 21 to 23, write an integer expression that describes the situation. Then answer the question.

21. Science and Medicine   The temperature at noon on a June day was 82°F. It fell by 12 degrees in the next 4 h. What was the temperature at 4:00 P.M.?

22. Business and Finance   Jason's checking account shows a balance of $853. He has discovered that a deposit of $70 was accidently recorded as a check for $70. What is the corrected balance?

23. Business and Finance   Ylena's checking account shows a balance of $947. She has discovered that a check for $86 was recorded as a deposit of $86. What is the corrected balance?

24. Technology   How long ago was the year 1250 B.C.E.? What year was 3,300 years ago? Make a number line and locate the following events, cultures, and objects on it. How long ago was each item in the list? Which two events are the closest to each other? You may want to learn more about some of the cultures in the list and the mathematics and science developed by each culture.

Inca culture in Peru—1400 A.D.
The *Ahmes Papyrus,* a mathematical text from Egypt—1650 B.C.E.
Babylonian arithmetic develops the use of a zero symbol—300 B.C.E.
First Olympic Games—776 B.C.E.
Pythagoras of Greece dies—500 B.C.E.
Mayans in Central America independently develop use of zero—500 A.D.
The *Chou Pei,* a mathematics classic from China—1000 B.C.E.
The *Aryabhatiya,* a mathematics work from India—499 A.D.
Trigonometry arrives in Europe via the Arabs and India—1464 A.D.
Arabs receive algebra from Greek, Hindu, and Babylonian sources and develop it into a new systematic form—850 A.D.
Development of calculus in Europe—1670 A.D.
Rise of abstract algebra—1860 A.D.
Growing importance of probability and development of statistics—1902 A.D.

1. _____

2. _____

3. _____

4. _____

5. _____

6. _____

7. _____

8. _____

9. _____

10. _____

11. _____

12. _____

13. _____

14. _____

15. _____

16. _____

17. _____

18. _____

19. _____

20. _____

21. _____

22. _____

23. _____

24.

25.

26.

27.

28.

29.

30.

31.

32.

33.

34.

35.

36.

37.

38.

25. Complete the following statement: "$3 - (-7)$ is the same as _____ because . . ."
Write a problem that might be answered by doing this subtraction.

26. Explain the difference between the two phrases: "a number subtracted from 5"
and "a number less than 5." Use algebra and English to explain the meaning of
these phrases. Write other ways to express subtraction in English. Which ones are
confusing?

27. Science and Medicine The greatest one-day temperature drop in the United
States happened on Christmas Eve, 1924, in Montana. The temperature went from
$63°F$ during the day to $-21°F$ at night. What was the total temperature drop?

28. Science and Medicine A similar one-day temperature drop happened in Alaska.
The temperature went from $47°F$ during the day to $-29°F$ at night. What was the
total temperature drop?

29. Science and Medicine The tide at the mouth of the Delaware River tends to vary
between a maximum of $+10$ ft and a minimum of -5 ft. What is the difference in
feet between the high tide and the low tide?

30. Science and Medicine The tide at the mouth of the Sacramento River tends to
vary between a maximum of $+7$ ft and a minimum of -2 ft. What is the difference
in feet between the high tide and the low tide?

Calculator Exercises

Use your calculator to perform the following operations.

31. $-789 + (-128)$ **32.** $-910 + (-567)$

33. $-349 + (-431)$ **34.** $-412 + (-367)$

35. $47 - (-25)$ **36.** $123 - (-219)$

37. $234 - (-456)$ **38.** $412 - (-123)$

Answers

1. 8   **3.** 37   **5.** -2 **7.** -21 **9.** -8 **11.** -23 **13.** 16
15. 19   **17.** -12 **19.** 8   **21.** $82° - 12° = 70°F$
23. $947 - 86 + (-86)$; $775 is the balance   **25.**   **27.** $84°F$

29. 15   **31.** -917 **33.** -780 **35.** 72   **37.** 690

2.4 Multiplication of Integers

2.4 OBJECTIVES

1. Find the product of two or more integers
2. Use the order of operations with integers

When you first considered multiplication in arithmetic, it was thought of as repeated addition. Now we look at what our work with the addition of integers can tell us about multiplication when integers are involved. For example,

$3 \cdot 4 = \underline{4 + 4 + 4} = 12$

> We interpret multiplication as repeated addition to find the product, 12.

Now, consider the product $(3)(-4)$:

$(3)(-4) = (-4) + (-4) + (-4) = -12$

Looking at this product suggests the first portion of our rule for multiplying integers. The product of a positive integer and a negative integer is negative.

Property: Multiplying Integers Case 1: Different Signs

The product of two integers with different signs is negative.

To use this rule in multiplying two integers with different signs, multiply their absolute values and attach a negative sign.

OBJECTIVE 1

Example 1 Multiplying Integers

Multiply.

(a) $(5)(-6) = -30$

> The product is negative.

(b) $(-10)(10) = -100$

(c) $(8)(-12) = -96$

CHECK YOURSELF 1

Multiply.

(a) $(-7)(5)$ **(b)** $(-12)(9)$ **(c)** $(-15)(8)$

The product of two negative integers is harder to visualize. The following pattern may help you see how we can determine the sign of the product.

$$(3)(-2) = -6$$
$$(2)(-2) = -4$$
$$(1)(-2) = -2$$
$$(0)(-2) = 0$$
$$(-1)(-2) = 2$$

NOTE This number is decreasing by 1.

Do you see that the product is *increasing* by 2 each time as you go down?

NOTE $(-1)(-2)$ is the opposite of -2.

What should the product $(-2)(-2)$ be? Continuing the pattern shown, we see that

$$(-2)(-2) = 4$$

This suggests that the product of two negative integers is positive, which is the case. We can extend our multiplication rule.

Property: Multiplying Integers Case 2: Same Sign

The product of two integers with the same sign is positive.

Example 2 Multiplying Integers

Multiply.

(a) $9 \cdot 7 = 63$ The product of two positive numbers (same sign, $+$) is positive.

(b) $(-8)(-5) = 40$ The product of two negative numbers (same sign, $-$) is positive.

CHECK YOURSELF 2

Multiply.

(a) $10 \cdot 12$ (b) $(-8)(-9)$

The *multiplicative identity property* and *multiplicative property of zero* studied in Section 1.5 can be applied to integers, as illustrated in Example 3.

Example 3 Multiplying Integers by One and Zero

Find each product.

(a) $(1)(-7) = -7$

(b) $(15)(1) = 15$

(c) $(-7)(0) = 0$

(d) $0 \cdot 12 = 0$

CHECK YOURSELF 3 _____

Multiply.

(a) $(-10)(1)$ **(b)** $(0)(-17)$

We can now extend the rules for the order of operations learned in Section 1.8 to simplify expressions containing integers. First, we will work with integers raised to a power.

OBJECTIVE 2

Example 4   Integers with Exponents

Evaluate each expression.

NOTE In part (b) of Example 4, we have a negative integer raised to a power.
   In part (c), only the 3 is raised to a power. We have the opposite of 3 squared.

(a) $(-3)^2 = (-3)(-3) = 9$

(b) $(-3)^3 = (-3)(-3)(-3) = -27$

(c) $-3^2 = -(3 \cdot 3) = -9$ Note that the negative is *not* squared.

CHECK YOURSELF 4 _____

Evaluate each expression.

(a) $(-4)^2$ **(b)** $(-4)^3$ **(c)** -4^2

In Example 5 we will apply the order of operations.

Example 5   Using Order of Operations with Integers

Evaluate each expression.

(a) $7(-9 + 12)$ Evaluate inside the parentheses first.
    $= 7(3) = 21$

(b) $(-8)(-7) - 40$ Multiply first, then subtract.
    $= 56 - 40$
    $= 16$

(c) $(-5)^2 - 3$ Evaluate the power first.
    $= (-5)(-5) - 3$ Note that $(-5)^2 = (-5)(-5)$
    $= 25 - 3$ $= 25$
    $= 22$

(d) $-5^2 - 3$ Note that $-5^2 = -25$. The power applies *only* to the 5.
    $= -25 - 3$
    $= -28$

✔ CHECK YOURSELF 5 ⎯⎯⎯⎯⎯⎯⎯⎯⎯⎯⎯⎯⎯⎯⎯⎯⎯

Evaluate each expression.

(a) $8(-9 + 7)$ **(b)** $(-3)(-5) + 7$

(c) $(-4)^2 - (-4)$ **(d)** $-4^2 - (-4)$

READING YOUR TEXT ⎯⎯⎯⎯⎯⎯⎯⎯⎯⎯⎯

The following fill-in-the-blank exercises are designed to assure that you understand the key vocabulary used in this section. Each sentence comes directly from the section. You will find the correct answers in Appendix C.

Section 2.4

(a)   The product of two integers with different signs is ⎯⎯⎯⎯⎯⎯⎯⎯⎯⎯.

(b)   The product of two integers with the same sign is ⎯⎯⎯⎯⎯⎯⎯⎯⎯⎯.

(c)   Given the expression -3^2, the ⎯⎯⎯⎯⎯⎯⎯⎯⎯⎯ is *not* squared.

(d)   The rules for order of operations were learned in Section ⎯⎯⎯⎯⎯⎯⎯.

CHECK YOURSELF ANSWERS ⎯⎯⎯⎯⎯⎯⎯⎯⎯⎯

1. (a) -35; **(b)** -108; **(c)** -120 **2. (a)** 120; **(b)** 72 **3. (a)** -10; **(b)** 0
4. (a) 16; **(b)** -64; **(c)** -16 **5. (a)** -16; **(b)** 22; **(c)** 20; **(d)** -12

2.4 Exercises

Name _____

Section _____ Date _____

ANSWERS

Multiply.

1. $4 \cdot 10$ **ALEKS**

2. $3 \cdot 14$ **ALEKS**

3. $(5)(-12)$ **ALEKS**

4. $(10)(-2)$ **ALEKS**

5. $(-8)(9)$ **ALEKS**

6. $(-12)(3)$ **ALEKS**

7. $(-8)(-7)$ **ALEKS**

8. $(-9)(-8)$ **ALEKS**

9. $(-5)(-12)$ **ALEKS**

10. $(-7)(-3)$ **ALEKS**

11. $(0)(-18)$ **ALEKS**

12. $(-17)(0)$ **ALEKS**

13. $(15)(0)$ **ALEKS**

14. $(0)(25)$ **ALEKS**

Do the indicated operations. Remember the rules for the order of operations.

15. $5(7 - 2)$

16. $7(8 - 5)$

17. $2(5 - 8)$

18. $6(14 - 16)$

19. $-3(9 - 7)$

20. $-6(12 - 9)$

21. $-3(-2 - 5)$ **VIDEO**

22. $-2(-7 - 3)$

23. $(-2)(3) - 5$

24. $(-6)(8) - 27$

25. $4(-7) - 5$

26. $(-3)(-9) - 11$

27. $(-5)(-2) - 12$

28. $(-7)(-3) - 25$

29. $(3)(-7) + 20$

30. $(2)(-6) + 8$

31. $-4 + (-3)(6)$

32. $-5 + (-2)(3)$

33. $7 - (-4)(-2)$

34. $9 - (-2)(-7)$

35. $(-7)^2 - 17$

36. $(-6)^2 - 20$

1. _____
2. _____
3. _____
4. _____
5. _____
6. _____
7. _____
8. _____
9. _____
10. _____
11. _____
12. _____

13. _____ 14. _____
15. _____ 16. _____
17. _____ 18. _____
19. _____ 20. _____
21. _____ 22. _____
23. _____ 24. _____
25. _____ 26. _____
27. _____ 28. _____
29. _____ 30. _____
31. _____ 32. _____
33. _____ 34. _____
35. _____ 36. _____

37. _____

38. _____

39. _____

40. _____

41. _____

42. _____

43. _____

44. _____

45. _____

46. _____

47. _____

48. _____

49. _____

50. _____

51. _____

52. _____

53. _____

37. $(-5)^2 + 18$ **38.** $(-2)^2 + 10$

39. $-6^2 - 4$ **40.** $-5^2 - 3$

41. $(-4)^2 - (-2)(-5)$ **42.** $(-3)^3 - (-8)(-2)$

43. $(-8)^2 - 5^2$ **44.** $(-6)^2 - 4^2$

45. $(-6)^2 - (-3)^2$ **46.** $(-8)^2 - (-4)^2$

47. $-8^2 - 5^2$ **48.** $-6^2 - 3^2$

49. $-8^2 - (-5)^2$ **50.** $-9^2 - (-6)^2$

51. Business and Finance Stores occasionally sell products at a loss in order to draw in customers or to reward good customers. The theory is that customers will buy other products along with the discounted item and the store will ultimately profit.

Beguhn Industries sells five different products. The company makes $18 on each product-A item sold, loses $4 on product-B items, earns $11 on product C, makes $38 on product D, and loses $15 on product E.

One month, Beguhn Industries sold 127 units of product A, 273 units of product B, 201 units of product C, 377 units of product D, and 43 units of product E. What was their profit or loss that month?

52. Statistics In Atlantic City, Nick played the slot machines for 12 h. He lost $45 an hour. Use integers to represent the change in Nick's financial status at the end of the 12 h.

53. Science and Medicine The temperature is $-6°F$ at 5:00 in the evening. If the temperature drops 2 degrees every hour, what is the temperature at 1:00 A.M.?

Answers

1. 40   **3.** -60 **5.** -72 **7.** 56   **9.** 60   **11.** 0   **13.** 0
15. 25   **17.** -6 **19.** -6 **21.** 21   **23.** -11 **25.** -33
27. -2 **29.** -1 **31.** -22 **33.** -1 **35.** 32   **37.** 43
39. -40 **41.** 6   **43.** 39   **45.** 27   **47.** -89 **49.** -89
51. $+\$17,086$ **53.** $-22°F$

2.5 Division of Integers

2.5 OBJECTIVES

1. Find the quotient of two integers
2. Use the order of operations with integers

You know from your work in arithmetic that multiplication and division are related operations. We can use that fact, and our work of Section 2.4, to determine rules for the division of integers. Every division problem can be stated as an equivalent multiplication problem. For instance,

$$\frac{15}{5} = 3 \qquad \text{because} \qquad 15 = 5 \cdot 3$$

$$\frac{-24}{6} = -4 \qquad \text{because} \qquad -24 = (6)(-4)$$

$$\frac{-30}{-5} = 6 \qquad \text{because} \qquad -30 = (-5)(6)$$

These examples illustrate that because the two operations are related, the rule of signs that we stated in Section 2.4 for multiplication is also true for division.

> **Property: Dividing Integers**
>
> 1. The quotient of two integers with different signs is negative.
> 2. The quotient of two integers with the same sign is positive.

Again, the rule is easy to use. To divide two integers, divide their absolute values. Then attach the proper sign according to the rule.

OBJECTIVE 1

Example 1 Dividing Integers

Divide.

(a) Positive $\longrightarrow \dfrac{28}{7} = 4 \longleftarrow$ Positive
Positive $\longrightarrow$

(b) Negative $\longrightarrow \dfrac{-36}{-4} = 9 \longleftarrow$ Positive
Negative $\longrightarrow$

(c) Negative $\longrightarrow \dfrac{-42}{7} = -6 \longleftarrow$ Negative
Positive $\longrightarrow$

(d) Positive $\longrightarrow \dfrac{75}{-3} = -25 \longleftarrow$ Negative
Negative $\longrightarrow$

> ✔ **CHECK YOURSELF 1**
>
> *Divide.*
>
> (a) $\dfrac{-55}{11}$ (b) $\dfrac{80}{20}$
>
> (c) $\dfrac{-48}{-8}$ (d) $\dfrac{144}{-12}$

As discussed in Section 1.6, we must be very careful when 0 is involved in a division problem. Remember that 0 divided by any nonzero number is just 0. This rule can be extended to include integers, so that

$$\frac{0}{-7} = 0 \qquad \text{because} \qquad 0 = (-7)(0)$$

However, if zero is the *divisor,* we have a special problem. Consider

$$\frac{-9}{0} = ?$$

This means that $-9 = 0 \cdot ?$.

Can 0 times a number ever be -9? No, so there is no solution.

Because $\dfrac{-9}{0}$ cannot be replaced by any number, we agree that *division by 0 is not allowed.* We say that division by 0 is *undefined.*

Example 2 Dividing Integers

Divide, if possible.

(a) $\dfrac{7}{0}$ is undefined.

(b) $\dfrac{-9}{0}$ is undefined.

(c) $\dfrac{0}{5} = 0$

(d) $\dfrac{0}{-8} = 0$

Note: The expression $\dfrac{0}{0}$ is called an **indeterminate form.** You will learn more about this in later mathematics classes.

✔ **CHECK YOURSELF 2**

Divide, if possible.

(a) $\dfrac{0}{3}$ (b) $\dfrac{5}{0}$ (c) $\dfrac{-7}{0}$ (d) $\dfrac{0}{-9}$

The fraction bar, like parentheses and the absolute value bars, serves as a *grouping symbol.* This means that all operations in the numerator and denominator should be performed separately. Then the division is done as the last step. Example 3 illustrates this property.

OBJECTIVE 2   **Example 3   Using Order of Operations**

Evaluate each expression.

(a) $\dfrac{(-6)(-7)}{3} = \dfrac{42}{3} = 14$ Multiply in the numerator, then divide.

(b) $\dfrac{3 + (-12)}{3} = \dfrac{-9}{3} = -3$ Add in the numerator, then divide.

(c) $\dfrac{-4 + (2)(-6)}{-6 - 2} = \dfrac{-4 + (-12)}{-6 - 2}$ Multiply in the numerator. Then add in the numerator and subtract in the denominator.

$\qquad\qquad = \dfrac{-16}{-8} = 2$ Divide as the last step.

✔ **CHECK YOURSELF 3**

Evaluate each expression.

(a) $\dfrac{-4 + (-8)}{6}$ (b) $\dfrac{3 - (2)(-6)}{-5}$ (c) $\dfrac{(-2)(-4) - (-6)(-5)}{(-2)(11)}$

Using Your Calculator to Multiply and Divide Integers

Finding the product of two integers using a calculator is relatively straightforward.

Example 4   Multiplying Integers

Find the product. $(457)(-734)$

$457 \;\boxed{\times}\; 734 \;\boxed{+/-}\; \boxed{=}$

Your display should read $-335{,}438$.

✔ CHECK YOURSELF 4 —————————————————————

Find the products.

(a) $(36)(-91)$ **(b)** $(-12)(-284)$

Finding the quotient of integers is also straightforward.

Example 5 Dividing Integers

Find the quotient. $\dfrac{-384}{16}$

384 $\boxed{+/-}$ $\boxed{\div}$ 16 $\boxed{=}$

Your display should read -24.

✔ CHECK YOURSELF 5 —————————————————————

Find the quotient.

$(-7{,}865) \div (-242)$

We can also use the calculator to raise an integer to a power.

Example 6 Raising a Number to a Power

Evaluate.

$(-3)^6$

$\boxed{(}$ 3 $\boxed{+/-}$ $\boxed{)}$ $\boxed{y^x}$ 6 $\boxed{=}$

NOTE The parentheses ensure that the negative is attached to the 3 *before* it is raised to a power.

or, on some calculators

$\boxed{(}$ $\boxed{(-)}$ 3 $\boxed{)}$ $\boxed{\wedge}$ 6 $\boxed{\text{Enter}}$

Either way, your display should read 729.

✔ CHECK YOURSELF 6 —————————————————————

Evaluate.

$(-2)^9$

READING YOUR TEXT

The following fill-in-the-blank exercises are designed to assure that you understand the key vocabulary used in this section. Each sentence comes directly from the section. You will find the correct answers in Appendix C.

Section 2.5

(a) The quotient of two integers with different signs is _____.

(b) The quotient of two integers with the same sign is _____.

(c) Division by _____ is *not* allowed.

(d) The fraction bar serves as a _____ symbol.

CHECK YOURSELF ANSWERS

1. (a) −5; **(b)** 4; **(c)** 6; **(d)** −12   **2. (a)** 0; **(b)** undefined; **(c)** undefined; **(d)** 0
3. (a) −2; **(b)** −3; **(c)** 1   **4. (a)** −3,276; **(b)** 3,408   **5.** 32.5   **6.** −512

Name _____

Section _____ Date _____

ANSWERS

1. _____
2. _____
3. _____
4. _____
5. _____
6. _____
7. _____
8. _____
9. _____
10. _____
11. _____
12. _____ 13. _____
14. _____
15. _____ 16. _____
17. _____
18. _____
19. _____ 20. _____
21. _____ 22. _____
23. _____ 24. _____
25. _____ 26. _____
27. _____ 28. _____
29. _____ 30. _____
31. _____ 32. _____
33. _____
34. _____

156 SECTION 2.5

2.5 Exercises

Divide.

1. $\dfrac{-20}{-4}$ **ALEKS**

2. $\dfrac{70}{14}$ **ALEKS**

3. $\dfrac{48}{6}$ **ALEKS**

4. $\dfrac{-24}{8}$ **ALEKS**

5. $\dfrac{50}{-5}$ **VIDEO** **ALEKS**

6. $\dfrac{-32}{-8}$ **ALEKS**

7. $\dfrac{-52}{4}$ **ALEKS**

8. $\dfrac{56}{-7}$ **ALEKS**

9. $\dfrac{-75}{-3}$ **ALEKS**

10. $\dfrac{-60}{15}$ **ALEKS**

11. $\dfrac{0}{-8}$ **VIDEO** **ALEKS**

12. $\dfrac{-125}{-25}$ **ALEKS**

13. $\dfrac{-9}{-1}$ **ALEKS**

14. $\dfrac{-10}{0}$ **ALEKS**

15. $\dfrac{-96}{-8}$ **ALEKS**

16. $\dfrac{-20}{2}$ **ALEKS**

17. $\dfrac{18}{0}$ **ALEKS**

18. $\dfrac{0}{8}$ **ALEKS**

19. $\dfrac{-17}{1}$ **ALEKS**

20. $\dfrac{-27}{-1}$ **ALEKS**

21. $\dfrac{-144}{-16}$ **ALEKS**

22. $\dfrac{-150}{6}$ **ALEKS**

Perform the indicated operations.

23. $\dfrac{(-6)(-3)}{2}$

24. $\dfrac{(-9)(5)}{-3}$

25. $\dfrac{(-8)(2)}{-4}$

26. $\dfrac{(7)(-8)}{-14}$

27. $\dfrac{24}{-4-8}$

28. $\dfrac{36}{-7+3}$

29. $\dfrac{-12-12}{-3}$

30. $\dfrac{-14-4}{-6}$

31. $\dfrac{55-19}{-12-6}$ **VIDEO**

32. $\dfrac{-11-7}{-14+8}$

33. $\dfrac{7-5}{2-2}$

34. $\dfrac{10-6}{4-4}$

For exercises 35 to 37, use integers to write an expression that represents the situation. Then answer the question.

35. **Business and Finance** Patrick worked all day mowing lawns and was paid $9 per hour. If he had $125 at the end of a 9-h day, how much did he have before he started working?

36. **Social Science** A woman lost 42 lb. If she lost 3 lb each week, how long has she been dieting?

37. **Business and Finance** Suppose that you and your two brothers bought equal shares of an investment for a total of $20,000 and sold it later for $16,232. How much did each person lose?

35. _____

36. _____

37. _____

38. _____

39. _____

40. _____

41. _____

42. _____

43. _____

44. _____

45. _____

Calculator Exercises

Use your calculator to multiply and divide.

38. $(15)(-45)$

39. $(78)(-12)$

40. $(-56)(31)$

41. $(-34)(-28)$

42. $(-71)(-19)$

43. $\dfrac{-28}{-14}$

44. $(-5)^4$

45. $(-4)^5$

Answers

1. 5   **3.** 8   **5.** -10 **7.** -13 **9.** 25   **11.** 0   **13.** 9
15. 12   **17.** Undefined   **19.** -17 **21.** 9   **23.** 9   **25.** 4
27. -2 **29.** 8   **31.** -2 **33.** Undefined   **35.** $125 - 9 \cdot 9 = \$44$
37. $\dfrac{20,000 - 16,232}{3} = \$1,256$ **39.** -936 **41.** 952   **43.** 2
45. $-1,024$

2.6 Introduction to Algebra: Variables and Expressions

2.6 OBJECTIVES

1. Represent addition, subtraction, multiplication, and division by using the symbols of algebra
2. Identify algebraic expressions

In arithmetic, you learned how to do calculations with numbers by using the basic operations of addition, subtraction, multiplication, and division.

In algebra, you will still use numbers and the same four operations. However, you will also use letters to represent numbers. Letters such as x, y, L, or W are called **variables** when they can represent different numerical values. If we need to represent the length and width of *any* rectangle, we can use the variables L and W.

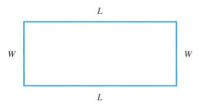

RECALL In arithmetic:
+ denotes addition
− denotes subtraction
· denotes multiplication
÷ denotes division.

You are familiar with the four symbols $(+, -, \cdot, \div)$ used to indicate the fundamental operations of arithmetic.

Next, we will look at how these operations are indicated in algebra. We begin by looking at addition.

OBJECTIVE 1

Example 1 Writing Expressions that Indicate Addition

(a) The *sum* of a and 3 is written as $a + 3$.

(b) L plus W is written as $L + W$.

(c) 5 more than m is written as $m + 5$.

(d) x increased by 7 is written as $x + 7$.

 CHECK YOURSELF 1

Write, using symbols.

(a) The sum of y and 4   (b) a plus b

(c) 3 more than x (d) n increased by 6

In Example 2, we look at how subtraction is indicated in algebra.

Example 2 Writing Expressions that Indicate Subtraction

(a) r minus s is written as $r - s$.

(b) The *difference* of m and 5 is written as $m - 5$.

(c) x decreased by 8 is written as $x - 8$.

(d) 4 less than a is written as $a - 4$.

CHECK YOURSELF 2 _____

Write, using symbols.

(a) *w* minus *z*
(c) *y* decreased by 3

(b) The difference of *a* and 7
(d) 5 less than *b*

You have seen that the operations of addition and subtraction are written exactly the same way in algebra as in arithmetic. This is not true in multiplication because the sign × looks like the letter *x*. So in algebra we use other symbols to show multiplication to avoid any confusion. Here are some ways to write multiplication.

NOTE *x* and *y* are called the **factors** of the product *xy*.

Definition: Multiplication

A centered dot	$x \cdot y$	
Parentheses	$(x)(y)$	These all indicate the *product* of *x* and *y* or *x* times *y*.
Writing the letters next to each other	xy	

Example 3   Writing Expressions that Indicate Multiplication

NOTE You can place letters next to each other or numbers and letters next to each other to show multiplication. But you *cannot* place numbers side by side to show multiplication: 37 means the number "thirty-seven," not 3 times 7.

(a) The product of 5 and *a* is written as $5 \cdot a$, $(5)(a)$, or $5a$. The last expression, $5a$, is the shortest and the most common way of writing the product.
(b) 3 times 7 can be written as $3 \cdot 7$, $(3)(7)$, or $3(7)$.
(c) Twice *z* is written as $2z$.
(d) The product of 2, *s*, and *t* is written as $2st$.
(e) 4 more than the product of 6 and *x* is written as $6x + 4$.

CHECK YOURSELF 3 _____

Write, using symbols.

(a) *m* times *n*
(c) The product of 8 and 9
(e) 3 more than the product of 8 and *a*

(b) The product of *h* and *b*
(d) The product of 5, *w*, and *y*

Before we move on to division, we look at how we can combine the symbols we have learned so far.

Definition: Expression

NOTE Not every collection of symbols is an expression.

An **expression** is a meaningful collection of numbers, variables, and signs of operation.

OBJECTIVE 2   **Example 4   Identifying Expressions**

(a) $2m + 3$ is an expression. It means that we multiply 2 and m and then add 3.

(b) $x + \cdot + 3$ is not an expression. The three operations in a row have no meaning.

(c) $y = 2x - 1$ is not an expression. The equals sign is not an operation sign.

(d) $3a + 5b - 4c$ is an expression. Its meaning is clear.

 CHECK YOURSELF 4 _____

Identify which are expressions and which are not.

(a) $7 - \cdot x$ (b) $6 + y = 9$

(c) $a + b - c$ (d) $3x - 5yz$

To write more complicated products in algebra, we need to use grouping symbols. Parentheses ( ) mean that an expression is to be thought of as a single quantity. Brackets [ ] are used in exactly the same way as parentheses in algebra. Look at Example 5, which shows the use of these signs of grouping.

Example 5   Expressions with More than One Operation

(a) 3 times the sum of a and b is written as

NOTE This can be read as "3 times the quantity a plus b."

$3(a + b)$

The sum of a and b is a single quantity, so it is enclosed in parentheses.

NOTE No parentheses are used in part (b) because the 3 multiplies *only* the a.

(b) The sum of 3 times a and b is written as $3a + b$.

(c) 2 times the difference of m and n is written as $2(m - n)$.

(d) The product of s plus t and s minus t is written as $(s + t)(s - t)$.

(e) The product of b and 3 less than b is written as $b(b - 3)$.

 CHECK YOURSELF 5 _____

Write, using symbols.

(a) Twice the sum of p and q

(b) The sum of twice p and q

(c) The product of a and the quantity $b - c$

(d) The product of x plus 2 and x minus 2

(e) The product of x and 4 more than x

NOTE In algebra, the fraction form is usually used.

Now we look at the operation of division.

Example 6   Writing Expressions that Indicate Division

(a) m divided by 3 is written as $\dfrac{m}{3}$.

(b) The quotient of a plus b and 5 is written as $\dfrac{a + b}{5}$.

(c) The sum p plus q divided by the difference p minus q is written as $\dfrac{p + q}{p - q}$.

 CHECK YOURSELF 6

Write, using symbols.

(a) r divided by s

(b) The quotient when x minus y is divided by 7

(c) The difference a minus 2 divided by the sum a plus 2

Notice that we can use many different letters to represent variables. In Example 6, the letters m, a, b, p, and q represented different variables. We often choose a letter that reminds us of what it represents, for example, L for *length* or W for *width*.

Example 7   Writing Geometric Expressions

(a) *Length* times *width* is written LW.

(b) One-half of *altitude* times *base* is written $\dfrac{1}{2}ab$.

(c) *Length* times *width* times *height* is written LWH.

(d) Pi (π) times *diameter* is written πd.

 CHECK YOURSELF 7

Write each geometric expression, using symbols.

(a) Two times *length* plus two times *width*   **(b)** Two times pi (π) times *radius*

READING YOUR TEXT

The following fill-in-the-blank exercises are designed to assure that you understand the key vocabulary used in this section. Each sentence comes directly from the section. You will find the correct answers in Appendix C.

Section 2.6

(a) Letters are called _____ when they can represent different numerical values.

(b) An _____ is a meaningful collection of numbers, variables, and signs of operations.

(c) A raised dot between two letters, x and y, indicates the _____ of x and y.

(d) $y = 2x - 1$ is not an expression because _____ is not an operation sign.

CHECK YOURSELF ANSWERS

1. **(a)** $y + 4$; **(b)** $a + b$; **(c)** $x + 3$; **(d)** $n + 6$ **2.** **(a)** $w - z$; **(b)** $a - 7$; **(c)** $y - 3$;
(d) $b - 5$ **3.** **(a)** mn; **(b)** hb; **(c)** $8 \cdot 9$ or $(8)(9)$; **(d)** $5wy$; **(e)** $8a + 3$
4. **(a)** Not an expression; **(b)** not an expression; **(c)** an expression; **(d)** an expression
5. **(a)** $2(p + q)$; **(b)** $2p + q$; **(c)** $a(b - c)$; **(d)** $(x + 2)(x - 2)$; **(e)** $x(x + 4)$
6. **(a)** $\dfrac{r}{s}$; **(b)** $\dfrac{x - y}{7}$; **(c)** $\dfrac{a - 2}{a + 2}$ **7.** **(a)** $2L + 2W$; **(b)** $2\pi r$

2.6   Exercises

Name _____

Section _____ Date _____

ANSWERS

Write each of the phrases, using symbols.

1. The sum of c and d

2. a plus 7

3. w plus z

4. The sum of m and n

5. x increased by 2

6. 3 more than b

7. 10 more than y

8. m increased by 4

9. a minus b

10. 5 less than s

11. b decreased by 7

12. r minus 3

13. 6 less than r

14. x decreased by 3

15. w times z

16. The product of 3 and c

17. The product of 5 and t

18. 8 times a

19. The product of 8, m, and n

20. The product of 7, r, and s

21. The product of 3 and the quantity p plus q

22. The product of 5 and the sum of a and b

23. Twice the sum of x and y

24. 3 times the sum of m and n

25. The sum of twice x and y

26. The sum of 3 times m and n

27. Twice the difference of x and y

28. 3 times the difference of c and d

29. The quantity a plus b times the quantity a minus b

1. _____	2. _____
3. _____	4. _____
5. _____	6. _____
7. _____	8. _____
9. _____	
10. _____	
11. _____	
12. _____	
13. _____	
14. _____	
15. _____	
16. _____	
17. _____	
18. _____	
19. _____	
20. _____	
21. _____	
22. _____	
23. _____	
24. _____	
25. _____	
26. _____	
27. _____	
28. _____	
29. _____	

30. The product of x plus y and x minus y

31. The product of m and 3 less than m **VIDEO**

32. The product of a and 7 more than a

33. x divided by 5

34. The quotient when b is divided by 8

35. The quotient of a plus b, and 7

36. The difference x minus y, divided by 9

37. The difference of p and q, divided by 4 **VIDEO**

38. The sum of a and 5, divided by 9

39. The sum of a and 3, divided by the difference of a and 3

40. The difference of m and n, divided by the sum of m and n

Write each of the phrases, using symbols. Use the variable x to represent the unknown number in each case.

41. 5 more than a number

42. A number increased by 8

43. 7 less than a number

44. A number decreased by 10

45. 9 times a number

46. Twice a number

47. 6 more than 3 times a number

48. 5 times a number, decreased by 10

49. Twice the sum of a number and 5

50. 3 times the difference of a number and 4

51. The product of 2 more than a number and 2 less than that same number

52. The product of 5 less than a number and 5 more than that same number

53. The quotient of a number and 7

54. A number divided by 3

55. The sum of a number and 5, divided by 8

56. The quotient when 7 less than a number is divided by 3

57. 6 more than a number divided by 6 less than that same number

58. The quotient when 3 less than a number is divided by 3 more than that same number

Write each of the following geometric expressions using symbols.

59. Four times the length of a side (s)

60. $\dfrac{4}{3}$ times π times the cube of the radius (r)

61. The radius (r) squared times the height (h) times π

62. Twice the length (L) plus twice the width (W)

63. One-half the product of the height (h) and the sum of two unequal sides (b_1 and b_2)

64. Six times the length of a side (s) squared

Identify which are expressions and which are not.

65. $2(x + 5)$

66. $4 + (x - 3)$

67. $4 + \div m$

68. $6 + a = 7$

69. $2b = 6$

70. $x(y + 3)$

71. $2a + 5b$

72. $4x + \cdot 7$

51. _____

52. _____

53. _____

54. _____

55. _____

56. _____

57. _____

58. _____

59. _____

60. _____

61. _____

62. _____

63. _____

64. _____

65. _____

66. _____

67. _____

68. _____

69. _____

70. _____

71. _____

72. _____

73. Social Science Earth's population has doubled in the last 40 years. If we let x represent Earth's population 40 years ago, what is the population today? **ALEKS**

74. Science and Medicine It is estimated that Earth is losing 4,000 species of plants and animals every year. If S represents the number of species living last year, how many species are on Earth this year? **ALEKS**

75. Business and Finance The simple interest (I) earned when a principal (P) is invested at a rate (r) for a time (t) is calculated by multiplying the principal times the rate times the time. Write a formula for the interest earned.

76. Science and Medicine The kinetic energy (KE) of a particle of mass m is found by taking one-half of the product of the mass and the square of the velocity (v). Write a formula for the kinetic energy of a particle.

77. Rewrite each algebraic expression as an English phrase. Exchange papers with another student to edit your writing. Be sure the meaning in English is the same as in algebra. These expressions are not complete sentences, so your English does not have to be in complete sentences. Here is an example.

Algebra: $2(x - 1)$

English: We could write, "One less than a number is doubled." Or we might write, "A number is diminished by one and then multiplied by two."

(a) $n + 3$ **(b)** $\dfrac{x + 2}{5}$ **(c)** $3(5 + a)$ **(d)** $3 - 4n$ **(e)** $\dfrac{x + 6}{x - 1}$

Answers

1. $c + d$ **3.** $w + z$ **5.** $x + 2$ **7.** $y + 10$ **9.** $a - b$

11. $b - 7$ **13.** $r - 6$ **15.** wz **17.** $5t$ **19.** $8mn$ **21.** $3(p + q)$

23. $2(x + y)$ **25.** $2x + y$ **27.** $2(x - y)$ **29.** $(a + b)(a - b)$

31. $m(m - 3)$ **33.** $\dfrac{x}{5}$ **35.** $\dfrac{a + b}{7}$ **37.** $\dfrac{p - q}{4}$ **39.** $\dfrac{a + 3}{a - 3}$

41. $x + 5$ **43.** $x - 7$ **45.** $9x$ **47.** $3x + 6$ **49.** $2(x + 5)$

51. $(x + 2)(x - 2)$ **53.** $\dfrac{x}{7}$ **55.** $\dfrac{x + 5}{8}$ **57.** $\dfrac{x + 6}{x - 6}$

59. $4s$ **61.** $\pi r^2 h$ **63.** $\dfrac{1}{2}h(b_1 + b_2)$ **65.** Expression

67. Not an expression   **69.** Not an expression   **71.** Expression

73. $2x$ **75.** $I = Prt$ **77.**

2.7 Evaluating Algebraic Expressions

2.7 OBJECTIVES

1. Substitute integer values for variables in an expression and evaluate
2. Interpret summation notation

In applying algebra to problem solving, you will often want to find the value of an algebraic expression when you know certain values for the letters (or variables) in the expression. Finding the value of an expression is called *evaluating the expression* and uses the following steps.

Step by Step: To Evaluate an Algebraic Expression

Step 1 Replace each variable by the given number value.

Step 2 Do the necessary arithmetic operations, following the rules for order of operations.

OBJECTIVE 1

Example 1 Evaluating Algebraic Expressions

Suppose that $a = 5$ and $b = 7$.

(a) To evaluate $a + b$, we replace a with 5 and b with 7.

$$a + b = (5) + (7) = 12$$

NOTE In this text, we will usually show such replacement using parentheses to hold the value of the variable.

(b) To evaluate $3ab$, we again replace a with 5 and b with 7.

$$3ab = 3(5)(7) = 105$$

 CHECK YOURSELF 1

If x = 6 and y = 7, evaluate.

(a) $y - x$ **(b)** $5xy$

We are now ready to evaluate algebraic expressions that require following the rules for the order of operations.

Example 2 Evaluating Algebraic Expressions

Evaluate the expressions if $a = 2$, $b = 3$, $c = 4$, and $d = 5$.

(a) $5a + 7b = 5(2) + 7(3)$ Multiply first.

             $= 10 + 21 = 31$ Then add.

CAUTION

This is different from
$(3c)^2 = [3 \cdot (4)]^2$
$ = 12^2 = 144$

(b) $3c^2 = 3(4)^2$ Evaluate the power.

       $= 3 \cdot 16 = 48$ Then multiply.

(c) $7(c + d) = 7[(4) + (5)]$ Add inside the grouping symbols.

           $= 7 \cdot 9 = 63$

(d) $5a^4 - 2d^2 = 5(2)^4 - 2(5)^2$ Evaluate the powers.

$\qquad\qquad = 5 \cdot 16 - 2 \cdot 25$ Multiply.

$\qquad\qquad = 80 - 50 = 30$ Subtract.

NOTE The parentheses attach the negative sign to the number **before** it is raised to a power.

(e) $(-a)^2 = (-2)^2$

$\qquad\qquad = (-2) \cdot (-2)$

$\qquad\qquad = 4$

(f) $-a^2 = -(2)^2$ Evaluate the powers.

$\qquad\qquad = -(2 \cdot 2)$

$\qquad\qquad = -4$

 CHECK YOURSELF 2

If x = 3, y = 2, z = 4, and w = 5, evaluate the expressions.

(a) $4x^2 + 2$ **(b)** $5(z + w)$ **(c)** $7(z^2 - y^2)$

(d) $-x^2$ **(e)** $(-x)^2$

To evaluate algebraic expressions when a fraction bar is used, do the following: Start by doing all the work in the numerator and then do the work in the denominator. Divide the numerator by the denominator as the last step.

Example 3   Evaluating Algebraic Expressions

If $p = 2$, $q = 3$, and $r = 4$, evaluate:

(a) $\dfrac{8p}{r}$

RECALL In Section 2.5, we mentioned that the fraction bar is a grouping symbol, like parentheses. Work first in the numerator and then in the denominator.

Replace p with 2 and r with 4.

$\dfrac{8p}{r} = \dfrac{8 \cdot (2)}{(4)} = \dfrac{16}{4} = 4$ Divide as the last step.

(b) $\dfrac{7q + r}{p + q} = \dfrac{7(3) + (4)}{(2) + (3)}$ Now evaluate the top and bottom separately.

$\qquad\qquad = \dfrac{21 + 4}{2 + 3} = \dfrac{25}{5} = 5$

 CHECK YOURSELF 3

Evaluate the expressions if c = 5, d = 8, and e = 3.

(a) $\dfrac{6c}{e}$ **(b)** $\dfrac{4d + e}{c}$ **(c)** $\dfrac{10d - e}{d + e}$

Example 4 shows how a scientific calculator can be used to evaluate algebraic expressions.

Example 4   Using a Calculator to Evaluate Expressions

Use a scientific calculator to evaluate the expressions.

(a) $\dfrac{4x + y}{z}$ if $x = 2$, $y = 1$, and $z = 3$

NOTE You must enclose the entire numerator in parentheses.

Replace x with 2, y with 1, and z with 3:

$$\frac{4x + y}{z} = \frac{4(2) + (1)}{(3)}$$

Now, use the following keystrokes:

(4 × 2 + 1) ÷ 3 =

The display will read 3.

NOTE Both, the numerator and the denominator, must be enclosed in parentheses.

(b) $\dfrac{7x - y}{3z - x}$ if $x = 2$, $y = 6$, and $z = 2$

$$\frac{7x - y}{3z - x} = \frac{7(2) - (6)}{3(2) - (2)}$$

Use the following keystrokes:

(7 × 2 − 6) ÷ (3 × 2 − 2) =

The display will read 2.

CHECK YOURSELF 4

Use a scientific calculator to evaluate the expressions if x = 2, y = 6, and z = 5.

(a) $\dfrac{2x + y}{z}$

(b) $\dfrac{4y - 2z}{x}$

Example 5   Evaluating Expressions

Evaluate $5a + 4b$ if $a = -2$ and $b = 3$.

Replace *a* with −2 and *b* with 3.

NOTE Remember the rules for the order of operations. Multiply first, then add.

$$5a + 4b = 5(-2) + 4(3)$$
$$= -10 + 12$$
$$= 2$$

CHECK YOURSELF 5

Evaluate 3x + 5y if x = −2 and y = −5.

We follow the same rules no matter how many variables are in the expression.

Example 6 Evaluating Expressions

Evaluate the expressions if $a = -4$, $b = 2$, $c = -5$, and $d = 6$.

This becomes $-(-20)$, or $+20$.

(a) $7a - 4c = 7(-4) - 4(-5)$

$= -28 + 20$

$= -8$

Evaluate the power first, then multiply by 7.

(b) $7c^2 = 7(-5)^2 = 7 \cdot 25$

$= 175$

(c) $b^2 - 4ac = (2)^2 - 4(-4)(-5)$

$= 4 - 4(-4)(-5)$

$= 4 - 80$

$= -76$

Add inside the brackets first.

(d) $b(a + d) = 2[(-4) + (6)]$

$= 2(2)$

$= 4$

CAUTION

When a squared variable is replaced by a negative number, square the negative.

$(-5)^2 = (-5)(-5) = 25$

The exponent applies to -5!

$-5^2 = -(5 \cdot 5) = -25$

The exponent applies only to 5!

✓ CHECK YOURSELF 6

Evaluate the expressions if $p = -4$, $q = 3$, and $r = -2$.

(a) $5p - 3r$ **(b)** $2p^2 + q$ **(c)** $p(q + r)$

(d) $-q^2$ **(e)** $(-q)^2$

As mentioned earlier, the fraction bar is a grouping symbol. Example 7 further illustrates this concept.

Example 7 Evaluating Expressions

Evaluate the expressions if $x = 4$, $y = -5$, $z = 2$, and $w = -3$.

(a) $\dfrac{z - 2y}{x} = \dfrac{(2) - 2(-5)}{(4)}$

$= \dfrac{2 - (-10)}{4}$

$= \dfrac{12}{4} = 3$

(b) $\dfrac{3x - w}{2x + w} = \dfrac{3(4) - (-3)}{2(4) + (-3)} = \dfrac{12 + 3}{8 + (-3)}$

$= \dfrac{15}{5} = 3$

CHECK YOURSELF 7

Evaluate the expressions if m = −6, n = 4, and p = −3.

(a) $\dfrac{m + 3n}{p}$ **(b)** $\dfrac{4m + n}{m + 4n}$

When an expression is evaluated by a calculator, the same order of operations that we introduced in Section 1.8 is followed.

	Algebraic Notation	Calculator Notation
Addition	$6 + 2$	6 $\boxed{+}$ 2
Subtraction	$4 - 8$	4 $\boxed{-}$ 8
Multiplication	$(3)(-5)$	3 $\boxed{\times}$ $\boxed{(-)}$ 5 or 3 $\boxed{\times}$ 5 $\boxed{+/-}$
Division	$\dfrac{8}{6}$	8 $\boxed{\div}$ 6
Exponential	3^4	3 $\boxed{\wedge}$ 4 or 3 $\boxed{y^x}$ 4

In many applications, you need to find the sum of a group of numbers you are working with. In mathematics, the shorthand symbol for "sum of" is the Greek letter Σ (capital sigma, the "S" of the Greek alphabet). The expression Σx, in which x refers to all the numbers in a given group, means the sum of all the numbers in that group.

OBJECTIVE 2   **Example 8   Summing a Group of Integers**

Find Σx for the group of integers.

$-2, -6, 3, 5, -4$

$$\Sigma x = -2 + (-6) + 3 + 5 + (-4)$$
$$= (-8) + 3 + 5 + (-4)$$
$$= (-8) + 8 + (-4)$$
$$= -4$$

CHECK YOURSELF 8

Find Σx for each group of integers.

(a) $-3, 4, -7, -9, 8$ **(b)** $-2, 6, -5, -3, 4, 7$

READING YOUR TEXT

The following fill-in-the-blank exercises are designed to assure that you understand the key vocabulary used in this section. Each sentence comes directly from the section. You will find the correct answers in Appendix C.

Section 2.7

(a) To evaluate an algebraic expression, first replace each _____ by the given number value.

(b) The _____ bar is a grouping symbol.

(c) When a squared variable is replaced with a negative number, square the _____.

(d) In mathematics, we use the _____ letter Σ as shorthand for "sum of."

CHECK YOURSELF ANSWERS

1. (a) 1; **(b)** 210   **2. (a)** 38; **(b)** 45; **(c)** 84; **(d)** -9; **(e)** 9   **3. (a)** 10; **(b)** 7; **(c)** 7
4. (a) 2; **(b)** 7   **5.** -31 **6. (a)** -14; **(b)** 35; **(c)** -4; **(d)** -9; **(e)** 9
7. (a) -2; **(b)** -2 **8. (a)** -7; **(b)** 7

Evaluate each of the expressions if $a = -2$, $b = 5$, $c = -4$, and $d = 6$.

1. $3c - 2b$ **ALEKS**

2. $4c - 2b$ **ALEKS**

3. $8b + 2c$ **ALEKS**

4. $7a - 2c$ **ALEKS**

5. $-b^2 + b$ **VIDEO**

6. $(-b)^2 + b$

7. $3a^2$

8. $6c^2$

9. $c^2 - 2d$

10. $3a^2 + 4c$

11. $2a^2 + 3b^2$

12. $4b^2 - 2c^2$

13. $2(a + b)$

14. $5(b - c)$

15. $4(2a - d)$

16. $6(3c - d)$

17. $a(b + 3c)$ **VIDEO**

18. $c(3a - d)$

19. $\dfrac{6d}{c}$ **ALEKS**

20. $\dfrac{8b}{5c}$ **ALEKS**

21. $\dfrac{3d + 2c}{b}$ **ALEKS**

22. $\dfrac{2b + 3d}{2a}$ **ALEKS**

23. $\dfrac{2b - 3a}{c + 2d}$ **ALEKS**

24. $\dfrac{3d - 2b}{5a + d}$ **ALEKS**

25. $d^2 - b^2$

26. $c^2 - a^2$

27. $(d - b)^2$

28. $(c - a)^2$

29. $(d - b)(d + b)$

30. $(c - a)(c + a)$

31. $d^3 - b^3$

32. $c^3 + a^3$

33. $(d - b)^3$

34. $(c + a)^3$

35. $(d - b)(d^2 + db + b^2)$

36. $(c + a)(c^2 - ac + a^2)$

37. $b^2 + a^2$

38. $d^2 - a^2$

39. $(b + a)^2$

40. $(d - a)^2$

41. $a^2 + 2ad + d^2$ ALEKS

42. $b^2 - 2bc + c^2$ ALEKS

For each group of integers, evaluate Σx.

43. 1, 2, 3, 7, 8, 9, 11

44. 2, 4, 5, 6, 10, 11, 12

45. $-5, -3, -1, 2, 3, 4, 8$

46. $-4, -2, -1, 5, 7, 8, 10$

47. $3, 2, -1, -4, -3, 8, 6$

48. $3, -4, 2, -1, 2, -7, 9$

For exercises 49 to 52, decide if the given values make the statement true or false.

49. $x - 7 = 2y + 5; x = 22, y = 5$

50. $3(x - y) = 6; x = 5, y = -3$

51. $2(x + y) = 2x + y; x = -4, y = -2$

52. $x^2 - y^2 = x - y; x = 4, y = -3$

53. Geometry The perimeter of a rectangle of length L and width W is sometimes given by the formula $P = 2L + 2W$. Find the perimeter when $L = 10$ in. and $W = 5$ in.

54. **Geometry** Using the equation given in exercise 53, find the perimeter of a sheet of paper that is 11 in. wide and 14 in. long.

55. **Statistics** A major-league pitcher throws a ball straight up into the air. The height of the ball (in feet) can be determined by substituting the number of seconds (s) that have passed after the throw into the expression $110s - 16s^2$.

(a) Determine the height after 2 s. (Hint: Replace s with 2 in the expression and then evaluate.)

(b) Find the height after 3 s.

56. **Statistics** Given the same equation used in exercise 55, find the height of the ball after 4 s. How does this compare to the answer for exercise 55? What has happened?

57. **Number Problem** Is $a^n + b^n = (a + b)^n$? Try a few numbers and decide if you think this is true for all numbers, for some numbers, or never true. Write an explanation of your findings and give examples.

58. Enjoyment of patterns in art, music, and language is common to all cultures, and many cultures also delight in and draw spiritual significance from patterns in numbers. One such set of patterns is that of the "magic" square. One of these squares appears in a famous etching by Albrecht Dürer, who lived from 1471 to 1528 in Europe. He was one of the first artists in Europe to use geometry to give perspective, a feeling of three dimensions, in his work.

The magic square in his work is this one:

16	3	2	13
5	10	11	8
9	6	7	12
4	15	14	1

Why is this square "magic"? It is magic because every row, every column, and both diagonals add to the same number. In this square, there are 16 spaces for the numbers 1 through 16.

Part 1: What number does each row and column add to?

Write the square that you obtain by adding −17 to each number. Is this still a magic square? If so, what number does each column and row add to? If you add 5 to each number in the original magic square, do you still have a magic square? You have been studying the operations of addition, multiplication, subtraction, and division with integers and with rational numbers. What operations can you perform on this magic square and still have a magic square? Try to find something that will not work. Use algebra to help you decide what will work and what won't. Write a description of your work and explain your conclusions.

Part 2: Here is the oldest published magic square. It is from China, about 250 B.C.E. Legend has it that it was brought from the River Lo by a turtle to the Emperor Yii, who was a hydraulic engineer.

4	9	2
3	5	7
8	1	6

Check to make sure that this is a magic square. Work together to decide what operation might be done to every number in the magic square to make the sum of each row, column, and diagonal the *opposite* of what it is now. What would you do to every number to cause the sum of each row, column, and diagonal to equal zero?

The formula used for converting a Fahrenheit temperature to a Celsius temperature is

$$C = \frac{5}{9}(F - 32)$$

Use this formula to convert each of the following Fahrenheit temperatures to its Celsius equivalent.

59. 32°F

60. 212°F

61. 68°F

62. 104°F

Answers

1. −22   **3.** 32   **5.** −20   **7.** 12   **9.** 4   **11.** 83   **13.** 6
15. −40   **17.** 14   **19.** −9   **21.** 2   **23.** 2   **25.** 11   **27.** 1
29. 11   **31.** 91   **33.** 1   **35.** 91   **37.** 29   **39.** 9   **41.** 16
43. 41   **45.** 8   **47.** 11   **49.** True   **51.** False   **53.** 30 in.
55. (a) 156 ft; **(b)** 186 ft   **57.**   **59.** 0°C   **61.** 20°C

2.8 Simplifying Algebraic Expressions

2.8 OBJECTIVES

1. Identify terms, like terms, and numerical coefficients
2. Simplify algebraic expressions by combining like terms

In Section 1.2, we found the perimeter of a rectangle by using the formula

$$P = L + W + L + W$$

There is another version of this formula that we can use. Because we are adding two times the length and two times the width, we can use the language of algebra to write this formula as

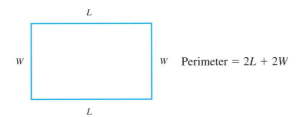

We call $2L + 2W$ an **algebraic expression,** or more simply an **expression.** Recall from Section 2.6 that an expression allows us to write a mathematical idea in symbols. It can be thought of as a meaningful collection of letters, numbers, and operation signs.

Some expressions are

$$5x^2 \qquad\qquad 3a + 2b \qquad\qquad 4x^3 + (-2y) + 1$$

In algebraic expressions, the addition and subtraction signs break the expressions into smaller parts called **terms.**

> **Definition:** Term
>
> A **term** is a number, or the product of a number and one or more variables, raised to a power.

In an expression, each sign (+ or −) is a part of the term that follows the sign.

OBJECTIVE 1

> **Example 1   Identifying Terms**

(a) $5x^2$ has one term.

(b) $\underline{3a} + \underline{2b}$ has two terms: $3a$ and $2b$.
   Term Term

NOTE This could also be written as $4x^3 - 2y + 1$

(c) $\underline{4x^3} + \underline{(-2y)} + \underline{1}$ has three terms: $4x^3$, $-2y$, and 1.
   Term   Term   Term

CHECK YOURSELF 1

List the terms of each expression.

(a) $2b^4$ **(b)** $5m + 3n$ **(c)** $2s^2 - 3t - 6$

Note that a term in an expression may have any number of factors. For instance, $5xy$ is a term. It has factors of 5, x, and y. The number factor of a term is called the **numerical coefficient.** So for the term $5xy$, the numerical coefficient is 5.

Example 2 Identifying the Numerical Coefficient

(a) $4a$ has the numerical coefficient 4.

(b) $6a^3b^4c^2$ has the numerical coefficient 6.

(c) $-7m^2n^3$ has the numerical coefficient -7.

(d) Because $1 \cdot x = x$, the numerical coefficient of x is understood to be 1.

CHECK YOURSELF 2

Give the numerical coefficient for each term.

(a) $8a^2b$ **(b)** $-5m^3n^4$ **(c)** y

If terms contain exactly the *same letters* (or variables) raised to the *same powers,* they are called **like terms.**

Example 3 Identifying Like Terms

(a) The following are like terms.

$6a$ and $7a$
$5b^2$ and b^2 Each pair of terms has the same letters, with each letter raised to the same power—the numerical coefficients can be any number.
$10x^2y^3z$ and $-6x^2y^3z$
$-3m^2$ and m^2

(b) The following are *not* like terms.

Different letters

$6a$ and $7b$

Different exponents

$5b^2$ and b^3

Different exponents

$3x^2y$ and $4xy^2$

CHECK YOURSELF 3 _____

Circle like terms.

$5a^2b \qquad ab^2 \qquad a^2b \qquad -3a^2 \qquad 4ab \qquad 3b^2 \qquad -7a^2b$

Like terms of an expression can always be combined into a single term. Look at the following:

$$\underbrace{2x}_{x+x} + \underbrace{5x}_{x+x+x+x+x} = \underbrace{7x}_{x+x+x+x+x+x+x}$$

Rather than having to write out all those *x*'s, try

RECALL Here we use the distributive property discussed first in Section 1.5.

$$2x + 5x = (2 + 5)x$$
$$= 7x$$

In the same way,

NOTE You don't have to write all this out—just do it mentally!

$$9b + 6b = (9 + 6)b$$
$$= 15b$$

and

$$10a - 4a = 10a + (-4a)$$
$$= (10 + (-4))a$$
$$= 6a$$

This leads us to a rule.

Step by Step: To Combine Like Terms

To combine like terms, use steps 1 and 2.

Step 1   Add or subtract the numerical coefficients.

Step 2   Attach the common variables.

OBJECTIVE 2   **Example 4   Combining Like Terms**

Combine like terms.

NOTE The exponents do not change when combining like terms.

(a) $8m + 5m = (8 + 5)m = 13m$

(b) $5pq^3 - 4pq^3 = 5pq^3 + (-4pq^3) = 1pq^3 = pq^3$

RECALL When any factor is multiplied by 0, the product is 0.

(c) $7a^3b^2 - 7a^3b^2 = 7a^3b^2 + (-7a^3b^2) = 0a^3b^2 = 0$

 CHECK YOURSELF 4 _____

Combine like terms.

(a) $6b + 8b$ **(b)** $12x^2 - 3x^2$ **(c)** $8xy^3 - 7xy^3$ **(d)** $9a^2b^4 - 9a^2b^4$

With expressions involving more than two terms, the idea is just the same.

Example 5   Combining Like Terms

Combine like terms.

(a) $5ab - 2ab + 3ab$

$= 5ab + (-2ab) + 3ab$

$= (5 + (-2) + 3)ab = 6ab$

NOTE The distributive property can be used over any number of like terms.

Only like terms can be combined.

(b) $8x - 2x \quad + 5y$

$(8 + (-2))\, x + 5y$

$= 6x \qquad + 5y$

Like terms   Like terms

(c) $5m + 8n \quad + 4m - 3n$

$= (5m + 4m) + (8n + (-3n))$

$= \qquad 9m \quad + \quad 5n$

Here we have used the associative and commutative properties of addition.

NOTE With practice, you will be doing this mentally, rather than writing out all the steps.

(d) $4x^2 + 2x - 3x^2 + x$

$= (4x^2 + (-3x^2)) + (2x + x)$

$= x^2 + 3x$

As these examples illustrate, combining like terms often means changing the grouping and the order in which the terms are written. Again all this is possible because of the properties of addition that we introduced in Section 1.2.

CHECK YOURSELF 5

Combine like terms.

(a) $4m^2 - 3m^2 + 8m^2$ **(b)** $9ab + 3a - 5ab$

(c) $4p + 7q + 5p - 3q$

As you have seen in arithmetic, subtraction can be performed directly. As this is the form used for most of mathematics, we will use that form throughout this text. Just remember, by using the additive inverse of numbers, you can always rewrite a subtraction problem as an addition problem.

Example 6   Combining Like Terms

Combine like terms.

(a) $2xy - 3xy + 5xy$

$= (2 - 3 + 5)xy$

$= 4xy$

(b) $5a - 2b + 7b - 8a$

$= 5a + (-2)b + 7b + (-8)a$

$= 5a + (-8)a + (-2)b + 7b$

$= -3a + 5b$

NOTE Note that x and x^2 are *not* like terms.

(c) $2x^2 + 3x + 4x^2 + x$

$= (2x^2 + 4x^2) + (3x + x)$

$= 6x^2 + 4x$

 CHECK YOURSELF 6 _____

Combine like terms.

(a) $4ab + 5ab - 3ab - 7ab$ **(b)** $2x - 7y - 8x - y$

The distributive property studied in Section 1.5 can be used to simplify expressions containing variables. This is illustrated in Example 7.

Example 7   Using the Distributive Property before Combining Like Terms

Use the distributive property to remove the parentheses. Then simplify by combining like terms.

(a) $2(x - 5) - 8x$

$= 2x - 2(5) - 8x$

$= 2x - 10 - 8x$

$= 2x - 8x - 10$

$= -6x - 10$

(b) $3a - 2b + 4(a + 3b)$

$= 3a - 2b + 4a + 4(3b)$

$= 3a - 2b + 4a + 12b$

$= 3a + 4a - 2b + 12b$

$= 7a + 10b$

 CHECK YOURSELF 7 _____

Use the distributive property to remove the parentheses. Then simplify by combining like terms.

(a) $7x + 3(x - 1)$ **(b)** $5(m + 2n) + 3(m - n)$

READING YOUR TEXT

The following fill-in-the-blank exercises are designed to assure that you understand the key vocabulary used in this section. Each sentence comes directly from the section. You will find the correct answers in Appendix C.

Section 2.8

(a) A _____ is a number or the product of a number and one or more variables raised to a power.

(b) The number factor of a term is called the numerical _____.

(c) If terms contain exactly the same variables raised to the same powers, they are called _____ terms.

(d) Like terms in an expression can always be combined into a _____ term.

CHECK YOURSELF ANSWERS

1. (a) $2b^4$; **(b)** $5m, 3n$; **(c)** $2s^2, -3t, -6$ **2. (a)** 8; **(b)** -5; **(c)** 1
3. The like terms are $5a^2b$, a^2b, and $-7a^2b$ **4. (a)** $14b$; **(b)** $9x^2$; **(c)** xy^3; **(d)** 0
5. (a) $9m^2$; **(b)** $4ab + 3a$; **(c)** $9p + 4q$ **6. (a)** $-ab$; **(b)** $-6x - 8y$
7. (a) $10x - 3$; **(b)** $8m + 7n$

2.8　Exercises

Name _____

Section _____ Date _____

ANSWERS

List the terms of each expression.

1. $5a + 2$

2. $7a - 4b$

3. $4x^3$

4. $3x^2$

5. $3x^2 + 3x - 7$

6. $2a^3 - a^2 + a$

Identify the like terms in each set of terms.

7. $5ab, 3b, 3a, 4ab$

8. $9m^2, 8mn, 5m^2, 7m$

9. $4xy^2, 2x^2y, 5x^2, -3x^2y, 5y, 6x^2y$

10. $8a^2b, 4a^2, 3ab^2, -5a^2b, 3ab, 5a^2b$

Combine like terms.

11. $3m + 7m$　**ALEKS**

12. $6a^2 + 8a^2$　**ALEKS**

13. $7b^3 + 10b^3$ 　**ALEKS**

14. $7rs + 13rs$　**ALEKS**

15. $21xyz + 7xyz$　**ALEKS**

16. $4mn^2 + 15mn^2$　**ALEKS**

17. $9z^2 - 3z^2$　**ALEKS**

18. $7m - 6m$　**ALEKS**

19. $5a^3 - 5a^3$　**ALEKS**

20. $13xy - 9xy$　**ALEKS**

21. $19n^2 - 18n^2$ 　**ALEKS**

22. $7cd - 7cd$　**ALEKS**

23. $21p^2q - 6p^2q$　**ALEKS**

24. $17r^3s^2 - 8r^3s^2$　**ALEKS**

25. $10x^2 - 7x^2 + 3x^2$　**ALEKS**

26. $13uv + 5uv - 12uv$　**ALEKS**

27. $9a - 7a + 4b$　**ALEKS**

28. $5m^2 - 3m + 6m^2$　**ALEKS**

29. $7x + 5y - 4x - 4y$　**ALEKS**

30. $6a^2 + 11a + 7a^2 - 9a$　**ALEKS**

31. $4a + 7b + 3 - 2a + 3b - 2$

32. $5p^2 + 2p + 8 + 4p^2 + 5p - 6$

1. _____

2. _____

3. _____

4. _____

5. _____

6. _____

7. _____

8. _____

9. _____

10. _____

11. _____	12. _____
13. _____	14. _____
15. _____	16. _____
17. _____	18. _____
19. _____	20. _____
21. _____	22. _____
23. _____	24. _____
25. _____	26. _____

27. _____

28. _____

29. _____

30. _____

31. _____

32. _____

33. _____

34. _____

35. _____

36. _____

37. _____

38. _____

39. _____

40. _____

41. _____

42. _____

43. _____

44. _____

45. _____

46. _____

47. _____

48. _____

49. _____

50. _____

Perform the indicated operations.

33. Find the sum of $5a^4$ and $8a^4$.

34. Find the sum of $9p^2$ and $12p^2$.

35. Subtract $12a^3$ from $15a^3$.

36. Subtract $5m^3$ from $18m^3$.

37. Subtract $4x$ from the sum of $8x$ and $3x$.

38. Subtract $8ab$ from the sum of $7ab$ and $5ab$.

39. Subtract $3mn^2$ from the sum of $9mn^2$ and $5mn^2$.

40. Subtract $4x^2y$ from the sum of $6x^2y$ and $12x^2y$.

Use the distributive property to remove the parentheses in each expression. Then simplify by combining like terms.

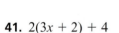

41. $2(3x + 2) + 4$ **42.** $3(4z + 5) - 9$

43. $5(6a - 2) + 12a$ **44.** $7(4w - 3) - 25w$

45. $4s + 2(s + 4) + 4$ **46.** $5p + 4(p + 3) - 8$

47. Write a paragraph explaining the difference between n^2 and $2n$.

48. Complete the explanation: "x^3 and $3x$ are not the same because . . ."

49. Allied Health The *ideal body weight*, in pounds, for a woman can be approximated by substituting her height, in inches, into the formula $105 + 5(h - 60)$. Use the distributive property to simplify the ideal body weight formula.

50. Allied Health Use the simplified formula you found in exercise 49 to approximate the ideal body weight of a woman who stands 5 ft 4 in. tall.

51. Work with another student to complete this exercise. Place $>$, $<$, or $=$ in the blank in these statements.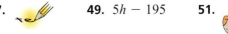

1^2 _____ 2^1 What happens as the table of numbers is extended? Try more examples.

2^3 _____ 3^2
What sign seems to occur the most in your table? $>$, $<$, or $=$?

3^4 _____ 4^3

4^5 _____ 5^4 Write an algebraic statement for the pattern of numbers in this table. Do you think this is a pattern that continues? Add more lines to the table and extend the pattern to the general case by writing the pattern in algebraic notation. Write a short paragraph stating your conjecture.

Answers

1. $5a$, 2   **3.** $4x^3$ **5.** $3x^2$, $3x$, -7 **7.** $5ab$, $4ab$ **9.** $2x^2y$, $-3x^2y$, $6x^2y$

11. $10m$ **13.** $17b^3$ **15.** $28xyz$ **17.** $6z^2$ **19.** 0   **21.** n^2

23. $15p^2q$ **25.** $6x^2$ **27.** $2a + 4b$ **29.** $3x + y$ **31.** $2a + 10b + 1$

33. $13a^4$ **35.** $3a^3$ **37.** $7x$ **39.** $11mn^2$ **41.** $6x + 8$

43. $42a - 10$ **45.** $6s + 12$ **47.**   **49.** $5h - 195$ **51.**

Introduction to Linear Equations

1. Determine whether a given number is a solution for an equation
2. Identify linear equations

In Chapter 1, we introduced one of the most important tools of mathematics, the equation. The ability to recognize and solve various types of equations is probably the most useful algebraic skill you will learn. We continue to build upon the methods of this chapter throughout the remainder of the text. We start by reviewing what we mean by an *equation*.

> **Definition: Equation**
>
> An **equation** is a mathematical statement that two expressions are equal.

Some examples are $3 + 4 = 7$, $x + 3 = 5$, and $P = 2L + 2W$.

Note that some of these equations contain letters. In algebra, we use letters to represent numerical values that we don't know or that could change. These letters are called variables. Recall that a *variable* is a letter used to represent a number. In the equation $x + 3 = 5$, x is a variable.

As you can see, an equals sign ($=$) separates the two equal expressions. These expressions are usually called the **left side** and the **right side** of the equation.

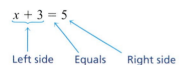

Left side   Equals   Right side

NOTE An equation such as

$x + 3 = 5$

is called a **conditional equation** because it can be either true or false depending on the value given to the variable.

Just as the balance scale may be in balance or out of balance, an equation may be either true or false. For instance, $3 + 4 = 7$ is true because both sides name the same number. What about an equation such as $x + 3 = 5$ that has a letter or variable on one side? Any number can replace x in the equation. However, in this case, only one number will make this equation a true statement.

$$\text{If } x = \begin{cases} 1 & 1 + 3 = 5 \text{ is false} \\ 2 & 2 + 3 = 5 \text{ is true} \\ 3 & 3 + 3 = 5 \text{ is false} \end{cases}$$

The number 2 is called the **solution** (or **root**) of the equation $x + 3 = 5$ because substituting 2 for x gives a true statement.

> **Definition: Solution**
>
> A **solution** for an equation is any value for the variable that makes the equation a true statement.

OBJECTIVE 1

Example 1   Verifying a Solution

(a) Is 3 a solution for the equation $2x + 4 = 10$?

To find out, replace x with 3 and evaluate $2x + 4$ on the left.

Left side		Right side
$2(3) + 4$	$\overset{?}{=}$	10
$6 + 4$	$\overset{?}{=}$	10
10	$=$	10

Because $10 = 10$ is a true statement, 3 is a solution of the equation.

(b) Is 5 a solution of the equation $3x - 2 = 2x + 1$?

To find out, replace x with 5 and evaluate each side separately.

RECALL We must follow the rules for the order of operations. Multiply first and then add or subtract.

Left side		Right side
$3(5) - 2$	$\overset{?}{=}$	$2(5) + 1$
$15 - 2$	$\overset{?}{=}$	$10 + 1$
13	$\neq$	11

Because the two sides do not name the same number, we do not have a true statement. Therefore, 5 is not a solution.

CHECK YOURSELF 1

For the equation

$2x - 1 = x + 5$

(a) Is 4 a solution?   **(b)** Is 6 a solution?

You may be wondering whether an equation can have more than one solution. It certainly can. For instance,

$x^2 = 9$

has two solutions. They are 3 and -3 because

$(3)^2 = 9$ and $(-3)^2 = 9$

In this chapter, however, we will always work with *linear equations in one variable.* These are equations that can be put into the form

$$ax + b = 0$$

in which the variable is x, a and b are any numbers, and a is not equal to 0. In a linear equation, the variable can appear only to the first power. No other power (x^2, x^3, etc.) can appear. Linear equations are also called **first-degree equations.** The degree of an equation in one variable is the highest power to which the variable appears.

> **Definition:** Linear Equations
>
> **Linear equations in one variable** are equations that can be written in the form
>
> $$ax + b = 0 \qquad a \neq 0$$
>
> Every such equation will have exactly one solution.

OBJECTIVE 2

Example 2 **Identifying Expressions and Equations**

Given the following

$$4x + 5$$
$$2x + 8 = 0$$
$$3x^2 - 9 = 0$$
$$5x = 15$$

Label each as an expression, a linear equation, or an equation that is not linear.

(a) $4x + 5$ is an expression.

(b) $2x + 8 = 0$ is a linear equation.

(c) $3x^2 - 9 = 0$ is an equation that is not linear.

(d) $5x = 15$ is a linear equation because this can be written as $5x - 15 = 0$.

CHECK YOURSELF 2

Label each as an expression, a linear equation, or an equation that is not linear.

(a) $2x^2 = 8$ **(b)** $2x - 3 = 0$
(c) $5x - 10$ **(d)** $2x + 1 = 7$

READING YOUR TEXT

The following fill-in-the-blank exercises are designed to assure that you understand the key vocabulary used in this section. Each sentence comes directly from the section. You will find the correct answers in Appendix C.

Section 2.9

(a) An equation is a mathematical statement that two _____ are equal.

(b) A _____ for an equation is any value for the variable that makes the statement true.

(c) _____ equations in one variable are equations that can be written in the form $ax + b = 0$.

(d) Linear equations are also called _____-degree equations.

CHECK YOURSELF ANSWERS

1. (a) 4 is not a solution; (b) 6 is a solution
2. (a) Nonlinear equation; (b) linear equation; (c) expression; (d) linear equation

Name _____

Section _____ Date _____

ANSWERS

1. _____	**2.** _____
3. _____	**4.** _____
5. _____	
6. _____	
7. _____	
8. _____	
9. _____	
10. _____	
11. _____	
12. _____	
13. _____	
14. _____	
15. _____	
16. _____	
17. _____	
18. _____	
19. _____	
20. _____	
21. _____	
22. _____	
23. _____	
24. _____	
25. _____	
26. _____	

Is the number shown in parentheses a solution for the given equation?

1. $x + 4 = 9$ (5)

2. $x + 2 = 11$ (8)

3. $x - 15 = 6$ (−21)

4. $x - 11 = 5$ (16)

5. $5 - x = 2$ (4)

6. $10 - x = 7$ (3)

7. $4 - x = 6$ (−2)

8. $5 - x = 6$ (−3)

9. $3x + 4 = 13$ (8)

10. $5x + 6 = 31$ (5)

11. $4x - 5 = 7$ (2)

12. $2x - 5 = 1$ (3)

13. $5 - 2x = 7$ (−1)

14. $4 - 5x = 9$ (−2)

15. $4x - 5 = 2x + 3$ (4)

16. $5x + 4 = 2x + 10$ (4)

17. $x + 3 + 2x = 5 + x + 8$ (5)

18. $5x - 3 + 2x = 3 + x - 12$ (−2)

Label each as an expression or a linear equation.

19. $2x + 1 = 9$

20. $7x + 14$

21. $2x - 8$

22. $5x - 3 = 12$

23. $7x + 2x + 8 - 3$

24. $x + 5 = 13$

25. $2x - 8 = 3$

26. $12x - 5x + 2 + 5$

27. An algebraic equation is a complete sentence. It has a subject, a verb, and a predicate. For example, $x + 2 = 5$ can be written in English as "Two more than a number is five." Or, "A number added to two is five." Write an English version of the following equations. Be sure you write complete sentences and that the sentences express the same idea as the equations. Exchange sentences with another student and see if your interpretation of each other's sentences results in the same equation.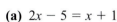

(a) $2x - 5 = x + 1$ **(b)** $2(x + 2) = 14$

(c) $n + 5 = \dfrac{n}{2} - 6$ **(d)** $7 - 3a = 5 + a$

28. Complete the following explanation in your own words: "The difference between $3(x - 1) + 4 - 2x$ and $3(x - 1) + 4 = 2x$ is . . . ."

Answers

1. Yes   **3.** No   **5.** No   **7.** Yes   **9.** No   **11.** No   **13.** Yes
15. Yes   **17.** Yes   **19.** Linear equation   **21.** Expression
23. Expression   **25.** Linear equation   **27.**

27. _____

28. _____

2.10 The Addition Property of Equality

1. Use the addition property to solve an equation
2. Combine like terms while solving an equation
3. Use the distributive property while solving an equation

It is not difficult to find the solution for an equation such as $x + 3 = 8$ by guessing the answer to the question "What plus 3 is 8?" Here the answer to the question is 5, and that is also the solution for the equation. But for more complicated equations, you are going to need something more than guesswork. A better method is to transform the given equation to an *equivalent equation* whose solution can be found by inspection.

> **Definition:** Equivalent Equations
>
> Equations that have the same solution(s) are called **equivalent equations.**

These expressions are all equivalent equations:

$$2x + 3 = 5 \qquad 2x = 2 \qquad \text{and} \qquad x = 1$$

They all have the same solution, 1. We say that a linear equation is *solved* when it is transformed to an equivalent equation of the form

NOTE In some cases, we'll write the equation in the form

$\square = x$.

The number will be our solution when the equation has the variable isolated on the left or on the right.

$x = \square$

The variable is alone on the left side.

The right side is some number, the solution.

The addition property of equality is the first property you will need to transform an equation to an equivalent form.

> **Property:** The Addition Property of Equality
>
> If $\quad a = b$
>
> then $\quad a + c = b + c$
>
> In words, adding the same quantity to both sides of an equation gives an equivalent equation.

RECALL An equation is a statement that the two sides are equal. Adding the same quantity to both sides does not change the equality or "balance."

We said that a true equation was like a scale in balance.

© 2010 McGraw-Hill Companies

The addition property is equivalent to adding the same weight to both sides of the scale. It remains in balance.

OBJECTIVE 1

Example 1   Using the Addition Property to Solve an Equation

Solve.

$$x - 3 = 9$$

Remember that our goal is to isolate x on one side of the equation. Because 3 is being subtracted from x, we can add 3 to remove it. We must use the addition property to add 3 to both sides of the equation.

NOTE To check, replace x with 12 in the original equation:

$x - 3 = 9$

$(12) - 3 \stackrel{?}{=} 9$

$9 = 9$

Because we have a true statement, 12 is the solution.

$$
\begin{array}{rl}
x - 3 = & 9 \\
+\ 3 & +3 \\
\hline
x\quad = & 12
\end{array}
$$

{ Adding 3 "undoes" the subtraction and leaves x alone on the left.

Because 12 is the solution for the equivalent equation $x = 12$, it is the solution for our original equation.

 CHECK YOURSELF 1

Solve and check.

$$x - 5 = 4$$

The addition property also allows us to add a negative number to both sides of an equation. This is really the same as subtracting the same quantity from both sides.

Example 2   Using the Addition Property to Solve an Equation

Solve.

NOTE Recall our comment that we could write an equation in the equivalent forms $x = \square$ or $\square = x$, in which $\square$ represents some number. Suppose we have an equation like

$12 = x + 7$

Adding -7 will isolate x *on the right:*

$12 = x + 7$

$\underline{-7 \qquad -7}$

$5 = x$

and the solution is 5.

$$x + 5 = 9$$

In this case, 5 is *added* to x on the left. We can use the addition property to add a -5 to both sides. Because $5 + (-5) = 0$, this will "undo" the addition and leave the variable x alone on one side of the equation.

$$
\begin{array}{rl}
x + 5 = & 9 \\
-\ 5 & -5 \\
\hline
x\quad = & 4
\end{array}
$$

The solution is 4. To check, replace x with 4 in the original equation.

$(4) + 5 = 9$ (True)

CHECK YOURSELF 2

Solve and check.

$x + 6 = 13$

What if the equation has a variable term on both sides? You can use the addition property to add or subtract a term involving the variable to get the desired result.

OBJECTIVE 2   **Example 3   Using the Addition Property to Solve an Equation**

Solve.

$5x = 4x + 7$

We start by adding $-4x$ to both sides of the equation. Do you see why? Remember that an equation is solved when we have an equivalent equation of the form $x = \square$.

RECALL Adding $-4x$ is identical to subtracting $4x$.

$$\begin{array}{rl} 5x = & 4x + 7 \\ -4x & -4x \\ \hline x = & 7 \end{array}$$ $\left\{ \begin{array}{l} \text{Adding } -4x \text{ to both} \\ \text{sides } \textit{removes } 4x \\ \text{from the right.} \end{array} \right.$

To check: Because 7 is a solution for the equivalent equation $x = 7$, it should be a solution for the original equation. To find out, replace x with 7:

$$5(7) \overset{?}{=} 4(7) + 7$$
$$35 \overset{?}{=} 28 + 7$$
$$35 = 35 \quad \text{(True)}$$

CHECK YOURSELF 3

Solve and check.

$7x = 6x + 3$

You may have to apply the addition property more than once to solve an equation as we see in Example 4.

Example 4   Using the Addition Property to Solve an Equation

Solve.

$7x - 8 = 6x$

We want all the variables on *one* side of the equation. If we choose the left, we add $-6x$ to both sides of the equation. This will remove $6x$ from the right:

$$\begin{array}{rl} 7x - 8 = & 6x \\ -6x & -6x \\ \hline x - 8 = & 0 \end{array}$$

We want the variable alone, so we add 8 to both sides. This isolates x on the left.

$$
\begin{array}{rcl}
x - 8 = & 0 \\
+\,8 & +8 \\
\hline
x \quad = & 8
\end{array}
$$

The solution is 8. We leave it to you to check this result.

CHECK YOURSELF 4

Solve and check.

$9x + 3 = 8x$

Often an equation has more than one variable term *and* more than one number. In this case, you have to apply the addition property twice in solving these equations.

Example 5   Using the Addition Property to Solve an Equation

Solve.

$5x - 7 = 4x + 3$

We would like the variable terms on the left, so we start by adding $-4x$ to remove the $4x$ term from the right side of the equation:

$$
\begin{array}{rcl}
5x - 7 = & 4x + 3 \\
-4x & -4x \\
\hline
x - 7 = & 3
\end{array}
$$

Now, to isolate the variable, we add 7 to both sides.

$$
\begin{array}{rcl}
x - 7 = & 3 \\
+\,7 & +7 \\
\hline
x \quad = & 10
\end{array}
$$

NOTE You could just as easily have added 7 to both sides and *then* added $-4x$. The result would be the same. In fact, some students prefer to combine the two steps.

The solution is 10. To check, replace x with 10 in the original equation:

$5(10) - 7 \overset{?}{=} 4(10) + 3$

$43 = 43$ (True)

CHECK YOURSELF 5

Solve and check.

(a) $4x - 5 = 3x + 2$ **(b)** $6x + 2 = 5x - 4$

RECALL By *simplify,* we mean to combine all like terms.

In solving an equation, you should always "simplify" each side as much as possible before using the addition property.

Example 6 **Combining Like Terms and Solving the Equation**

Solve.

Like terms   Like terms

$$5 + 8x - 2 = 2x - 3 + 5x$$

Because like terms appear on each side of the equation, we start by combining the numbers on the left (5 and -2). Then we combine the like terms ($2x$ and $5x$) on the right. We have

$$3 + 8x = 7x - 3$$

Now we can apply the addition property, as before:

$$
\begin{array}{rll}
3 + 8x = & 7x - 3 & \\
\underline{-7x = -7x} & & \text{Add } -7x. \\
3 + x = & -3 & \\
\underline{-3 \qquad\qquad -3} & & \text{Add } -3. \\
x = & -6 & \text{Isolate } x.
\end{array}
$$

The solution is -6. To check, always return to the original equation. That will catch any possible errors in simplifying. Replacing x with -6 gives

$$5 + 8(-6) - 2 \stackrel{?}{=} 2(-6) - 3 + 5(-6)$$
$$5 - 48 - 2 \stackrel{?}{=} -12 - 3 - 30$$
$$-45 = -45 \quad \text{(True)}$$

 CHECK YOURSELF 6 _____

Solve and check.

(a) $3 + 6x + 4 = 8x - 3 - 3x$ **(b)** $5x + 21 + 3x = 20 + 7x - 2$

We may have to apply some of the properties discussed in Section 1.5 in solving equations. Example 7 illustrates the use of the distributive property to clear an equation of parentheses.

OBJECTIVE 3   **Example 7** **Using the Distributive Property and Solving Equations**

Solve.

NOTE $2(3x + 4) =$ $2(3x) + 2(4) = 6x + 8.$

$$2(3x + 4) = 5x - 6$$

Applying the distributive property on the left, we have

$$6x + 8 = 5x - 6$$

We can then proceed as before:

$$
\begin{array}{rll}
6x + 8 = & 5x - 6 & \\
\underline{-5x \qquad\qquad -5x} & & \text{Add } -5x. \\
x + 8 = & -6 & \\
\underline{-8 \qquad\qquad -8} & & \text{Add } -8. \\
x = & -14 &
\end{array}
$$

NOTE Remember that $x = -14$ and $-14 = x$ are equivalent equations.

The solution is -14. We leave the checking of this result to the reader.
     Remember: Always return to the original equation to check.

CHECK YOURSELF 7 _____

Solve and check each of the equations.

(a) $4(5x - 2) = 19x + 4$ (b) $3(5x + 1) = 2(7x - 3) - 4$

Given an expression such as

$-2(x - 5)$

the distributive property can be used to create the equivalent expression.

$-2x + 10$

The distribution of a negative number is used in Example 8.

Example 8   Distributing a Negative Number

Solve each of the equations.

(a) $-2(x - 5) = -3x + 2$

$$
\begin{array}{rll}
-2x + 10 = -3x + 2 & \text{Distribute the } -2. \\
\underline{+3x \qquad\quad +3x} & \text{Add } 3x. \\
x + 10 = \qquad\quad 2 & \\
\underline{-10 = \quad -10} & \text{Add } -10. \\
x \quad = \quad -8 &
\end{array}
$$

(b) $-3(3x + 5) = -5(2x - 2)$

$$
\begin{array}{rll}
-9x - 15 = -5(2x - 2) & \text{Distribute the } -3. \\
-9x - 15 = -10x + 10 & \text{Distribute the } -5. \\
\underline{+10x \qquad\quad +10x} & \text{Add } 10x. \\
x - 15 = \qquad\quad 10 & \\
\underline{+15 \qquad\quad +15} & \text{Add } 15. \\
x \quad = \quad 25 &
\end{array}
$$

 CHECK YOURSELF 8 _____

Solve each of the equations.

(a) $-2(x - 3) = -x + 5$ (b) $-4(2x - 1) = -3(3x + 2)$

When parentheses are preceded only by a negative, or by the minus sign, we say that we have a silent negative one. Example 9 illustrates this case.

Example 9 **Distributing the Silent Negative One**

Solve.

$$-(2x + 3) = -3x + 7$$
$$-1(2x + 3) = -3x + 7$$
$$(-1)(2x) + (-1)(3) = -3x + 7$$

$$
\begin{array}{rcl}
-2x - 3 &=& -3x + 7 \\
\underline{+3x \qquad\quad +3x} & & \qquad \text{Add } 3x. \\
x - 3 &=& 7 \\
\underline{\quad +3 \qquad\quad +3} & & \qquad \text{Add } 3. \\
x &=& 10
\end{array}
$$

 CHECK YOURSELF 9 _____

Solve.

$$-(3x + 2) = -2x - 6$$

READING YOUR TEXT _____

The following fill-in-the-blank exercises are designed to assure that you understand the key vocabulary used in this section. Each sentence comes directly from the section. You will find the correct answers in Appendix C.

Section 2.10

(a) Equations that have the same solution are called _____ equations.

(b) The addition property is equivalent to adding the same _____ to both sides of a scale.

(c) In solving an equation, you should always _____ each side as much as possible before using the addition property.

(d) **Remember:** Always return to the original equation to _____ the result.

CHECK YOURSELF ANSWERS _____

1. 9   **2.** 7   **3.** 3   **4.** -3 **5. (a)** 7; **(b)** -6 **6. (a)** -10; **(b)** -3
7. (a) 12; **(b)** -13 **8. (a)** 1; **(b)** -10 **9.** 4

2.10 Exercises

Solve and check each equation.

1. $x + 9 = 11$ **ALEKS**

2. $x - 4 = 6$ **ALEKS**

3. $x - 8 = 3$ **ALEKS**

4. $x + 11 = 15$ **ALEKS**

5. $x - 8 = -10$ **VIDEO** **ALEKS**

6. $x + 5 = 2$ **ALEKS**

7. $x + 4 = -3$ **ALEKS**

8. $x - 5 = -4$ **ALEKS**

9. $11 = x + 5$ **ALEKS**

10. $x + 7 = 0$ **ALEKS**

11. $4x = 3x + 4$

12. $7x = 6x - 8$

13. $11x = 10x - 10$

14. $9x = 8x + 5$

15. $6x + 3 = 5x$

16. $12x - 6 = 11x$

17. $8x - 4 = 7x$

18. $9x - 7 = 8x$

19. $2x + 3 = x + 5$

20. $3x - 2 = 2x + 1$

21. $3x - 5 + 2x - 7 + x = 5x + 2$

22. $5x + 8 + 3x - x + 5 = 6x - 3$

23. $3(7x + 2) = 5(4x + 1) + 17$ **VIDEO**

24. $5(5x + 3) = 3(8x - 2) + 4$

25. Which equation is equivalent to the equation $5x - 7 = 4x - 12$?

   (a) $9x = 19$ **(b)** $9x - 7 = -12$ **(c)** $x = -18$ **(d)** $x - 7 = -12$

26. Which equation is equivalent to the equation $12x - 6 = 8x + 14$?

   (a) $4x - 6 = 14$ **(b)** $x = 20$ **(c)** $20x = 20$ **(d)** $4x = 8$

27. Which equation is equivalent to the equation $7x + 5 = 12x - 10$?

   (a) $5x = -15$ **(b)** $7x - 5 = 12x$ **(c)** $-5 = 5x$ **(d)** $7x + 15 = 12x$

Name _____

Section _____ Date _____

ANSWERS

1. _____
2. _____
3. _____
4. _____
5. _____
6. _____
7. _____
8. _____
9. _____
10. _____
11. _____
12. _____
13. _____
14. _____
15. _____
16. _____
17. _____
18. _____
19. _____
20. _____   21. _____
22. _____   23. _____
24. _____   25. _____
26. _____   27. _____

True or false?

28. Every linear equation with one variable has exactly one solution.

29. Isolating the variable on the right side of the equation will result in a negative solution.

30. "Surprising Results!" Work with other students to try this experiment. Each person should do the following six steps mentally, not telling anyone else what their calculations are:

(a) Think of a number. (b) Add 7.
(c) Multiply by 3. (d) Add 3 more than the original number.
(e) Divide by 4. (f) Subtract the original number.

What number do you end up with? Compare your answer with everyone else's. Does everyone have the same answer? Make sure that everyone followed the directions accurately. How do you explain the results? Algebra makes the explanation clear. Work together to do the problem again, using a variable for the number. Make up another series of computations that give "surprising results."

Answers

1. 2   **3.** 11   **5.** −2   **7.** −7   **9.** 6   **11.** 4   **13.** −10
15. −3   **17.** 4   **19.** 2   **21.** 14   **23.** 16   **25.** d   **27.** d
29. False

ACTIVITY 2: CHARTING TEMPERATURES

Each chapter in this text includes an activity. The activity is related to the vignette you encountered in the chapter opening. The activity provides you with the opportunity to apply the math you studied in the chapter.

Your instructor will determine how best to use this activity in your class. You may find yourself working in class or outside of class; you may find yourself working alone or in small groups; or you may even be asked to perform research in a library or on the Internet.

In Section 2.1, we looked at a number line. The figure below is the typical representation of the number line.

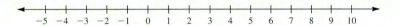

A thermometer can be thought of as a number line. Compare the thermometer below with the number line that follows.

We can use the number line above to locate a series of temperatures. For example, a series of low temperatures for the first seven days of December in Anchorage, Alaska, is presented below.

12°F, −5°F, −7°F, −2°F, 6°F, 9°F, 25°F

Plotting these seven points on the number line, we get

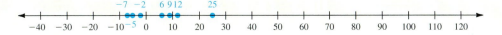

Although this presentation of the data gives us some information about the temperatures, it is really of limited use. Perhaps the most important thing we are looking for when we examine a series of temperatures is a trend. We cannot detect a trend from this number line.

In order to help us see trends, we usually plot temperatures on a time chart. In a time chart, we can look at temperature changes over regular intervals—sometimes hours and sometimes days. The following chart has one point for each of the first seven days in December. Compare the points to the data listed on p. 201.

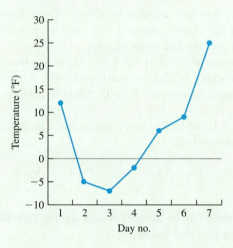

Do you see the advantages of this kind of presentation? It is called a **line graph** because of the lines that connect the temperature points. The trend in temperatures over the most recent five days is definitely increasing. This guarantees nothing, but it is encouraging!

Make a similar presentation for each of the following data sets.

1. The high temperatures during the first seven days in June of 2004 in Death Valley were

    102°F, 98°F, 101°F, 105°F, 106°F, 109°F, 112°F

2. Using the U.S. weather service forecast, plot the predicted high for the next seven days in your area.

3. The record low temperature in the U.S. was in Prospect Camp Creek, Alaska, during Christmas week of 1924. Research the low temperatures in Alaska that week and plot the temperatures over a seven day period.

4. Find a tide chart that gives the high and low tides for some location in the U.S. Use a chart similar to the one used for temperatures to plot the high tides (use a red line) and the low tides (use a blue line) on the same chart.

2 Summary

DEFINITION/PROCEDURE	EXAMPLE	REFERENCE
Introduction to Integers		**Section 2.1**
Positive Integers Integers used to name whole numbers to the right of the origin on the number line. *Negative Integers* Integers used to name the opposites of whole numbers. Negatives are found to the left of the origin on the number line. *Integers* Whole numbers and their opposites. The integers are $$\{\ldots, -3, -2, -1, 0, 1, 2, 3, \ldots\}$$	The origin $-3\ -2\ -1\ \ 0\ \ 1\ \ 2\ \ 3$ Negative integers · Positive integers	*p. 122*
Absolute Value The distance (on the number line) between the point named by a signed number and the origin. The absolute value of x is written $\lvert x \rvert$.	$\lvert 7 \rvert = 7$ $\lvert -10 \rvert = 10$	*p. 124*
Addition of Integers		**Section 2.2**
Adding Integers 1. If two integers have the same sign, add their absolute values. Give the result the sign of the original integers. 2. If two integers have different signs, subtract their absolute values, the smaller from the larger. Give the result the sign of the integer with the larger absolute value.	$9 + 7 = 16$ $(-9) + (-7) = -16$ $15 + (-10) = 5$ $(-12) + 9 = -3$	*p. 133–134*
Subtraction of Integers		**Section 2.3**

Subtracting Integers
 1. Rewrite the subtraction problem as an addition problem by
   a. Changing the subtraction symbol to an addition symbol
   b. Replacing the integer being subtracted with its opposite
 2. Add the resulting integers as before.
	$16 - 8 = 16 + (-8)$
	 $= 8$
	 $8 - 15 = 8 + (-15)$
	 $= -7$
	 $-9 - (-7) = -9 + 7$
	 $= -2$
		p. 139

Multiplication of Integers		**Section 2.4**
Multiplying Integers Multiply the absolute values of the two integers. 1. If the integers have different signs, the product is negative. 2. If the integers have the same sign, the product is positive.	$5(-7) = -35$ $(-10)(9) = -90$ $8 \cdot 7 = 56$ $(-9)(-8) = 72$ $(-2)^2 = (-2)(-2) = 4$ $-2^2 = -(2 \cdot 2) = -4$	*p. 145–146*

Continued

DEFINITION/PROCEDURE	EXAMPLE	REFERENCE
Division of Integers		Section 2.5
Dividing Integers Divide the absolute values of the two integers. **1.** If the integers have different signs, the quotient is negative. **2.** If the integers have the same sign, the quotient is positive.	$\dfrac{-32}{4} = -8$ $\dfrac{-18}{-9} = 2$	p. 151
Introduction to Algebra: Variables and Expressions		Section 2.6
Multiplication $x \cdot y$ $(x)(y)$ These all mean the **product** of x **and** y or x **times** y. xy	The product of m and n is mn. The product of 2 and the sum of a and b is $2(a + b)$.	p. 159
Evaluating Algebraic Expressions		Section 2.7
Evaluating Algebraic Expressions To evaluate an algebraic expression: **1.** Replace each variable or letter with its number value. **2.** Do the necessary arithmetic, following the rules for the order of operations.	Evaluate $2x + 3y$ if $x = 5$ and $y = -2$. $2x + 3y$ $= 2(5) + 3(-2)$ $= 10 + (-6) = 4$	p. 167
Simplifying Algebraic Expressions		Section 2.8
Term A number or the product of a number and one or more variables.	$3xy$ is a term	p. 177
Combining Like Terms To combine like terms: **1.** Add or subtract the coefficients (the numbers multiplying the variables). **2.** Attach the common variable.	$5x + 2x = 7x$ $8a - 5a = 3a$	p. 179
Introduction to Linear Equations		Section 2.9
Equation A statement that two expressions are equal.	$2x - 3 = 5$ is an equation	p. 186
Solution A value for a variable that makes an equation a true statement.	4 is a solution for the above equation because $2(4) - 3 = 5$	p. 187
The Addition Property of Equality		Section 2.10
Equivalent Equations Equations that have exactly the same solutions.	$2x - 3 = 5$ and $x = 4$ are equivalent equations	p. 192
The Addition Property of Equality If $a = b$, then $a + c = b + c$	If $2x - 3 = 7$, then $2x - 3 + 3 = 7 + 3$	p. 192

Summary Exercises

This summary exercise set is provided to give you practice with each of the objectives of this chapter. Each exercise is keyed to the appropriate chapter section. When you are finished, you can check your answers to the odd-numbered exercises against those presented in the back of the text. If you have difficulty with any of these questions, go back and reread the examples from that section. The answers to the even-numbered exercises appear in the *Instructor's Solutions Manual.* Your instructor will give you guidelines on how to best use these exercises in your instructional setting.

[2.1] Represent the integers on the number line shown.

1. $6, -18, -3, 2, 15, -9$

Place each of the groups of integers in ascending order.

2. $4, -3, 6, -7, 0, 1, -2$

3. $-7, 8, -8, 1, 2, -3, 3, 0, 7$

Find the opposite of each number.

4. 17

5. -63

Evaluate.

6. $|9|$

7. $|-9|$

8. $-|9|$

9. $-|-9|$

10. $|12 - 8|$

11. $|8| + |-12|$

12. $-|8 + 12|$

13. $|-18| - |-12|$

14. $|-7| - |-3|$

15. $|-9| + |-5|$

[2.2] Add.

16. $-3 + (-8)$

17. $10 + (-4)$

18. $6 + (-6)$

19. $-16 + (-16)$

20. $-18 + 0$

[2.3] Subtract.

21. $8 - 13$

22. $-7 - 10$

23. $10 - (-7)$

24. $-5 - (-1)$

25. $-9 - (-9)$

26. $0 - (-2)$

[2.4] Multiply.

27. $(10)(-7)$

28. $(-8)(-5)$

29. $(-3)(-15)$

30. $(1)(-15)$

31. $(0)(-8)$

Evaluate each of the expressions.

32. $18 - 3 \cdot 5$

33. $(18 - 3) \cdot 5$

34. $5 \cdot 4^2$

35. $(5 \cdot 4)^2$

36. $5 \cdot 3^2 - 4$

37. $5(3^2 - 4)$

38. $5(4 - 2)^2$

39. $5 \cdot 4 - 2^2$

40. $(5 \cdot 4 - 2)^2$

41. $3(5 - 2)^2$

42. $3 \cdot 5 - 2^2$

43. $(3 \cdot 5 - 2)^2$

[2.5] Divide.

44. $\dfrac{80}{16}$

45. $\dfrac{-63}{7}$

46. $\dfrac{-81}{-9}$

47. $\dfrac{0}{-5}$

48. $\dfrac{32}{-8}$

49. $\dfrac{-7}{0}$

Perform the indicated operations.

50. $\dfrac{-8 + 6}{-8 - (-10)}$

51. $\dfrac{-6 - 1}{5 - (-2)}$

52. $\dfrac{25 - 4}{-5 - (-2)}$

[2.6] Write, using symbols.

53. 5 more than y

54. c decreased by 10

55. The product of 8 and a

56. The quotient when y is divided by 3

57. 5 times the product of m and n

58. The product of a and 5 less than a

59. 3 more than the product of 17 and x

60. The quotient when a plus 2 is divided by a minus 2

Identify which are expressions and which are not.

61. $4(x + 3)$

62. $7 \div \cdot 8$

63. $y + 5 = 9$

64. $11 + 2(3x - 9)$

[2.7] Evaluate the expressions if $x = -3$, $y = 6$, $z = -4$, and $w = 2$.

65. $3x + w$

66. $5y - 4z$

67. $x + y - 3z$

68. $5z^2$

69. $3x^2 - 2w^2$

70. $3x^3$

71. $5(x^2 - w^2)$

72. $\dfrac{6z}{2w}$

73. $\dfrac{2x - 4z}{y - z}$

74. $\dfrac{3x - y}{w - x}$

75. $\dfrac{x(y^2 - z^2)}{(y + z)(y - z)}$

76. $\dfrac{y(x - w)^2}{x^2 - 2xw + w^2}$

[2.8] List the terms of the expressions.

77. $4a^3 - 3a^2$

78. $5x^2 - 7x + 3$

Circle like terms.

79. $5m^2, -3m, -4m^2, 5m^3, m^2$

80. $4ab^2, 3b^2, -5a, ab^2, 7a^2, -3ab^2, 4a^2b$

Combine like terms.

81. $5c + 7c$

82. $2x + 5x$

83. $4a - 2a$

84. $6c - 3c$

85. $9xy - 6xy$

86. $5ab^2 + 2ab^2$

87. $7a + 3b + 12a - 2b$

88. $6x - 2x + 5y - 3x$

89. $5x^3 + 17x^2 - 2x^3 - 8x^2$

90. $3a^3 + 5a^2 + 4a - 2a^3 - 3a^2 - a$

[2.9] Tell whether the number shown in parentheses is a solution for the given equation.

91. $7x + 2 = 16$ (2)

92. $5x - 8 = 3x + 2$ (4)

93. $7x - 2 = 2x + 8$ (2)

94. $4x + 3 = 2x - 11$ (-7)

95. $x + 5 + 3x = 2 + x + 23$ (6)

96. $\dfrac{1}{3}x + 2 = 10$ (21)

[2.10] Solve the equations and check your results.

97. $x + 5 = 7$

98. $x - 9 = 3$

99. $5x = 4x - 5$

100. $3x - 9 = 2x$

101. $5x - 3 = 4x + 2$

102. $9x + 2 = 8x - 7$

103. $7x - 5 = 6x - 4$

104. $3 + 4x - 1 = x - 7 + 2x$

105. $4(2x + 3) = 7x + 5$

Self-Test for Chapter 2

Name _____

Section _____ Date _____

ANSWERS

The purpose of this self-test is to help you check your progress so that you can find sections and concepts that you need to review before the next in-class exam. Allow yourself about an hour to take this test. At the end of that hour, check your answers against those given in the back of this text. If you missed a question, notice the section reference that accompanies the answer. Go back to that section and reread the examples until you have mastered that particular concept.

Represent the integers on the number line shown.

1. $5, -12, 4, -7, 18, -17$

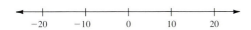

2. Place the following group of integers in ascending order: $4, -3, -6, 5, 0, 2, -2$

Evaluate.

3. $|7|$

4. $|-7|$

5. $|18 - 7|$

6. $|18| - |-7|$

7. $-|24 - 5|$

Find the opposite of each integer.

8. 40

9. -19

Add.

10. $-8 + (-5)$

11. $6 + (-9)$

12. $(-9) + (-12)$

Subtract.

13. $9 - 15$

14. $-9 - 15$

15. $5 - (-4)$

16. $-7 - (-7)$

Multiply.

17. $(-8)(5)$

18. $(-9)(-7)$

19. $(6)(-4)$

1. _____

2. _____

3. _____

4. _____

5. _____

6. _____

7. _____

8. _____

9. _____

10. _____

11. _____

12. _____

13. _____

14. _____

15. _____

16. _____

17. _____

18. _____

19. _____

20.

21.

22.

23.

24.

25.

26.

27.

28.

29.

30.

31.

32.

33.

34.

35.

36.

37.

38.

39.

40.

41.

42.

43.

44.

Evaluate each expression.

20. $\dfrac{75}{-3}$

21. $\dfrac{-36 + 9}{-9}$

22. $\dfrac{(-15)(-3)}{-9}$

23. $\dfrac{9}{0}$

24. $23 - 4 \cdot 5$

25. $4 \cdot 5^2 - 35$

26. $4(2 + 4)^2$

27. $16 \div (-4) + (-5)$

28. If $x = 2$, $y = -1$, and $z = 3$, evaluate the expression $\dfrac{9x^2 y}{3z}$.

Write, using symbols.

29. 5 less than a

30. The product of 6 and m

31. 4 times the sum of m and n

32. The quotient when the sum of a and b is divided by 3

Identify which are expressions and which are not.

33. $5x + 6 = 4$

34. $4 + (6 + x)$

Combine like terms.

35. $8a + 7a$

36. $10x + 8y + 9x - 3y$

37. Subtract $9a^2$ from the sum of $12a^2$ and $5a^2$.

38. Number Problem Tom is 8 years younger than twice Moira's age. Write an expression for Tom's age. Let x represent Moira's age.

39. Geometry The length of a rectangle is 4 more than twice the width. Write an expression for the length of the rectangle.

Tell whether the number shown in parentheses is a solution for the given equation.

40. $7x - 3 = 25$ (5)

41. $8x - 3 = 5x + 9$ (4)

Solve the equations and check your results.

42. $x - 7 = 4$

43. $7x - 12 = 6x$

44. $9x - 2 = 8x + 5$

Cumulative Review
for Chapters 1 and 2

The following exercises are presented to help you review concepts from earlier chapters that you may have forgotten. This section is meant as review material and not as a comprehensive exam. The answers are presented in the back of the text. If you have difficulty with any of these exercises, be certain to at least read through the summary related to that section.

Name _____

Section _____ Date _____

ANSWERS

1. Give the place value of 7 in 3,738,500.

2. Give the word name for 302,525.

3. Write two million, four hundred thirty thousand as a numeral.

In exercises 4 to 6, name the property of addition that is illustrated.

4. $5 + 12 = 12 + 5$ **5.** $9 + 0 = 9$

6. $(7 + 3) + 8 = 7 + (3 + 8)$

In exercises 7 and 8, perform the indicated operations.

7.
$$\begin{array}{r} 593 \\ 275 \\ +\ \ 98 \\ \hline \end{array}$$

8. Find the sum of 58, 673, 5,325, and 17,295.

In exercises 9 and 10, round the numbers to the indicated place value.

9. 5,873 to the nearest hundred

10. 953,150 to the nearest ten thousand

In exercise 11, estimate the sum by rounding to the nearest hundred.

11.
$$\begin{array}{r} 943 \\ 3,281 \\ 778 \\ 2,112 \\ +\ \ 570 \\ \hline \end{array}$$

ANSWERS

1. _____
2. _____
3. _____
4. _____
5. _____
6. _____
7. _____
8. _____
9. _____
10. _____
11. _____

12. _____

13. _____

14. _____

15. _____

16. _____

17. _____

18. _____

19. _____

20. _____

21. _____

22. _____

23. _____

24. _____

25. _____

26. _____

27. _____

28. _____

29. _____

30. _____

31. _____

32. _____

Evaluate.

12. $|5 - 14|$ **13.** $|5| - |14|$

14. What is the opposite of -3?

In exercises 15 to 24, evaluate each expression.

15. $-12 + (-6)$ **16.** $-7 + 7$

17. $5 + (-7)$ **18.** $9 + (-4) + (-7)$

19. $-8 - (-8)$ **20.** $5 - (-5)$

21. $(-8)(-12)$ **22.** $(-6)(15)$

23. $14 \div (-7)$ **24.** $-25 \div 0$

In exercises 25 to 28, evaluate the expressions if $a = 5$, $b = -3$, $c = 4$, and $d = -2$.

25. $6ad$ **26.** $3b^2$

27. $3(c - 2d)$ **28.** $\dfrac{2a - 7d}{a - b}$

In exercises 29 and 30, combine like terms.

29. $6x + 14 - 3x + 5$ **30.** $4x + 8y - 2x - 7y$

In exercises 31 and 32, solve each equation and check your solution.

31. $x - 5 = 17$ **32.** $3x + 5 = 2x - 3$

FRACTIONS AND EQUATIONS

CHAPTER 3   OUTLINE

Section 3.1   Introduction to Fractions   page 215

Section 3.2   Prime Numbers and Factorization   page 223

Section 3.3   Equivalent Fractions   page 235

Section 3.4   Multiplication and Division of Fractions   page 250

Section 3.5   The Multiplication Property of Equality   page 264

Section 3.6   Linear Equations in One Variable   page 272

INTRODUCTION

There's a saying, "You are what you eat." Unfortunately, many Americans eat too much or perhaps eat too many highly processed foods. Consider the following:

- $\frac{3}{5}$ of all U.S. adults are classified as overweight.

- $\frac{1}{2}$ of all U.S. adults are classified as obese.

- $\frac{1}{5}$ of all U.S. citizens have high blood pressure that can be directly attributed to a high-sodium diet.

Researchers say that the two most likely causes for these conditions are consuming too many calories and not participating in physical activity. Clearly, monitoring the quantity and types of foods we eat and getting regular exercise are important issues for today's society.

In this chapter, we develop many of the skills needed in handling fractions, which are frequently used to answer questions related to the areas of diet and exercise. In Sections 3.1, 3.3, 3.4, and 3.6, you will find exercises that deal with these topics. At the end of the chapter, but before the chapter summary, you will find a more in-depth activity involving diet and exercise.

Pretest Chapter 3

This pretest will provide a preview of the types of exercises you will encounter in each section of this chapter. The answers for these exercises can be found in the back of the text. If you are working on your own or are ahead of the class, this pretest can help you identify the sections in which you should focus more of your time.

[3.2]   **1.** List all the factors of 42.

2. Write the prime factorization of 350.

3. Find the greatest common factor (GCF) of 24, 36, and 52.

[3.3]   **4.** Are $\dfrac{8}{12}$ and $\dfrac{20}{30}$ equivalent fractions?

5. Reduce $\dfrac{-18}{30}$ to lowest terms.

[3.4]   Carry out the indicated operations.

6. $\dfrac{-4}{5} \cdot \dfrac{5}{7}$ **7.** $\dfrac{6}{25} \div \dfrac{3}{10}$

8. $\dfrac{2}{3} \div 7$

[3.5] and [3.6]   Solve the following equations.

9. $7x + 3 = 2x - 2$ **10.** $\dfrac{x-3}{8} = \dfrac{x-2}{10}$

3.1 OBJECTIVES

1. Identify the numerator and denominator of a fraction
2. Use fractions to name parts of a whole
3. Identify proper and improper fractions

Earlier chapters dealt with integers and the operations that are performed on them. Now we will define *fractions,* learn the language of fractions, and learn to perform operations on fractions.

> **Definition:** Fraction
>
> Whenever a unit or a whole quantity is divided into parts, we call those parts **fractions** of the unit.

NOTE The word *fraction* comes from the Latin stem *fractio,* which means "breaking into pieces."

In Figure 1, the whole has been divided into five equivalent parts. We use the symbol $\frac{2}{5}$ to represent the shaded portion of the whole.

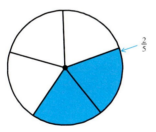

$\frac{2}{5}$

Figure 1

NOTE *Common fraction* is technically the correct term. We will normally just use *fraction* in the text if there is no room for confusion.

The symbol $\frac{2}{5}$ is called a **common fraction,** or more simply a fraction. A fraction is written in the form $\frac{a}{b}$, in which a is an integer and b is a natural number.

We give the numbers a and b special names. The **denominator,** b, is the number on the bottom. This tells us into how many equal parts the unit or whole has been divided. The **numerator,** a, is the number on the top. This tells us how many parts of the unit are being considered.

In Figure 1, the *denominator* is 5; the unit or whole (the circle) has been divided into five equal parts. The *numerator* is 2. We have shaded two parts of the unit.

$\frac{2}{5}$ ← Numerator
← Denominator

OBJECTIVES 1 and 2   **Example 1**   **Labeling Fraction Components**

The fraction $\frac{4}{7}$ names the shaded part of the rectangle in Figure 2.

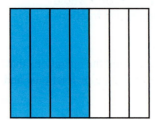

Figure 2

The unit or whole is divided into seven equal parts, so the denominator is 7. We have shaded four of those parts, and so we have a numerator of 4.

CHECK YOURSELF 1

What fraction names the shaded part of this diagram? Identify the numerator and denominator.

Fractions can also be used to name a part of a collection or a set of identical objects.

Example 2   **Naming a Fractional Part**

The fraction $\frac{5}{6}$ names the shaded part of Figure 3. We have shaded five of the six identical objects.

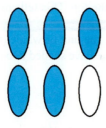

Figure 3

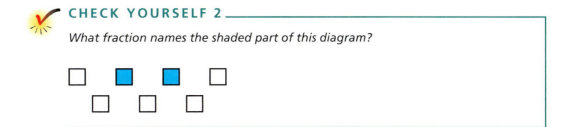

CHECK YOURSELF 2

What fraction names the shaded part of this diagram?

Example 3 Naming a Fractional Part

NOTE Of course, the fraction $\frac{8}{23}$ names the part of the class that is not women.

In a class of 23 students, 15 are women. We can name the part of the class that is women as $\frac{15}{23}$.

CHECK YOURSELF 3

Seven replacement parts out of a shipment of 50 were faulty. What fraction names the portion of the shipment that was faulty?

NOTE $\frac{a}{b}$ names the *quotient* when *a* is divided by *b*. Of course, *b* cannot be 0.

A fraction can also be thought of as indicating division. The symbol $\frac{a}{b}$ also means $a \div b$.

Example 4 Writing a Fraction as Division

$\frac{2}{3}$ is the quotient when 2 is divided by 3. So $\frac{2}{3} = 2 \div 3$.

Note: $\frac{2}{3}$ can be read as "two-thirds" or as "2 divided by 3."

CHECK YOURSELF 4

Using the numbers 5 and 9, write $\frac{5}{9}$ using division.

We can use the relative size of the numerator and denominator of a fraction to separate fractions into different categories.

Definition: Proper Fraction

A fraction that names a number less than 1 (the numerator is *less than* the denominator) is called a **proper fraction.**

> **Definition:** Improper Fraction
>
> A fraction that names a number greater than or equal to 1 (the numerator is *greater than* or *equal to* the denominator) is called an **improper fraction.**

OBJECTIVE 3   **Example 5   Categorizing Fractions**

(a) $\frac{2}{3}$ is a proper fraction because the numerator is less than the denominator (Figure 4).

(b) $\frac{4}{3}$ is an improper fraction because the numerator is larger than the denominator (Figure 5).

(c) Also, $\frac{6}{6}$ is an improper fraction because it names exactly 1 unit; the numerator is equal to the denominator (Figure 6).

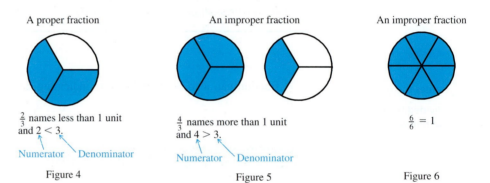

A proper fraction

$\frac{2}{3}$ names less than 1 unit and $2 < 3$.

Numerator   Denominator

Figure 4

An improper fraction

$\frac{4}{3}$ names more than 1 unit and $4 > 3$.

Numerator   Denominator

Figure 5

An improper fraction

$\frac{6}{6} = 1$

Figure 6

 CHECK YOURSELF 5

List the proper fractions and the improper fractions in the group:

$$\frac{5}{4}, \frac{10}{11}, \frac{3}{4}, \frac{8}{5}, \frac{6}{6}, \frac{13}{10}, \frac{7}{8}, \frac{15}{8}$$

One special kind of improper fraction should be mentioned at this point: a fraction with a denominator of 1.

> **Definition:** Fractions with a Denominator of 1
>
> Any fraction with a denominator of 1 is simply equal to the numerator.
>
> For example, $\frac{5}{1} = 5 \div 1 = 5$

READING YOUR TEXT

The following fill-in-the-blank exercises are designed to assure that you understand the key vocabulary used in this section. Each sentence usually comes directly from the section. You will find the correct answers in Appendix C.

Section 3.1

(a) In a fraction, the _____ is the number that tells us into how many parts the whole has been divided.

(b) In a fraction, the _____ is the number that tells us how many parts of the whole are used.

(c) A fraction that names a number less than 1 is called a _____ fraction.

(d) A fraction that names a number greater than or equal to 1 is called an _____ fraction.

CHECK YOURSELF ANSWERS

1. $\dfrac{3}{8}$ $\begin{array}{l}\leftarrow \text{Numerator}\\ \leftarrow \text{Denominator}\end{array}$

2. $\dfrac{2}{7}$ 3. $\dfrac{7}{50}$ 4. $5 \div 9$

5. Proper fractions: Improper fractions:

$\dfrac{10}{11}, \dfrac{3}{4}, \dfrac{7}{8}$ $\dfrac{5}{4}, \dfrac{8}{5}, \dfrac{6}{6}, \dfrac{13}{10}, \dfrac{15}{8}$

Name _____

Section _____ Date _____

ANSWERS

1. _____

2. _____

3. _____

4. _____

5. _____

6. _____

7. _____

8. _____

9. _____

10. _____

11. _____

12. _____

3.1 Exercises

Identify the numerator and denominator of each fraction.

1. $\dfrac{6}{11}$

2. $\dfrac{5}{12}$

3. $\dfrac{5}{3}$

4. $\dfrac{7}{2}$

What fraction names the shaded part of each of the figures?

5.

6.

7.

8.

9.

10.

Solve the applications.

11. Statistics You missed 7 questions on a 20-question test. What fraction names the part you got correct? The part you got wrong?

12. Statistics Of the five starters on a basketball team, two fouled out of a game. What fraction names the part of the starting team that fouled out?

13. **Business and Finance** A used-car dealer sold 11 of the 17 cars in stock. What fraction names the portion sold? What fraction names the portion *not* sold?

14. **Business and Finance** At lunch, five people out of a group of nine had hamburgers. What fraction names the part of the group that had hamburgers? What fraction names the part that did *not* have hamburgers?

Identify each number as a proper fraction or an improper fraction.

15. $\dfrac{3}{5}$

16. $\dfrac{9}{5}$

17. $\dfrac{6}{6}$

18. $\dfrac{7}{9}$

19. $\dfrac{11}{8}$ VIDEO

20. $\dfrac{8}{8}$

21. $\dfrac{13}{17}$

22. $\dfrac{16}{15}$

For exercises 23 and 24, use the data in the table below. It shows robbery offenses committed in one year.

Location of Crime	Street or Highway	Commercial Building	Gas Station	Convenience Store	Residence	Bank	Total
Number of Offenses (in thousands)	219	61	10	26	54	9	379

Source: U.S. Federal Bureau of Investigation.

23. What fraction names the portion of robberies committed at residences?

24. What fraction names the portion of robberies committed on streets or highways?

For exercises 25 and 26, fill in the blank with "less" or "greater."

25. Suppose we have two fractions with the same numerator, but the denominator of the first is greater than the denominator of the second. Then the portion represented by the first fraction is _____ than the portion represented by the second.

26. Suppose we have two fractions with the same denominator, but the numerator of the first is greater than the numerator of the second. Then the portion represented by the first fraction is _____ than the portion represented by the second.

13.

14.

15.

16.

17.

18.

19.

20.

21.

22.

23.

24.

25.

26.

27. _____

28. _____

29. _____

30. _____

27. Suppose that a rectangle is divided into four parts as shown.

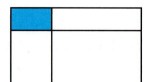

If one part is shaded as shown, should the fraction naming the shaded region be $\frac{1}{4}$? Why or why not?

28. Based on a 2000-calorie diet, consult with your local health officials and determine what fraction of your daily diet each of the following should contribute.

(a) Fat   **(b)** Carbohydrate   **(c)** Unsaturated fat   **(d)** Protein

Using this information, work with your classmates to plan a daily menu that will satisfy the requirements.

29. Suppose the U.S. national debt had to be paid by individuals.

(a) How would the amount each individual owed be determined?
(b) Would this be a proper or improper fraction?

30. In statistics, we describe the probability of an event by dividing the number of ways the event can happen by the total number of possibilities. Using this definition, determine the probability of each of the following:

(a) Obtaining a head on a single toss of a coin
(b) Obtaining a prime number (in this case 2, 3, or 5) when a die is tossed

Answers

1. 6 is the numerator; 11 is the denominator
3. 5 is the numerator; 3 is the denominator   **5.** $\frac{5}{6}$ **7.** $\frac{5}{5}$ **9.** $\frac{5}{8}$

11. $\frac{13}{20}, \frac{7}{20}$ **13.** $\frac{11}{17}, \frac{6}{17}$ **15.** Proper   **17.** Improper   **19.** Improper

21. Proper   **23.** $\frac{54}{379}$ **25.** less   **27.**   **29.**

3.2 Prime Numbers and Factorization

3.2 OBJECTIVES

1. Find the factors of a natural number
2. Determine whether a number is prime, composite, or neither
3. Find the prime factorization for any number
4. Find the greatest common factor (GCF) of two or more numbers

Overcoming Math Anxiety

Working Together

How many of your classmates do you know? Whether you are by nature outgoing or shy, you have much to gain by getting to know your classmates.

1. It is important to have someone to call when you have missed class or if you are unclear on an assignment.

2. Working with another person is almost always beneficial to both people. If you don't understand something, it helps to have someone to ask about it. If you do understand something, nothing will cement that understanding more than explaining the idea to another person.

3. Sometimes we need to share our feelings of distress. If an assignment is particularly frustrating, it is reassuring to find that it is also frustrating for other students.

4. Have you ever thought you had the right answer, but it doesn't match the answer in the text? Frequently the answers are equivalent, but that's not always easy to see. A different perspective can help you see that. Occasionally there is an error in a textbook (here, we are talking about *other* textbooks). In such cases, it is wonderfully reassuring to find that someone else has the same answer as you do.

NOTE 2 and 5 can also be called **divisors** of 10. They divide 10 exactly.

In Section 1.5 we said that because $2 \cdot 5 = 10$, we call 2 and 5 the **factors** of 10.

RECALL A natural number, or counting number, is any number in the set {1, 2, 3, . . .}.

Definition: Factor

A **factor** of a natural number is another natural number that will *divide* exactly into that number. This means that the division will have a remainder of 0.

OBJECTIVE 1

Example 1 Finding Factors

List all factors of 18.

$1 \cdot 18 = 18$ 1 and 18 are factors of 18.

$2 \cdot 9 = 18$ 2 and 9 are also factors of 18.

$3 \cdot 6 = 18$ 3 and 6 are factors (or divisors) of 18.

NOTE This is a complete list of the factors. There are no other whole numbers that divide 18 exactly. Note that the factors of 18, except for 18 itself, are *smaller* than 18.

1, 2, 3, 6, 9, and 18 are all the factors of 18.

✔ CHECK YOURSELF 1

List all the factors of 24.

Listing factors leads us to an important classification of natural numbers. Any natural number other than 1 will be either a *prime* or a *composite* number. Look at the following definitions.

A natural number other than 1 will always have itself and 1 as factors. Sometimes these will be the *only* factors. For instance, 1 and 3 are the only factors of 3.

> **Definition:** Prime Number
>
> A **prime number** is any natural number (other than 1) that has exactly two factors, 1 and itself.

NOTE How large can a prime number be? There is no largest prime number. To date, the largest *known* prime is $2^{25,964,951} - 1$. This is a number with 7,816,230 digits, if you are curious. Of course, a computer had to be used to verify that a number of this size is prime. By the time you read this, someone may very well have found an even larger prime number.

As examples, 2, 3, 5, and 7 are prime numbers. Their only factors are 1 and themselves.

To check whether a number is prime, one approach is simply to divide the smaller primes, 2, 3, 5, 7, and so on, into the given number. If no factors other than 1 and the given number are found, the number is prime.

Here is the method known as the **sieve of Eratosthenes** for identifying prime numbers.

1. Write down a sequence of natural numbers, starting with the number 2. In this example, we stop at 50.

2. Start at the number 2. Delete every second number after the 2.

3. Move to the number 3. Delete every third number after 3 (some numbers will be deleted twice).

4. Continue this process, deleting every fourth number after 4, every fifth number after 5, and so on.

5. When you have finished, the undeleted numbers are the prime numbers.

	2	3	4̶	5	6̶	7	8̶	9̶	10̶
11	12̶	13	14̶	15̶	16̶	17	18̶	19	20̶
21̶	22̶	23	24̶	25̶	26̶	27̶	28̶	29	30̶
31	32̶	33̶	34̶	35̶	36̶	37	38̶	39̶	40̶
41	42̶	43	44̶	45̶	46̶	47	48̶	49̶	50̶

The prime numbers less than 50 are 2, 3, 5, 7, 11, 13, 17, 19, 23, 29, 31, 37, 41, 43, 47.

OBJECTIVE 2   **Example 2   Identifying Prime Numbers**

Which of the numbers are prime?

17 is a prime number.   1 and 17 are the only factors.

29 is a prime number.   1 and 29 are the only factors.

33 is *not* prime.   1, 3, 11, and 33 are all factors of 33.

Note: If a number less than 100 is *not* a prime, it will have one or more of the numbers 2, 3, 5, or 7 as factors.

CHECK YOURSELF 2

Which of the following numbers are prime numbers?

2, 6, 9, 11, 15, 19, 23, 35, 41

We can now define a second class of natural numbers.

Definition: Composite Number

NOTE This definition tells us that a composite number *does* have factors other than 1 and itself.

A **composite number** is any natural number greater than 1 that is not prime. Every composite number has more than two factors.

Example 3   Identifying Composite Numbers

Which of the numbers are composite?

18 is a composite number.	1, 2, 3, 6, 9, and 18 are all factors of 18.
23 is not a composite number.	1 and 23 are the only factors. This means that 23 is a *prime number.*
25 is a composite number.	1, 5, and 25 are factors.
38 is a composite number.	1, 2, 19, and 38 are factors.

CHECK YOURSELF 3

Which of the following numbers are composite?

2, 6, 10, 13, 16, 17, 22, 27, 31, 35

By the definitions of prime and composite numbers:

Property: One

The natural number 1 is neither prime nor composite.

This is simply a matter of the way in which prime and composite numbers are defined in mathematics. The number 1 is the *only* natural number that cannot be classified as one or the other.

To **factor a number** means to write the number as a product of its natural-number factors.

Example 4   Factoring a Composite Number

Factor the number 10.

$10 = 2 \cdot 5$ The order in which you write the factors does not matter, so $10 = 5 \cdot 2$ would also be correct.

Of course, $10 = 10 \cdot 1$ is also a correct statement. However, in this section we are interested in factors other than 1 and the given number.

Factor the number 21.

$21 = 3 \cdot 7$

CHECK YOURSELF 4

Factor the number 35.

In writing composite numbers as a product of factors, there can be a number of different possible factorizations.

Example 5 Factoring a Composite Number

Find three ways to factor 72.

NOTE There have to be at least two different factorizations because a composite number has factors other than 1 and itself.

$$72 = 8 \cdot 9 \quad (1)$$
$$= 6 \cdot 12 \quad (2)$$
$$= 3 \cdot 24 \quad (3)$$

CHECK YOURSELF 5

Find three ways to factor 42.

We now want to write composite numbers as a product of their *prime factors.* Look again at the first factored line of Example 5. The process of factoring can be continued until all the factors are prime numbers.

OBJECTIVE 3   **Example 6   Factoring a Composite Number**

NOTE This is often called a **factor tree.**

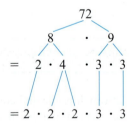

4 is still not prime, and so we continue by factoring 4.

72 is now written as a product of prime factors.

NOTE Finding the prime factorization of a number will be important in our later work in adding fractions.

When we write 72 as $2 \cdot 2 \cdot 2 \cdot 3 \cdot 3$, no further factorization is possible. This is called the **prime factorization** of 72.

Now, what if we start with the second factored line from the same example, $72 = 6 \cdot 12$?

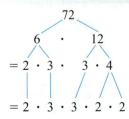

Continue to factor 6 and 12.

Continue again to factor 4. Other choices for the factors of 12 are possible. As we shall see, the end result will be the same.

© 2010 McGraw-Hill Companies

No matter which pair of factors you start with, you will find the same prime factorization. In Example 6, 2 is a factor three times, and 3 is a factor twice. Because *multiplication is commutative,* the order in which we write the factors does not matter.

 CHECK YOURSELF 6 _____

We could also write

$72 = 2 \cdot 36$

Continue the factorization.

Property: The Fundamental Theorem of Arithmetic

There is exactly one prime factorization for any composite number.

The method of Example 6 will always work. However, another method for factoring composite numbers exists. This method is particularly useful when numbers get large, in which case, factoring with a number tree becomes difficult.

Property: Factoring by Division

To find the prime factorization of a number, divide the number by a series of primes until the final quotient is a prime number.

NOTE The prime factorization is then the product of all the prime divisors and the final quotient.

Example 7   Finding Prime Factors

To write 60 as a product of prime factors, divide 2 into 60 for a quotient of 30. Continue to divide by 2 again for the quotient 15. Because 2 won't divide evenly into 15, we try 3. Because the quotient 5 is prime, we are done.

$$\frac{30}{2)60} \searrow \frac{15}{2)30} \searrow \frac{5}{3)15} \quad \text{Prime}$$

Our factors are the prime divisors and the final quotient. We have

$60 = 2 \cdot 2 \cdot 3 \cdot 5$

 CHECK YOURSELF 7 _____

Complete the process to find the prime factorization of 90.

$$\frac{45}{2)90} \searrow \frac{?}{?)45}$$

Remember to continue until the final quotient is prime.

Writing composite numbers in their completely factored form can be simplified if we use a format called **continued division.**

Example 8   Finding Prime Factors Using Continued Division

Use the continued-division method to divide 60 by a series of prime numbers.

NOTE In each short division, we write the quotient *below* rather than above the dividend. This is just a convenience for the next division.

Primes

Stop when the final quotient is prime.

To write the factorization of 60, we list each divisor used and the final prime quotient. In our example, we have

$$60 = 2 \cdot 2 \cdot 3 \cdot 5$$

 CHECK YOURSELF 8

Find the prime factorization of 234.

We know that a factor or a divisor of a whole number divides that number exactly.
    The factors or divisors of 20 are

NOTE Again the factors of 20, other than 20 itself, are less than 20.

1, 2, 4, 5, 10, 20

Each of these numbers divides 20 exactly, that is, with no remainder.
    Our work in this section involves common factors or divisors. A **common factor** or **divisor** for two numbers is any factor that divides both the numbers exactly.

Example 9   Finding Common Factors

Look at the numbers 20 and 30. Is there a common factor for the two numbers?
    First, we list the factors. Then we circle the ones that appear in both lists.

Factors

20:    ①, ②, 4, ⑤, ⑩, 20

30:    ①, ②, 3, ⑤, 6, ⑩, 15, 30

We see that 1, 2, 5, and 10 are common factors of 20 and 30. Each of these numbers divides both 20 and 30 exactly.
    Our later work with fractions will require that we find the greatest common factor (GCF) of a group of numbers.

Definition: Greatest Common Factor

The **greatest common factor** (GCF) of a group of numbers is the *largest* number that will divide each of the given numbers exactly.

 CHECK YOURSELF 9

List the factors of 30 and 36 and then find the GCF.

To find the GCF of a group of numbers, we use the prime factorization of each number. We will outline the process.

> **Step by Step:**   Finding the Greatest Common Factor
>
> **Step 1**   Write the prime factorization for each of the numbers in the group.
> **Step 2**   Locate the prime factors that are *common* to all the numbers.
> **Step 3**   The greatest common factor will be the *product* of all the common prime factors.

NOTE If there are no common prime factors, the GCF is 1.

OBJECTIVE 4

Example 10   **Finding the Greatest Common Factor**

Find the GCF of 20 and 30.

Step 1   Write the prime factorization of 20 and 30.

$20 = 2 \cdot 2 \cdot 5$
$30 = 2 \cdot 3 \cdot 5$

Step 2   Find the prime factors common to each number.

$20 = ②\cdot 2 \cdot ⑤$ 2 and 5 are the common prime factors.
$30 = ②\cdot 3 \cdot ⑤$

Step 3   Form the product of the common prime factors.

$2 \cdot 5 = 10$

10 is the GCF.

CHECK YOURSELF 10

Find the GCF of 30 and 36 using prime factorizations.

To find the GCF of a group of more than two numbers, we use the same process.

Example 11   **Finding the Greatest Common Factor**

Find the GCF of 24, 30, and 36.

$24 = ②\cdot 2 \cdot 2 \cdot ③$
$30 = ②\cdot ③\cdot 5$
$36 = ②\cdot 2 \cdot ③\cdot 3$

2 and 3 are the prime factors common to *all three numbers.*

$2 \cdot 3 = 6$ is the GCF.

CHECK YOURSELF 11

Find the GCF of 15, 30, and 45.

Example 12   Finding the Greatest Common Factor

NOTE If two numbers, such as 15 and 28, have no common factor other than 1, they are called **relatively prime**.

Find the GCF of 15 and 28.

$15 = 3 \cdot 5$ There are no common prime factors listed. But remember that 1 is a factor of every natural number.
$28 = 2 \cdot 2 \cdot 7$

The GCF of 15 and 28 is 1.

CHECK YOURSELF 12

Find the GCF of 30 and 49.

READING YOUR TEXT

The following fill-in-the-blank exercises are designed to assure that you understand the key vocabulary used in this section. Each sentence usually comes directly from the section. You will find the correct answers in Appendix C.

Section 3.2

(a)   A _____ of a natural number is another natural number that will divide exactly into the first number.

(b)   A _____ number is any natural number (other than 1) that has exactly two factors, 1 and itself.

(c)   Every _____ number has more than two factors.

(d)   To factor a number means to write the number as a _____ of its natural-number factors.

CHECK YOURSELF ANSWERS

1. 1, 2, 3, 4, 6, 8, 12, and 24   **2.** 2, 11, 19, 23, and 41 are prime numbers
3. 6, 10, 16, 22, 27, and 35 are composite numbers   **4.** $5 \cdot 7$
5. $2 \cdot 21, 3 \cdot 14, 6 \cdot 7$ **6.** $2 \cdot 2 \cdot 2 \cdot 3 \cdot 3$
7. $\begin{array}{ccc} 45 & 15 & 5 \\ 2\overline{)90} & 3\overline{)45} & 3\overline{)15} \end{array}$ **8.** $2 \cdot 3 \cdot 3 \cdot 13$
     $90 = 2 \cdot 3 \cdot 3 \cdot 5$
9. 30: ①, ②, ③, 5, ⑥, 10, 15, 30
     36: ①, ②, ③, 4, ⑥, 9, 12, 18, 36
     6 is the GCF.
10. $30 = ② \cdot ③ \cdot 5$
     $36 = ② \cdot 2 \cdot ③ \cdot 3$
     The GCF is $2 \cdot 3 = 6$.
11. 15   **12.** 1

3.2 Exercises

List the factors of each of the numbers.

1. 4   ALEKS

2. 6   ALEKS

3. 24   ALEKS

4. 32   ALEKS

5. 64   VIDEO   ALEKS

6. 66   ALEKS

7. 11   ALEKS

8. 37   ALEKS

Use the given list of numbers for exercises 9 and 10.

0, 1, 15, 19, 23, 31, 49, 55, 59, 87, 91, 97, 103, 105

9. Which of the given numbers are prime?   ALEKS

10. Which of the given numbers are composite?

Find the prime factorization of each number.

11. 18   ALEKS

12. 22   ALEKS

13. 30   ALEKS

14. 35   ALEKS

15. 70   VIDEO   ALEKS

16. 90   ALEKS

17. 66   ALEKS

18. 100   ALEKS

19. 130   ALEKS

20. 88   ALEKS

21. 315   ALEKS

22. 400   ALEKS

In the study of algebra, you often will want to find factors of a number with a given sum or difference. The exercises 23 to 26 use this technique.

23. Find two factors of 24 with a sum of 10.

24. Find two factors of 15 with a difference of 2.

25. Find two factors of 30 with a difference of 1.

26. Find two factors of 28 with a sum of 11.

Find the GCF for each of the groups of numbers.

27. 4 and 6   ALEKS

28. 6 and 9   ALEKS

29. 20 and 21   ALEKS

30. 28 and 42   ALEKS

31. 18 and 24   ALEKS

32. 35 and 36   ALEKS

33. 18 and 54   VIDEO   ALEKS

34. 12 and 48   ALEKS

35. 12, 36, and 60

36. 15, 45, and 90

37. 25, 75, and 150

38. 36, 72, and 144

39. Suppose that two quantities of screws, 70 round head and 98 flat head, are to be packaged in small containers of the same size. The two types are not to be mixed, and the same number of screws is to be placed into each container, with none left over. What is the greatest number of screws that can be put into each container?

40. Suppose that two quantities of candies, 72 dark chocolate and 90 white chocolate, are to be packaged in small containers of the same size. The two types are not to be mixed, and the same number of candies is to be placed into each container, with none left over. What is the greatest number of candies that can be put into each container?

For exercises 41 and 42, determine whether the given statement is true or false.

41. The only even prime number is 2.

42. If a natural number is odd, then it is a prime number.

Solve the applications.

43. A natural number is said to be "perfect" if it is equal to the sum of its natural number divisors, except itself.

   (a) Show that 28 is a perfect number.

   (b) Identify another perfect number less than 28.

44. Find the smallest natural number that is divisible by all of the numbers. 2, 3, 4, 6, 8, 9.

45. Tom and Dick both work the night shift at the steel mill. Tom has every sixth night off, and Dick has every eighth night off. If they both have August 1 off, when will they both be off together again?

46. Mercury, Venus, and Earth revolve around the sun once every 3, 7, and 12 months, respectively. If the three planets are now in a straight line, what is the smallest number of months that must pass before they line up again?

47. Use the *sieve of Eratosthenes* to determine all the prime numbers less than 100.

	2	3	4	5	6	7	8	9	10
11	12	13	14	15	16	17	18	19	20
21	22	23	24	25	26	27	28	29	30
31	32	33	34	35	36	37	38	39	40
41	42	43	44	45	46	47	48	49	50
51	52	53	54	55	56	57	58	59	60
61	62	63	64	65	66	67	68	69	70
71	72	73	74	75	76	77	78	79	80
81	82	83	84	85	86	87	88	89	90
91	92	93	94	95	96	97	98	99	100

48. Prime numbers that differ by 2 are called **twin primes.** Examples are 3 and 5, and 5 and 7. Find one pair of twin primes between 85 and 105.

49. Parts **(a)** to **(c)** refer to "twin primes" (see exercise 48).

   (a) Search for and make a list of several pairs of twin primes in which the primes are greater than 3.
   (b) What do you notice about each number that lies *between* a pair of twin primes?
   (c) Write an explanation for your observation in part **(b).**

50. Obtain (or imagine that you have) a quantity of square tiles. Six tiles can be arranged in the shape of a rectangle in two different ways:

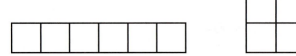

(a) Record the dimensions of the given rectangles.

(b) If you use 7 tiles, how many different rectangles can you form?

(c) If you use 10 tiles, how many different rectangles can you form?

(d) What kind of number (of tiles) permits *only one* arrangement into a rectangle? *More than one* arrangement?

51. The number 10 has 4 factors: 1, 2, 5, and 10. We can say that 10 has an even number of factors. Investigate several numbers to determine which numbers have an *even number* of factors and which numbers have an *odd number* of factors.

52. Suppose that a school has 1,000 lockers and that they are all closed. A person passes through, opening every other locker, beginning with locker #2. Then another person passes through, changing every third locker (closing it if it is open and opening it if it is closed), starting with locker #3. Yet another person passes through, changing every fourth locker, beginning with locker #4. This process continues until 1,000 people pass through.

(a) At the end of this process, which locker numbers are closed?

(b) Write an explanation for your answer to part **(a)**.
    (*Hint:* It may help to attempt exercise 51 first.)

Answers

1. 1, 2, 4   **3.** 1, 2, 3, 4, 6, 8, 12, 24   **5.** 1, 2, 4, 8, 16, 32, 64   **7.** 1, 11

9. 19, 23, 31, 59, 97, 103   **11.** $2 \cdot 3 \cdot 3$ **13.** $2 \cdot 3 \cdot 5$ **15.** $2 \cdot 5 \cdot 7$

17. $2 \cdot 3 \cdot 11$ **19.** $2 \cdot 5 \cdot 13$ **21.** $3 \cdot 3 \cdot 5 \cdot 7$ **23.** 4, 6   **25.** 5, 6

27. 2   **29.** 1   **31.** 6   **33.** 18   **35.** 12   **37.** 25   **39.** 14

41. True   **43.**   **45.** August 25   **47.** 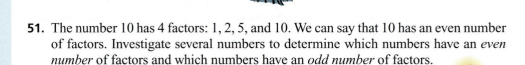   **49.**

51.

3.3 Equivalent Fractions

3.3 OBJECTIVES

1. Determine whether two fractions are equivalent
2. Use the fundamental principle to simplify fractions
3. Use the fundamental principle to build fractions

It is possible to represent the same portion of the whole by different fractions. Look at Figure 1, representing $\frac{3}{6}$ and $\frac{1}{2}$. The two fractions are simply different names for the same number. They are called **equivalent fractions** for this reason.

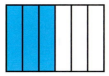

Figure 1

Any fraction has infinitely many equivalent fractions. For instance, $\frac{2}{3}$, $\frac{4}{6}$, and $\frac{6}{9}$ are all equivalent fractions because they name the same part of a unit. This is illustrated in Figure 2.

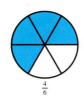

Figure 2

Many more fractions are equivalent to $\frac{2}{3}$. All these fractions can be used interchangeably. An easy way to find out if two fractions are equivalent is to use cross products.

$$\frac{a}{b} = \frac{c}{d}$$ We call $a \cdot d$ and $b \cdot c$ the **cross products.**

Property: Testing for Equivalence

If the cross products for two fractions are equal, the two fractions are equivalent.

OBJECTIVE 1

Example 1 Identifying Equivalent Fractions Using Cross Products

(a) Are $\frac{3}{24}$ and $\frac{4}{32}$ equivalent fractions?

The cross products are $3 \cdot 32 = 96$ and $24 \cdot 4 = 96$. Because the cross products are equal, the fractions are equivalent.

(b) Are $\frac{2}{5}$ and $\frac{3}{7}$ equivalent fractions?

The cross products are $2 \cdot 7$ and $5 \cdot 3$.

$2 \cdot 7 = 14$ and $5 \cdot 3 = 15$

Because $14 \neq 15$, the fractions are *not* equivalent.

NOTE Where would we find $\dfrac{-3}{11}$ on the number line? It is between 0 and −1.

(c) Are $\dfrac{-3}{11}$ and $\dfrac{-6}{22}$ equivalent?

The cross products are

$$(-3) \cdot 22 = -66 \qquad \text{and} \qquad 11 \cdot (-6) = -66$$

Because the cross products are equal, the fractions are equivalent.

 CHECK YOURSELF 1 _____

(a) Are $\dfrac{3}{8}$ and $\dfrac{9}{24}$ equivalent fractions?   **(b)** Are $\dfrac{7}{8}$ and $\dfrac{8}{9}$ equivalent fractions?

In writing equivalent fractions, we use an important principle.

> **Property:** **The Fundamental Principle of Fractions, Part 1**
>
> For the fraction $\dfrac{a}{b}$ and any nonzero number c,
>
> $$\dfrac{a}{b} = \dfrac{a \div c}{b \div c}$$

The Fundamental Principle of Fractions tells us that we can divide the numerator and denominator by the same nonzero number. The result is an equivalent fraction. For instance,

NOTE Divide the numerator and denominator by 2, 3, and 4, respectively.

$$\dfrac{2}{4} = \dfrac{2 \div 2}{4 \div 2} = \dfrac{1}{2} \qquad \dfrac{3}{6} = \dfrac{3 \div 3}{6 \div 3} = \dfrac{1}{2} \qquad \dfrac{4}{8} = \dfrac{4 \div 4}{8 \div 4} = \dfrac{1}{2}$$

NOTE Divide the numerator and denominator by 5, 6, and 7, respectively.

$$\dfrac{5}{10} = \dfrac{5 \div 5}{10 \div 5} = \dfrac{1}{2} \qquad \dfrac{6}{12} = \dfrac{6 \div 6}{12 \div 6} = \dfrac{1}{2} \qquad \dfrac{7}{14} = \dfrac{7 \div 7}{14 \div 7} = \dfrac{1}{2}$$

We will see how this is applied.

Simplifying a fraction or **reducing a fraction to lower terms** means finding an equivalent fraction with a *smaller* numerator and denominator than those of the original fraction. Dividing the numerator and denominator by the same nonzero number will do exactly that.

Consider Example 2.

OBJECTIVE 2 | **Example 2**   **Simplifying Fractions**

Simplify each fraction.

NOTE We apply the fundamental principle to divide the numerator and denominator by 5.

(a) $\dfrac{5}{15} = \dfrac{5 \div 5}{15 \div 5} = \dfrac{1}{3}$

$\dfrac{5}{15}$ and $\dfrac{1}{3}$ are equivalent fractions.   *Check this by finding the cross products.*

NOTE We divide the numerator and denominator by 2.

(b) $\dfrac{4}{8} = \dfrac{4 \div 2}{8 \div 2} = \dfrac{2}{4}$ *$\dfrac{2}{4}$ can be further simplified!*

$\dfrac{4}{8}$ and $\dfrac{2}{4}$ are equivalent fractions.

CHECK YOURSELF 2 _____

Write two fractions that are equivalent to $\frac{30}{45}$.

(a) Divide the numerator and denominator by 5.

(b) Divide the numerator and denominator by 15.

We say that a fraction is in **simplest form,** or in **lowest terms,** if the numerator and denominator have no common factors other than 1. This means that the fraction has the smallest possible numerator and denominator.

In Example 2, $\frac{1}{3}$ is in simplest form because the numerator and denominator have no common factors other than 1. The fraction is in lowest terms.

NOTE In this case, the numerator and denominator are *not* as small as possible. The numerator and denominator have a common factor of 2.

$\frac{2}{4}$ is *not* in simplest form.   *Do you see that* $\frac{2}{4}$ *can also be written as* $\frac{1}{2}$?

To write a fraction in simplest form or to reduce a fraction to lowest terms, divide the numerator and denominator by their GCF.

Example 3   Simplifying Fractions

Write $\frac{10}{15}$ in simplest form.

From our work in this chapter, we know that the GCF of 10 and 15 is 5. To write $\frac{10}{15}$ in simplest form, divide the numerator and denominator by 5.

$$\frac{10}{15} = \frac{10 \div 5}{15 \div 5} = \frac{2}{3}$$

The resulting fraction, $\frac{2}{3}$, is in lowest terms.

CHECK YOURSELF 3 _____

Write $\frac{12}{18}$ *in simplest form by dividing the numerator and denominator by the GCF.*

Many students prefer another method of reducing fractions, which uses the prime factorizations of the numerator and denominator. Example 4 uses this method.

Example 4   Factoring to Simplify a Fraction

(a) Simplify $\frac{24}{42}$.

To simplify $\frac{24}{42}$, factor.

NOTE From the prime factorization of 24 and 42, we divide by the common factors of 2 and 3.

$$\frac{24}{42} = \frac{\overset{1}{\cancel{2}} \cdot 2 \cdot 2 \cdot \overset{1}{\cancel{3}}}{\underset{1}{\cancel{2}} \cdot \underset{1}{\cancel{3}} \cdot 7} = \frac{4}{7}$$ *Each time we divide the numerator and denominator by a common factor, we write 1s in place of the factors.*

Note: The numerator and denominator of the simplified fraction are the *product* of the prime factors remaining after the division by 2 and 3.

(b) Simplify $\dfrac{120}{180}$.

To reduce $\dfrac{120}{180}$ to lowest terms, write the prime factorizations of the numerator and denominator. Then divide by any common factors.

$$\frac{120}{180} = \frac{\overset{1}{\cancel{2}} \cdot \overset{1}{\cancel{2}} \cdot 2 \cdot \overset{1}{\cancel{3}} \cdot \overset{1}{\cancel{5}}}{\underset{1}{\cancel{2}} \cdot \underset{1}{\cancel{2}} \cdot \underset{1}{\cancel{3}} \cdot 3 \cdot \underset{1}{\cancel{5}}} = \frac{2}{3}$$

CHECK YOURSELF 4

Write each of the following fractions in simplest form.

(a) $\dfrac{60}{75}$

(b) $\dfrac{210}{252}$

There is another way to organize your work in simplifying fractions. It again uses the fundamental principle to divide the numerator and denominator by any common factors. We will illustrate with the fractions considered in Example 4.

Example 5   Using Common Factors to Simplify Fractions

(a) $\dfrac{24}{42} = \dfrac{\overset{12}{\cancel{24}}}{\underset{21}{\cancel{42}}} = \dfrac{\overset{4}{\cancel{12}}}{\underset{7}{\cancel{21}}} = \dfrac{4}{7}$

Divide by the common factor of 2.   Divide by the common factor of 3.

The original numerator and denominator are divisible by 2, and so we divide by that factor to arrive at $\dfrac{12}{21}$. A common factor of 3 still exists. Divide again for the result $\dfrac{4}{7}$, which is in lowest terms.

Note: If we had seen the GCF of 6 at first, we could have divided by 6 and arrived at the same result in one step.

(b) $\dfrac{120}{180} = \dfrac{\overset{\overset{2}{\cancel{20}}}{\cancel{120}}}{\underset{\underset{3}{\cancel{30}}}{\cancel{180}}} = \dfrac{2}{3}$

Our first step is to divide by the common factor of 6. We then have $\dfrac{20}{30}$. There is still a common factor of 10, so we again divide.

Again, we could have divided by the GCF of 60 in one step if we had recognized it.

CHECK YOURSELF 5

Using the method of Example 5, write each of the fractions in simplest form.

(a) $\dfrac{60}{75}$

(b) $\dfrac{84}{196}$

Most of the fractions we have looked at up to this point have been positive numbers. Fractions can certainly be negative numbers. The standard form for a negative fraction uses a negative number in the numerator and a positive number in the denominator.

Example 6   Writing Negative Fractions in Standard Form

Rewrite each fraction in standard form.

(a) $\dfrac{2}{-5}$

Recall that a positive number divided by a negative number results in a negative number. Here the fraction is negative. The standard form has the negative in the numerator, so we write $\dfrac{-2}{5}$.

(b) $\dfrac{-3}{-8}$

A negative number divided by a negative number results in a positive number. In standard form, we simply write this fraction as $\dfrac{3}{8}$.

 CHECK YOURSELF 6

Rewrite each fraction in standard form.

(a) $\dfrac{-4}{-5}$ 　　　　　　 **(b)** $\dfrac{2}{-9}$ 　　　　　　 **(c)** $\dfrac{5}{-8}$

The fundamental principle is also used to simplify negative fractions.

Example 7   Simplifying Negative Fractions

Simplify $\dfrac{42}{-56}$.

First, we rewrite the fraction in standard form.

$$\frac{42}{-56} = \frac{-42}{56}$$

Then we may divide by the common factor of 14 in the numerator and denominator.

$$\frac{-42}{56} = \frac{-42 \div 14}{56 \div 14} = \frac{-3}{4}$$

 CHECK YOURSELF 7

Simplify.

(a) $\dfrac{26}{-52}$ 　　　　　　　　　　　 **(b)** $\dfrac{-18}{-64}$

To this point in this section, we have used the Fundamental Principle of Fractions to simplify fractions. It can also be used to find an equivalent fraction with a larger denominator.

> **Property:** The Fundamental Principle of Fractions, Part 2
>
> For the fraction $\dfrac{a}{b}$ and any nonzero number c,
>
> $$\frac{a}{b} = \frac{a \cdot c}{b \cdot c}$$

OBJECTIVE 3 **Example 8** **Finding an Equivalent Fraction**

Find the fraction equivalent to $\dfrac{2}{5}$ with a denominator of 25.

We are looking to find the new numerator where

$$\frac{2}{5} = \frac{?}{25}$$

Note that the original denominator, 5, must be multiplied by 5 to get the new denominator, 25. By the fundamental principle, we must multiply the numerator and denominator by the same number to get an equivalent fraction, so we multiply each by 5.

$$\frac{2}{5} = \frac{2 \cdot 5}{5 \cdot 5} = \frac{10}{25}$$

 CHECK YOURSELF 8 _____

Find the fraction equivalent to $\dfrac{5}{7}$ with a denominator of 42.

We can also use the fundamental principle to find equivalent negative fractions.

Example 9 **Finding Equivalent Negative Fractions**

Find the fraction equivalent to $\dfrac{-3}{7}$ with a denominator of 35.

We are looking to find the new numerator where

$$\frac{-3}{7} = \frac{?}{35}$$

Note that the original denominator, 7, must be multiplied by 5 to get the new denominator, 35. By the fundamental principle, we must multiply the numerator and denominator by the same number to get an equivalent fraction, so we multiply each by 5.

$$\frac{-3}{7} = \frac{-3 \cdot 5}{7 \cdot 5} = \frac{-15}{35}$$

 CHECK YOURSELF 9 _____

Find the fraction equivalent to $\dfrac{-2}{7}$ with a denominator of 21.

Using Your Calculator to Simplify Fractions

If you have a calculator that supports fraction arithmetic, you should learn to use it to check your work. We'll look at two different types of these calculators.

 ## Scientific Calculator

Scientific calculators include the TI-34, the Casio fx-250 or fx-115, and the Sharp 506h or 509h.

Before doing Example 10, find the button on your scientific calculator that is labeled $\boxed{\textbf{a b/c}}$. This is the button that will be used to enter fractions.

Example 10   Using a Scientific Calculator to Simplify Fractions

Simplify the fraction $\dfrac{-24}{68}$.

There are four steps in simplifying fractions using a scientific calculator.

RECALL Pressing the $\boxed{+/-}$ key changes 24 to −24.

Step 1   Enter the numerator, 24 $\boxed{+/-}$.

Step 2   Press the $\boxed{\textbf{a b/c}}$ key.

Step 3   Enter the denominator, 68.

Step 4   Press $\boxed{=}$.

The calculator will display the simplified fraction, $\dfrac{-6}{17}$.

 CHECK YOURSELF 10 _____

Simplify the fraction $\dfrac{-51}{81}$.

 ## Graphing Calculator

To simplify the same fraction, $\dfrac{-24}{68}$, using a graphing calculator, such as the TI-83 or TI-84:

Step 1   Enter the fraction as a division problem: $\boxed{(-)}$ 24 $\boxed{\div}$ 68.

Step 2   Press the $\boxed{\textbf{MATH}}$ key.

Step 3   Select $\boxed{\textbf{1: ▶ Frac}}$.

Step 4   Press $\boxed{\text{Enter}}$.

NOTE Some scientific calculators cannot handle denominators larger than 999.

The calculator displays the simplified fraction, $\dfrac{-6}{17}$, as −6/17.

The graphing calculator is particularly useful for simplifying fractions with large values in the numerator and denominator.

Example 11   Using a Graphing Calculator to Simplify Fractions

Simplify $\dfrac{-546}{637}$.

Using our calculator, we find that

$$\frac{-546}{637} = \frac{-6}{7}$$

CHECK YOURSELF 11 _____

Simplify $\dfrac{-649}{885}$.

READING YOUR TEXT

The following fill-in-the-blank exercises are designed to assure that you understand the key vocabulary used in this section. Each sentence usually comes directly from the section. You will find the correct answers in Appendix C.

Section 3.3

(a) Given the statement $\dfrac{a}{b} = \dfrac{c}{d}$, we call $a \cdot d$ and $b \cdot c$ the _____.

(b) If the cross-products for two fractions are equal, the two fractions are _____.

(c) The _____ Principle of Fractions tells us that we can divide the numerator and denominator by the same nonzero number.

(d) A fraction is in simplest form if the numerator and denominator have no _____ factors other than 1.

CHECK YOURSELF ANSWERS

1. (a) Yes; **(b)** No   **2. (a)** $\dfrac{6}{9}$; **(b)** $\dfrac{2}{3}$

3. 6 is the GCF of 12 and 18, so $\dfrac{12}{18} = \dfrac{12 \div 6}{18 \div 6} = \dfrac{2}{3}$

4. (a) $\dfrac{60}{75} = \dfrac{2 \cdot 2 \cdot \overset{1}{\cancel{3}} \cdot \overset{1}{\cancel{5}}}{\underset{1}{\cancel{3}} \cdot \underset{1}{\cancel{5}} \cdot 5} = \dfrac{4}{5}$; **(b)** $\dfrac{210}{252} = \dfrac{\overset{1}{\cancel{2}} \cdot \overset{1}{\cancel{3}} \cdot 5 \cdot \overset{1}{\cancel{7}}}{\underset{1}{\cancel{2}} \cdot 2 \cdot 3 \cdot \underset{1}{\cancel{3}} \cdot \underset{1}{\cancel{7}}} = \dfrac{5}{6}$

5. (a) Divide by the common factors of 3 and 5, $\dfrac{60}{75} = \dfrac{4}{5}$;

(b) Divide by the common factors of 4 and 7, $\dfrac{84}{196} = \dfrac{3}{7}$

6. (a) $\dfrac{4}{5}$; **(b)** $\dfrac{-2}{9}$; **(c)** $\dfrac{-5}{8}$ **7. (a)** $\dfrac{-1}{2}$; **(b)** $\dfrac{9}{32}$ **8.** $\dfrac{30}{42}$ **9.** $\dfrac{-6}{21}$ **10.** $\dfrac{-17}{27}$

11. $\dfrac{-11}{15}$

3.3 Exercises

Are the pairs of fractions equivalent?

1. $\dfrac{1}{3}, \dfrac{3}{5}$

2. $\dfrac{3}{5}, \dfrac{9}{15}$

3. $\dfrac{1}{7}, \dfrac{4}{28}$

4. $\dfrac{2}{3}, \dfrac{3}{5}$

5. $\dfrac{-5}{6}, \dfrac{-15}{18}$

6. $\dfrac{-3}{4}, \dfrac{-16}{20}$

7. $\dfrac{2}{21}, \dfrac{4}{25}$

8. $\dfrac{20}{24}, \dfrac{5}{6}$

9. $\dfrac{-2}{7}, \dfrac{-3}{11}$

10. $\dfrac{-12}{15}, \dfrac{-36}{45}$

11. $\dfrac{16}{24}, \dfrac{40}{60}$

12. $\dfrac{15}{20}, \dfrac{20}{25}$

Solve the given applications.

13. **Statistics** On a test of 72 questions, Sam answered 54 correctly. On another test Sam answered 66 correct out of 88. Did Sam get the same portion of each test correct?

14. **Statistics** Jeff Bagwell of the Houston Astros had 104 hits in 325 times at bat. Matt Williams of the Arizona Diamondbacks had 88 hits in 275 times at bat. Did they have the same batting average?

Boost your GRADE at ALEKS.com!

ALEKS®

- Practice Problems
- Self-Tests
- NetTutor
- e-Professors
- Videos

Name _____

Section _____ Date _____

ANSWERS

1. _____
2. _____
3. _____
4. _____
5. _____
6. _____
7. _____
8. _____
9. _____
10. _____
11. _____
12. _____
13. _____
14. _____

15. _____

16. _____

17. _____

18. _____

19. _____

20. _____

21. _____

22. _____

23. _____

24. _____

25. _____

26. _____

27. _____

28. _____

29. _____

30. _____

31. _____

32. _____

33. _____

34. _____

35. _____

36. _____

Write each fraction in simplest form.

15. $\dfrac{8}{12}$ **ALEKS**

16. $\dfrac{12}{15}$ **ALEKS**

17. $\dfrac{10}{14}$ **ALEKS**

18. $\dfrac{15}{50}$ **ALEKS**

19. $\dfrac{-12}{18}$ *VIDEO*

20. $\dfrac{-28}{35}$

21. $\dfrac{35}{40}$ **ALEKS**

22. $\dfrac{21}{24}$ **ALEKS**

23. $\dfrac{-11}{-44}$

24. $\dfrac{-10}{-25}$

25. $\dfrac{12}{36}$ *VIDEO* **ALEKS**

26. $\dfrac{18}{48}$ **ALEKS**

27. $\dfrac{24}{-27}$

28. $\dfrac{30}{-50}$

29. $\dfrac{32}{40}$ **ALEKS**

30. $\dfrac{17}{51}$ **ALEKS**

31. $\dfrac{-75}{105}$

32. $\dfrac{-62}{93}$

33. $\dfrac{48}{60}$ **ALEKS**

34. $\dfrac{48}{66}$ **ALEKS**

35. $\dfrac{-105}{-135}$ *VIDEO*

36. $\dfrac{-54}{-126}$

37. $\dfrac{66}{110}$ ALEKS

38. $\dfrac{280}{320}$ ALEKS

39. $\dfrac{16}{-21}$

40. $\dfrac{21}{-32}$

41. $\dfrac{31}{52}$ ALEKS

42. $\dfrac{42}{55}$ ALEKS

Write each fraction as an equivalent fraction using the given denominator.

43. $\dfrac{1}{3}$; 9 ALEKS

44. $\dfrac{1}{5}$; 20 ALEKS

45. $\dfrac{1}{8}$; 128 ALEKS

46. $\dfrac{1}{10}$; 200 ALEKS

47. $\dfrac{-1}{4}$; 16

48. $\dfrac{-1}{6}$; 36

49. $\dfrac{-1}{7}$; 35

50. $\dfrac{-1}{9}$; 72

51. $\dfrac{2}{3}$; 27 ALEKS

52. $\dfrac{3}{5}$; 45 ALEKS

53. $\dfrac{4}{7}$; 28 ALEKS

54. $\dfrac{3}{20}$; 100 ALEKS

55. $\dfrac{-4}{5}$; 35 VIDEO

56. $\dfrac{-5}{9}$; 99

57. $\dfrac{-7}{24}$; 96

58. $\dfrac{-6}{17}$; 68

59. $\dfrac{27}{31}$; 93 ALEKS

60. $\dfrac{-17}{50}$; 200

ANSWERS

37. _____

38. _____

39. _____

40. _____

41. _____

42. _____

43. _____

44. _____

45. _____

46. _____

47. _____

48. _____

49. _____

50. _____

51. _____

52. _____

53. _____

54. _____

55. _____

56. _____

57. _____ 58. _____

59. _____ 60. _____

ANSWERS

61. _____

62. _____

63. _____

64. _____

65. _____

66. _____

67. _____

68. _____

69. _____

For exercises 61 to 64, use the following survey data concerning the blood types of 500 individuals.

Blood Type	A	B	AB	O
Number of Individuals	210	40	25	225

61. Science and Medicine Write, in simplest form, the fraction of individuals with type A blood.

62. Science and Medicine Write, in simplest form, the fraction of individuals with type B blood.

63. Science and Medicine Write, in simplest form, the fraction of individuals with type AB blood.

64. Science and Medicine Write, in simplest form, the fraction of individuals with type O blood.

Solve the applications.

65. Business and Finance A quarter is what fractional part of a dollar? Simplify your result.

66. Business and Finance A dime is what fractional part of a dollar? Simplify your result.

67. Science and Medicine What fractional part of an hour is 15 min? Simplify your result. VIDEO

68. Science and Medicine What fractional part of a day is 6 h? Simplify your result.

69. Science and Medicine A meter is equal to 100 cm. What fractional part of a meter is 70 cm? Simplify your result.

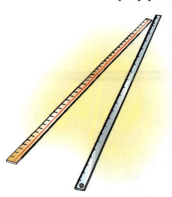

70. **Science and Medicine**   A kilometer is equal to 1,000 m. What fractional part of a kilometer is 300 m? Simplify your result.

71. **Business and Finance**   Susan did a tune-up on her automobile. She found that two of her eight spark plugs were fouled. What fraction represents the number of fouled plugs? Reduce to lowest terms.

72. **Statistics**   Samantha answered 18 of 20 problems correctly on a test. What fractional part did she answer correctly? Reduce your answer to lowest terms.

73. **Statistics**   The local baseball team won 36 of the 58 games they played. What fractional part did they win? Reduce your answer to lowest terms.

74. **Business and Finance**   Sharon earned $250 at her after-school job. A new bike costs $120. What fractional part of her money will remain after she purchases the bike? Reduce your answer to lowest terms.

For exercises 75 and 76, indicate whether the given statement is always true, sometimes true, or never true.

75. If the numerator of a fraction is a prime number, then the fraction is in simplified form.

76. If you add a nonzero number to both the numerator and denominator of a fraction, you will obtain an equivalent fraction.

77. A student is attempting to reduce the fraction $\dfrac{8}{12}$ to lowest terms. He produces the following argument:

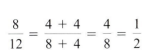

$$\frac{8}{12} = \frac{4+4}{8+4} = \frac{4}{8} = \frac{1}{2}$$

What is the fallacy in this argument? What is the correct answer?

78. _____

79. _____

80. _____

81. _____

82. _____

83. _____

84. _____

78. Can any of the following fractions be simplified?

(a) $\dfrac{824}{73}$ (b) $\dfrac{59}{11}$ (c) $\dfrac{135}{17}$

What characteristic do you notice about the denominator of each fraction? What rule would you make up based on your observations?

79. Consider Figure (a).

(a) Give the fraction that represents the shaded region.

(a)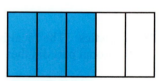

(b) Draw a horizontal line through the figure, as shown in Figure (b). Now give the fraction representing the shaded region.

(b)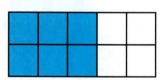

80. Repeat exercise 79 using these figures.

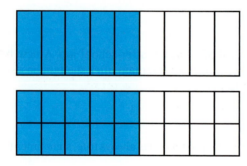

81. **Science and Medicine** Ernesto got advice from a personal trainer regarding his diet. The trainer recommended that Ernesto should consume approximately 2,000 calories per day, and of these calories, 1,200 should be from carbohydrates. What fraction of his calories should come from carbohydrates?

82. **Science and Medicine** Lisa's trainer advised her to consume approximately 1,800 calories per day and that 360 calories should be from protein sources. What fraction of her daily caloric intake does this represent?

 Calculator Exercises

Use your calculator to simplify the fractions.

83. $\dfrac{-28}{40}$ 84. $\dfrac{-121}{132}$

85. $\dfrac{96}{-144}$

86. $\dfrac{385}{-605}$

87. $\dfrac{-445}{-623}$

88. $\dfrac{-153}{-255}$

89. $\dfrac{-299}{391}$

90. $\dfrac{-152}{209}$

91. $\dfrac{-289}{459}$

ANSWERS

85. _____

86. _____

87. _____

88. _____

89. _____

90. _____

91. _____

Answers

1. $1 \cdot 5 = 5$; $3 \cdot 3 = 9$. The fractions are not equivalent.   **3.** Yes
5. Yes   **7.** No   **9.** No
11. $16 \cdot 60 = 960$, and $24 \cdot 40 = 960$. The fractions are equivalent.   **13.** Yes
15. $\dfrac{2}{3}$ **17.** $\dfrac{5}{7}$ **19.** $\dfrac{-2}{3}$ **21.** $\dfrac{7}{8}$ **23.** $\dfrac{1}{4}$ **25.** $\dfrac{1}{3}$ **27.** $\dfrac{-8}{9}$
29. $\dfrac{4}{5}$ **31.** $\dfrac{-5}{7}$ **33.** $\dfrac{4}{5}$ **35.** $\dfrac{7}{9}$ **37.** $\dfrac{3}{5}$ **39.** $\dfrac{-16}{21}$ **41.** $\dfrac{31}{52}$
43. $\dfrac{3}{9}$ **45.** $\dfrac{16}{128}$ **47.** $\dfrac{-4}{16}$ **49.** $\dfrac{-5}{35}$ **51.** $\dfrac{18}{27}$ **53.** $\dfrac{16}{28}$ **55.** $\dfrac{-28}{35}$
57. $\dfrac{-28}{96}$ **59.** $\dfrac{81}{93}$ **61.** $\dfrac{21}{50}$ **63.** $\dfrac{1}{20}$ **65.** $\dfrac{1}{4}$ **67.** $\dfrac{1}{4}$ **69.** $\dfrac{7}{10}$
71. $\dfrac{1}{4}$ **73.** $\dfrac{18}{29}$ **75.** Sometimes   **77.** 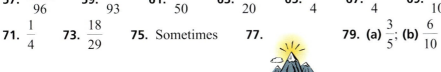   **79.** (a) $\dfrac{3}{5}$; (b) $\dfrac{6}{10}$
81. $\dfrac{3}{5}$ **83.** $\dfrac{-7}{10}$ **85.** $\dfrac{-2}{3}$ **87.** $\dfrac{5}{7}$ **89.** $\dfrac{-13}{17}$ **91.** $\dfrac{-17}{27}$

3.4 Multiplication and Division of Fractions

3.4 OBJECTIVES

1. Multiply two fractions
2. Determine the reciprocal of a number
3. Divide two fractions

Multiplication is the easiest of the four operations with fractions. We can illustrate multiplication by picturing fractions as parts of a whole or unit. Using this idea, we show the fractions $\frac{4}{5}$ and $\frac{2}{3}$.

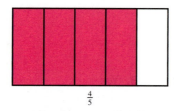

NOTE A fraction followed by the word *of* means that we want to multiply by that fraction.

Suppose now that we wish to find $\frac{2}{3}$ *of* $\frac{4}{5}$. We can combine the diagrams.

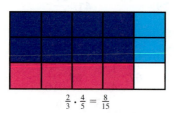

The part of the whole representing the product $\frac{2}{3} \cdot \frac{4}{5}$ is the purple region in the figure.

The unit has been divided into 15 parts and 8 of those parts are used, so $\frac{2}{3} \cdot \frac{4}{5}$ must be $\frac{8}{15}$.

The following rule is suggested by the diagrams.

Step by Step: To Multiply Fractions

Step 1	Multiply the numerators to find the numerator of the product.
Step 2	Multiply the denominators to find the denominator of the product.
Step 3	Simplify the resulting fraction if possible.

Example 1 will require using steps 1 and 2.

OBJECTIVE 1

Example 1 Multiplying Two Fractions

Multiply.

NOTE We multiply fractions in this way *not* because it is easy, but because it works!

(a) $\dfrac{1}{4} \cdot \dfrac{1}{7} = \dfrac{1 \cdot 1}{4 \cdot 7} = \dfrac{1}{28}$

(b) $\dfrac{2}{3} \cdot \dfrac{4}{5} = \dfrac{2 \cdot 4}{3 \cdot 5} = \dfrac{8}{15}$

(c) $\dfrac{5}{8} \cdot \dfrac{7}{9} = \dfrac{5 \cdot 7}{8 \cdot 9} = \dfrac{35}{72}$

 CHECK YOURSELF 1 _____

Multiply.

(a) $\dfrac{7}{8} \cdot \dfrac{3}{10}$ **(b)** $\dfrac{5}{7} \cdot \dfrac{3}{4}$

Step 3 indicates that the product of fractions should always be simplified to lowest terms. Consider Example 2.

> **Example 2   Multiplying Two Fractions**

Multiply and write the result in lowest terms.

(a) $\dfrac{3}{4} \cdot \dfrac{2}{9} = \dfrac{3 \cdot 2}{4 \cdot 9}$

$= \dfrac{6}{36}$ Noting that $\dfrac{6}{36}$ is not in simplest form,

$= \dfrac{1}{6}$ we divide numerator and denominator by 6 to write the product in lowest terms.

(b) $\dfrac{4}{7} \cdot \dfrac{3}{8} = \dfrac{4 \cdot 3}{7 \cdot 8}$

$= \dfrac{12}{56}$

$= \dfrac{3}{14}$

 CHECK YOURSELF 2 _____

Multiply and write the result in lowest terms.

$\dfrac{5}{7} \cdot \dfrac{3}{10}$

To find the product of a fraction and a whole number, write the whole number as a fraction (the whole number divided by 1) and apply the multiplication rule as before. Example 3 illustrates this approach.

Example 3   Multiplying a Whole Number and a Fraction

Do the indicated multiplication and simplify.

Remember that $5 = \dfrac{5}{1}$.

(a) $5 \cdot \dfrac{3}{4} = \dfrac{5}{1} \cdot \dfrac{3}{4} = \dfrac{5 \cdot 3}{1 \cdot 4}$

$\qquad = \dfrac{15}{4}$

(b) $\dfrac{5}{12} \cdot 6 = \dfrac{5}{12} \cdot \dfrac{6}{1}$

$\qquad = \dfrac{5 \cdot 6}{12 \cdot 1}$

NOTE Leave the product as an improper fraction unless instructed to write as a mixed number.

$\qquad = \dfrac{30}{12} = \dfrac{5}{2}$

(c) $\dfrac{21}{2} \cdot \dfrac{4}{7} = \dfrac{84}{14}$

$\qquad = 6$

🖊 **CHECK YOURSELF 3**

Multiply.

(a) $\dfrac{3}{16} \cdot 8$ 　　　　　　　　(b) $4 \cdot \dfrac{5}{7}$

When multiplying fractions, it is usually easier to simplify, that is, remove any common factors in the numerator and denominator, *before multiplying*. Remember that to simplify means to *divide* by the same common factor.

Example 4   Simplifying before Multiplying Two Fractions

Simplify and then multiply.

NOTE Once again we are applying the fundamental principle to divide the numerator and denominator by 3.

NOTE Because we divide by any common factors before we multiply, the resulting product *is in simplest form.*

$$\dfrac{3}{5} \cdot \dfrac{4}{9} = \dfrac{\overset{1}{\cancel{3}} \cdot 4}{5 \cdot \underset{3}{\cancel{9}}}$$

To simplify, we divide the *numerator* and *denominator* by the common factor 3. Remember that $\overset{1}{\cancel{3}}$ means $3 \div 3 = 1$, and $\underset{3}{\cancel{9}}$ means $9 \div 3 = 3$.

$$= \dfrac{1 \cdot 4}{5 \cdot 3}$$

$$= \dfrac{4}{15}$$

CHECK YOURSELF 4

Simplify and then multiply.

$$\frac{7}{8} \cdot \frac{5}{21}$$

Our work in Example 4 leads to the following general rule about simplifying fractions in multiplication.

> **Property:** Simplifying Fractions before Multiplying
>
> In multiplying two or more fractions, we can divide any factor of the numerator and any factor of the denominator by the same nonzero number to simplify the product.

The same rule can be applied when we have more than two fractions, as in Example 5.

Example 5   Multiplying More than Two Fractions

Find the product.

$$\frac{2}{5} \cdot \frac{5}{4} \cdot \frac{8}{9} = \frac{2 \cdot \overset{1}{\cancel{5}} \cdot \overset{2}{\cancel{8}}}{\underset{1}{\cancel{5}} \cdot \underset{1}{\cancel{4}} \cdot 9} = \frac{4}{9}$$

 CHECK YOURSELF 5

Find the product.

$$\frac{3}{5} \cdot \frac{10}{13} \cdot \frac{26}{27}$$

If there are negative fractions involved in the product, it is best to first determine the sign of the product, then simplify.

Example 6   Multiplying More than Two Fractions

Find the product.

$$\frac{-3}{5} \cdot \frac{5}{7} \cdot \frac{-2}{9}$$

Note that the numerator will be positive (the product of two negatives). Because we will have a positive divided by a positive, our answer will be positive. We can now continue.

$$\frac{-3}{5} \cdot \frac{5}{7} \cdot \frac{-2}{9} = \frac{\overset{1}{\cancel{3}} \cdot \overset{1}{\cancel{5}} \cdot 2}{\cancel{5} \cdot 7 \cdot \cancel{9}} = \frac{2}{21}$$

 CHECK YOURSELF 6

Find the product.

$$\frac{-3}{7} \cdot \frac{-12}{5} \cdot \frac{-2}{9}$$

We are now ready to look at the operation of division on fractions. Before we do so, we will need a new concept, the *reciprocal* of a fraction.

NOTE In general, the reciprocal of the fraction $\frac{a}{b}$ is $\frac{b}{a}$. Neither *a* nor *b* can be 0.

> **Definition:** The Reciprocal of a Fraction
>
> We invert, or interchange, the numerator and denominator of a fraction to write its **reciprocal.**

OBJECTIVE 2   **Example 7   Finding the Reciprocal of a Fraction**

Find the reciprocal of (a) $\frac{3}{4}$, (b) 5, and (c) $\frac{-2}{7}$.

(a) The reciprocal of $\frac{3}{4}$ is $\frac{4}{3}$.   Just invert, or turn over, the fraction.

(b) The reciprocal of 5, or $\frac{5}{1}$, is $\frac{1}{5}$.   Write 5 as $\frac{5}{1}$ and then turn over the fraction.

(c) The reciprocal of $\frac{-2}{7}$ is $\frac{7}{-2}$, which we write in standard form as $\frac{-7}{2}$.

 CHECK YOURSELF 7

Find the reciprocal of (a) $\frac{5}{8}$, (b) 3, and (c) $\frac{-1}{6}$.

An important property relating a number and its reciprocal is given here.

NOTE In symbols, this says:
$$\frac{a}{b} \cdot \frac{b}{a} = 1$$
Neither *a* nor *b* can be 0.

> **Property:** Reciprocal Products
>
> The product of any number and its reciprocal is 1. (Every number except zero has a reciprocal.)

We are now ready to use the reciprocal to find a rule for dividing fractions. Recall that we can represent the operation of division in several ways. We used the symbol ÷ earlier. Remember that a fraction also indicates division. For instance,

RECALL $3 \div 5$ and $\frac{3}{5}$ both mean "3 divided by 5."

$$3 \div 5 = \frac{3}{5}$$

In this statement, 5 is called the *divisor*. It follows the division sign ÷ and is written *below* the fraction bar.

Using this information, we can write a statement involving fractions and division as a **complex fraction,** which may contain fractions in the numerator and denominator. Example 8 illustrates this. We will study more about complex fractions in Section 4.6.

Example 8   Writing a Quotient as a Complex Fraction

Write $\frac{2}{5} \div \frac{3}{4}$ as a complex fraction.

The numerator is $\frac{2}{5}$.

$$\frac{\frac{2}{5}}{\frac{3}{4}}$$

A *complex fraction* is written by placing the dividend in the numerator and the divisor in the denominator.

The denominator is $\frac{3}{4}$.

 CHECK YOURSELF 8 _____

Write $\frac{2}{7} \div \frac{4}{5}$ as a complex fraction.

We will continue with the same division problem.

OBJECTIVE 3   **Example 9   Dividing Two Fractions**

(1) $\dfrac{2}{5} \div \dfrac{3}{4} = \dfrac{\dfrac{2}{5}}{\dfrac{3}{4}}$ Write the original quotient as a complex fraction.

$= \dfrac{\dfrac{2}{5} \cdot \dfrac{4}{3}}{\dfrac{3}{4} \cdot \dfrac{4}{3}}$ Multiply the numerator and denominator by $\frac{4}{3}$, the reciprocal of the denominator. This does *not* change the value of the fraction.

$= \dfrac{\dfrac{2}{5} \cdot \dfrac{4}{3}}{1}$ The denominator becomes 1.

(2) $= \dfrac{2}{5} \cdot \dfrac{4}{3}$ Recall that a number divided by 1 is just that number.

We see from lines (1) and (2) that

$$\frac{2}{5} \div \frac{3}{4} = \frac{2}{5} \cdot \frac{4}{3}$$

We would certainly like to be able to divide fractions easily without all the work of Example 9. Look carefully at the example. A rule to divide fractions is suggested.

Property: To Divide Fractions

To divide one fraction by another, multiply the dividend by the reciprocal of the divisor.

 CHECK YOURSELF 9

Write $\frac{2}{7} \div \frac{4}{5}$ as a multiplication problem.

Example 10 applies the rule for dividing fractions.

Example 10   Dividing Two Fractions

Divide.

RECALL The number inverted is the divisor. It *follows* the division sign.

$$\frac{1}{3} \div \frac{4}{7} = \frac{1}{3} \cdot \frac{7}{4}$$ We invert the divisor, $\frac{4}{7}$, then multiply.

$$= \frac{1 \cdot 7}{3 \cdot 4} = \frac{7}{12}$$

 CHECK YOURSELF 10

Divide.

$$\frac{2}{7} \div \frac{4}{5}$$

Now we will look at an example that contains a negative fraction.

Example 11   Dividing Two Fractions

Divide.

NOTE If there is only one negative in a multiplication or division, the result will be negative.

$$\frac{-5}{8} \div \frac{3}{5} = \frac{-5}{8} \cdot \frac{5}{3} = \frac{-5 \cdot 5}{8 \cdot 3} = \frac{-25}{24}$$

CHECK YOURSELF 11 _____

Divide.

$$\frac{-5}{6} \div \frac{3}{7}$$

Simplifying will also be useful in dividing fractions. Consider Example 12.

Example 12   Dividing Two Fractions

CAUTION

We must invert the divisor *before any simplification.*

Divide.

$$\frac{3}{5} \div \frac{-6}{7} = \frac{3}{5} \cdot \frac{-7}{6}$$ Invert the divisor *first!* Then you can divide by the common factor of 3.

$$= \frac{(\cancel{3}^{1})(-7)}{(5)(\cancel{6}_{2})} = \frac{-7}{10}$$

CHECK YOURSELF 12 _____

Divide.

$$\frac{4}{9} \div \frac{-8}{15}$$

Using Your Calculator to Multiply and Divide Fractions

Scientific Calculator

To multiply fractions on a scientific calculator, you enter the first fraction, using the $\boxed{\textbf{a b/c}}$ key, and then press the multiplication sign. Next enter the second fraction and then press the equals sign.

Example 13   Multiplying Two Fractions

Find the product.

$$\frac{-7}{15} \cdot \frac{-5}{21}$$

The keystroke sequence is

7 $\boxed{+/-}$ $\boxed{\textbf{a b/c}}$ 15 $\boxed{\times}$ 5 $\boxed{+/-}$ $\boxed{\textbf{a b/c}}$ 21 $\boxed{=}$

The result is $\frac{1}{9}$.

CHECK YOURSELF 13 _____

Find the product.

$$\frac{-24}{33} \cdot \frac{22}{39}$$

Graphing Calculator

When using a graphing calculator, you must choose the fraction option $\boxed{1:\blacktriangleright \textbf{Frac}}$ from the $\boxed{\textbf{MATH}}$ menu before pressing $\boxed{\text{Enter}}$.

For the fraction problem in Example 13, $\dfrac{-7}{15} \cdot \dfrac{-5}{21}$, the keystroke sequence is

$\boxed{(-)}$ 7 $\boxed{\div}$ 15 $\boxed{\times}$ $\boxed{(-)}$ 5 $\boxed{\div}$ 21 $\boxed{\textbf{MATH}}$ $\boxed{1:\blacktriangleright \textbf{Frac}}$ $\boxed{\text{Enter}}$

Again, the result will be $\dfrac{1}{9}$.

Scientific Calculator

Dividing fractions on a scientific calculator requires only that you enter the problem followed by the equal sign.

Example 14   Dividing Two Fractions

Find the quotient.

$$\frac{-24}{23} \div \frac{-16}{13}$$

The keystroke sequence is

24 $\boxed{+/-}$ $\boxed{\text{a b/c}}$ 23 $\boxed{\div}$ 16 $\boxed{+/-}$ $\boxed{\text{a b/c}}$ 13 $\boxed{=}$

The result is $\dfrac{39}{46}$, a positive number (we divided two negative numbers).

CHECK YOURSELF 14 _____

Find the quotient.

$$\frac{-17}{12} \div \frac{-16}{3}$$

Graphing Calculator

When using a graphing calculator, you must choose the fraction option $\boxed{1:\blacktriangleright \textbf{Frac}}$ from the $\boxed{\textbf{MATH}}$ menu before pressing $\boxed{\text{Enter}}$.

The keystroke sequence for the fraction problem in Example 14, $\dfrac{-24}{23} \div \dfrac{-16}{13}$, is

NOTE The divison fraction is inside the parentheses.

$\boxed{(-)}$ 24 $\boxed{\div}$ 23 $\boxed{\div}$ $\boxed{(}$ $\boxed{(-)}$ 16 $\boxed{\div}$ 13 $\boxed{)}$ $\boxed{\textbf{MATH}}$ $\boxed{1:\blacktriangleright \textbf{Frac}}$ $\boxed{\text{Enter}}$

READING YOUR TEXT

The following fill-in-the-blank exercises are designed to assure that you understand the key vocabulary used in this section. Each sentence usually comes directly from the section. You will find the correct answers in Appendix C.

Section 3.4

(a) A fraction followed by the word *of* means that we want to _____ by that fraction.

(b) We interchange the numerator and denominator of a fraction to write its _____.

(c) The _____ of any nonzero number and its reciprocal is 1.

(d) To divide one fraction by another, multiply the dividend by the _____ of the divisor.

CHECK YOURSELF ANSWERS

1. (a) $\dfrac{7}{8} \cdot \dfrac{3}{10} = \dfrac{7 \cdot 3}{8 \cdot 10} = \dfrac{21}{80}$; (b) $\dfrac{5}{7} \cdot \dfrac{3}{4} = \dfrac{5 \cdot 3}{7 \cdot 4} = \dfrac{15}{28}$

2. $\dfrac{5}{7} \cdot \dfrac{3}{10} = \dfrac{5 \cdot 3}{7 \cdot 10} = \dfrac{15}{70} = \dfrac{3}{14}$ **3.** (a) $\dfrac{3}{2}$; (b) $\dfrac{20}{7}$

4. $\dfrac{7}{8} \cdot \dfrac{5}{21} = \dfrac{\overset{1}{7} \cdot 5}{8 \cdot \underset{3}{21}} = \dfrac{5}{24}$ **5.** $\dfrac{4}{9}$ **6.** $\dfrac{-8}{35}$ **7.** (a) $\dfrac{8}{5}$; (b) $\dfrac{1}{3}$; (c) -6

8. $\dfrac{\frac{2}{7}}{\frac{4}{5}}$ **9.** $\dfrac{2}{7} \cdot \dfrac{5}{4}$ **10.** $\dfrac{5}{14}$ **11.** $\dfrac{-35}{18}$

12. $\dfrac{4}{9} \div \dfrac{-8}{15} = \dfrac{4}{9} \cdot \dfrac{-15}{8} = \dfrac{-5}{6}$ **13.** $\dfrac{-16}{39}$ **14.** $\dfrac{17}{64}$

Name _____

Section _____ Date _____

ANSWERS

1. _____
2. _____
3. _____
4. _____
5. _____
6. _____
7. _____
8. _____

9. _____	10. _____
11. _____	12. _____
13. _____	14. _____
15. _____	16. _____
17. _____	18. _____
19. _____	20. _____
21. _____	22. _____
23. _____	24. _____
25. _____	26. _____
27. _____	28. _____
29. _____	30. _____

Multiply. Be sure to write each answer in simplest form.

1. $\dfrac{3}{4} \cdot \dfrac{5}{11}$ ALEKS

2. $\dfrac{2}{7} \cdot \dfrac{5}{9}$ ALEKS

3. $\dfrac{3}{4} \cdot \dfrac{7}{11}$ ALEKS

4. $\dfrac{2}{5} \cdot \dfrac{3}{7}$ ALEKS

5. $\dfrac{-3}{5} \cdot \dfrac{5}{7}$

6. $\dfrac{-6}{11} \cdot \dfrac{8}{6}$

7. $\dfrac{6}{13} \cdot \dfrac{4}{9}$ ALEKS

8. $\dfrac{5}{9} \cdot \dfrac{6}{11}$ ALEKS

9. $\dfrac{-3}{11} \cdot \dfrac{-7}{9}$

10. $\dfrac{-7}{9} \cdot \dfrac{-3}{5}$

11. $\dfrac{3}{10} \cdot \dfrac{5}{9}$ ALEKS

12. $\dfrac{5}{21} \cdot \dfrac{14}{25}$ ALEKS

13. $\dfrac{7}{9} \cdot \dfrac{6}{5}$ ALEKS

14. $\dfrac{8}{13} \cdot \dfrac{26}{5}$ ALEKS

15. $\dfrac{-3}{4} \cdot \dfrac{6}{7}$

16. $\dfrac{-3}{7} \cdot 14$ ALEKS

17. $9 \cdot \dfrac{5}{6}$ ALEKS

18. $15 \cdot \dfrac{5}{6}$ ALEKS

19. $\dfrac{-12}{25} \cdot \dfrac{-11}{18}$

20. $\dfrac{-10}{12} \cdot \dfrac{-16}{25}$

21. Find $\dfrac{2}{3}$ of $\dfrac{3}{7}$

22. What is $\dfrac{5}{6}$ of $\dfrac{9}{10}$?

Find the product.

23. $\dfrac{2}{3} \cdot \dfrac{3}{5} \cdot \dfrac{5}{7}$

24. $\dfrac{1}{7} \cdot \dfrac{7}{12} \cdot \dfrac{12}{13}$

25. $\dfrac{2}{5} \cdot \dfrac{10}{11} \cdot \dfrac{22}{25}$

26. $\dfrac{3}{8} \cdot \dfrac{1}{3} \cdot \dfrac{16}{17}$

27. $\dfrac{-1}{3} \cdot \dfrac{3}{8} \cdot \dfrac{12}{17}$ ALEKS

28. $\dfrac{4}{7} \cdot \dfrac{-3}{8} \cdot \dfrac{14}{15}$ ALEKS

29. $\dfrac{-1}{5} \cdot \dfrac{7}{9} \cdot \dfrac{-15}{14}$ ALEKS

30. $\dfrac{-2}{7} \cdot \dfrac{-5}{4} \cdot \dfrac{-14}{15}$ ALEKS

Find the quotient. Write each result in simplest form.

31. $\dfrac{1}{5} \div \dfrac{3}{4}$ **ALEKS**

32. $\dfrac{2}{5} \div \dfrac{1}{3}$ **ALEKS**

33. $\dfrac{-2}{5} \div \dfrac{3}{4}$

34. $\dfrac{-5}{8} \div \dfrac{3}{4}$

35. $\dfrac{8}{9} \div \dfrac{4}{3}$ **VIDEO** **ALEKS**

36. $\dfrac{5}{9} \div \dfrac{8}{11}$ **ALEKS**

37. $\dfrac{-7}{10} \div \dfrac{-5}{9}$

38. $\dfrac{-8}{9} \div \dfrac{-11}{15}$

39. $\dfrac{8}{15} \div \dfrac{2}{5}$ **ALEKS**

40. $\dfrac{5}{27} \div \dfrac{15}{54}$ **ALEKS**

41. $\dfrac{5}{27} \div \dfrac{25}{36}$ **ALEKS**

42. $\dfrac{9}{28} \div \dfrac{27}{35}$ **ALEKS**

43. $\dfrac{-4}{5} \div 4$

44. $-27 \div \dfrac{3}{7}$

45. $12 \div \dfrac{2}{3}$

46. $\dfrac{5}{8} \div 5$

47. $\dfrac{-12}{17} \div -6$ **VIDEO**

48. $\dfrac{-3}{4} \div -9$

49. **Construction** The dimensions of a rectangular parcel of land are $\dfrac{5}{8}$ of a mile by $\dfrac{3}{10}$ of a mile. Find the area of the parcel. (Recall that the area of a rectangle is found by multiplying the length by the width.)

50. **Construction** The dimensions of a rectangular parcel of land are $\dfrac{7}{16}$ of a mile by $\dfrac{4}{5}$ of a mile. Find the area of the parcel. (Recall that the area of a rectangle is found by multiplying the length by the width.)

51. **Science and Medicine** The recommended daily allowance (RDA) for carbohydrates is 300 grams. If a PowerBar has 45 grams of carbohydrates, first find the fraction of the RDA for carbohydrates in the bar and then find the fraction of the RDA if three bars are consumed.

52. **Science and Medicine** The recommended daily allowance (RDA) for sodium is 22 grams. If a can of CherryCoke has 42 grams of sodium, first find the fraction of the RDA for sodium in a can and then find the fraction of the RDA if two cans are consumed.

ANSWERS

31. _____

32. _____

33. _____

34. _____

35. _____

36. _____

37. _____

38. _____

39. _____

40. _____

41. _____

42. _____

43. _____

44. _____

45. _____

46. _____

47. _____

48. _____

49. _____

50. _____

51. _____

52. _____

53. _____

54. _____

55. _____

56. _____

57. _____

58. _____

59. _____

60. _____

61. _____

62. _____

63. _____

64. _____

65. _____

66. _____

67. _____

68. _____

53. Science and Medicine To determine the upper limit for a person's heart rate during aerobic training, subtract the person's age from 220 and then multiply by $\frac{9}{10}$. Find this upper limit for a 30-year-old.

54. Science and Medicine Using the information in exercise 53, find the upper limit heart rate for a 40-year-old.

55. Science and Medicine To determine the lower limit for a person's heart rate during aerobic training, subtract the person's age from 220 and then multiply by $\frac{3}{5}$. Find this lower limit for a 30-year-old.

56. Science and Medicine Using the information in exercise 55, find the lower limit heart rate for a 40-year-old.

In exercises 57 to 60, indicate whether the given statement is always true, sometimes true, or never true.

57. The product of two proper fractions is itself a proper fraction.

58. The quotient of two proper fractions is itself a proper fraction.

59. The product of two proper fractions is smaller than either of the two original fractions.

60. The reciprocal of a proper fraction is itself a proper fraction.

Calculator Exercises

Find the products using your calculator.

61. $\dfrac{-15}{20} \cdot \dfrac{-8}{12}$

62. $\dfrac{-7}{8} \cdot \dfrac{-4}{21}$

63. $\dfrac{-18}{84} \cdot \dfrac{36}{27}$

64. $\dfrac{-6}{35} \cdot \dfrac{20}{12}$

65. $\dfrac{7}{12} \cdot \dfrac{-36}{63}$

66. $\dfrac{8}{27} \cdot \dfrac{-45}{64}$

67. $\dfrac{-27}{72} \cdot \dfrac{-24}{45}$

68. $\dfrac{-81}{136} \cdot \dfrac{-84}{135}$

Find the quotients using your calculator.

69. $\dfrac{1}{5} \div \dfrac{2}{15}$

70. $\dfrac{13}{17} \div \dfrac{39}{34}$

71. $\dfrac{20}{27} \div \dfrac{-35}{36}$

72. $\dfrac{13}{15} \div \dfrac{-39}{5}$

73. $\dfrac{15}{18} \div \dfrac{45}{27}$

74. $\dfrac{2}{3} \div \dfrac{10}{9}$

75. $\dfrac{-25}{45} \div \dfrac{-100}{135}$

76. $\dfrac{-19}{63} \div \dfrac{-38}{9}$

Answers

1. $\dfrac{15}{44}$ **3.** $\dfrac{21}{44}$ **5.** $\dfrac{-3}{7}$ **7.** $\dfrac{8}{39}$ **9.** $\dfrac{7}{33}$ **11.** $\dfrac{1}{6}$ **13.** $\dfrac{14}{15}$

15. $\dfrac{-9}{14}$ **17.** $\dfrac{15}{2}$ **19.** $\dfrac{22}{75}$ **21.** $\dfrac{2}{7}$ **23.** $\dfrac{2}{7}$ **25.** $\dfrac{8}{25}$ **27.** $\dfrac{-3}{34}$

29. $\dfrac{1}{6}$ **31.** $\dfrac{4}{15}$ **33.** $\dfrac{-8}{15}$ **35.** $\dfrac{2}{3}$ **37.** $\dfrac{63}{50}$ **39.** $\dfrac{4}{3}$ **41.** $\dfrac{4}{15}$

43. $\dfrac{-1}{5}$ **45.** 18   **47.** $\dfrac{2}{17}$ **49.** $\dfrac{3}{16}$ mi^2 **51.** $\dfrac{3}{20}, \dfrac{9}{20}$ **53.** 171

55. 114   **57.** Always   **59.** Always   **61.** $\dfrac{1}{2}$ **63.** $\dfrac{-2}{7}$ **65.** $\dfrac{-1}{3}$

67. $\dfrac{1}{5}$ **69.** $\dfrac{3}{2}$ **71.** $\dfrac{-16}{21}$ **73.** $\dfrac{1}{2}$ **75.** $\dfrac{3}{4}$

ANSWERS

69.

70.

71.

72.

73.

74.

75.

76.

3.5 The Multiplication Property of Equality

3.5 OBJECTIVE

1. Use the multiplication property of equality to solve an equation

In Chapter 2, we learned how to solve certain equations using the addition property. Now we will look at a different type of equation. For instance, what if we want to solve a multiplication equation? Here is an example:

$6x = 18$

Using the addition property won't help. We need a second property for solving equations. First, we will revisit the idea of a reciprocal.

Example 1 Finding the Reciprocal of a Fraction

RECALL In general, the reciprocal of the fraction $\frac{a}{b}$ is $\frac{b}{a}$. Neither a nor b can be 0.

Find the reciprocal of (a) $\frac{4}{5}$ and (b) 7.

(a) The reciprocal of $\frac{4}{5}$ is $\frac{5}{4}$.

(b) The reciprocal of 7, or $\frac{7}{1}$, is $\frac{1}{7}$.   Write 7 as $\frac{7}{1}$ and then turn over the fraction.

 CHECK YOURSELF 1 _____

Find the reciprocal of (a) $\frac{3}{7}$ and (b) 17.

The relationship of a number and its reciprocal is shown in Example 2.

Example 2 Multiplying a Number by Its Reciprocal

Multiply.

(a) $\left(\dfrac{3}{7}\right)\left(\dfrac{7}{3}\right) = 1$

(b) $(8)\left(\dfrac{1}{8}\right) = 1$

(c) $(-17)\left(\dfrac{-1}{17}\right) = 1$

CHECK YOURSELF 2

Find the products.

(a) $\left(\dfrac{1}{4}\right)(4)$

(b) $\left(\dfrac{-2}{11}\right)\left(\dfrac{-11}{2}\right)$

This relationship between a number and its reciprocal will be very important in applying the multiplication property of equality for solving equations.

NOTE The *c* cannot be zero. Multiplying by 0 gives 0 = 0. We have lost the variable!

Property: The Multiplication Property of Equality

If $a = b$ then $ac = bc$

In words, multiplying both sides of an equation by the same nonzero number gives an equivalent equation.

Again, we return to the image of the balance scale. We start with the assumption that *a* and *b* have the same weight.

The multiplication property tells us that the scale will be in balance as long as we have the same number of "*a* weights" as we have of "*b* weights."

We will work through some examples, using this second rule.

OBJECTIVE 1

> **Example 3** **Solving Equations Using the Multiplication Property**
>
> Solve.
>
> $6x = 18$
>
> Here the variable x is multiplied by 6. So we apply the multiplication property and multiply both sides by the reciprocal of 6, $\frac{1}{6}$. Keep in mind that we want an equation of the form

NOTE

$\frac{1}{6}(6x) = \left(\frac{1}{6} \cdot 6\right)x$

$\qquad = 1 \cdot x$, or x

We then have x alone on one side of the equation, which is what we want.

> $x = \square$
>
> $\frac{1}{6}(6x) = \frac{1}{6}(18)$
>
> We can now simplify.
>
> $1 \cdot x = 3 \qquad$ or $\qquad x = 3$
>
> The solution is 3. To check, replace x with 3:
>
> $6 \cdot 3 \overset{?}{=} 18$
>
> $\quad 18 = 18 \qquad$ (True)

CHECK YOURSELF 3 _____

Solve and check.

$8x = 32$

In Example 3, we solved the equation by multiplying both sides by the reciprocal of the coefficient of x.

Example 4 illustrates a slightly different approach to solving an equation using the multiplication property.

> **Example 4** **Solving Equations Using the Multiplication Property**
>
> Solve.
>
> $5x = -35$

NOTE Because division is defined in terms of multiplication, we can also divide both sides of an equation by the same nonzero number.

> The variable x is multiplied by 5. We *divide* both sides by 5 to "undo" that multiplication:
>
> $\dfrac{5x}{5} = \dfrac{-35}{5}$
>
> $x = -7 \qquad$ ⎰ Note that the right side reduces to -7. Be careful with the rules for signs.
>
> We will leave it to you to check the solution.

CHECK YOURSELF 4

Solve and check.

$7x = -42$

Example 5   Solving Equations Using the Multiplication Property

Solve.

$-9x = 54$

In this case, x is multiplied by -9, so we divide both sides by -9 to isolate x on the left:

$$\frac{-9x}{-9} = \frac{54}{-9}$$

$$x = -6$$

The solution is -6. To check:

$$(-9)(-6) \overset{?}{=} 54$$

$$54 = 54 \quad \text{(True)}$$

CHECK YOURSELF 5

Solve and check.

$-10x = -60$

Example 6   Solving Equations Using the Multiplication Property

Solve.

$$\frac{-1}{3}x = 6$$

Here x is multiplied by $\dfrac{-1}{3}$, so we multiply both sides by -3, the reciprocal of $\dfrac{-1}{3}$, to isolate x on the left:

$$(-3)\left(\frac{-1}{3}x\right) = (-3)(6)$$

We can now simplify.

$$1 \cdot x = -18 \qquad \text{or} \qquad x = -18$$

The solution is -18. To check, replace x with -18:

$$\frac{-1}{3}(-18) \overset{?}{=} 6$$

Rewriting the left side as $\dfrac{(-1)(-18)}{3} = \dfrac{18}{3}$, we have

$$\dfrac{18}{3} \overset{?}{=} 6$$

$$6 = 6 \qquad \text{(True)}$$

 CHECK YOURSELF 6 _____

Solve and check.

$$\dfrac{1}{7}x = 9$$

You may sometimes have to simplify an equation before applying the methods of this section. Example 7 illustrates this property.

Example 7   Combining Like Terms and Solving Equations

Solve and check.

$$3x + 5x = 40$$

Using the distributive property, we can combine the like terms on the left to write

$$8x = 40$$

We can now proceed as before.

$$\dfrac{8x}{8} = \dfrac{40}{8} \qquad \text{Divide by 8.}$$

$$x = 5$$

The solution is 5. To check, we return to the original equation. Substituting 5 for x yields

$$3 \cdot 5 + 5 \cdot 5 \overset{?}{=} 40$$
$$15 + 25 \overset{?}{=} 40$$
$$40 = 40 \qquad \text{(True)}$$

The solution is verified.

 CHECK YOURSELF 7 _____

Solve and check.

$$7x + 4x = -66$$

READING YOUR TEXT

The following fill-in-the-blank exercises are designed to assure that you understand the key vocabulary used in this section. Each sentence usually comes directly from the section. You will find the correct answers in Appendix C.

Section 3.5

(a) The _____ of the fraction $\dfrac{a}{b}$ is $\dfrac{b}{a}$.

(b) In the equation $5x = -35$, we _____ both sides by 5 to "undo" the multiplication.

(c) In the equation $\dfrac{-1}{3}x = 6$, we _____ both sides by -3.

(d) To solve $3x + 5x = 40$, we use the _____ property so we can combine like terms on the left.

CHECK YOURSELF ANSWERS

1. (a) $\dfrac{7}{3}$; (b) $\dfrac{1}{17}$ 2. (a) 1; (b) 1   3. 4   4. -6 5. 6   6. 63   7. -6

Name _____

Section _____ Date _____

ANSWERS

1. _____   2. _____

3. _____   4. _____

5. _____   6. _____

7. _____   8. _____

9. _____   10. _____

11. _____   12. _____

13. _____

14. _____

15. _____

16. _____

17. _____

18. _____

19. _____

20. _____

21. _____

22. _____

23. _____

24. _____

25. _____

26. _____

27. _____

28. _____

29. _____

30. _____

3.5 Exercises

Find the reciprocal.

1. $\dfrac{7}{8}$

2. $\dfrac{9}{5}$

3. $\dfrac{1}{2}$

4. 6

5. $\dfrac{-2}{3}$

6. -5

7. 1

8. $\dfrac{1}{8}$

Find each product.

9. $\left(\dfrac{12}{5}\right)\left(\dfrac{5}{12}\right)$

10. $\left(\dfrac{-2}{3}\right)\left(\dfrac{-3}{2}\right)$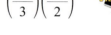

11. $(10)\left(\dfrac{1}{10}\right)$

12. $(-5)\left(\dfrac{-1}{5}\right)$

13. $\left(\dfrac{-1}{9}\right)(-9)$

14. $\left(\dfrac{1}{14}\right)(14)$

Solve for x and check your result.

15. $5x = 20$

16. $6x = 30$

17. $9x = 54$

18. $6x = -42$

19. $63 = 9x$

20. $66 = 6x$

21. $4x = -16$

22. $-3x = 27$

23. $-9x = 72$

24. $10x = -100$

25. $6x = -54$

26. $-7x = 49$

27. $-4x = -12$

28. $52 = -4x$

29. $-42 = 6x$

30. $-7x = -35$

31. $-6x = -54$

32. $-4x = -24$

33. $\dfrac{1}{2}x = 4$

34. $\dfrac{1}{3}x = 2$

35. $\dfrac{1}{5}x = 3$

36. $\dfrac{1}{8}x = 5$

37. $6 = \dfrac{1}{7}x$

38. $6 = \dfrac{1}{3}x$

39. $\dfrac{1}{5}x = -4$

40. $\dfrac{1}{7}x = -5$

41. $\dfrac{-1}{3}x = 8$

42. $5x + 4x = 36$

43. $16x - 9x = -35$

44. $4x - 2x + 7x = 36$

In exercises 45 and 46, write and solve an equation that models the given situation.

45. Business and Finance James bought five battery packs and Marc purchased seven of the same packs. They spent a total of $48. How much did each pack cost?

46. Business and Finance Bea worked 8 hours one day and 6 hours the next. She was paid $126. How much is Bea's hourly wage?

In exercises 47 and 48, indicate whether the given statement is always true, sometimes true, or never true.

47. The reciprocal of a proper fraction is an improper fraction.

48. The reciprocal of a whole number is an improper fraction.

31. _____
32. _____
33. _____
34. _____
35. _____
36. _____
37. _____
38. _____
39. _____
40. _____
41. _____
42. _____
43. _____
44. _____
45. _____
46. _____
47. _____
48. _____

Answers

1. $\dfrac{8}{7}$ **3.** 2 **5.** $\dfrac{-3}{2}$ **7.** 1 **9.** 1 **11.** 1 **13.** 1 **15.** 4

17. 6 **19.** 7 **21.** -4 **23.** -8 **25.** -9 **27.** 3 **29.** -7

31. 9 **33.** 8 **35.** 15 **37.** 42 **39.** -20 **41.** -24 **43.** -5

45. $4 **47.** Always

3.6 Linear Equations in One Variable

3.6 OBJECTIVES

1. Combine the addition and multiplication properties to solve an equation
2. Use the order of operations when solving an equation
3. Recognize identities
4. Recognize equations with no solutions

In all our examples thus far, either the addition property or the multiplication property was used in solving an equation. Often, finding a solution will require the use of both properties. If this is the case, we apply the addition property first and then use the multiplication property.

OBJECTIVE 1

Example 1 Solving Equations

(a) Solve.

$$4x - 5 = 7$$

NOTE If we were to begin by dividing by 4, we would need to distribute and divide:

$$\frac{4x}{4} - \frac{5}{4} = \frac{7}{4}$$

This is a valid operation but makes our work more difficult by introducing fractions.

It is easier to begin by adding 5 to both sides.

Here x is *multiplied* by 4. The result, $4x$, then has 5 subtracted from it (or -5 added to it) on the left side of the equation. These two operations mean that both properties must be applied in solving the equation.

Because the variable term is already on the left, we start by adding 5 to both sides:

$$
\begin{array}{rl}
4x - 5 = & 7 \\
+ 5 & +5 \\
\hline
4x = & 12
\end{array}
$$

We now divide both sides by 4:

$$\frac{4x}{4} = \frac{12}{4}$$

$$x = 3$$

The solution is 3. To check, replace x with 3 in the original equation. Be careful to follow the rules for the order of operations.

$$4 \cdot 3 - 5 \overset{?}{=} 7$$
$$12 - 5 \overset{?}{=} 7$$
$$7 = 7 \quad \text{(True)}$$

(b) Solve.

$$
\begin{array}{rl}
3x + 8 = & -4 \\
- 8 & -8 \qquad \text{Add } -8 \text{ to both sides.} \\
\hline
3x = & -12
\end{array}
$$

Now divide both sides by 3 to isolate x on the left.

$$\frac{3x}{3} = \frac{-12}{3}$$

$$x = -4$$

The solution is -4. We'll leave it to you to check this result.

CHECK YOURSELF 1 _____

Solve and check.

(a) $6x + 9 = -15$ **(b)** $5x - 8 = 7$

The variable may appear in any position in an equation. Just apply the rules carefully as you try to write an equivalent equation, and you will find the solution. Example 2 illustrates this property.

Example 2   Solving Equations

 C A U T I O N

The sign in front of a term belongs to that term.

NOTE $\dfrac{-2}{-2} = 1$, so we divide by -2 to isolate x on the left.

Solve.

$$\begin{array}{rr} 3 - 2x = & 9 \\ -3 \quad\quad & -3 \\ \hline -2x = & 6 \end{array}$$ First add -3 to both sides.

Now divide both sides by -2. This will leave x alone on the left.

$$\frac{-2x}{-2} = \frac{6}{-2}$$
$$x = -3$$

The solution is -3. We'll leave it to you to check this result.

 CHECK YOURSELF 2 _____

Solve and check.

$10 - 3x = 1$

You may also have to combine multiplication with addition or subtraction to solve an equation. Consider Example 3.

Example 3   Solving Equations

(a) Solve.

$$\frac{1}{5}x - 3 = 4$$

To get the x term alone, we first add 3 to both sides.

$$\begin{array}{rr} \frac{1}{5}x - 3 = & 4 \\ + 3 \quad & +3 \\ \hline \frac{1}{5}x \quad\; = & 7 \end{array}$$

Now, multiply both sides of the equation by 5, which is the reciprocal of $\frac{1}{5}$.

$$5\left(\frac{1}{5}x\right) = 5 \cdot 7$$
$$x = 35$$

The solution is 35. Just return to the original equation to check the result.

$$\frac{1}{5}(35) - 3 \overset{?}{=} 4$$
$$7 - 3 \overset{?}{=} 4$$
$$4 = 4 \quad \text{(True)}$$

(b) Solve.

$$5 - \frac{1}{4}x = 2$$

To get the x term alone, we first add -5 to both sides.

$$\begin{array}{rcr} 5 - \dfrac{1}{4}x &=& 2 \\ -5 && -5 \\ \hline \dfrac{-1}{4}x &=& -3 \end{array}$$

Now multiply both sides by -4, the reciprocal of $\frac{-1}{4}$.

$$(-4)\left(\frac{-1}{4}x\right) = (-4)(-3)$$

or

$$x = 12$$

The solution is 12. We'll leave it to you to check this result.

CHECK YOURSELF 3

Solve and check.

(a) $\frac{1}{6}x + 5 = 3$ **(b)** $-8 - \frac{1}{4}x = 10$

In Section 2.10, you learned how to solve certain equations when the variable appeared on both sides. Example 4 will show you how to extend that work by using the multiplication property of equality.

Example 4   Solving an Equation

Solve.

$$6x - 14 = 3x - 2$$

First add 14 to both sides. This will undo the subtraction on the left.

$$
\begin{array}{rcl}
6x - 14 & = & 3x - 2 \\
+\ 14 & & +\ 14 \\
\hline
6x & = & 3x + 12
\end{array}
$$

Now add $-3x$ so that the terms in x will be on the left only.

$$
\begin{array}{rcl}
6x = & 3x + 12 \\
-3x & -3x \\
\hline
3x = & 12
\end{array}
$$

Finally divide by 3.

$$\frac{3x}{3} = \frac{12}{3}$$

$$x = 4$$

Check:

$$6(4) - 14 \stackrel{?}{=} 3(4) - 2$$
$$24 - 14 \stackrel{?}{=} 12 - 2$$
$$10 = 10 \quad \text{(True)}$$

As you know, the basic idea is to use our two properties to form an equivalent equation with the x isolated. Here we added 14 and then subtracted $3x$. You can do these steps in either order. Try it for yourself the other way. In either case, the multiplication property is then used as the *last step* in finding the solution.

 CHECK YOURSELF 4 _____

Solve and check.

$7x - 5 = 3x + 15$

We will look at two approaches to solving equations in which the coefficient on the right side is greater than the coefficient on the left side.

Example 5   Solving an Equation (Two Methods)

Solve $4x - 8 = 7x + 7$.

Method 1

$$
\begin{array}{rcrr}
4x - 8 =& & 7x +& 7 \\
-7x & & -7x & \\
\hline
-3x - 8 =& & & 7 \\
+8 & & +& 8 \\
\hline
-3x \quad =& & & 15
\end{array}
$$

Adding $-7x$ will get the variable terms on the left.

Adding 8 will leave the x term alone on the left.

$$\frac{-3x}{-3} = \frac{15}{-3}$$

Dividing by -3 will isolate x on the left.

$$x = -5$$

We'll let you check this result.

To avoid a negative coefficient (in this example, -3), some students prefer a different approach.

This time we'll work toward having the number on the *left* and the x term on the *right*, or

$\square = x$.

Method 2

NOTE It is usually easier to isolate the variable term on the side that will result in a positive coefficient.

$$
\begin{array}{rcr}
4x - 8 =& & 7x + 7 \\
-4x & & -4x \\
\hline
-8 =& & 3x + 7 \\
-7 & & -7 \\
\hline
-15 =& & 3x
\end{array}
$$

Add $-4x$ to get the variables on the right.

Add -7 to both sides.

$$\frac{-15}{3} = \frac{3x}{3}$$

Divide by 3 to isolate x on the right.

$$-5 = x$$

Because $-5 = x$ and $x = -5$ are equivalent equations, it really makes no difference; the solution is still -5! You can use whichever approach you prefer.

✔ CHECK YOURSELF 5

Solve $5x + 3 = 9x - 21$ by finding equivalent equations of the form $x = \square$ and $\square = x$ to compare the two methods of finding the solution.

It may also be necessary to remove grouping symbols in solving an equation.

OBJECTIVE 2 | **Example 6   Solving Equations that Contain Parentheses**

Solve and check.

<table>
<tr><td>$5(x - 3) - 2x = x + 7$</td><td>First, apply the distributive property.</td></tr>
<tr><td>$5x - 15 - 2x = x + 7$</td><td>Combine like terms.</td></tr>
<tr><td>$3x - 15 = x + 7$</td><td></td></tr>
<tr><td>$+ 15 \quad\quad + 15$</td><td>Add 15.</td></tr>
<tr><td>$3x = x + 22$</td><td></td></tr>
<tr><td>$-x \quad\quad -x$</td><td>Add $-x$.</td></tr>
<tr><td>$2x = 22$</td><td>Divide by 2.</td></tr>
<tr><td>$x = 11$</td><td></td></tr>
</table>

RECALL

$5(x - 3)$
$= 5(x + (-3))$
$= 5x + 5(-3)$
$= 5x + (-15)$
$= 5x - 15$

The solution is 11. To check, substitute 11 for x in the original equation. Again note the use of our rules for the order of operations.

$5(11 - 3) - 2 \cdot 11 \overset{?}{=} 11 + 7$ Simplify terms in parentheses.

$5 \cdot 8 - 2 \cdot 11 \overset{?}{=} 11 + 7$ Multiply.

$40 - 22 \overset{?}{=} 11 + 7$ Add and subtract.

$18 = 18$ A true statement.

 CHECK YOURSELF 6 _____

Solve and check.

$7(x + 5) - 3x = x - 7$

An equation that is true for any value of x is called an **identity.**

OBJECTIVE 3 | **Example 7   Solving an Equation**

Solve the equation $2(x - 3) = 2x - 6$.

$$2(x - 3) = 2x - 6$$
$$2x - 6 = 2x - 6$$
$$-2x \quad\quad -2x$$
$$-6 = -6$$

NOTE We could ask the question "For what values of x does $-6 = -6$?"

The statement $-6 = -6$ is true for any value of x. The original equation is an identity.

 CHECK YOURSELF 7 _____

Solve the equation $3(x - 4) - 2x = x - 12$.

There are also equations for which there are no solutions.

OBJECTIVE 4 | **Example 8   Solving an Equation**

Solve the equation $3(2x - 5) - 4x = 2x + 1$.

$$3(2x - 5) - 4x = 2x + 1$$
$$6x - 15 \quad - 4x = 2x + 1$$
$$2x - 15 = 2x + 1$$
$$\underline{-2x \qquad\qquad -2x}$$
$$- 15 = \qquad 1$$

NOTE We could ask the question "For what values of x does $-15 = 1$?"

These two numbers are never equal. The original equation has no solutions.

CHECK YOURSELF 8

Solve the equation $2(x - 5) + x = 3x - 3$.

Step by Step: Solving Linear Equations

NOTE Such an outline of steps is sometimes called an **algorithm** for the process.

Step 1   Use the distributive property to remove any grouping symbols. Then simplify by combining like terms on each side of the equation.

Step 2   Add or subtract the same term on each side of the equation until the variable term is on one side and a number is on the other.

Step 3   Multiply or divide both sides of the equation by the same nonzero number so that the variable is alone on one side of the equation. If no variable remains, determine whether the original equation is an identity or whether it has no solutions.

Step 4   Check the solution in the original equation.

READING YOUR TEXT

The following fill-in-the-blank exercises are designed to assure that you understand the key vocabulary used in this section. Each sentence usually comes directly from the section. You will find the correct answers in Appendix C.

Section 3.6

(a)   Often, in solving an equation, both the _____ property and the multiplication property must be used.

(b)   When both addition and multiplication are used in solving an equation, the _____ property is used as the last step.

(c)   It is usually easier to isolate the variable term on the side that will result in a _____ coefficient.

(d)   An equation that is true for any value of x is called an _____.

CHECK YOURSELF ANSWERS

1. (a) -4; **(b)** 3   **2.** 3   **3. (a)** -12; **(b)** -72 **4.** 5   **5.** 6   **6.** -14
7. The equation is an identity; x is any real number.   **8.** There are no solutions.

3.6 Exercises

Name _____

Section _____ Date _____

Solve for x and check your result.

1. $2x + 1 = 9$ **ALEKS**

2. $3x - 1 = 17$ **ALEKS**

3. $3x - 2 = 7$ **ALEKS**

4. $5x + 3 = 23$ **ALEKS**

5. $4x + 7 = 35$ **ALEKS**

6. $7x - 8 = 13$ **ALEKS**

7. $2x + 9 = 5$ **ALEKS**

8. $6x + 25 = -5$ **ALEKS**

9. $4 - 7x = 18$

10. $8 - 5x = -7$

11. $3 - 4x = -9$

12. $5 - 4x = 25$

13. $\frac{1}{2}x + 1 = 5$

14. $\frac{1}{3}x - 2 = 3$

15. $\frac{1}{4}x - 5 = 3$

16. $\frac{1}{5}x + 3 = 8$

17. $5x = 2x + 9$

18. $7x = 18 - 2x$

19. $3x = 10 - 2x$

20. $11x = 7x + 20$

21. $9x + 2 = 3x + 38$

22. $8x - 3 = 4x + 17$

23. $4x - 8 = x - 14$

24. $6x - 5 = 3x - 29$

25. $7x - 3 = 9x + 5$

26. $5x - 2 = 8x - 11$

27. $5x + 4 = 7x - 8$

28. $2x + 23 = 6x - 5$

29. $2x - 3 + 5x = 7 + 4x + 2$

30. $8x - 7 - 2x = 2 + 4x - 5$

31. $6x + 7 - 4x = 8 + 7x - 26$

32. $7x - 2 - 3x = 5 + 8x + 13$

33. $9x - 2 + 7x + 13 = 10x - 13$

34. $5x + 3 + 6x - 11 = 8x + 25$

35. $7(2x - 1) - 5x = x + 25$ **ALEKS**

36. $9(3x + 2) - 10x = 12x - 7$ **ALEKS**

ANSWERS

1. _____	**2.** _____
3. _____	**4.** _____
5. _____	**6.** _____
7. _____	**8.** _____
9. _____	**10.** _____
11. _____	**12.** _____
13. _____	**14.** _____
15. _____	**16.** _____
17. _____	**18.** _____
19. _____	**20.** _____
21. _____	**22.** _____

23. _____

24. _____

25. _____

26. _____

27. _____

28. _____

29. _____

30. _____

31. _____

32. _____

33. _____

34. _____

35. _____

36. _____

37. _____

38. _____

39. _____

40. _____

41. _____

42. _____

43. _____

44. _____

45. _____

46. _____

47. _____

48. _____

49. _____

50. _____

37. $4(x + 5) = 4x + 20$

38. $-3(2x - 4) - 12 = -6x$

39. $5(x + 1) - 4x = x - 5$

40. $-4(2x - 3) = -8x + 5$

41. $6x - 4x + 1 = 12 + 2x - 11$

42. $-2x + 5x - 9 = 3(x - 4) - 5$

43. $-4(x + 2) - 11 = 2(-2x - 3) - 13$

44. $4(-x - 2) + 5 = -2(2x + 7)$

45. Create an equation of the form $ax + b = c$ that has 2 as a solution.

46. Create an equation of the form $ax + b = c$ that has 7 as a solution.

47. The equation $3x = 3x + 5$ has no solution, whereas the equation $7x + 8 = 8$ has zero as a solution. Explain the difference between a solution of zero and no solution.

48. Construct an equation for which every real number is a solution.

In exercises 49 and 50, write and solve an equation that models the given situation.

49. Science and Medicine   To determine the upper limit for a person's heart rate during aerobic training, we subtract the person's age from 220 and then multiply by $\dfrac{9}{10}$. How old is a person whose upper limit heart rate is 153?

50. Science and Medicine   To determine the lower limit for a person's heart rate during aerobic training, we subtract the person's age from 220 and then multiply by $\dfrac{3}{5}$. How old is a person whose lower limit heart rate is 111?

Answers

1. 4   3. 3   5. 7   7. -2 9. -2 11. 3   13. 8   15. 32
17. 3   19. 2   21. 6   23. -2 25. -4 27. 6   29. 4
31. 5   33. -4 35. 4   37. Identity   39. No solution
41. Identity   43. Identity   45.   47.   49. 50

ACTIVITY 3: FRACTIONS IN DIET AND EXERCISE

Each chapter in this text includes an activity. The activity is related to the vignette you encountered in the chapter opening. The activity provides you with the opportunity to apply the math you studied in the chapter.

Your instructor will determine how best to use this activity in your class. You may find yourself working in class or outside of class; you may find yourself working alone or in small groups; or you may even be asked to perform research in a library or on the Internet.

Part 1: Recommended Daily Allowance

According to the Food and Drug Administration (FDA), the following table represents the recommended daily allowance (RDA) for each food type.

	RDA
Total fat	75 grams
Saturated fat	40 grams
Cholesterol	5 grams
Sodium	22 grams
Carbohydrates	300 grams
Dietary fiber	25 grams

A high-performance energy bar made by PowerBar has the following amounts of each food type.

	Amount
Total fat	$\frac{5}{2}$ gram
Saturated fat	$\frac{1}{2}$ gram
Cholesterol	0 grams
Sodium	1 gram
Carbohydrates	45 grams
Dietary fiber	3 grams

Use the tables to answer each of the following questions.

1. What fraction of the RDA for sodium is contained in the PowerBar?

2. What fraction of the RDA for carbohydrates is contained in the PowerBar?

3. What fraction of the RDA for dietary fiber is contained in the PowerBar?

4. What fraction of the RDA for total fat is contained in the PowerBar?

5. What fraction of the RDA for saturated fat is contained in the PowerBar?

Part 2: Aerobic Exercise

According to the website www.Fitnesszone.be, **aerobic training zone** refers to the training intensity range that will produce improvement in your level of aerobic fitness without overtaxing your cardiorespiratory system. Your aerobic training zone is based on a percentage of your maximal heart rate (MHR). As a general rule, your maximal heart rate is measured directly or estimated by subtracting your age from 220. Depending upon how physically fit you are, the lower and upper limits of your aerobic training zone are then based on a fraction of the maximal heart rate, approximately $\frac{3}{5}$ and $\frac{9}{10}$, respectively.

The chart below uses this information to determine maximal heart rate by subtracting the age from 220. Complete the table.

Age	Maximal Heart Rate
20	
25	
30	
35	
40	
45	
50	
55	

Using the information provided in the first paragraph, complete the table below to find the lower and upper limits for the training zone for each age. Note that when we compute the upper limit of the training zone, we round up to the next integer. For example, we compute the upper limit for age 25 as

$$195 \times \frac{9}{10} = \frac{195}{1} \times \frac{9}{10} = \frac{39}{1} \times \frac{9}{2} = \frac{351}{2} = 175\frac{1}{2}$$

Rounding up, we get an upper limit of 176.

Age	Maximal Heart Rate	Lower Limits of Zone $\left(\frac{3}{5}\text{ MHR}\right)$	Upper Limit of Zone $\left(\frac{9}{10}\text{ MHR}\right)$
20			
25			
30			
35			
40			
45			
50			
55			

3 Summary

DEFINITION/PROCEDURE	EXAMPLE	REFERENCE
Introduction to Fractions		**Section 3.1**
Fraction Fractions name a number of equal parts of a unit or whole. A fraction is written in the form $\frac{a}{b}$, in which a is an integer and b is a natural number.	$\frac{5}{8}$ is a fraction.	*p. 215*
Numerator The number of parts of the whole that are being considered. **Denominator** The number of equal parts into which the whole is divided.	Numerator $\frac{5}{8}$ Denominator	*p. 215*
Proper Fraction A fraction whose numerator is less than its denominator. It names a number less than 1.	$\frac{2}{3}$ and $\frac{11}{15}$ are proper fractions.	*p. 217*
Improper Fraction A fraction whose numerator is greater than or equal to its denominator. It names a number greater than or equal to 1.	$\frac{7}{5}, \frac{21}{20}$, and $\frac{8}{8}$ are improper fractions.	*p. 217*
Prime Numbers and Factorization		**Section 3.2**
Prime Number Any natural number greater than 1 that has exactly two factors, 1 and itself.	7, 13, 29, and 73 are prime numbers.	*p. 224*
Composite Number Any natural number greater than 1 that is not prime.	8, 15, 42, and 65 are composite numbers.	*p. 225*
Prime Factorization To find the prime factorization of a number, divide the number by a series of primes until the final quotient is a prime number. The prime factors include each prime divisor and the final quotient.	$630 = 2 \cdot 3 \cdot 3 \cdot 5 \cdot 7$	*p. 226*
Greatest Common Factor (GCF) The GCF is the *largest* number that is a factor of each of a group of numbers.		*p. 228*
To Find the GCF **Step 1** Write the prime factorization for each of the numbers in the group. **Step 2** Locate the prime factors that are common to all the numbers. **Step 3** The greatest common factor (GCF) will be the product of all the common prime factors. If there are no common prime factors, the GCF is 1.	To find the GCF of 24, 30, and 36: $24 = ②\cdot 2 \cdot 2 \cdot ③$ $30 = ② \cdot ③ \cdot 5$ $36 = ② \cdot 2 \cdot ③ \cdot 3$ The GCF is $2 \cdot 3 = 6$	*p. 229*

Continued

DEFINITION/PROCEDURE	EXAMPLE	REFERENCE
Equivalent Fractions		Section 3.3
The Fundamental Principle of Fractions For the fraction $\frac{a}{b}$, and any nonzero number c, $$\frac{a}{b} = \frac{a \div c}{b \div c}$$ **In words:** We can divide the numerator and denominator of a fraction by the same nonzero number. This is used to simplify (or reduce) a fraction.	$$\frac{8}{12} = \frac{8 \div 4}{12 \div 4} = \frac{2}{3}$$ $\frac{8}{12}$ and $\frac{2}{3}$ are equivalent fractions.	p. 236
Equivalent Fractions The fundamental principle can also be written as $$\frac{a}{b} = \frac{a \cdot c}{b \cdot c} \qquad c \neq 0$$ This is used to build up an equivalent fraction.	$$\frac{2}{3} = \frac{2 \cdot 5}{3 \cdot 5} = \frac{10}{15}$$	p. 240
Multiplication and Division of Fractions		Section 3.4
To Multiply Two Fractions 1. Multiply numerator by numerator. This gives the numerator of the product. 2. Multiply denominator by denominator. This gives the denominator of the product. 3. Simplify the resulting fraction if possible. In multiplying fractions, it is usually easiest to divide by any common factors in the numerator and denominator *before* multiplying.	$$\frac{5}{8} \cdot \frac{3}{7} = \frac{5 \cdot 3}{8 \cdot 7} = \frac{15}{56}$$ $$\frac{5}{9} \cdot \frac{3}{10} = \frac{\overset{1}{5} \cdot \overset{1}{3}}{\underset{3}{9} \cdot \underset{2}{10}} = \frac{1}{6}$$	p. 250
To Divide Two Fractions Replace the divisor by its reciprocal and multiply.	$$\frac{3}{7} \div \frac{4}{5} = \frac{3}{7} \cdot \frac{5}{4} = \frac{15}{28}$$	p. 256
The Multiplication Property of Equality		Section 3.5
The Multiplication Property of Equality If $a = b$ then $a \cdot c = b \cdot c$	If $\frac{1}{2} x = 7$ then $2\left(\frac{1}{2} x\right) = 2(7)$	p. 265
Linear Equations in One Variable		Section 3.6
Solving Linear Equations The steps for solving a linear equation are as follows: 1. Use the distributive property to remove any grouping symbols. Then simplify by combining like terms. 2. Add or subtract the same term on both sides of the equation until the variable term is on one side and a number is on the other. 3. Multiply or divide both sides of the equation by the same nonzero number so that the variable is alone on one side of the equation. 4. Check the solution in the original equation.	Solve: $$\begin{aligned} 3(x-2) + 4x &= 3x + 14 \\ 3x - 6 + 4x &= 3x + 14 \\ 7x - 6 &= 3x + 14 \\ +6 &\quad +6 \\ \hline 7x &= 3x + 20 \\ -3x &\quad -3x \\ \hline 4x &= 20 \end{aligned}$$ $$\frac{4x}{4} = \frac{20}{4}$$ $$x = 5$$	p. 277

Summary Exercises

This summary exercise set is provided to give you practice with each of the objectives of this chapter. Each exercise is keyed to the appropriate chapter section. When you are finished, you can check your answers to the odd-numbered exercises against those presented in the back of the text. If you have difficulty with any of these questions, go back and reread the examples from that section. The answers to the even-numbered exercises appear in the *Instructor's Manual*. Your instructor will give you guidelines on how to best use these exercises in your class.

[3.1] Identify the numerator and denominator of each fraction.

1. $\dfrac{5}{9}$

2. $\dfrac{17}{23}$

Give the fractions that name the shaded portions of the diagrams. Identify the numerator and the denominator.

3.

Fraction _____

Numerator _____

Denominator _____

4.

Fraction: _____

Numerator: _____

Denominator _____

5. From the group of numbers:

$$\frac{2}{3}, \frac{5}{4}, \frac{45}{8}, \frac{7}{7}, \frac{9}{1}, \frac{7}{10}, \frac{12}{5}$$

List the proper fractions. _____

List the improper fractions _____

[3.2] In exercises 6 and 7, list all the factors of the given numbers.

6. 52

7. 41

In exercise 8, use the group of numbers 2, 5, 7, 11, 14, 17, 21, 23, 27, 39, and 43.

8. List the prime numbers; then list the composite numbers.

In exercises 9 and 10, determine which, if any, of the numbers 2, 3, and 5 are factors of the given numbers.

9. 2,350

10. 33,451

In exercises 11 to 14, find the prime factorization for the given numbers.

11. 48

12. 420

13. 2,640

14. 2,250

In exercises 15 to 20, find the greatest common factor (GCF).

15. 15 and 20

16. 30 and 31

17. 24 and 40

18. 39 and 65

19. 49, 84, and 119

20. 77, 121, and 253

[3.3] Determine whether each of the pairs of fractions are equivalent.

21. $\dfrac{5}{8}, \dfrac{7}{12}$

22. $\dfrac{8}{15}, \dfrac{32}{60}$

Write each fraction in simplest form.

23. $\dfrac{24}{36}$

24. $\dfrac{45}{75}$

25. $\dfrac{140}{180}$

26. $\dfrac{16}{21}$

Find the missing numerators.

27. $\dfrac{15}{25} = \dfrac{?}{5}$

28. $\dfrac{32}{40} = \dfrac{?}{5}$

29. $\dfrac{2}{7} = \dfrac{?}{28}$

30. $\dfrac{4}{5} = \dfrac{?}{30}$

[3.4] Multiply.

31. $\dfrac{7}{15} \cdot \dfrac{-5}{21}$

32. $\dfrac{-10}{27} \cdot \dfrac{-9}{20}$

33. $4 \cdot \dfrac{3}{8}$

Divide.

34. $\dfrac{5}{12} \div \dfrac{5}{8}$

35. $\dfrac{-7}{15} \div \dfrac{14}{25}$

[3.5] and [3.6] Solve each equation.

36. $7x = 147$

37. $5x = 35$

38. $7x = -28$

39. $-6x = 24$

40. $-9x = -63$

41. $\dfrac{1}{4}x = 8$

42. $\dfrac{-1}{5}x = -3$

43. $5x - 3 = 12$

44. $4x + 3 = -13$

45. $7x + 8 = 3x$

46. $3 - 5x = -17$

47. $\dfrac{1}{3}x - 5 = 1$

48. $\dfrac{1}{4}x + 4 = 7$

49. $6x - 5 = 3x + 13$

50. $3x + 7 = x - 9$

51. $2x + 7 = 4x - 5$

52. $3x - 2 + 5x = 7 + 2x + 21$

53. $x - 5 - 6x = 3(9 + x)$

Self-Test for Chapter 3

The purpose of this self-test is to help you check your progress so that you can find sections and concepts that you need to review before the next in-class exam. Allow yourself about an hour to take this test. At the end of that hour, check your answers against those given in the back of this text. If you missed a question, notice the section reference that accompanies the question. Go back to that section and reread the examples until you have mastered that particular concept.

Name _____

Section _____ Date _____

ANSWERS

For exercises 1 to 3, what fraction names the shaded part of each diagram? Identify the numerator and denominator.

1.   **2.**   **3.**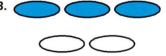

4. Identify the proper fractions and improper fractions in the following group.

$$\frac{10}{11}, \frac{9}{5}, \frac{7}{7}, \frac{8}{1}, \frac{1}{8}$$

Proper   Improper

5. Which of the numbers 5, 9, 13, 17, 22, 27, 31, and 45 are prime numbers? Which are composite numbers?

6. Determine which, if any, of the numbers 2, 3, and 5 are factors of 54,204.

7. Find the prime factorization for 264.

In exercises 8 and 9, find the greatest common factor (GCF) for the given numbers.

8. 36 and 84   **9.** 16, 24, and 72

In exercises 10 to 12, use the cross-product method to determine whether the pair of fractions is equivalent.

10. $\dfrac{2}{7}, \dfrac{8}{28}$ **11.** $\dfrac{8}{20}, \dfrac{12}{30}$ **12.** $\dfrac{3}{20}, \dfrac{2}{15}$

In exercises 13 to 15, write the fractions in simplest form.

13. $\dfrac{21}{27}$ **14.** $\dfrac{36}{84}$ **15.** $\dfrac{8}{23}$

1. _____
2. _____
3. _____
4. _____
5. _____
6. _____
7. _____
8. _____
9. _____
10. _____
11. _____
12. _____
13. _____
14. _____
15. _____

16. _____	
17. _____	
18. _____	
19. _____	
20. _____	
21. _____	
22. _____	
23. _____	
24. _____	
25. _____	
26. _____	
27. _____	
28. _____	
29. _____	
30. _____	
31. _____	
32. _____	
33. _____	
34. _____	
35. _____	
36. _____	
37. _____	

In exercises 16 to 18, find the missing numerators.

16. $\dfrac{28}{35} = \dfrac{?}{5}$ **17.** $\dfrac{42}{98} = \dfrac{?}{7}$ **18.** $\dfrac{105}{120} = \dfrac{?}{8}$

In exercises 19 and 20, multiply.

19. $\dfrac{9}{10} \cdot \dfrac{5}{8}$ **20.** $\dfrac{16}{35} \cdot \dfrac{-14}{24}$

In exercises 21 to 25, divide.

21. $\dfrac{7}{12} \div \dfrac{14}{15}$ **22.** $\left(\dfrac{-2}{5}\right) \div \dfrac{8}{3}$

23. $\left(\dfrac{-16}{15}\right) \div \left(\dfrac{-8}{3}\right)$ **24.** $\left(\dfrac{-7}{3}\right) \cdot \left(\dfrac{-3}{7}\right)$

25. $\left(\dfrac{-7}{3}\right) \div \left(\dfrac{-3}{7}\right)$

Solve the equations and check your results.

26. $7x = 49$ **27.** $\dfrac{1}{4}x = -3$

28. $\dfrac{-1}{5}x = -5$ **29.** $7x - 5 = 16$

30. $10 - 3x = -2$ **31.** $7x + 10 = 4x - 5$

32. $2x - 7 = 5x + 8$ **33.** $\dfrac{x}{3} - 5 = 2$

34. $\dfrac{2x}{5} + 1 = \dfrac{6}{5}$ **35.** $\dfrac{2x}{3} + 5 = \dfrac{1}{4}$

Solve the following applications.

36. Business and Finance Ellen estimates she will need $600 for books next term. She has $350 saved for this purpose. What fraction of her book costs does she already have? Simplify your answer.

37. Social Science There are 1,680 Hispanic students enrolled at a certain college. If the total number of students is 7,000, what fraction are Hispanic? Simplify your answer.

Cumulative Review
for Chapters 1 to 3

Name _____

Section _____ Date _____

ANSWERS

The following exercises are presented to help you review concepts from earlier chapters that you may have forgotten. This section is meant as review material and not as a comprehensive exam. The answers are presented in the back of the text. If you have difficulty with any of these exercises, be certain to at least read through the summary related to that section.

In exercises 1 to 15, perform the indicated operations.

1. $\begin{array}{r} 1,369 \\ + 5,804 \\ \hline \end{array}$

2. $\begin{array}{r} 489 \\ 562 \\ 613 \\ + 254 \\ \hline \end{array}$

3. $\begin{array}{r} 357 \\ 28 \\ + 2,346 \\ \hline \end{array}$

4. $\begin{array}{r} 13 \\ 2,543 \\ + 10,547 \\ \hline \end{array}$

5. $-289 + 54$

6. $53,294 - 41,074$

7. $503 - 74$

8. $5,731 - 2,492$

9. $58 \cdot (-3)$

10. Find the product of -273 and -7.

11. $\begin{array}{r} 89 \\ \times 56 \\ \hline \end{array}$

12. $\begin{array}{r} 538 \\ \times 103 \\ \hline \end{array}$

13. $281\overline{)6,935}$

14. $571\overline{)12,583}$

15. $293\overline{)61,382}$

In exercises 16 to 21, evaluate each of the expressions.

16. $12 \div 6 + 3$

17. $4 + 12 \div 4$

18. $3^3 \div 9$

19. $28 \div 7 \cdot 4$

20. $5 \cdot 8 \div 2$

21. $36 \div (3^2 + 3)$

ANSWERS

1. _____
2. _____
3. _____
4. _____
5. _____
6. _____
7. _____
8. _____
9. _____
10. _____
11. _____
12. _____
13. _____
14. _____
15. _____
16. _____
17. _____
18. _____
19. _____
20. _____
21. _____

22. _____

23. _____

24. _____

25. _____

26. _____

27. _____

28. _____

29. _____

30. _____

31. _____

32. _____

33. _____

34. _____

35. _____

36. _____

37. _____

38. _____

39. _____

40. _____

In exercise 22, identify the proper fractions and improper fractions from the group.

$$\frac{5}{7}, \frac{15}{9}, \frac{8}{8}, \frac{11}{1}, \frac{2}{5}$$

22. Proper: ____   Improper: _____

In exercises 23 and 24, find out whether the pair of fractions is equivalent.

23. $\dfrac{8}{32}, \dfrac{9}{36}$

24. $\dfrac{6}{11}, \dfrac{7}{9}$

Perform the indicated operations.

25. $\dfrac{7}{15} \cdot \dfrac{5}{21}$

26. $\dfrac{-10}{27} \cdot \dfrac{-9}{20}$

27. $-4 \cdot \dfrac{3}{8}$

28. $3 \cdot \dfrac{-5}{8}$

29. $\dfrac{-5}{6} \div \dfrac{-1}{3}$

30. $\dfrac{12}{13} \div (-4)$

Simplify each expression.

31. $3x - 5x + 7x$

32. $4(x + 2) - 5(x - 3)$

33. $7 - 2(x + 3)$

34. $2x - 5y - 4x + 8y$

35. $3(x - y) + 5(2x + 3y)$

Solve the equations and check your results.

36. $9x - 5 = 8x$

37. $\dfrac{-3}{4}x = 18$

38. $6x - 8 = 2x - 3$

39. $2x + 3 = 7x + 5$

40. $\dfrac{4}{3}x - 6 = 4 - \dfrac{2}{3}x$

APPLICATIONS OF FRACTIONS AND EQUATIONS

CHAPTER 4 OUTLINE

Section 4.1   Addition and Subtraction of Fractions   page 293

Section 4.2   Operations on Mixed Numbers   page 313

Section 4.3   Applications Involving Fractions   page 330

Section 4.4   Equations Containing Fractions   page 351

Section 4.5   Applications of Linear Equations in
One Variable   page 360

Section 4.6   Complex Fractions   page 373

chapter

4

INTRODUCTION

Home remodeling, once attempted by few people, has become popular and accessible to many. Thanks to several large discount stores, those interested in do-it-yourself projects can tackle them. Whether the project is small or large, you can find the necessary materials, tools, and even expert advice at such stores.

One thing is certain, however. The planning of a project (and its execution!) will involve mathematics and almost certainly the use of fractions. Suppose, for example, you want to put a new floor in your kitchen. You may well be faced with building up the subfloor to a certain height. The necessary flooring material comes in a variety of thicknesses, so you will have choices and decisions to make.

In this chapter, we will work with the ideas needed to make decisions like these. In Sections 4.1, 4.2, 4.3, and 4.5, you will find exercises that relate to the topics of home remodeling and construction. At the end of the chapter, but before the chapter summary, you will find a more in-depth activity involving a kitchen flooring project.

ANSWERS

1. _____

2. _____

3. _____

4. _____

5. _____

6. _____

7. _____

8. _____

9. _____

10. _____

11. _____

12. _____

13. _____

14. _____

15. _____

16. _____

Pretest Chapter 4

This pretest will provide a preview of the types of exercises you will encounter in each section of this chapter. The answers for these exercises can be found in the back of the text. If you are working on your own or are ahead of the class, this pretest can help you identify the sections in which you should focus more of your time.

[4.1]   **1. (a)** $\dfrac{3}{7} + \dfrac{2}{7}$ **(b)** $\dfrac{8}{9} - \dfrac{3}{9}$

2. Find the least common denominator for fractions with the denominators 16 and 24.

3. Find the least common multiple of each group of numbers.

   (a) 25 and 40   **(b)** 20, 24, and 30

Add or subtract. Simplify when possible.

4. $\dfrac{4}{5} + \dfrac{7}{8} + \dfrac{9}{20}$ **5.** $\dfrac{7}{24} + \dfrac{-2}{9}$

[4.2]   **6.** Convert $\dfrac{37}{4}$ to a mixed number.

7. Convert $6\dfrac{4}{7}$ to an improper fraction.

Carry out the indicated operation.

8. $2\dfrac{2}{5} \cdot 1\dfrac{3}{4}$ **9.** $\dfrac{2}{3} \div 7$

10. $3\dfrac{5}{6} + 2\dfrac{7}{8}$ **11.** $2\dfrac{1}{9} - 1\dfrac{1}{12}$

[4.3]   **12.** A piece of wood that is $15\dfrac{3}{4}$ in. long is to be cut into blocks $1\dfrac{1}{8}$ in. long. How many blocks can be cut?

13. A house plan calls for $12\dfrac{3}{4}$ yd^2 of carpeting in the living room and $5\dfrac{1}{2}$ yd^2 in the hallway. How much carpeting will be needed?

[4.4]   **14.** Solve for x. $\dfrac{x}{4} - \dfrac{x}{5} = 2$

[4.5]   **15.** If two-fifths of a number is added to one-half of that number, the sum is 27. Find the number.

[4.6]   **16.** Simplify. $\dfrac{\frac{3}{8}}{\frac{5}{6}}$

4.1 Addition and Subtraction of Fractions

4.1 OBJECTIVES

1. Add and subtract fractions with like denominators
2. Find a least common multiple (LCM)
3. Order fractions with unlike denominators
4. Add and subtract fractions with unlike denominators

Recall from our work in Chapter 1 that adding can be thought of as combining groups of the *same kind* of objects. This is also true when you think about adding fractions.

Fractions can be added only if they represent the *same parts* of a whole. This means that we can add fractions only when they are **like fractions,** that is, when they have the *same* (*a common*) denominator.

As long as we are dealing with like fractions, addition is an easy matter. Just use this rule:

NOTE For instance, we can add two nickels and three nickels to get five nickels. We *cannot* directly add two nickels and three dimes!

> **Step by Step:** To Add Like Fractions
>
> **Step 1** Add the numerators.
> **Step 2** Place the sum over the common denominator.
> **Step 3** Simplify the resulting fraction when necessary.

Example 1 illustrates the use of this rule.

OBJECTIVE 1

> **Example 1** Adding Like Fractions

Add.

$$\frac{1}{5} + \frac{3}{5}$$

Step 1 Add the numerators.

$$1 + 3 = 4$$

Step 2 Write that sum over the common denominator, 5. We are done at this point because the answer, $\frac{4}{5}$, is in the simplest possible form.

$$\frac{1}{5} + \frac{3}{5} = \overset{\text{Step 1}}{\frac{1 + 3}{5}} = \overset{\text{Step 2}}{\frac{4}{5}}$$

We will illustrate with a diagram.

NOTE Combining 1 of the 5 parts with 3 of the 5 parts gives a total of 4 of the 5 equal parts.

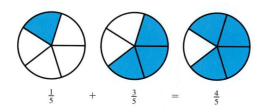

$$\frac{1}{5} \qquad + \qquad \frac{3}{5} \qquad = \qquad \frac{4}{5}$$

CHECK YOURSELF 1

Add.

$$\frac{2}{9} + \frac{5}{9}$$

Be Careful! In adding fractions, *do not* follow the rule for multiplying fractions. To multiply $\frac{1}{5} \cdot \frac{3}{5}$, you would multiply both the numerators and the denominators:

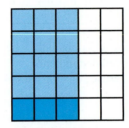

$$\frac{1}{5} \cdot \frac{3}{5} = \frac{3}{25}$$

$\frac{1}{5}$ of $\frac{3}{5} = \frac{3}{25}$

But to add fractions, you must have a common denominator that becomes the denominator of the sum.

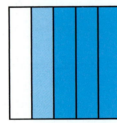

$$\frac{1}{5} + \frac{3}{5} \neq \frac{4}{10}$$

$\frac{1}{5} + \frac{3}{5} = \frac{4}{5}$

Step 3 of the addition rule for like fractions tells us to *simplify* the sum. Sums of fractions are usually written in lowest terms. Consider Example 2.

Example 2   Adding Like Fractions that Require Simplifying

Add and simplify.

Step 3

$$\frac{3}{12} + \frac{5}{12} = \frac{8}{12} = \frac{2}{3}$$

The sum $\frac{8}{12}$ is *not* in lowest terms.
Divide the numerator and denominator by 4 to simplify the result.

CHECK YOURSELF 2

Add.

$$\frac{4}{15} + \frac{6}{15}$$

We can also easily extend our addition rule to find the sum of more than two fractions as long as they all have the same denominator. This is shown in Example 3.

Example 3   Adding a Group of Like Fractions

Add.

$$\frac{2}{7} + \frac{3}{7} + \frac{6}{7} = \frac{11}{7}$$ Add the numerators: $2 + 3 + 6 = 11$.

 CHECK YOURSELF 3

Add.

$$\frac{1}{8} + \frac{3}{8} + \frac{5}{8}$$

RECALL Like fractions have the same denominator.

If a problem involves like fractions, then subtraction, as in addition, is not difficult.

NOTE Compare this to our rule for adding like fractions.

Step by Step: To Subtract Like Fractions

Step 1   Subtract the numerators.
Step 2   Place the difference over the common denominator.
Step 3   Simplify the resulting fraction when necessary.

Example 4   Subtracting Like Fractions

Subtract.

Step 1   Step 2   Subtract the numerators: $4 - 2 = 2$. Write the difference over the common denominator, 5. Step 3 is not necessary because the difference is in simplest form.

(a) $\dfrac{4}{5} - \dfrac{2}{5} = \dfrac{4 - 2}{5} = \dfrac{2}{5}$

Illustrating with a diagram:

NOTE Subtracting 2 of the 5 parts from 4 of the 5 parts leaves 2 of the 5 equal parts.

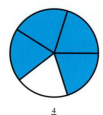  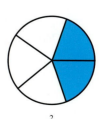

$$\frac{4}{5} \qquad - \qquad \frac{2}{5} \qquad = \qquad \frac{2}{5}$$

NOTE Always write the result in lowest terms.

(b) $\dfrac{5}{8} - \dfrac{3}{8} = \dfrac{5 - 3}{8} = \dfrac{2}{8} = \dfrac{1}{4}$

CHECK YOURSELF 4

Subtract.

$$\frac{11}{12} - \frac{5}{12}$$

As was the case with integers, we should determine the sign of the result first when adding fractions.

Example 5 Addition Involving Negative Fractions

Find the sum.

(a) $\dfrac{-3}{5} + \dfrac{2}{5}$

NOTE You may think of the magnitude of a number as its distance from 0 on the number line. Another term for *magnitude* is *absolute value*.

Because the signs are opposite and $\dfrac{-3}{5}$ has greater magnitude than $\dfrac{2}{5}$, the sum will be negative.

$$\frac{-3}{5} + \frac{2}{5} = \frac{-1}{5}$$

(b) $\dfrac{-3}{5} + \dfrac{-2}{5}$

The signs are the same (both are negative), so we know that the sum will be negative.

$$\frac{-3}{5} + \frac{-2}{5} = \frac{-5}{5} = -1$$

CHECK YOURSELF 5

Find the sum.

(a) $\dfrac{-4}{9} + \dfrac{7}{9}$ **(b)** $\dfrac{-6}{35} + \dfrac{-12}{35}$

Recall that, when subtracting integers, we changed the exercise to an equivalent addition exercise. We will do the same thing with fractions when negative numbers are involved.

Example 6 Subtraction Involving Negative Fractions

Find the difference.

(a) $\dfrac{-3}{7} - \dfrac{2}{7}$

First, we change the operation to addition, also replacing $\dfrac{2}{7}$ with its opposite $\left(\dfrac{-2}{7}\right)$.

$$\frac{-3}{7} - \frac{2}{7} = \frac{-3}{7} + \frac{-2}{7} = \frac{-5}{7}$$

(b) $\dfrac{-3}{7} - \dfrac{-4}{7}$

Again, we change to an addition exercise.

$$\frac{-3}{7} - \frac{-4}{7} = \frac{-3}{7} + \frac{4}{7}$$

We are now finding the sum of two numbers with opposite signs. First we find the difference of $\dfrac{4}{7}$ and $\dfrac{3}{7}$. Then, because $\dfrac{4}{7}$ has greater magnitude than $\dfrac{-3}{7}$, we note that the result is positive.

$$\frac{-3}{7} + \frac{4}{7} = \frac{1}{7}$$

CHECK YOURSELF 6

Find the difference.

(a) $\dfrac{-4}{11} - \dfrac{7}{11}$ **(b)** $\dfrac{-6}{7} - \dfrac{-12}{7}$

In this section, we are discussing the process used for adding or subtracting two fractions. One of the most important concepts we use in the addition and subtraction of fractions is that of *multiples*.

> **Definition: Multiples**
>
> The **multiples** of a number are the product of that number with the natural numbers 1, 2, 3, 4, 5, ...

> **Example 7   Listing Multiples**

List the multiples of 3.
   The multiples of 3 are

$$3 \cdot 1, 3 \cdot 2, 3 \cdot 3, 3 \cdot 4, \ldots$$

or

NOTE The multiples, except for 3 itself, are *larger* than 3.

$3, 6, 9, 12, \ldots$ The three dots indicate that the list continues indefinitely.

An easy way of listing the multiples of 3 is to think of *counting by threes*.

CHECK YOURSELF 7

List the first nine multiples of 4.

Sometimes we need to find common multiples of two or more numbers.

> **Definition:** Common Multiples
>
> If a number is a multiple of each of a group of numbers, it is called a **common multiple** of the numbers; that is, it is a number that is evenly divisible by all the numbers in the group.

Example 8 **Finding Common Multiples**

Find three common multiples of 3 and 5.
  Some multiples of 3 are

3, 6, 9, 12, 15, 18, 21, 24, 27, 30, 33, 36, 39, 42, 45

Some multiples of 5 are

5, 10, 15, 20, 25, 30, 35, 40, 45, 50, 55, 60

So, common multiples of 3 and 5 include

NOTE 15, 30, and 45 are multiples of *both* 3 and 5.

15, 30, 45

 CHECK YOURSELF 8

List the first six multiples of 6. Then look at your list from Check Yourself 7 and list some common multiples of 4 and 6.

For our later work, we will use the *least common multiple* of a group of numbers.

> **Definition:** Least Common Multiple
>
> The **least common multiple (LCM)** of a group of numbers is the *smallest* number that is a multiple of each number in the group.

It is possible to simply list the multiples of each number and then find the LCM by inspection.

OBJECTIVE 2   **Example 9** **Finding the Least Common Multiple**

Find the least common multiple of 6 and 8.
  Multiples

NOTE 48 is also a common multiple of 6 and 8, but we are looking for the *smallest* common multiple.

6: 6, 12, 18, ⟨24⟩, 30, 36, 42, 48, . . .

8: 8, 16, ⟨24⟩, 32, 40, 48, . . .

We see that 24 is the smallest number common to both lists. So 24 is the LCM of 6 and 8.

 CHECK YOURSELF 9

Find the least common multiple of 20 and 30 by listing the multiples of each number.

The technique of Example 9 will work for any group of numbers. However, it becomes tedious for larger numbers. We can outline a different approach.

Step by Step: Finding the Least Common Multiple

NOTE For instance, if a number appears three times in the factorization of a number, it must be included at least three times in forming the least common multiple.

Step 1   Write the prime factorization for each of the numbers in the group.
Step 2   Find all the prime factors that appear in any one of the prime factorizations.
Step 3   Form the product of those prime factors using each factor the greatest number of times it occurs in any one factorization.

Some students prefer the following method of lining up the factors to find the LCM of a group of numbers.

Example 10   Finding the Least Common Multiple

To find the LCM of 10 and 18, factor:

NOTE Line up the *like* factors vertically.

$$10 = 2 \qquad \cdot 5$$
$$\underline{18 = 2 \cdot 3 \cdot 3}$$
$$2 \cdot 3 \cdot 3 \cdot 5 \qquad \text{Bring down the factors.}$$

2 and 5 appear, at most, one time in any one factorization. And 3 appears two times in one factorization.

$$2 \cdot 3 \cdot 3 \cdot 5 = 90$$

So 90 is the LCM of 10 and 18.

 CHECK YOURSELF 10 _____

Use the method of Example 10 to find the LCM of 24 and 36.

The procedure is the same for a group of more than two numbers.

Example 11   Finding the Least Common Multiple

To find the LCM of 12, 18, and 20, we factor:

$$12 = 2 \cdot 2 \cdot 3$$
$$18 = 2 \qquad \cdot 3 \cdot 3$$

NOTE The different factors that appear are 2, 3, and 5.

$$\underline{20 = 2 \cdot 2 \qquad \cdot 5}$$
$$2 \cdot 2 \cdot 3 \cdot 3 \cdot 5$$

2 and 3 appear twice in one factorization, and 5 appears just once.

$$2 \cdot 2 \cdot 3 \cdot 3 \cdot 5 = 180$$

So 180 is the LCM of 12, 18, and 20.

✔ **CHECK YOURSELF 11** _____

Find the LCM of 3, 4, and 6.

The process of finding the least common multiple is very useful when adding, sub-tracting, or comparing unlike fractions (fractions with different denominators).

Suppose you are asked to compare the sizes of the fractions $\frac{3}{7}$ and $\frac{4}{7}$. Because each unit in the diagram is divided into seven parts, it is easy to see that $\frac{4}{7}$ is larger than $\frac{3}{7}$.

$\frac{4}{7}$ $\frac{3}{7}$

Four parts of seven are a greater portion than three parts. Now compare the size of the fractions $\frac{2}{5}$ and $\frac{1}{3}$.

$\frac{2}{5}$ $\frac{1}{3}$

We cannot readily compare fifths with thirds, say $\frac{2}{5}$ and $\frac{1}{3}$, because they name differ-ent ways of dividing the whole. Deciding which fraction is larger is not nearly so easy.

To compare the sizes of fractions, we change them to equivalent fractions having a *common denominator*. This common denominator must be a common multiple of the orig-inal denominators.

OBJECTIVE 3   **Example 12   Finding Common Denominators to Order Fractions**

Compare the sizes of $\frac{2}{5}$ and $\frac{1}{3}$.

The original denominators are 5 and 3. Because 15 is a common multiple of 5 and 3, we will use 15 as our common denominator.

NOTE $\frac{2}{5}$ and $\frac{6}{15}$ are **equivalent fractions.** They name the same portion of a whole.

$$\frac{2}{5} = \frac{6}{15}$$

Think, "What must we multiply 5 by to get 15?" The answer is 3. Multiply the numerator and denominator by that number.

$$\frac{1}{3} = \frac{5}{15}$$

Multiply the numerator and denominator by 5.

NOTE 6 of 15 parts represents a greater portion of the whole than 5 of 15 parts.

Because $\dfrac{2}{5} = \dfrac{6}{15}$ and $\dfrac{1}{3} = \dfrac{5}{15}$, we see that $\dfrac{2}{5}$ is larger than $\dfrac{1}{3}$.

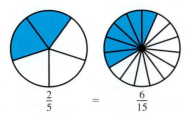

$$\frac{2}{5} \quad = \quad \frac{6}{15}$$

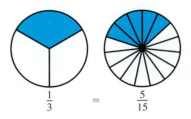

$$\frac{1}{3} \quad = \quad \frac{5}{15}$$

 CHECK YOURSELF 12

Which is larger, $\dfrac{1}{2}$ or $\dfrac{4}{7}$?

Now we consider an example that uses the inequality notation.

Example 13   Using an Inequality Symbol with Two Fractions

NOTE The inequality symbol "points" to the smaller quantity.

Use the inequality symbol $<$ or $>$ to complete the statement.

$$\frac{5}{8} \underline{} \frac{3}{5}$$

Once again we must compare the sizes of the two fractions, and this is done by converting the fractions to equivalent fractions with a common denominator. Here we will use 40 as that denominator.

NOTE The LCM of 8 and 5 is 40.

$$\overset{\times 5}{\frac{5}{8}} \underset{\times 5}{\frac{25}{40}} \qquad \overset{\times 8}{\frac{3}{5}} \underset{\times 8}{\frac{24}{40}}$$

Because $\dfrac{5}{8}\left(\text{or } \dfrac{25}{40}\right)$ is larger than $\dfrac{3}{5}\left(\text{or } \dfrac{24}{40}\right)$, we write

$$\frac{5}{8} > \frac{3}{5}$$

CHECK YOURSELF 13 _____

Use the symbol < or > to complete the statement.

$$\frac{5}{9} \underline{\qquad} \frac{6}{11}$$

Thus far, we have dealt only with finding sums and differences of like fractions (fractions with a common denominator). What about a sum that deals with *unlike fractions,* such as $\frac{1}{3} + \frac{1}{4}$?

RECALL Only *like* fractions can be added.

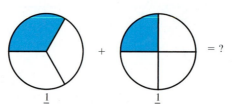

We cannot add unlike fractions because they have different denominators.

To add unlike fractions, write them as equivalent fractions with a common denominator. In this case, we will use 12 as the denominator.

NOTE We can now add because we have like fractions.

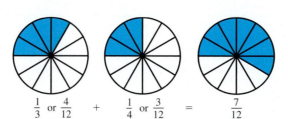

$\frac{1}{3}$ or $\frac{4}{12}$ + $\frac{1}{4}$ or $\frac{3}{12}$ = $\frac{7}{12}$

We have chosen 12 because it is a *multiple* of 3 and 4.
$\frac{1}{3}$ is equivalent to $\frac{4}{12}$.
$\frac{1}{4}$ is equivalent to $\frac{3}{12}$.

Any common multiple of the denominators will work in forming equivalent fractions. For instance, we can write $\frac{1}{3}$ as $\frac{8}{24}$ and $\frac{1}{4}$ as $\frac{6}{24}$. Our work is simplest, however, if we use the smallest possible number for the common denominator. This is called the **least common denominator (LCD).**

The LCD is the least common multiple of the denominators of the fractions. This is the *smallest* number that is a multiple of all the denominators. For example, the LCD for $\frac{1}{3}$ and $\frac{1}{4}$ is 12, *not* 24. Fractions can be added using any common denominator, but if the LCD is not used, the result will always need to be simplified.

NOTE This is virtually identical to the Step by Step on page 299 for finding the LCM.

Step by Step: To Find the Least Common Denominator

Step 1 Write the prime factorization for each of the denominators.
Step 2 Find all the prime factors that appear in any one of the prime factorizations.
Step 3 Form the product of those prime factors using each factor the greatest number of times it occurs in any one factorization.

We are now ready to add unlike fractions. In this case, the fractions must be renamed as equivalent fractions that have the same denominator. We will use this rule:

> **Step by Step: To Add Unlike Fractions**
>
> **Step 1**   Find the LCD of the fractions.
> **Step 2**   Change each unlike fraction to an equivalent fraction with the LCD as a common denominator.
> **Step 3**   Add the resulting like fractions as before.

Example 14 shows the use of this rule.

OBJECTIVE 4   **Example 14   Adding Unlike Fractions**

Add the fractions $\dfrac{1}{6}$ and $\dfrac{3}{8}$.

Step 1   We find that the LCD for fractions with denominators of 6 and 8 is 24.

Step 2   Convert the fractions so that they have the denominator 24.

$$\frac{1}{6} = \frac{4}{24}$$

How many sixes are in 24? There are 4. So multiply the numerator and denominator by 4.

$$\frac{3}{8} = \frac{9}{24}$$

How many eights are in 24? There are 3. So multiply the numerator and denominator by 3.

Step 3   We can now add the equivalent like fractions.

$$\frac{1}{6} + \frac{3}{8} = \frac{4}{24} + \frac{9}{24} = \frac{13}{24}$$ Add the numerators and place that sum over the common denominator.

CHECK YOURSELF 14 _____

Add.

$$\frac{3}{5} + \frac{1}{3}$$

Following is a similar example. Remember that the sum should always be written in simplest form.

Example 15 Adding Unlike Fractions that Require Simplifying

Add the fractions $\dfrac{-7}{10}$ and $\dfrac{-2}{15}$.

Step 1   The LCD for fractions with denominators of 10 and 15 is 30.

Step 2 $\dfrac{-7}{10} = \dfrac{-21}{30}$

$\dfrac{-2}{15} = \dfrac{-4}{30}$ Do you see how the equivalent fractions are formed?

Step 3 $\dfrac{-7}{10} + \dfrac{-2}{15} = \dfrac{-21}{30} + \dfrac{-4}{30}$

$= \dfrac{-25}{30} = \dfrac{-5}{6}$ Add the resulting like fractions. Be sure the sum is in simplest form.

 CHECK YOURSELF 15 _____

Add.

$\dfrac{1}{6} + \dfrac{7}{12}$

We can easily add more than two fractions by using the same procedure. Example 16 illustrates this approach.

Example 16 Adding a Group of Unlike Fractions

Add $\dfrac{5}{6} + \dfrac{2}{9} + \dfrac{4}{15}$.

Step 1   The LCD is 90.

Step 2 $\dfrac{5}{6} = \dfrac{75}{90}$ Multiply the numerator and denominator by 15.

$\dfrac{2}{9} = \dfrac{20}{90}$ Multiply the numerator and denominator by 10.

$\dfrac{4}{15} = \dfrac{24}{90}$ Multiply the numerator and denominator by 6.

Step 3 $\dfrac{75}{90} + \dfrac{20}{90} + \dfrac{24}{90} = \dfrac{119}{90}$ Now add.

 CHECK YOURSELF 16 _____

Add.

$\dfrac{2}{5} + \dfrac{3}{8} + \dfrac{7}{20}$

To subtract unlike fractions, which are fractions that do not have the same denominator, we have this rule:

> ### Step by Step: To Subtract Unlike Fractions
>
> **Step 1** Find the LCD of the fractions.
> **Step 2** Change each unlike fraction to an equivalent fraction with the LCD as a common denominator.
> **Step 3** Subtract the resulting like fractions as before.

NOTE Of course, this is the same as our rule for adding fractions. We just subtract instead of add!

Example 17   Subtracting Unlike Fractions

NOTE This example could also be written as $\dfrac{5}{8} + \dfrac{-1}{6}$

Subtract $\dfrac{5}{8} - \dfrac{1}{6}$.

Step 1 The LCD is 24.

Step 2 Convert the fractions so that they have the common denominator 24.

$$\frac{5}{8} = \frac{15}{24}$$
$$\frac{1}{6} = \frac{4}{24}$$

The first two steps are exactly the same as if we were adding.

NOTE You can use your calculator to check your result.

Step 3 Subtract the equivalent like fractions.

$$\frac{5}{8} - \frac{1}{6} = \frac{15}{24} - \frac{4}{24} = \frac{11}{24}$$

 CAUTION

Be Careful! You *cannot* simply subtract the numerators and the denominators.

$$\frac{5}{8} - \frac{1}{6} \quad \text{is not} \quad \frac{4}{2}$$

 **CHECK YOURSELF 17** _____

Subtract.

$$\frac{7}{10} - \frac{1}{4}$$

 # Using Your Calculator to Add and Subtract Fractions

Adding and subtracting fractions on the calculator is very much like the multiplication and division you did in Chapter 3. The only thing that changes is the operation.

 ## Scientific Calculator

Here's where the fraction calculator is a great tool for checking your work. No muss, no fuss, no searching for a common denominator. Just enter the fractions and get the right answer!

Example 18   Adding Fractions

Find the sum or difference.

(a) $\dfrac{-3}{14} + \dfrac{-7}{12}$

The keystroke sequence is

3 [+/−] [a b/c] 14 [+] 7 [+/−] [a b/c] 12 [=]

The result is $\dfrac{-67}{84}$.

(b) $\dfrac{-5}{8} - \dfrac{-7}{18}$

5 [+/−] [a b/c] 8 [−] 7 [+/−] [a b/c] 18 [=]

The result is $\dfrac{-17}{72}$.

Graphing Calculator

Use your graphing calculator to find the sum.

$$\dfrac{-7}{24} + \dfrac{-17}{42}$$

The keystroke sequence is

[(−)] 7 [÷] 24 [+] [(−)] 17 [÷] 42 [Math] [1: ▶ Frac] [Enter]

The result is $\dfrac{-39}{56}$.

CHECK YOURSELF 18

Find the sum or difference.

(a) $\dfrac{-5}{24} + \dfrac{-11}{18}$

(b) $\dfrac{-9}{11} - \dfrac{-1}{3}$

READING YOUR TEXT

The following fill-in-the-blank exercises are designed to assure that you understand the key vocabulary used in this section. Each sentence usually comes directly from the section. You will find the correct answers in Appendix C.

Section 4.1

(a) We can add fractions only when they have the same _____.

(b) We may think of the _____ of a number as its distance from 0 on the number line.

(c) The _____ of a number are the product of that number with the natural numbers.

(d) If a number is a multiple of each of a group of numbers, it is called a _____ multiple of the numbers.

CHECK YOURSELF ANSWERS

1. $\dfrac{2}{9} + \dfrac{5}{9} = \dfrac{2+5}{9} = \dfrac{7}{9}$ 2. $\dfrac{2}{3}$ 3. $\dfrac{9}{8}$ 4. $\dfrac{1}{2}$ 5. (a) $\dfrac{3}{9} = \dfrac{1}{3}$; (b) $\dfrac{-18}{35}$

6. (a) $\dfrac{-11}{11} = -1$; (b) $\dfrac{6}{7}$

7. The first seven multiples of 4 are 4, 8, 12, 16, 20, 24, 28, 32, and 36.

8. 6, 12, 18, 24, 30, 36; some common multiples of 4 and 6 are 12, 24, and 36.

9. The multiples of 20 are 20, 40, 60, 80, 100, 120, . . . ; the multiples of 30 are 30, 60, 90, 120, 150, . . . ; the least common multiple of 20 and 30 is 60, the smallest number common to both lists.

10. $2 \cdot 2 \cdot 2 \cdot 3 \cdot 3 = 72$ 11. 12   12. $\dfrac{4}{7}$ is larger   13. $\dfrac{5}{9} > \dfrac{6}{11}$

14. $\dfrac{14}{15}$ 15. $\dfrac{1}{6} + \dfrac{7}{12} = \dfrac{2}{12} + \dfrac{7}{12} = \dfrac{9}{12} = \dfrac{3}{4}$ 16. $\dfrac{9}{8}$ 17. $\dfrac{9}{20}$

18. (a) $\dfrac{-59}{72}$; (b) $\dfrac{-16}{33}$

Name _____

Section _____ Date _____

ANSWERS

1. _____
2. _____
3. _____
4. _____
5. _____
6. _____
7. _____
8. _____
9. _____
10. _____
11. _____
12. _____
13. _____
14. _____
15. _____
16. _____
17. _____ 18. _____
19. _____ 20. _____
21. _____ 22. _____
23. _____ 24. _____

308 SECTION 4.1

4.1 Exercises

Add or subtract as indicated. Write all answers in lowest terms.

1. $\dfrac{3}{5} + \dfrac{1}{5}$ ALEKS

2. $\dfrac{4}{7} + \dfrac{1}{7}$ ALEKS

3. $\dfrac{4}{11} + \dfrac{6}{11}$ ALEKS

4. $\dfrac{5}{16} + \dfrac{4}{16}$ ALEKS

5. $\dfrac{2}{10} + \dfrac{3}{10}$ ALEKS

6. $\dfrac{5}{12} + \dfrac{1}{12}$ ALEKS

7. $\dfrac{3}{7} + \dfrac{4}{7}$ ALEKS

8. $\dfrac{3}{20} + \dfrac{7}{20}$ ALEKS

9. $\dfrac{-9}{30} + \dfrac{-11}{30}$

10. $\dfrac{-4}{9} + \dfrac{-5}{9}$

11. $\dfrac{-13}{48} + \dfrac{-23}{48}$

12. $\dfrac{-17}{60} + \dfrac{-31}{60}$

13. $\dfrac{3}{7} + \dfrac{6}{7}$ ALEKS

14. $\dfrac{3}{5} + \dfrac{4}{5}$ ALEKS

15. $\dfrac{7}{10} - \dfrac{-9}{10}$

16. $\dfrac{5}{8} - \dfrac{-7}{8}$

17. $\dfrac{11}{12} - \dfrac{-10}{12}$

18. $\dfrac{13}{18} - \dfrac{-11}{18}$

19. $\dfrac{1}{8} + \dfrac{1}{8} + \dfrac{3}{8}$

20. $\dfrac{1}{10} + \dfrac{3}{10} + \dfrac{3}{10}$

21. $\dfrac{1}{9} + \dfrac{4}{9} + \dfrac{5}{9}$

22. $\dfrac{7}{12} + \dfrac{11}{12} + \dfrac{1}{12}$

23. $\dfrac{3}{5} + \dfrac{-1}{5}$

24. $\dfrac{5}{7} + \dfrac{-2}{7}$

25. $\dfrac{7}{9} + \dfrac{-4}{9}$

26. $\dfrac{7}{10} + \dfrac{-3}{10}$

27. $\dfrac{13}{20} - \dfrac{3}{20}$ ALEKS

28. $\dfrac{19}{30} - \dfrac{17}{30}$ ALEKS

29. $\dfrac{19}{24} - \dfrac{5}{24}$ ALEKS

30. $\dfrac{25}{36} - \dfrac{13}{36}$ ALEKS

31. $\dfrac{11}{12} - \dfrac{7}{12}$ ALEKS

32. $\dfrac{9}{10} - \dfrac{6}{10}$ ALEKS

33. $\dfrac{-8}{9} - \dfrac{-3}{9}$

34. $\dfrac{-5}{8} - \dfrac{-1}{8}$

Evaluate. Write all answers in lowest terms.

35. $\dfrac{7}{12} - \dfrac{4}{12} + \dfrac{3}{12}$ VIDEO

36. $\dfrac{8}{9} + \dfrac{3}{9} - \dfrac{5}{9}$

37. $\dfrac{6}{13} - \dfrac{3}{13} + \dfrac{11}{13}$

38. $\dfrac{9}{11} - \dfrac{3}{11} + \dfrac{7}{11}$

39. $\dfrac{18}{23} - \dfrac{13}{23} - \dfrac{3}{23}$

40. $\dfrac{17}{18} - \dfrac{11}{18} - \dfrac{5}{18}$

Find the least common multiple for each group of numbers. Use whichever method you wish.

41. 2 and 3 ALEKS

42. 3 and 5 ALEKS

43. 4 and 6   ALEKS

44. 6 and 9 ALEKS

45. 10 and 20 ALEKS

46. 12 and 36 ALEKS

47. 9 and 12 ALEKS

48. 20 and 30 ALEKS

49. 18, 21, and 28

50. 8, 15, and 20

51. 20, 30, and 45

52. 12, 20, and 35

ANSWERS

25. _____

26. _____

27. _____

28. _____

29. _____

30. _____

31. _____

32. _____

33. _____

34. _____

35. _____

36. _____

37. _____

38. _____

39. _____

40. _____

41. _____

42. _____

43. _____

44. _____

45. _____ 46. _____

47. _____ 48. _____

49. _____ 50. _____

51. _____ 52. _____

In exercises 53 and 54, indicate whether the given statement is always true, sometimes true, or never true.

53. The least common multiple of two numbers is smaller than each of the two numbers.

54. The greatest common factor of two numbers is smaller than each of the two numbers.

Arrange the given fractions from smallest to largest.

55. $\dfrac{11}{12}, \dfrac{4}{5}, \dfrac{5}{6}$

56. $\dfrac{5}{8}, \dfrac{9}{16}, \dfrac{13}{32}$

Complete the statements, using the symbol $<$ or $>$.

57. $\dfrac{5}{6} \underline{\hspace{1cm}} \dfrac{2}{5}$ **ALEKS**

58. $\dfrac{3}{4} \underline{\hspace{1cm}} \dfrac{10}{11}$ **ALEKS**

59. $\dfrac{5}{16} \underline{\hspace{1cm}} \dfrac{7}{20}$ **ALEKS**

60. $\dfrac{7}{12} \underline{\hspace{1cm}} \dfrac{9}{15}$ **ALEKS**

Find the least common denominator for fractions with the given denominators.

61. 48 and 80

62. 60 and 84

63. 3, 4, and 5

64. 3, 4, and 6

65. 8, 10, and 15

66. 6, 22, and 33

Add or subtract as indicated.

67. $\dfrac{2}{3} + \dfrac{1}{4}$ **ALEKS**

68. $\dfrac{-3}{5} + \dfrac{-1}{3}$ **ALEKS**

69. $\dfrac{-1}{5} + \dfrac{-3}{10}$ **ALEKS**

70. $\dfrac{1}{3} + \dfrac{1}{18}$ **ALEKS**

71. $\dfrac{5}{8} - \dfrac{-1}{12}$ **VIDEO**

72. $\dfrac{5}{12} - \dfrac{-3}{10}$

73. $\dfrac{1}{5} + \dfrac{7}{10} + \dfrac{4}{15}$

74. $\dfrac{2}{3} + \dfrac{1}{4} + \dfrac{3}{8}$

75. $\dfrac{4}{5} + \dfrac{-1}{3}$ **ALEKS**

76. $\dfrac{7}{9} + \dfrac{-1}{6}$ **ALEKS**

77. $\dfrac{11}{15} - \dfrac{3}{5}$ **ALEKS**

78. $\dfrac{5}{6} - \dfrac{2}{7}$ **ALEKS**

79. $\dfrac{-3}{8} - \dfrac{-1}{4}$

80. $\dfrac{-9}{10} - \dfrac{-4}{5}$

81. $\dfrac{15}{16} + \dfrac{5}{8} - \dfrac{1}{4}$

82. $\dfrac{9}{10} - \dfrac{1}{5} + \dfrac{1}{2}$

Solve the following applications.

83. Construction Find the combined thickness of subflooring that is made of $\dfrac{1}{8}$-inch plywood laid on $\dfrac{3}{4}$-inch plywood. **(4)**

84. Construction Find the combined thickness of subflooring that is made of $\dfrac{1}{4}$-inch plywood laid on $\dfrac{3}{8}$-inch plywood. **(4)**

Calculator Exercises

Find the following sums or differences using your calculator.

85. $\dfrac{2}{3} + \dfrac{-1}{2}$

86. $\dfrac{11}{12} + \dfrac{-5}{6}$

87. $\dfrac{3}{4} + \dfrac{-7}{9}$

88. $\dfrac{7}{11} + \dfrac{-5}{6}$

89. $\dfrac{5}{12} - \dfrac{-1}{6}$

90. $\dfrac{2}{7} - \dfrac{-3}{8}$

91. $\dfrac{15}{17} + \dfrac{-9}{11}$

92. $\dfrac{31}{42} + \dfrac{-18}{51}$

93. $\dfrac{-4}{9} - \dfrac{-2}{5}$

94. $\dfrac{-11}{13} - \dfrac{-2}{3}$

75. _____

76. _____

77. _____

78. _____

79. _____

80. _____

81. _____

82. _____

83. _____

84. _____

85. _____

86. _____

87. _____

88. _____

89. _____

90. _____

91. _____

92. _____

93. _____

94. _____

Answers

1. $\dfrac{4}{5}$ **3.** $\dfrac{10}{11}$ **5.** $\dfrac{2}{10} + \dfrac{3}{10} = \dfrac{5}{10} = \dfrac{1}{2}$ **7.** $\dfrac{3}{7} + \dfrac{4}{7} = \dfrac{7}{7} = 1$

9. $\dfrac{-2}{3}$ **11.** $\dfrac{-3}{4}$ **13.** $\dfrac{9}{7}$ **15.** $\dfrac{7}{10} + \dfrac{9}{10} = \dfrac{16}{10} = \dfrac{8}{5}$ **17.** $\dfrac{7}{4}$

19. $\dfrac{5}{8}$ **21.** $\dfrac{10}{9}$ **23.** $\dfrac{2}{5}$ **25.** $\dfrac{1}{3}$ **27.** $\dfrac{1}{2}$ **29.** $\dfrac{7}{12}$ **31.** $\dfrac{1}{3}$

33. $\dfrac{-5}{9}$ **35.** $\dfrac{1}{2}$ **37.** $\dfrac{14}{13}$ **39.** $\dfrac{2}{23}$ **41.** 6   **43.** 12   **45.** 20

47. 36   **49.** 252   **51.** 180   **53.** Never   **55.** $\dfrac{4}{5}, \dfrac{5}{6}, \dfrac{11}{12}$

57. $>$ **59.** $<$ **61.** 240   **63.** 60   **65.** 120   **67.** $\dfrac{11}{12}$ **69.** $\dfrac{-1}{2}$

71. $\dfrac{17}{24}$ **73.** $\dfrac{1}{5} + \dfrac{7}{10} + \dfrac{4}{15} = \dfrac{6}{30} + \dfrac{21}{30} + \dfrac{8}{30} = \dfrac{35}{30} = \dfrac{7}{6}$

75. $\dfrac{4}{5} - \dfrac{1}{3} = \dfrac{12}{15} - \dfrac{5}{15} = \dfrac{7}{15}$ **77.** $\dfrac{2}{15}$ **79.** $\dfrac{-1}{8}$ **81.** $\dfrac{21}{16}$ **83.** $\dfrac{7}{8}$ in.

85. $\dfrac{1}{6}$ **87.** $\dfrac{-1}{36}$ **89.** $\dfrac{7}{12}$ **91.** $\dfrac{12}{187}$ **93.** $\dfrac{-2}{45}$

4.2 Operations on Mixed Numbers

4.2 OBJECTIVES

1. Write improper fractions as mixed numbers
2. Write mixed numbers as improper fractions
3. Multiply two mixed numbers
4. Estimate products by rounding
5. Divide two mixed numbers
6. Add two mixed numbers
7. Subtract two mixed numbers

Another way to write a fraction whose absolute value is larger than 1 is called a *mixed number*.

> **Definition:** Mixed Number
>
> A **mixed number** is the sum of a whole number and a proper fraction.

> **Example 1** Identifying a Mixed Number

NOTE $2\frac{3}{4}$ means $2 + \frac{3}{4}$. In fact, we read the mixed number as "two *and* three-fourths." The addition sign is usually not written.

$2\frac{3}{4}$ is a mixed number. It represents the sum of the whole number 2 and the fraction $\frac{3}{4}$. Look at the diagram, which illustrates $2\frac{3}{4}$.

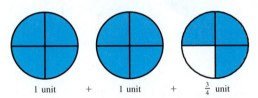

1 unit   +   1 unit   + $\frac{3}{4}$ unit

Notice that there are eleven *quarters*, and hence the improper fraction $\frac{11}{4}$.

✔ CHECK YOURSELF 1

Give the mixed number that names the shaded portion of the given diagram.

For our later work, it will be important to be able to change back and forth between improper fractions and mixed numbers. Because an improper fraction represents a number with absolute value greater than or equal to 1, we have this rule:

NOTE In subsequent courses, you will find that improper fractions are preferred to mixed numbers.

> **Property:** Improper Fractions to Mixed Numbers
>
> An improper fraction can always be written as either a mixed number or as a whole number.

RECALL You can write the fraction $\frac{7}{5}$ as $7 \div 5$. We divide the numerator by the denominator.

NOTE In step 1, the quotient gives the whole-number portion of the mixed number. Step 2 gives the fractional portion of the mixed number.

To do this, remember that you can think of a fraction as indicating division. The numerator is divided by the denominator. This leads us to the next rule:

Step by Step: To Change an Improper Fraction to a Mixed Number

Step 1 Divide the numerator by the denominator.

Step 2 If there is a remainder, write the remainder over the original denominator.

OBJECTIVE 1

Example 2 **Converting a Fraction to a Mixed Number**

Convert $\frac{17}{5}$ to a mixed number.

Divide 17 by 5.

$$\frac{17}{5} = 3\frac{2}{5}$$

Remainder — 2
Original denominator — 5
Quotient — 3

In diagram form:

$$\frac{17}{5} = 3\frac{2}{5}$$

 CHECK YOURSELF 2

Convert $\frac{32}{5}$ to a mixed number.

Example 3 **Converting a Fraction to a Mixed Number**

Convert $\frac{21}{7}$ to a mixed or a whole number.

Divide 21 by 7.

NOTE If there is *no* remainder, the improper fraction is equal to some whole number, in this case, the number is 3.

$$\begin{array}{r} 3 \\ 7)\overline{21} \\ \underline{21} \\ 0 \end{array} \qquad \frac{21}{7} = 3$$

 CHECK YOURSELF 3

Convert $\frac{48}{6}$ to a mixed or a whole number.

It is also easy to convert mixed numbers to improper fractions. Just use this rule:

Step by Step: To Change a Mixed Number to an Improper Fraction

Step 1   Multiply the denominator of the fraction by the whole-number portion of the mixed number.

Step 2   Add the numerator of the fraction to that product.

Step 3   Write that sum over the original denominator to form the improper fraction.

OBJECTIVE 2   **Example 4   Converting Mixed Numbers to Improper Fractions**

(a) Convert $3\frac{2}{5}$ to an improper fraction.

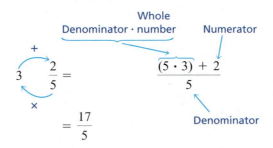

$$3\frac{2}{5} = \frac{(5 \cdot 3) + 2}{5}$$

Multiply the denominator by the whole number ($5 \cdot 3 = 15$). Add the numerator. We now have 17.

$$= \frac{17}{5}$$

Write 17 over the original denominator.

In diagram form:

Each of the three units has 5 fifths, so the whole-number portion is $5 \cdot 3$, or 15, fifths. Then add the 2 fifths from the fractional portion for 17 fifths.

(b) Convert $4\frac{5}{7}$ to an improper fraction.

NOTE Multiply the denominator, 7, by the whole number, 4, and add the numerator, 5.

$$4\frac{5}{7} = \frac{(7 \cdot 4) + 5}{7} = \frac{33}{7}$$

 CHECK YOURSELF 4

Convert $5\frac{3}{8}$ to an improper fraction.

When mixed numbers are involved in multiplication, the problem requires an additional step. First, change any mixed numbers to improper fractions. Then apply our multiplication rule for fractions.

> **Example 5** Multiplying a Mixed Number and a Fraction

$$1\frac{1}{2} \cdot \frac{3}{4} = \frac{3}{2} \cdot \frac{3}{4}$$ Change the mixed number to an improper fraction.

Here $1\frac{1}{2} = \frac{3}{2}$.

$$= \frac{3 \cdot 3}{2 \cdot 4}$$ Multiply as before.

$$= \frac{9}{8} = 1\frac{1}{8}$$ When the original problem includes mixed numbers, the product is usually written in mixed-number form.

 CHECK YOURSELF 5 _____

Multiply.

$$\frac{5}{8} \cdot 3\frac{1}{2}$$

If two mixed numbers are involved, change both of the mixed numbers to improper fractions. Example 6 illustrates this.

OBJECTIVE 3

> **Example 6** Multiplying Two Mixed Numbers

Multiply.

$$3\frac{2}{3} \cdot 2\frac{1}{2} = \frac{11}{3} \cdot \frac{5}{2}$$ Change the mixed numbers to improper fractions.

$$= \frac{11 \cdot 5}{3 \cdot 2} = \frac{55}{6} = 9\frac{1}{6}$$

CAUTION

NOTE This incorrect pattern leads to an answer of $6\frac{1}{3}$. The correct answer is $9\frac{1}{6}$.

Be Careful! Students sometimes think of

$$3\frac{2}{3} \cdot 2\frac{1}{2} \qquad \text{as} \qquad (3 \cdot 2) + \left(\frac{2}{3} \cdot \frac{1}{2}\right)$$

This is *not* the correct multiplication pattern. You must first change the mixed numbers to improper fractions before multiplying.

 CHECK YOURSELF 6 _____

Multiply.

$$2\frac{1}{3} \cdot 3\frac{1}{2}$$

The method of simplifying before multiplying two fractions that we studied in Section 3.4 can also be used when mixed numbers are involved. Consider Example 7.

Example 7   **Simplifying before Multiplying Two Mixed Numbers**

Multiply.

$$2\frac{2}{3} \cdot 2\frac{1}{4} = \frac{8}{3} \cdot \frac{9}{4}$$ First, convert the mixed numbers to improper fractions.

$$= \frac{\overset{2}{\cancel{8}} \cdot \overset{3}{\cancel{9}}}{\underset{1}{\cancel{3}} \cdot \underset{1}{\cancel{4}}}$$ To simplify, divide by the common factors of 3 and 4.

$$= \frac{2 \cdot 3}{1 \cdot 1}$$ Multiply as before.

$$= \frac{6}{1} = 6$$

 CHECK YOURSELF 7

Simplify and then multiply.

$$3\frac{1}{3} \cdot 2\frac{2}{5}$$

The ideas of Examples 1 to 7 also allow us to find the product of more than two fractions.

Example 8   **Simplifying before Multiplying Three Numbers**

Simplify and then multiply.

RECALL We can divide *any* factor of the numerator and *any* factor of the denominator by the same nonzero number.

$$\frac{2}{3} \cdot 1\frac{4}{5} \cdot \frac{5}{8} = \frac{2}{3} \cdot \frac{9}{5} \cdot \frac{5}{8}$$ Write any mixed or whole numbers as improper fractions.

$$= \frac{\overset{1}{\cancel{2}} \cdot \overset{3}{\cancel{9}} \cdot \overset{1}{\cancel{5}}}{\underset{1}{\cancel{3}} \cdot \underset{1}{\cancel{5}} \cdot \underset{4}{\cancel{8}}}$$ To simplify, divide by the common factors in the numerator and denominator.

$$= \frac{3}{4}$$

 CHECK YOURSELF 8

Simplify and then multiply.

$$\frac{5}{8} \cdot 4\frac{4}{5} \cdot \frac{1}{6}$$

We encountered estimation by rounding in our earlier work with whole numbers. Estimation can also be used to check the reasonableness of an answer when we are working with fractions or mixed numbers.

OBJECTIVE 4

NOTE In general, we round a mixed number to the nearest whole number. If the fractional part is less than $\frac{1}{2}$, we use the whole number part. If the fractional part is $\frac{1}{2}$ or more, we increase the whole number part by one.

Example 9 **Estimating the Product of Two Mixed Numbers**

Estimate the product of

$$3\frac{1}{8} \cdot 5\frac{5}{6}$$

Round each mixed number to the nearest whole number.

$$3\frac{1}{8} \rightarrow 3$$

$$5\frac{5}{6} \rightarrow 6$$

Our estimate of the product is then

$$3 \cdot 6 = 18$$

Note: The actual product in this case is $18\frac{11}{48}$, which certainly seems reasonable in view of our estimate.

 CHECK YOURSELF 9 _____

Estimate the product.

$$2\frac{7}{8} \cdot 8\frac{1}{3}$$

We learned how to divide fractions in Section 3.4. When mixed or whole numbers are involved, the process is similar. Simply change the mixed or whole numbers to improper fractions as the first step. Then proceed with the division rule. Example 10 illustrates this approach.

OBJECTIVE 5

Example 10 **Dividing Two Mixed Numbers**

Divide.

$$2\frac{3}{8} \div 1\frac{3}{4} = \frac{19}{8} \div \frac{7}{4}$$ Write the mixed numbers as improper fractions.

$$= \frac{19}{\underset{2}{\cancel{8}}} \cdot \frac{\overset{1}{\cancel{4}}}{7}$$ Invert the divisor and multiply as before.

$$= \frac{19}{14} = 1\frac{5}{14}$$

 CHECK YOURSELF 10 _____

Divide.

$$3\frac{1}{5} \div 2\frac{2}{5}$$

Example 11 illustrates the division process when a whole number is involved.

Example 11   Dividing a Mixed Number and a Whole Number

Divide and simplify.

NOTE Write the whole number 6 as $\dfrac{6}{1}$.

$$1\frac{4}{5} \div 6 = \frac{9}{5} \div \frac{6}{1}$$

$$= \frac{\overset{3}{\cancel{9}}}{5} \cdot \frac{1}{\underset{2}{\cancel{6}}} \qquad \text{Invert the divisor, then divide by the common factor of 3.}$$

$$= \frac{3}{10}$$

 CHECK YOURSELF 11 _____

Divide.

$$8 \div 4\frac{4}{5}$$

We learned how to add and subtract fractions in Section 4.1. Once you know how to add fractions, adding mixed numbers should be no problem if you keep in mind that addition involves combining groups of the *same kind* of objects. Because mixed numbers consist of two parts—a whole number and a fraction—we could work with the whole numbers and the fractions separately. Generally, it is easier to rewrite mixed numbers as improper fractions and then do the addition.

Consider $1\frac{1}{5} + 2\frac{2}{5}$.

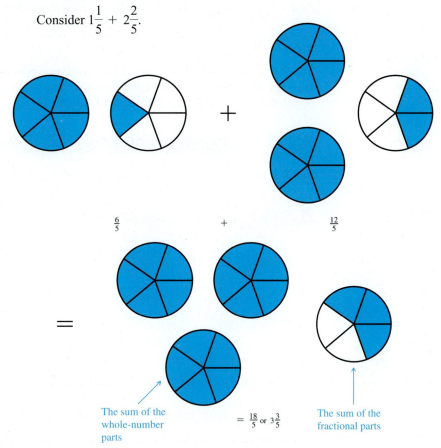

$\dfrac{6}{5}$ + $\dfrac{12}{5}$

The sum of the whole-number parts $= \dfrac{18}{5}$ or $3\dfrac{3}{5}$ The sum of the fractional parts

This suggests a general rule.

© 2010 McGraw-Hill Companies

> **Step by Step:** To Add Mixed Numbers
>
> **Step 1** Change the mixed numbers to improper fractions.
> **Step 2** Add the fractions.
> **Step 3** Rewrite the result as a mixed number.

NOTE Step 2 requires that the fractional parts have the same denominator.

Example 12 illustrates the use of this rule when adding two negative mixed numbers.

OBJECTIVE 6

Example 12 Adding Mixed Numbers

Add and write the result as a mixed number.

$$-3\frac{1}{5} + -4\frac{2}{5} = \frac{-16}{5} + \frac{-22}{5} \qquad \text{Rewrite as improper fractions.}$$

$$= \frac{-38}{5} \qquad \text{Add the numerators.}$$

$$= -7\frac{3}{5} \qquad \text{Rewrite as a mixed number.}$$

 CHECK YOURSELF 12 _____

Add $-2\frac{3}{10} + -3\frac{4}{10}$. Write the result as a mixed number.

When the fractional portions of the mixed numbers have different denominators, we must rename these fractions as equivalent fractions with the least common denominator to perform the addition in step 2. Consider Example 13.

Example 13 Adding Mixed Numbers with Different Denominators

Add and write the result as a mixed number.

$$3\frac{1}{6} + 2\frac{3}{8} = \frac{19}{6} + \frac{19}{8} \qquad \text{The LCD of the fractions is 24. Rename them with that denominator.}$$

$$= \frac{76}{24} + \frac{57}{24} \qquad \text{Then add as before.}$$

$$= \frac{133}{24}$$

$$= 5\frac{13}{24}$$

NOTE
$$\begin{array}{r} 5 \\ 24\overline{)133} \\ \underline{120} \\ 13 \end{array}$$

CHECK YOURSELF 13 _____

Add $5\frac{7}{10} + 3\frac{5}{6}$. Write the result as a mixed number.

You follow the same procedure if more than two mixed numbers are involved in the problem.

Example 14   Adding Mixed Numbers with Different Denominators

Add.

NOTE The LCD of the three denominators is 40. Convert to equivalent fractions.

$$2\frac{1}{5} + 3\frac{3}{4} + 4\frac{1}{8} = \frac{11}{5} + \frac{15}{4} + \frac{33}{8}$$

$$= \frac{88}{40} + \frac{150}{40} + \frac{165}{40}$$

$$= \frac{403}{40}$$

$$= 10\frac{3}{40}$$

 CHECK YOURSELF 14 _____

Add $5\frac{1}{2} + 4\frac{2}{3} + 3\frac{3}{4}$.

We can use a similar technique for *subtracting* mixed numbers. The rule is similar to that stated earlier for adding mixed numbers.

Step by Step: To Subtract Mixed Numbers

Step 1   Change the mixed numbers to improper fractions.
Step 2   Subtract the fractions.
Step 3   Rewrite the result as a mixed number.

Example 15 illustrates the use of this rule.

OBJECTIVE 7   **Example 15   Subtracting Mixed Numbers with Like Denominators**

Subtract.

$$5\frac{7}{12} - 3\frac{5}{12} = \frac{67}{12} - \frac{41}{12}$$

$$= \frac{26}{12}$$

$$= \frac{13}{6}$$

$$= 2\frac{1}{6}$$

CHECK YOURSELF 15 _____

Subtract $8\dfrac{7}{8} - 5\dfrac{3}{8}$.

Again, we must rename the fractions if different denominators are involved. This approach is shown in Example 16.

Example 16   Subtracting Mixed Numbers with Different Denominators

Subtract.

NOTE This example could also have been written as

$$8\dfrac{7}{10} + \left(-3\dfrac{3}{8}\right)$$

or

$$-3\dfrac{3}{8} + 8\dfrac{7}{10}$$

$$8\dfrac{7}{10} - 3\dfrac{3}{8} = \dfrac{87}{10} - \dfrac{27}{8} \qquad \text{Write the fractions with denominator 40.}$$

$$= \dfrac{348}{40} - \dfrac{135}{40} \qquad \text{Subtract as before.}$$

$$= \dfrac{213}{40}$$

$$= 5\dfrac{13}{40}$$

CHECK YOURSELF 16 _____

Subtract $7\dfrac{11}{12} - 3\dfrac{5}{8}$.

To subtract a mixed number from a whole number, we use the same techniques.

Example 17   Subtracting Mixed Numbers

Subtract.

$$6 - 2\dfrac{3}{4}$$

NOTE

$$6 = \dfrac{6}{1} = \dfrac{24}{4}$$

Multiply the numerator and denominator by 4 to form a common denominator.

$$6 - 2\dfrac{3}{4} = \dfrac{24}{4} - \dfrac{11}{4} \qquad \text{Write both the whole number and the mixed number as improper fractions with a common denominator.}$$

$$= \dfrac{13}{4}$$

$$= 3\dfrac{1}{4}$$

CHECK YOURSELF 17 _____

Subtract $7 - 3\dfrac{2}{5}$.

Using Your Calculator to Add and Subtract Mixed Numbers

We have already seen how to add, subtract, multiply, and divide fractions using our calculators. Now we will use our calculators to add and subtract mixed numbers.

Scientific Calculator

To enter a mixed number on a scientific calculator, press the fraction key between both the whole number and the numerator and denominator. For example, to enter $3\frac{7}{12}$, press

3 $\boxed{\textbf{a b/c}}$ 7 $\boxed{\textbf{a b/c}}$ 12

Example 18   Adding Mixed Numbers

Add.

$$-3\frac{7}{12} + 2\frac{11}{16}$$

The keystroke sequence is

3 $\boxed{\textbf{a b/c}}$ 7 $\boxed{\textbf{a b/c}}$ 12 $\boxed{+/-}$ $\boxed{+}$ 2 $\boxed{\textbf{a b/c}}$ 11 $\boxed{\textbf{a b/c}}$ 16 $\boxed{=}$

The result is $\dfrac{-43}{48}$.

Graphing Calculator

As with multiplying and dividing fractions, when using a graphing calculator, you must choose the fraction option from the math menu before pressing $\boxed{\text{Enter}}$.

For the problem in Example 18, $-3\frac{7}{12} + 2\frac{11}{16}$, the keystroke sequence is

$\boxed{(-)}$ $\boxed{(}$ 3 $\boxed{+}$ 7 $\boxed{\div}$ 12 $\boxed{)}$ $\boxed{+}$ 2 $\boxed{+}$ 11 $\boxed{\div}$ 16 $\boxed{\text{Math}}$ $\boxed{\blacktriangleright\textbf{Frac}}$ $\boxed{\text{Enter}}$

The display will read $\dfrac{-43}{48}$.

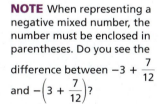

NOTE When representing a negative mixed number, the number must be enclosed in parentheses. Do you see the difference between $-3 + \dfrac{7}{12}$ and $-\left(3 + \dfrac{7}{12}\right)$?

✔ CHECK YOURSELF 18

Find the sum.

$$-8\frac{3}{7} + 4\frac{5}{6}$$

In this section, we learned to work with mixed numbers. In actual practice, and throughout the remainder of this text, mixed numbers are used almost exclusively when working with applications. For example, if, through the use of some formula, we discover that we need $\dfrac{13}{3}$ gal of paint, it makes far more sense to express that as $4\dfrac{1}{3}$ gal. On the other hand, if we solve an algebraic equation and find that $x = \dfrac{13}{3}$, we would leave the result as an improper fraction.

READING YOUR TEXT

The following fill-in-the-blank exercises are designed to assure that you understand the key vocabulary used in this section. Each sentence usually comes directly from the section. You will find the correct answers in Appendix C.

Section 4.2

(a) A _____ number is the sum of a whole number and a proper fraction.

(b) An improper fraction can always be written as either a mixed number or as a _____ number.

(c) If two mixed numbers are involved in multiplication, change both of the mixed numbers to _____ fractions.

(d) Adding mixed numbers requires that the fractional parts have the same _____.

CHECK YOURSELF ANSWERS

1. $3\dfrac{5}{6}$ **2.** $6\dfrac{2}{5}$ **3.** 8 **4.** $\dfrac{43}{8}$ **5.** $\dfrac{5}{8} \cdot 3\dfrac{1}{2} = \dfrac{5}{8} \cdot \dfrac{7}{2} = \dfrac{35}{16} = 2\dfrac{3}{16}$ **6.** $8\dfrac{1}{6}$

7. $3\dfrac{1}{3} \cdot 2\dfrac{2}{5} = \dfrac{10}{3} \cdot \dfrac{12}{5} = \dfrac{\overset{2}{\cancel{10}} \cdot \overset{4}{\cancel{12}}}{\underset{1}{\cancel{3}} \cdot \underset{1}{\cancel{5}}} = \dfrac{8}{1} = 8$ **8.** $\dfrac{1}{2}$ **9.** 24

10. $3\dfrac{1}{5} \div 2\dfrac{2}{5} = \dfrac{16}{5} \div \dfrac{12}{5} = \dfrac{\overset{4}{\cancel{16}}}{\underset{1}{\cancel{5}}} \cdot \dfrac{\overset{1}{\cancel{5}}}{\underset{3}{\cancel{12}}} = \dfrac{4}{3} = 1\dfrac{1}{3}$ **11.** $1\dfrac{2}{3}$

12. $-5\dfrac{7}{10}$ **13.** $5\dfrac{7}{10} + 3\dfrac{5}{6} = \dfrac{57}{10} + \dfrac{23}{6} = \dfrac{171}{30} + \dfrac{115}{30} = \dfrac{286}{30} = 9\dfrac{16}{30} = 9\dfrac{8}{15}$

14. $13\dfrac{11}{12}$ **15.** $3\dfrac{1}{2}$ **16.** $7\dfrac{11}{12} - 3\dfrac{5}{8} = \dfrac{95}{12} - \dfrac{29}{8} = \dfrac{190}{24} - \dfrac{87}{24} = \dfrac{103}{24} = 4\dfrac{7}{24}$

17. $3\dfrac{3}{5}$ **18.** $-3\dfrac{25}{42}$

4.2 Exercises

Identify each number as a proper fraction, an improper fraction, or a mixed number.

1. $\dfrac{3}{5}$

2. $\dfrac{-9}{5}$

3. $2\dfrac{3}{5}$

4. $\dfrac{7}{9}$

5. $\dfrac{6}{6}$

6. $-1\dfrac{1}{5}$

7. $\dfrac{11}{8}$

8. $\dfrac{8}{8}$

9. $4\dfrac{5}{7}$

10. $-5\dfrac{3}{7}$

11. $\dfrac{-13}{17}$

12. $\dfrac{16}{15}$

Give the mixed number that names the shaded portion of each diagram.

13.

14.

15.

16.

Change to a mixed or whole number.

17. $\dfrac{22}{5}$

18. $\dfrac{27}{8}$

19. $\dfrac{-34}{5}$

20. $\dfrac{-25}{6}$

Name _____

Section _____ Date _____

ANSWERS

1. _____
2. _____
3. _____
4. _____
5. _____
6. _____
7. _____
8. _____
9. _____
10. _____
11. _____
12. _____
13. _____
14. _____
15. _____
16. _____
17. _____
18. _____
19. _____
20. _____

21. _____

22. _____

23. _____

24. _____

25. _____

26. _____

27. _____

28. _____

29. _____

30. _____

31. _____

32. _____

33. _____

34. _____

35. _____

36. _____

37. _____

38. _____

39. _____

40. _____

41. _____

42. _____

Change to an improper fraction.

21. 8

22. $-4\dfrac{5}{8}$

23. $-6\dfrac{2}{9}$

24. 7

25. $3\dfrac{3}{7}$ ALEKS

26. $2\dfrac{2}{9}$ ALEKS

Multiply. Write each answer as a mixed number, whole number, or proper fraction.

27. $\dfrac{4}{9} \cdot 3\dfrac{3}{5}$ ALEKS

28. $5\dfrac{1}{3} \cdot \dfrac{7}{8}$ ALEKS

29. $\dfrac{-10}{27} \cdot 3\dfrac{3}{5}$

30. $-1\dfrac{1}{3} \cdot 1\dfrac{1}{5}$ ALEKS

31. $4\dfrac{1}{5} \cdot \dfrac{10}{21} \cdot \dfrac{9}{20}$

32. $\dfrac{7}{8} \cdot 5\dfrac{1}{3} \cdot \dfrac{-5}{14}$

Estimate the products.

33. $3\dfrac{1}{5} \cdot 4\dfrac{2}{3}$

34. $5\dfrac{1}{7} \cdot 2\dfrac{2}{13}$

35. $11\dfrac{3}{4} \cdot 5\dfrac{1}{4}$

36. $3\dfrac{4}{5} \cdot 5\dfrac{6}{7}$

37. $8\dfrac{2}{9} \cdot 7\dfrac{11}{12}$

38. $\dfrac{9}{10} \cdot 2\dfrac{2}{7}$

Find the reciprocal.

39. $2\dfrac{1}{3}$

40. $4\dfrac{3}{5}$

41. $9\dfrac{3}{4}$

42. $5\dfrac{1}{8}$

Divide. Write each result in simplest form.

43. $5\dfrac{3}{5} \div \dfrac{7}{15}$ **ALEKS**

44. $\dfrac{-7}{18} \div 5\dfrac{5}{6}$

45. $-1\dfrac{1}{3} \div 1\dfrac{1}{7}$

46. $3\dfrac{1}{2} \div 2\dfrac{4}{5}$ **ALEKS**

47. $3\dfrac{3}{4} \div 1\dfrac{3}{8}$ **ALEKS**

48. $5\dfrac{1}{3} \div 2\dfrac{2}{5}$ **ALEKS**

Do the indicated operations.

49. $2\dfrac{2}{9} + 3\dfrac{5}{9}$ **ALEKS**

50. $5\dfrac{2}{9} + 6\dfrac{4}{9}$ **ALEKS**

51. $-7\dfrac{7}{8} + 3\dfrac{3}{8}$

52. $-3\dfrac{5}{6} + 1\dfrac{1}{6}$

53. $3\dfrac{2}{5} - 1\dfrac{4}{5}$ **ALEKS**

54. $5\dfrac{3}{7} - 2\dfrac{1}{7}$ **ALEKS**

55. $3\dfrac{2}{3} - 2\dfrac{1}{4}$ **ALEKS**

56. $5\dfrac{4}{5} - 1\dfrac{1}{6}$ **ALEKS**

57. $-7\dfrac{5}{12} + 3\dfrac{11}{18}$

58. $-9\dfrac{3}{7} + 2\dfrac{13}{21}$

59. $5 - 2\dfrac{1}{4}$

60. $4 - 1\dfrac{2}{3}$

61. $3\dfrac{3}{4} + 5\dfrac{1}{2} - 2\dfrac{3}{8}$

62. $1\dfrac{5}{6} + 3\dfrac{5}{12} - 2\dfrac{1}{4}$

63. $2\dfrac{3}{8} + 2\dfrac{1}{4} - 1\dfrac{5}{6}$

64. $1\dfrac{1}{15} + 3\dfrac{3}{10} - 2\dfrac{4}{5}$

In exercises 65 and 66, indicate whether the given statement is always true, sometimes true, or never true.

65. When multiplying mixed numbers, we need to rewrite them with common denominators.

66. A mixed number can be written as an improper fraction.

ANSWERS

43. _____

44. _____

45. _____

46. _____

47. _____

48. _____

49. _____

50. _____

51. _____

52. _____

53. _____

54. _____

55. _____

56. _____

57. _____

58. _____

59. _____

60. _____

61. _____

62. _____

63. _____   64. _____

65. _____

66. _____

67. _____

68. _____

69. _____

70. _____

71. _____

72. _____

73. _____

74. _____

75. _____

76. _____

77. _____

78. _____

67. Is there any number that is equal to its own reciprocal? Find all such numbers.

68. Complete the following.
When multiplying or dividing mixed numbers, the first step is always to:

Solve the following applications.

69. Construction In building a new kitchen countertop, a top piece that is $\frac{3}{4}$-inch thick is to be laid on a base piece that is $\frac{5}{8}$-inch thick. What will be the total thickness of the countertop? ⟨4⟩

70. Construction In building a new kitchen countertop, a top piece that is $\frac{7}{8}$-inch thick is to be laid on two base pieces that are each $\frac{3}{8}$-inch thick. What will be the total thickness of the countertop? ⟨4⟩

71. Construction A bolt that is $2\frac{1}{4}$ inches long is placed through two sheets of plywood that are each $\frac{7}{8}$-inch thick. How much of the bolt protrudes through the sheets of plywood? ⟨4⟩

72. Construction A bolt that is $1\frac{7}{8}$ inches long is placed through two sheets of plywood that are $\frac{5}{8}$-inch thick and $\frac{1}{2}$-inch thick. How much of the bolt protrudes through the sheets of plywood? ⟨4⟩

Calculator Exercises

Add or subtract.

73. $3\frac{2}{3} + 2\frac{1}{4}$

74. $6\frac{1}{6} + 8\frac{2}{3}$

75. $-5\frac{4}{9} + \left(-2\frac{2}{3}\right)$

76. $-2\frac{3}{7} + \left(-4\frac{9}{14}\right)$

77. $11\frac{2}{3} + 5\frac{1}{4}$

78. $6\frac{3}{8} + 14\frac{5}{12}$

79. $14\frac{13}{18} + 22\frac{23}{27}$

80. $3\frac{41}{45} + 5\frac{25}{27}$

81. $-4\frac{7}{9} + 2\frac{11}{18}$

82. $-7\frac{8}{11} + 4\frac{13}{22}$

83. $5\frac{11}{16} - 2\frac{5}{12}$

84. $18\frac{5}{24} - 11\frac{3}{40}$

85. $-6\frac{2}{3} + 1\frac{5}{6}$

86. $-131\frac{43}{45} + 99\frac{27}{60}$

87. $10\frac{2}{3} + 4\frac{1}{5} + 7\frac{2}{15}$

88. $7\frac{1}{5} + 3\frac{2}{3} + 1\frac{1}{5}$

79.

80.

81.

82.

83.

84.

85.

86.

87.

88.

Answers

1. Proper   **3.** Mixed number   **5.** Improper   **7.** Improper

9. Mixed number   **11.** Proper   **13.** $1\frac{3}{4}$ **15.** $3\frac{5}{8}$ **17.** $4\frac{2}{5}$

19. $-6\frac{4}{5}$ **21.** $\frac{8}{1}$ **23.** $\frac{-56}{9}$ **25.** $3\frac{3}{7} = \frac{(7 \cdot 3) + 3}{7} = \frac{24}{7}$

27. $1\frac{3}{5}$ **29.** $-1\frac{1}{3}$ **31.** $\frac{9}{10}$ **33.** 15   **35.** 60   **37.** 64

39. $\frac{3}{7}$ **41.** $\frac{4}{39}$ **43.** 12   **45.** $-1\frac{1}{6}$ **47.** $2\frac{8}{11}$ **49.** $5\frac{7}{9}$

51. $-4\frac{1}{2}$ **53.** $1\frac{3}{5}$ **55.** $1\frac{5}{12}$ **57.** $-3\frac{29}{36}$ **59.** $2\frac{3}{4}$ **61.** $6\frac{7}{8}$

63. $2\frac{19}{24}$ **65.** Never   **67.** Yes; 1   **69.** $1\frac{3}{8}$ in.   **71.** $\frac{1}{2}$ in.

73. $5\frac{11}{12}$ **75.** $-8\frac{1}{9}$ **77.** $16\frac{11}{12}$ **79.** $37\frac{31}{54}$ **81.** $-2\frac{1}{6}$

83. $3\frac{13}{48}$ **85.** $-4\frac{5}{6}$ **87.** 22

4.3 Applications Involving Fractions

4.3 OBJECTIVE

1. Solve applications involving fractions

Our knowledge of fractions is critical in solving many applied problems. For example, you probably do many conversions between mixed and whole numbers without even thinking about the process that you follow, as Example 1 illustrates.

OBJECTIVE 1

Example 1 Converting Quarter Dollars to Dollars

Maritza has 53 quarters in her bank. How many dollars does she have?

Because there are 4 quarters in each dollar, 53 quarters can be written as

$$\frac{53}{4}$$

Converting the amount to dollars is the same as rewriting it as a mixed number.

$$\frac{53}{4} = 13\frac{1}{4}$$

She has $13\frac{1}{4}$ dollars, which you would probably write as $13.25. (*Note:* We will discuss decimal point usage in Chapter 5.)

 CHECK YOURSELF 1 _____

Kevin is doing the inventory in the convenience store in which he works. He finds there are 11 half gallons of milk. Write the amount of milk as a mixed number of gallons.

RECALL A **denominate number** has a unit of measure attached to it.

Units Analysis

When dividing two denominate numbers, the units are also divided. This yields a unit in fraction form.

Examples

$$250 \text{ mi} \div 10 \text{ gal} = \frac{250 \text{ mi}}{10 \text{ gal}} = \frac{25 \text{ mi}}{1 \text{ gal}} = 25 \, \frac{\text{mi}}{\text{gal}} \text{ (read "miles per gallon")}$$

$$360 \text{ ft} \div 30 \text{ s} = \frac{360 \text{ ft}}{30 \text{ s}} = 12 \, \frac{\text{ft}}{\text{s}} \text{ ("feet per second")}$$

When we multiply denominate numbers that have these units in fraction form, they behave just like fractions.

Examples

$$25 \, \frac{\text{mi}}{\text{gal}} \cdot 12 \text{ gal} = \frac{25 \text{ mi}}{1 \text{ gal}} \cdot \frac{12 \text{ gal}}{1} = 300 \text{ mi}$$

(If we look at the units, we see that the gallons essentially "cancel" when one is in the numerator and the other is in the denominator.)

$$12 \, \frac{\text{ft}}{\text{s}} \cdot 60 \, \frac{\text{s}}{\text{min}} = \frac{12 \text{ ft}}{1 \text{ s}} \cdot \frac{60 \text{ s}}{1 \text{ min}} = \frac{720 \text{ ft}}{1 \text{ min}} = 720 \, \frac{\text{ft}}{\text{min}}$$

(Again, the *seconds* cancel, leaving *feet* in the numerator and *minutes* in the denominator.)

Now we will look at some applications of our work with fraction operations. In solving these word problems, we will use the same approach we used earlier with whole numbers. We will review the four-step process introduced in Section 1.2.

Step by Step: Solving Applications Involving Fractions

Step 1   Read the problem carefully to determine the given information and what you are asked to find.

Step 2   Decide upon the operation or operations to be used.

Step 3   Write down the complete statement necessary to solve the problem and do the calculations.

Step 4   Check to make sure that you have answered the question of the problem and that your answer seems reasonable.

Now we will work through some examples, using these steps.

Example 2   An Application Involving Multiplication

Lisa worked $10\frac{1}{4}\frac{h}{day}$ for 5 days. How many hours did she work?

Step 1   We are looking for the total hours Lisa worked.

Step 2   We will multiply the hours per day by the days.

Step 3 $10\frac{1}{4}\frac{h}{day} \cdot 5 \text{ days} = \frac{41}{4}\frac{h}{day} \cdot 5 \text{ days} = \frac{205}{4} h = 51\frac{1}{4} h$

Step 4   Note the *days* cancel, leaving only the unit *hours*. The units should always be compared to the desired units from step 1. The answer also seems reasonable. An answer like 5 h or 500 h would not seem reasonable.

 CHECK YOURSELF 2 _____

Carlos worked $8\frac{1}{2}$ h per day for 6 days. How many hours did he work?

In Example 3, we will follow the four steps for solving applications, but we won't label the steps. You should still think about these steps as we solve the problem.

Example 3   An Application Involving the Multiplication of Mixed Numbers

A sheet of notepaper is $6\frac{3}{4}$ in. wide by $8\frac{2}{3}$ in. long. Find the area of the paper.

Multiply the length by the width. This will give the desired area. First, we will estimate the area.

$7 \text{ in.} \cdot 9 \text{ in.} = 63 \text{ in.}^2$

Now, we will find the exact area.

$$8\frac{2}{3} \text{ in.} \cdot 6\frac{3}{4} \text{ in.} = \frac{26}{3} \text{ in.} \cdot \frac{27}{4} \text{ in.}$$

$$= \frac{117}{2} \text{ in.}^2$$

$$= 58\frac{1}{2} \text{ in.}^2$$

RECALL The area of a rectangle is the product of its length and its width.

The units (square inches) are units of area. Note from our estimate that the result is reasonable.

 CHECK YOURSELF 3

A window is $4\frac{1}{2}$ ft high by $2\frac{1}{3}$ ft wide. What is its area?

> **Example 4** **An Application Involving the Multiplication of a Mixed Number and a Fraction**

A state park contains $38\frac{2}{3}$ acres. According to the plan for the park, $\frac{3}{4}$ of the park is to be left as a wildlife preserve. How many acres will this be?

RECALL The word *of* indicates multiplication.

We want to find $\frac{3}{4}$ *of* $38\frac{2}{3}$ acres. We then multiply as shown:

$$\frac{3}{4} \cdot 38\frac{2}{3} = \frac{\overset{1}{\cancel{3}}}{\underset{1}{\cancel{4}}} \cdot \frac{\overset{29}{\cancel{116}}}{\underset{1}{\cancel{3}}} \text{ acres} = 29 \text{ acres}$$

 CHECK YOURSELF 4

A backyard has $25\frac{3}{4}$ yd² of open space. If Patrick wants to build a vegetable garden covering $\frac{2}{3}$ of the open space, how many square yards will this be?

Units Analysis

When dividing by denominate numbers that have fractional units, we multiply by the reciprocal of the number *and its units*.

Examples

$$500 \text{ mi} \div \frac{25 \text{ mi}}{1 \text{ gal}} = 500 \text{ mi} \cdot \frac{1 \text{ gal}}{25 \text{ mi}} = 20 \text{ gal}$$

$$\$24,000 \div \frac{\$400}{1 \text{ yr}} = \$24,000 \cdot \frac{1 \text{ yr}}{\$400} = 60 \text{ yr}$$

(Note that in each case, the arithmetic of the units produces the final units.)

As was the case with multiplication, our work with the division of fractions will be used in the solution of a variety of applications. The steps of the problem-solving process remain the same.

> **Example 5   An Application Involving the Division of Mixed Numbers**

NOTE A kilometer, abbreviated km, is a metric unit of distance. It is about $\frac{3}{5}$ mi.

Jack traveled 140 km in $2\frac{1}{3}$ h. What was his average speed?

NOTE The important formula is speed = distance ÷ time.

$$\text{distance} \qquad \text{time}$$

$$\text{speed} = 140 \text{ km} \div 2\frac{1}{3} \text{ h}$$

We know the distance traveled and the time for that travel. To find the *average* speed, we must use division. Do you remember why?

$$= \frac{140}{1} \text{ km} \div \frac{7}{3} \text{ h}$$

$$= \frac{\overset{20}{\cancel{140}}}{1} \cdot \frac{3}{\underset{1}{\cancel{7}}} \frac{\text{km}}{\text{h}}$$

$\dfrac{\text{km}}{\text{h}}$ is read "kilometers per hour." This is a unit of speed.

$$= 60 \frac{\text{km}}{\text{h}}$$

 CHECK YOURSELF 5

A light plane flew 280 mi in $1\frac{3}{4}$ h. What was its average speed?

> **Example 6   An Application Involving the Division of Mixed Numbers**

An electrician needs pieces of wire $2\frac{3}{5}$ in. long. If she has a $20\frac{4}{5}$-in. piece of wire, how many of the shorter pieces can she cut?

NOTE We must divide the length of the longer piece by the desired length of the shorter piece.

$$20\frac{4}{5} \div 2\frac{3}{5} = \frac{104}{5} \div \frac{13}{5}$$

$$= \frac{\overset{8}{\cancel{104}}}{\underset{1}{\cancel{5}}} \cdot \frac{\overset{1}{\cancel{5}}}{\underset{1}{\cancel{13}}}$$

$$= 8 \text{ pieces}$$

 CHECK YOURSELF 6

A piece of plastic water pipe 63 in. long is to be cut into lengths of $3\frac{1}{2}$ in. How many of the shorter pieces can be cut?

Many applications can be solved by adding fractions.

Example 7 **An Application Involving the Addition of Like Fractions**

Noel walked $\frac{9}{10}$ mi to Jensen's house and then walked $\frac{7}{10}$ mi to school. How far did Noel walk?

$$\frac{9}{10} \text{ mi} \qquad \frac{7}{10} \text{ mi}$$

To find the total distance Noel walked, add the two distances.

$$\frac{9}{10} + \frac{7}{10} = \frac{16}{10} = 1\frac{6}{10} = 1\frac{3}{5}$$

Noel walked $1\frac{3}{5}$ mi.

 CHECK YOURSELF 7

Emir bought $\frac{7}{16}$ lb of candy at one store and $\frac{11}{16}$ lb at another store. How much candy did Emir buy?

Many of the measurements you deal with in everyday life involve fractions. Now we will look at some typical situations.

Example 8 **An Application Involving the Addition of Unlike Fractions**

Jack's doctor wants him to run at least 2 mi per week. Jack ran $\frac{1}{2}$ mi on Monday, $\frac{2}{3}$ mi on Wednesday, and $\frac{3}{4}$ mi on Friday. How far did he run during the week? Did he meet the goal set by his doctor?

The three distances that Jack ran are the given information in the problem. We want to find a total distance, so we must add for the solution.

$$\frac{1}{2} + \frac{2}{3} + \frac{3}{4} = \frac{6}{12} + \frac{8}{12} + \frac{9}{12}$$

Because we have no common denominator, we must convert to equivalent fractions before we can add.

$$= \frac{23}{12} = 1\frac{11}{12} \text{ mi}$$

Jack ran $1\frac{11}{12}$ mi during the week. He did not quite meet his goal.

 CHECK YOURSELF 8

Susan is designing an office complex. She needs $\frac{2}{5}$ acre for buildings, $\frac{1}{3}$ acre for driveways and parking, and $\frac{1}{6}$ acre for walks and landscaping. How much land does she need?

Example 9   An Application Involving the Addition of Unlike Fractions

Sam bought three packages of spices weighing $\frac{1}{4}$, $\frac{5}{8}$, and $\frac{1}{2}$ lb. What was the total weight?

We need to find the total weight, so we must add.

NOTE The abbreviation for pounds is *lb* from the Latin *libra*, meaning "balance" or "scales."

$$\frac{1}{4} + \frac{5}{8} + \frac{1}{2} = \frac{2}{8} + \frac{5}{8} + \frac{4}{8}$$ Write each fraction with the denominator 8.

$$= \frac{11}{8} = 1\frac{3}{8} \text{ lb}$$

The total weight was $1\frac{3}{8}$ lb.

CHECK YOURSELF 9 _____

For three different recipes, Max needs $\frac{3}{8}$, $\frac{1}{2}$, and $\frac{5}{8}$ gal tomato sauce. How many gallons should he buy altogether?

We will next look at an example that applies our work in subtracting unlike fractions.

Example 10   An Application Involving the Subtraction of Unlike Fractions

You have $\frac{7}{8}$ yd of a handwoven linen. A pattern for a placemat calls for $\frac{1}{2}$ yd. Will you have enough left for two napkins that will use a total of $\frac{1}{3}$ yd?

Find out how much fabric is left over after the placemat is made.

$$\frac{7}{8} \text{ yd} - \frac{1}{2} \text{ yd} = \frac{7}{8} \text{ yd} - \frac{4}{8} \text{ yd} = \frac{3}{8} \text{ yd}$$

Now compare the size of $\frac{1}{3}$ and $\frac{3}{8}$.

NOTE Remember that $\frac{3}{8}$ yd is left over and that $\frac{1}{3}$ yd is needed.

$$\frac{3}{8} \text{ yd} = \frac{9}{24} \text{ yd} \qquad \text{and} \qquad \frac{1}{3} \text{ yd} = \frac{8}{24} \text{ yd}$$

Because $\frac{3}{8}$ yd is *more than* the $\frac{1}{3}$ yd that is needed, there is enough material for the placemat *and* two napkins.

CHECK YOURSELF 10 _____

A concrete walk will require $\frac{3}{4}$ yd³ of concrete. If you have mixed $\frac{8}{9}$ yd³, will enough

concrete remain to do a project that will use $\frac{1}{6}$ yd³? (Note: yd³ is a measure of volume.)

Our next application involves measurement in inches. Note that on a ruler or yardstick, the marks divide each inch into $\frac{1}{2}$-in., $\frac{1}{4}$-in., and $\frac{1}{8}$-in. sections, and on some rulers, $\frac{1}{16}$-in. sections. We will use denominators of 2, 4, 8, and 16 in our measurement applications.

> **Example 11 An Application Involving the Subtraction of Unlike Fractions**

Alexei is cutting two slats that are each to be $\frac{3}{16}$ in. wide from a piece of wood that is $1\frac{3}{4}$ in. across. How much will be left?

The two $\frac{3}{16}$-in. pieces will total

$$2 \cdot \frac{3}{16} = \frac{6}{16} = \frac{3}{8} \text{ in.}$$

$$1\frac{3}{4} = \frac{7}{4} = \frac{14}{8}$$

$$\frac{14}{8} - \frac{3}{8} = \frac{11}{8}$$

The remaining strip will be $\frac{11}{8}$ in. or $1\frac{3}{8}$ in. wide.

CHECK YOURSELF 11 _____

Ricardo is cutting three strips from a piece of metal with a width of 1 in. Each strip has a width of $\frac{3}{16}$ in. How much metal will remain after the cuts?

Often we will have to use more than one operation to find the solution to a problem. Consider Example 12.

Example 12   An Application Involving Mixed Numbers

A rectangular poster is to have a total length of $12\frac{1}{4}$ in. We want a $1\frac{3}{8}$-in. border on the top and a 2-in. border on the bottom. What is the length of the printed part of the poster?

First, we will draw a sketch of the poster:

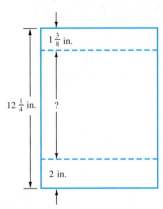

Now, we will use that sketch to find the total width of the top and bottom borders.

$$1\frac{3}{8} + 2 = \frac{11}{8} + \frac{16}{8} = \frac{27}{8} \text{ in.}$$

Now *subtract* that sum (the top and bottom borders) from the total length of the poster.

$$12\frac{1}{4} - \frac{27}{8} = \frac{49}{4} - \frac{27}{8} = \frac{98}{8} - \frac{27}{8}$$

$$= \frac{71}{8} = 8\frac{7}{8} \text{ in.}$$

The length of the printed part is $8\frac{7}{8}$ in.

CHECK YOURSELF 12 _____

You cut one shelf $3\frac{3}{4}$ ft long and one $4\frac{1}{2}$ ft long from a 12-ft piece of lumber. Can you cut another shelf 4 ft long?

READING YOUR TEXT _____

The following fill-in-the-blank exercises are designed to assure that you understand the key vocabulary used in this section. Each sentence usually comes directly from the section. You will find the correct answers in Appendix C.

Section 4.3

(a) A _____ number has a unit of measure attached to it.

(b) In problem solving, the last step is to make sure that you have answered the question of the problem and that your answer seems _____.

(c) The _____ of a rectangle is the product of its length and its width.

(d) An important formula is _____ = distance ÷ time.

CHECK YOURSELF ANSWERS

1. $\dfrac{11}{2} = 5\dfrac{1}{2}$ gal   2. 51 h   3. $10\dfrac{1}{2}$ ft^2 4. $17\dfrac{1}{6}$ yd^2 5. $160\dfrac{\text{mi}}{\text{h}}$

6. 18 pieces   7. $1\dfrac{1}{8}$ lb   8. $\dfrac{9}{10}$ acre   9. $1\dfrac{1}{2}$ gal

10. $\dfrac{5}{36}$ yd^3 will remain. You do *not* have enough concrete for both projects.

11. $\dfrac{7}{16}$ in.   12. No, only $3\dfrac{3}{4}$ ft is "left over."

4.3 Exercises

Solve the applications.

Name _____

Section _____ Date _____

1. **Business and Finance**   Clayton has 64 quarters in his bank. How many dollars does he have?

2. **Business and Finance**   Amy has 19 quarters in her purse. How many dollars does she have?

3. **Business and Finance**   Manuel counted 35 half gallons of orange juice in his store. Write the amount of orange juice as a mixed number of gallons.

4. **Business and Finance**   Sarah has 19 half gallons of turpentine in her paint store. Write the amount of turpentine as a mixed number of gallons.

5. **Crafts**   A recipe calls for $\frac{2}{3}$ cup of sugar for each serving. How much sugar is needed for six servings?

6. **Crafts**   Mom's French toast requires $\frac{3}{4}$ cup of batter for each serving. If five people are expected for breakfast, how much batter is needed?

7. **Construction**   A patch of dirt needs $3\frac{5}{6}$ ft^2 of sod to cover it. If Nick decides to cover only $\frac{3}{4}$ of the dirt, how much sod does he need?

8. **Construction**   A driveway requires $4\frac{5}{6}$ yd^3 of concrete to cover it. If Sheila wants to enlarge her driveway to $2\frac{1}{2}$ times its current size, how much concrete will she need?

1. _____

2. _____

3. _____

4. _____

5. _____

6. _____

7. _____

8. _____

ANSWERS

9. _____

10. _____

11. _____

12. _____

13. _____

14. _____

15. _____

16. _____

17. _____

9. **Social Science** The scale on a map is 1 in. = 200 mi. What actual distance, in miles, does $\frac{3}{8}$ in. represent?

10. **Business and Finance** You make $90 a day on a job. What will you receive for working $\frac{3}{4}$ of a day?

11. **Construction** A lumberyard has a stack of 80 sheets of plywood. If each sheet is $\frac{3}{4}$ in. thick, how high will the stack be?

12. **Business and Finance** A family uses $\frac{2}{5}$ of its monthly income for housing and utilities on average. If the family's monthly income is $1,750, what is spent for housing and utilities? What amount remains?

13. **Social Science** Of the eligible voters in an election, $\frac{3}{4}$ were registered. Of those registered, $\frac{5}{9}$ actually voted. What fraction of those people who were eligible voted? **ALEKS**

14. **Statistics** A survey has found that $\frac{7}{10}$ of the people in a city own pets. Of those who own pets, $\frac{2}{3}$ have dogs. What fraction of those surveyed own dogs? **ALEKS**

15. **Construction** A kitchen has dimensions $3\frac{1}{3}$ yd by $3\frac{3}{4}$ yd. How many square yards of linoleum must be bought to cover the floor?

16. **Science and Medicine** If you drive at an average speed of $52 \frac{\text{mi}}{\text{h}}$ for $1\frac{3}{4}$ h, how far will you travel?

17. **Science and Medicine** A jet flew at an average speed of $540 \frac{\text{mi}}{\text{h}}$ on a $4\frac{2}{3}$-h flight. What was the distance flown?

18. Construction   A piece of land that has $11\dfrac{2}{3}$ acres is being subdivided for home lots. It is estimated that $\dfrac{2}{7}$ of the area will be used for roads. What amount remains to be used for lots?

19. Geometry   To find the approximate circumference or distance around a circle, we multiply its diameter by $\dfrac{22}{7}$. What is the circumference of a circle with a diameter of 21 in.?

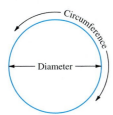

20. Geometry   The length of a rectangle is $\dfrac{6}{7}$ yd, and its width is $\dfrac{21}{26}$ yd. What is its area in square yards?

The formula for the area of a triangle is

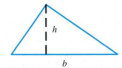

$$A = \frac{1}{2} \cdot h \cdot b$$

in which h is the height of the triangle and b is the base.

21. Find the area of a triangle with a height of $2\dfrac{2}{5}$ in. and a base of $3\dfrac{1}{3}$ in.   ALEKS

22. Find the area of a triangle with a height of $1\dfrac{7}{8}$ in. and a base of $2\dfrac{2}{5}$ in.   ALEKS

23. A recipe calls for the following ingredients:

$\dfrac{7}{8}$ cup of flour, $\dfrac{3}{4}$ cup of sugar, $\dfrac{2}{3}$ cup of milk, and $\dfrac{5}{6}$ teaspoon of salt. This recipe makes eight servings. What amount of each quantity would you use if you wanted two servings?

24. Obtain a map of your state and, using the legend provided, determine the distance between your state capital and any other city. Would this be the actual distance you would travel by car if you made the journey? Why or why not?

18. _____

19. _____

20. _____

21. _____

22. _____

23.  _____

24.  _____

25. Construction   A wire $5\frac{1}{4}$ ft long is to be cut into 7 pieces of the same length. How long will each piece be?

26. Crafts   A potter uses $\frac{2}{3}$ lb of clay in making a bowl. How many bowls can be made from 16 lb of clay?

27. Science and Medicine   Virginia made a trip of 95 mi in $1\frac{1}{4}$ h. What was her average speed?

28. Business and Finance   A piece of land measures $3\frac{3}{4}$ acres and is for sale at $60,000. What is the price per acre?

29. Crafts   A roast weighs $3\frac{1}{4}$ lb. How many $\frac{1}{4}$-lb servings will the roast provide?

30. Construction   A bookshelf is 55 in. long. If the books have an average thickness of $1\frac{1}{4}$ in., how many books can be put on the shelf?

31. Business and Finance   A butcher wants to wrap $\frac{3}{8}$-lb packages of ground beef from a cut of meat weighing $19\frac{1}{8}$ lb. How many packages can be prepared?

32. Business and Finance   A manufacturer has $45\frac{1}{2}$ yd of imported cotton fabric. A shirt pattern uses $1\frac{3}{4}$ yd. How many shirts can be made?

33. Construction   A stack of $\frac{3}{4}$-in.-thick plywood is 48 in. high. How many sheets of plywood are in the stack?

34. Construction   A landfill occupies land that measures $10\frac{2}{3}$ mi by $6\frac{3}{4}$ mi. If there are 144 cells of equal area in the landfill, what is the area of each cell?

35. Crafts   Manuel has $7\frac{1}{2}$ yd of cloth. He wants to cut it into strips $1\frac{3}{4}$ yd long. How many strips will he have? How much cloth remains, if any?

ANSWERS

36. _____

37. _____

38. _____

39. _____

40. _____

41. _____

36. Crafts Evette has $41\frac{1}{2}$ ft of string. She wants to cut it into pieces $3\frac{3}{4}$ ft long. How many pieces of string will she have? How much string remains, if any?

37. In squeezing oranges for fresh juice, three oranges yield about $\frac{1}{3}$ of a cup.

    (a) How much juice could you expect to obtain from a bag containing 24 oranges?

    (b) If you needed 8 cups of orange juice, how many bags of oranges should you buy?

38. A farmer died and left 17 cows to be divided among three workers. The first worker was to receive $\frac{1}{2}$ of the cows, the second worker was to receive $\frac{1}{3}$ of the cows, and the third worker was to receive $\frac{1}{9}$ of the cows. The executor of the farmer's estate realized that 17 cows could not be divided into halves, thirds, or ninths and so added a neighbor's cow to the farmer's. With 18 cows, the executor gave 9 cows to the first worker, 6 cows to the second worker, and 2 cows to the third worker. This accounted for the 17 cows, so the executor returned the borrowed cow to the neighbor. Explain why this works.

39. Division of fractions is not commutative.

    For example, $\frac{3}{4} \div \frac{5}{6} \neq \frac{5}{6} \div \frac{3}{4}$.

There could be an exception. Can you think of a situation in which division of fractions would be commutative?

40. Josephine's boss tells her that her salary is to be divided by $\frac{1}{3}$. Should she quit?

41. Compare the English phrases: "divide in half" and "divide by one-half." Do they say the same thing? Create examples to support your answer.

42. (a) Compute: $5 \div \dfrac{1}{10}$; $5 \div \dfrac{1}{100}$; $5 \div \dfrac{1}{1,000}$; $5 \div \dfrac{1}{10,000}$.

   (b) As the divisor gets smaller (approaches 0), what happens to the quotient?

   (c) What does this say about the answer to $5 \div 0$?

Solve the applications. Write each answer in lowest terms.

43. **Business and Finance**   You collect 3 dimes, 2 dimes, and then 4 dimes. How much money do you have as a fraction of a dollar?

44. **Business and Finance**   You collect 7 nickels, 4 nickels, and then 5 nickels. How much money do you have as a fraction of a dollar?

45. **Business and Finance**   You work 7 h one day, 5 h the second day, and 6 h the third day. How long did you work as a fraction of a 24-h day?

46. **Business and Finance**   One task took 7 min, a second task took 12 min, and a third task took 21 min. How long did the three tasks take as a fraction of an hour?

47. **Geometry**   What is the perimeter of a rectangle if the length is $\dfrac{7}{10}$ in. and the width is $\dfrac{2}{10}$ in.?

48. **Geometry**   Find the perimeter of a rectangular picture frame if the width is $\dfrac{7}{9}$ yd and the length is $\dfrac{5}{9}$ yd.

49. Statistics Patrick spent $\frac{4}{9}$ of an hour in the batting cages on Friday, and $\frac{7}{9}$ of an hour on Saturday. He wants to spend 2 h over 3 days. How much time should he spend on Sunday to accomplish this goal?

50. Construction Maria, a road inspector, must inspect $\frac{17}{30}$ of a mile of road. If she has already inspected $\frac{11}{30}$ of a mile, how much more does she need to inspect?

51. Geometry Find the perimeter of the figure.

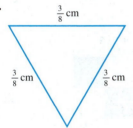

Find the perimeter of each triangles

52.

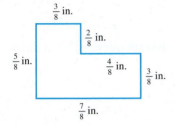

53.

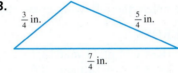

54.

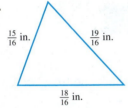

55.

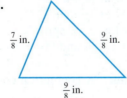

Find the perimeter of each polygon

56.

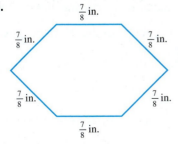

57.

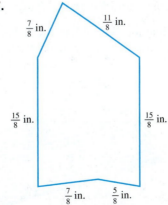

ANSWERS

49. _____

50. _____

51. _____

52. _____

53. _____

54. _____

55. _____

56. _____

57. _____

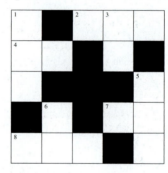
58. **Construction** Bolts can be purchased with diameters of $\frac{3}{8}, \frac{1}{4}$, or $\frac{3}{16}$ in. Which is smallest?

59. **Construction** Plywood comes in thicknesses of $\frac{5}{8}, \frac{3}{4}, \frac{1}{2}$, and $\frac{3}{8}$ in. Which size is thickest?

60. **Construction** Doweling is sold with diameters of $\frac{1}{2}, \frac{9}{16}, \frac{5}{8}$, and $\frac{3}{8}$ in. Which size is smallest?

61. Elian is asked to create a fraction equivalent to $\frac{1}{4}$. His answer is $\frac{4}{7}$. What did he do wrong? What would be a correct answer?

62. A sign on a busy highway says Exit 5A is $\frac{3}{4}$ mi away and Exit 5B is $\frac{5}{8}$ mi away. Which exit is first?

63. Complete the following cross number puzzle.

    ACROSS

    2. The LCM of 11 and 13
    4. The GCF of 120 and 300
    7. The GCF of 13 and 52
    8. The GCF of 360 and 540

    DOWN

    1. The LCM of 8, 14, and 21
    3. The LCM of 16 and 12
    5. The LCM of 2, 5, and 13
    6. The GCF of 54 and 90

64. **Construction** A countertop consists of a board $\frac{3}{4}$ in. thick and tile $\frac{3}{8}$ in. thick. What is the overall thickness?

65. **Business and Finance** Amy budgets $\frac{2}{5}$ of her income for housing and $\frac{1}{6}$ of her income for food. What fraction of her income is budgeted for these two purposes? What fraction of her income remains?

66. Social Science A person spends $\frac{3}{8}$ day at work and $\frac{1}{3}$ day sleeping. What fraction of a day do these two activities use? What fraction of the day remains?

66. _____

67. _____

68. _____

69. _____

70. _____

71. _____

72. _____

73. _____

67. Statistics Jose walked $\frac{3}{4}$ mi to the store, $\frac{1}{2}$ mi to a friend's house, and then $\frac{2}{3}$ mi home. How far did he walk?

68. Geometry Find the perimeter of, or the distance around, the figure.

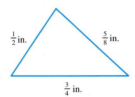

69. Business and Finance A budget guide states that you should spend $\frac{1}{4}$ of your salary for housing, $\frac{3}{16}$ for food, $\frac{1}{16}$ for clothing, and $\frac{1}{8}$ for transportation. What total portion of your salary will these four expenses account for?

70. Business and Finance Deductions from your paycheck are made roughly as follows: $\frac{1}{8}$ for federal tax, $\frac{1}{20}$ for state tax, $\frac{1}{20}$ for social security, and $\frac{1}{40}$ for a savings withholding plan. What portion of your pay is deducted?

For exercises 71 and 72, find the missing dimension (?) in the given figure.

71.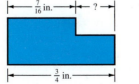

72.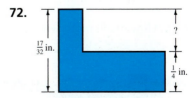

73. Crafts A roll of paper contains $30\frac{1}{4}$ yd. If $16\frac{7}{8}$ yd is cut from the roll, how much paper remains?

ANSWERS

74. _____

75. _____

76. _____

77. _____

78. _____

79. _____

74. Geometry Find the missing dimension in the figure.

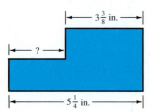

75. Construction A $4\frac{1}{4}$-in. bolt is placed through a board that is $3\frac{1}{2}$ in. thick. How far does the bolt extend beyond the board?

76. Business and Finance Ben can work 20 h per week on a part-time job. He works $5\frac{1}{2}$ h on Monday and $3\frac{3}{4}$ h on Tuesday. How many more hours can he work during the week?

77. Geometry Find the missing dimension in the figure.

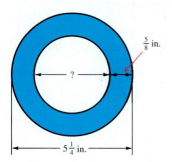

78. Construction The Whites used $20\frac{3}{4}$ yd^2 of carpet for their living room, $15\frac{1}{2}$ yd^2 for the dining room, and $6\frac{1}{4}$ yd^2 for a hallway. How much will remain if they began with a 50-yd^2 roll of carpeting?

79. Construction A construction company has bids for paving roads of $1\frac{1}{2}$, $\frac{3}{4}$, and $3\frac{1}{3}$ mi for the month of July. With their present equipment, they can pave 8 mi in 1 month. How much more work can they take on in July?

ANSWERS

80. _____

81. _____

82. _____

83. _____

84. _____

80. Statistics On an 8-h trip, Jack drives $2\frac{3}{4}$ h and Pat drives $2\frac{1}{2}$ h. How many hours are left to drive?

81. Statistics A runner has told herself that she will run 20 mi each week. She runs $5\frac{1}{2}$ mi on Sunday, $4\frac{1}{4}$ mi on Tuesday, $4\frac{3}{4}$ mi on Wednesday, and $2\frac{1}{8}$ mi on Friday. How far must she run on Saturday to meet her goal?

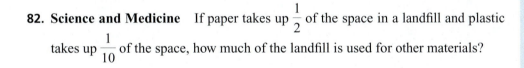

82. Science and Medicine If paper takes up $\frac{1}{2}$ of the space in a landfill and plastic takes up $\frac{1}{10}$ of the space, how much of the landfill is used for other materials?

83. Science and Medicine If paper takes up $\frac{1}{2}$ of the space in a landfill and organic waste takes up $\frac{1}{8}$ of the space, how much of the landfill is used for other materials?

84. Business and Finance The interest rate on an auto loan in May was $12\frac{3}{8}$%. By September the rate was up to $14\frac{1}{4}$%. How much did the interest rate increase over the period?

Answers

1. $16   **3.** $17\frac{1}{2}$ gal   **5.** 4 cups   **7.** $2\frac{7}{8}$ ft^2 **9.** 75 mi   **11.** 60 in.

13. $\frac{5}{12}$ **15.** $12\frac{1}{2}$ yd^2 **17.** 2,520 mi   **19.** 66 in.   **21.** 4 in.^{2}

23.   **25.** $\frac{3}{4}$ ft   **27.** $76\dfrac{\text{mi}}{\text{h}}$ **29.** 13 servings   **31.** 51 packages

33. 64 sheets **35.** 4; $\dfrac{1}{2}$ yd **37.** $2\dfrac{2}{3}$ cups; 3 bags **39.**

41.   **43.** $\dfrac{9}{10}$ of a dollar **45.** $\dfrac{3}{4}$ day **47.** $\dfrac{9}{5}$ in. or $1\dfrac{4}{5}$ in.

49. $\dfrac{7}{9}$ h **51.** 3 in. **53.** $\dfrac{15}{4}$ in. $= 3\dfrac{3}{4}$ in. **55.** $\dfrac{25}{8}$ in. $= 3\dfrac{1}{8}$ in.

57. $\dfrac{60}{8}$ in. $= 7\dfrac{1}{2}$ in. **59.** $\dfrac{3}{4}$ in. **61.** 🍴

63.

1	■	1	4	3
6	0	■	8	■
8	■	■	■	1
■	1	■	1	3
1	8	0	■	0

65. $\dfrac{17}{30}, \dfrac{13}{30}$ **67.** $1\dfrac{11}{12}$ mi **69.** $\dfrac{5}{8}$

71. $\dfrac{5}{16}$ in. **73.** $13\dfrac{3}{8}$ yd **75.** $\dfrac{3}{4}$ in. **77.** 4 in. **79.** $2\dfrac{5}{12}$ mi

81. $3\dfrac{3}{8}$ mi **83.** $\dfrac{3}{8}$

4.4 Equations Containing Fractions

4.4 OBJECTIVES

1. Solve equations containing fractions
2. Distinguish between solving fractional equations and simplifying fractional expressions

Recall from Section 3.6 that to solve an equation such as $\frac{1}{7}x = 9$ we multiply both sides of the equation by 7, the reciprocal of $\frac{1}{7}$. From our work with fractions, we know that $\frac{1}{7}x = 9$ is equivalent to the equation $\frac{x}{7} = 9$ since $\frac{1}{7}x = \frac{1}{7} \cdot \frac{x}{1} = \frac{x}{7}$. This means that we may solve the equation $\frac{x}{7} = 9$ in the same way we solved $\frac{1}{7}x = 9$, by multiplying both sides of the equation by the reciprocal of $\frac{1}{7}$ which is 7.

If we rewrite $\frac{x}{7} = 9$ as $\frac{x}{7} = \frac{9}{1}$, we observe that 7, the number by which we multiplied both sides of the equation, is the LCD of the fractions $\frac{x}{7}$ and $\frac{9}{1}$. This observation leads us to a method for solving **fractional equations,** which are equations that contain fractions as one or more of their terms.

To solve a fractional equation, we multiply each term of the equation by the LCD of all of the fractions. The resulting equation should be equivalent to the original equation and be cleared of all fractions.

OBJECTIVE 1

Example 1 Solving Fractional Equations

(a) Solve.

$$\frac{x}{3} = 6$$

$\frac{x}{3} = 6$ is equivalent to $\frac{x}{3} = \frac{6}{1}$. The LCD for $\frac{x}{3}$ and $\frac{6}{1}$ is 3. Multiply both sides of the equation by 3.

$$3\left(\frac{x}{3}\right) = 3 \cdot 6$$
$$x = 18$$

This leaves x alone on the left because

$$3\left(\frac{x}{3}\right) = \frac{3}{1} \cdot \frac{x}{3} = \frac{x}{1} = x$$

The solution is 18. To check, replace x with 18 in the *original* equation:

$$\frac{18}{3} \stackrel{?}{=} 6$$

$$6 = 6 \qquad \text{(True)}$$

The solution is verified.

(b) Solve.

$$\frac{x}{5} = -9$$

$\dfrac{x}{5} = -9$ is equivalent to $\dfrac{x}{5} = \dfrac{-9}{1}$. The LCD for $\dfrac{x}{5}$ and $\dfrac{-9}{1}$ is 5. Multiply both sides of the equation by 5.

$$5\left(\frac{x}{5}\right) = 5(-9)$$

$$x = -45$$

The solution is -45. To check, we replace x with -45:

$$\frac{-45}{5} \overset{?}{=} -9$$

$$-9 = -9 \qquad \text{(True)}$$

The solution is verified.

 CHECK YOURSELF 1

Solve and check.

(a) $\dfrac{x}{7} = 3$

(b) $\dfrac{x}{4} = -8$

When the variable is multiplied by a fraction that has a numerator other than 1, there are two approaches to finding the solution.

Example 2   Solving Fractional Equations

RECALL $\dfrac{3}{5}x$ can be written as $\dfrac{3x}{5}$.

Solve.

$$\frac{3}{5}x = 9$$

One approach is to multiply by 5, the LCD of $\dfrac{3x}{5}$ and $\dfrac{9}{1}$, as the first step.

$$5\left(\frac{3}{5}x\right) = 5 \cdot 9$$

$$3x = 45$$

Now we divide by 3.

$$\frac{3x}{3} = \frac{45}{3}$$

$$x = 15$$

To check:

$$\frac{3}{5} \cdot 15 \overset{?}{=} 9$$

$$9 = 9 \qquad \text{(True)}$$

The solution is verified.

A second approach uses our knowledge of reciprocals and is generally a bit more efficient. We multiply both sides of the equation by $\frac{5}{3}$.

RECALL $\frac{5}{3}$ is the *reciprocal* of $\frac{3}{5}$, and the product of a number and its reciprocal is just 1! So

$$\left(\frac{5}{3}\right)\left(\frac{3}{5}\right) = 1$$

$$\frac{5}{3}\left(\frac{3}{5}x\right) = \frac{5}{3} \cdot 9$$

$$x = \frac{5}{\cancel{3}_1} \cdot \frac{\cancel{9}^3}{1} = 15$$

So $x = 15$, as before.

✔ **CHECK YOURSELF 2**

Solve and check.

$$\frac{2}{3}x = 18$$

When an equation has more than one fraction, we first multiply by the LCD to clear the denominators.

Example 3   Solving an Equation with Two Fractions

Solve each equation.

(a) $\dfrac{x}{3} = \dfrac{5}{8}$

The LCD is 24. Multiplying both sides of the equation by 24, we get

NOTE To check, you should verify that $\dfrac{\frac{15}{8}}{3} = \dfrac{5}{8}$.

It may help to write the left hand side as

$$\frac{15}{8} \div 3$$

$$\frac{x}{3} \cdot 24 = \frac{5}{8} \cdot 24$$

$$8x = 15$$

$$x = \frac{15}{8}$$

(b) $\dfrac{3x}{10} = \dfrac{7}{12}$

The LCD is 60. Multiplying both sides by 60, we get

NOTE Again, verify that

$$\frac{3\left(\frac{35}{18}\right)}{10} = \frac{7}{12}$$

$$\frac{3x}{10} \cdot 60 = \frac{7}{12} \cdot 60$$

$$18x = 35$$

$$x = \frac{35}{18}$$

CHECK YOURSELF 3 _____

Solve each equation.

(a) $\dfrac{x}{10} = \dfrac{7}{2}$

(b) $\dfrac{2x}{5} = \dfrac{3}{10}$

Caution! Remember that, when we are solving an equation, we are looking for a value for the variable (usually x) that makes the equation true. When we are simplifying an expression, we are finding an equivalent expression.

OBJECTIVE 2

Example 4 **Identifying Equations**

For each, decide whether you are given an equation or an expression. Then state whether you could look for a solution.

RECALL Equations contain equal signs; expressions do not.

(a) $\dfrac{3x}{5} = \dfrac{3}{12}$ This is an equation. You can look for a solution.

(b) $\dfrac{3x}{10} + \dfrac{7}{12}$ This is an expression. You cannot solve an expression!

(c) $\dfrac{x}{3} - \dfrac{1}{5}$ This is an expression. You cannot solve an expression.

(d) $\dfrac{x}{6} + 6 = \dfrac{4}{9}$ This is an equation. We look at solving equations of this form in the remainder of this section.

 CHECK YOURSELF 4 _____

For each, decide whether you are given an equation or an expression. Then state whether you could look for a solution.

(a) $\dfrac{x}{15} - \dfrac{3}{12}$

(b) $\dfrac{3x}{10} = \dfrac{7}{12}$

(c) $\dfrac{x}{2} - 6$

(d) $\dfrac{x}{5} + 6 = \dfrac{4}{9}$

When solving an equation with more than one term on one side, multiply by the LCD to clear fractions. In the next example, we see that the distributive property plays an important role.

Example 5 **Solving Fractional Equations**

NOTE This equation has three terms: $\dfrac{x}{2}$, $-\dfrac{1}{3}$, and $\dfrac{2x + 3}{6}$. The sign of the term is not used to find the LCD.

Solve.

$$\dfrac{x}{2} - \dfrac{1}{3} = \dfrac{2x + 3}{6}$$

The LCD for $\dfrac{x}{2}$, $\dfrac{1}{3}$, and $\dfrac{2x+3}{6}$ is 6. Multiply both sides of the equation by 6. Using the distributive property, we multiply *each* term by 6.

NOTE By the multiplication property of equality, this equation is equivalent to the original equation.

$$6 \cdot \frac{x}{2} - 6 \cdot \frac{1}{3} = 6\left(\frac{2x+3}{6}\right) \qquad \text{or} \qquad 3x - 2 = 2x + 3$$

Solving as before, we have

$$3x - 2x = 3 + 2 \qquad \text{or} \qquad x = 5$$

To check, substitute 5 for x in the *original* equation:

$$\frac{5}{2} - \frac{1}{3} \overset{?}{=} \frac{2 \cdot 5 + 3}{6}$$

$$\frac{15}{6} - \frac{2}{6} \overset{?}{=} \frac{10+3}{6}$$

$$\frac{13}{6} = \frac{13}{6} \qquad \text{(True)}$$

The solution is verified.

CHECK YOURSELF 5

Solve and check.

$$\frac{x}{4} - \frac{1}{6} = \frac{4x-5}{12}$$

 **CAUTION**

Caution! Sometimes, one or more of the terms in an equation will be integers. It is important to remember that every term must be multiplied by the LCD. That includes the integer terms.

> **Example 6   Solving Equations Involving Both Fractions and Integers**

Solve the equation.

$$\frac{x}{2} + 6 = \frac{3}{4}$$

The LCD is 4. Multiplying both sides by 4, we get

$$4 \cdot \left(\frac{x}{2} + 6\right) = \frac{3}{4} \cdot 4$$

$$4 \cdot \frac{x}{2} + 4 \cdot 6 = \frac{3}{4} \cdot 4$$

$$2x + 24 = 3$$

$$2x = -21$$

NOTE Be sure to check this result.

$$x = \frac{-21}{2}$$

CHECK YOURSELF 6 _____

Solve the equation.

$$\frac{x}{3} - 4 = \frac{1}{4}$$

READING YOUR TEXT _____

The following fill-in-the-blank exercises are designed to assure that you understand the key vocabulary used in this section. Each sentence usually comes directly from the section. You will find the correct answers in Appendix C.

Section 4.4

(a) To solve a fractional equation, we multiply each term of the equation by the _____ of all the fractions.

(b) $\frac{5}{3}$ is the _____ of $\frac{3}{5}$.

(c) When we are solving an equation, we are looking for a value of the variable that makes the equation _____.

(d) When we are simplifying an expression, we are finding an _____ expression.

CHECK YOURSELF ANSWERS _____

1. **(a)** 21; **(b)** -32 **2.** 27   **3.** **(a)** $x = 35$; **(b)** $x = \frac{3}{4}$

4. **(a)** An expression; cannot be solved; **(b)** an equation; can be solved; **(c)** an expression; cannot be solved; **(d)** an equation; can be solved.

5. 3   **6.** $x = \frac{51}{4}$

4.4 Exercises

Solve for x and check your result.

1. $\dfrac{x}{4} = 8$ **ALEKS**

2. $\dfrac{x}{3} = 12$ **ALEKS**

3. $\dfrac{x}{8} = 5$ **ALEKS**

4. $\dfrac{x}{7} = 2$ **ALEKS**

5. $\dfrac{x}{6} = -3$ **ALEKS**

6. $\dfrac{x}{2} = -20$ **ALEKS**

7. $\dfrac{x}{5} = -4$ **ALEKS**

8. $\dfrac{x}{9} = -3$ **ALEKS**

9. $\dfrac{3}{4}x = 15$

10. $\dfrac{4}{5}x = 12$

11. $\dfrac{2}{3}x = 16$

12. $\dfrac{2}{5}x = 8$

13. $\dfrac{1}{6}x = -3$

14. $\dfrac{1}{4}x = 2$

15. $\dfrac{1}{5}x = 5$

16. $\dfrac{1}{3}x = -12$

17. $\dfrac{3}{2}x = 24$

18. $\dfrac{4}{3}x = 12$

19. $\dfrac{5}{3}x = -15$

20. $\dfrac{5}{2}x = -20$

In exercises 21 to 34, determine the smallest multiplier to use in order to clear the equation of fractions. Do not solve.

21. $\dfrac{x}{6} = \dfrac{5}{3}$

22. $\dfrac{x}{4} = \dfrac{7}{8}$

23. $\dfrac{x}{2} = \dfrac{-7}{3}$

24. $\dfrac{x}{4} = \dfrac{-5}{12}$

25. $\dfrac{x}{5} = \dfrac{2}{3}$

26. $\dfrac{x}{7} = \dfrac{3}{4}$

Name _____

Section _____ Date _____

ANSWERS

1. _____
2. _____
3. _____
4. _____
5. _____
6. _____
7. _____
8. _____
9. _____
10. _____
11. _____
12. _____
13. _____
14. _____
15. _____
16. _____
17. _____
18. _____
19. _____
20. _____
21. _____
22. _____
23. _____
24. _____
25. _____
26. _____

27. $\dfrac{x}{6} = \dfrac{-9}{4}$

28. $\dfrac{x}{8} = \dfrac{-7}{12}$

29. $\dfrac{x}{12} = \dfrac{5}{9}$

30. $\dfrac{x}{15} = \dfrac{3}{10}$

31. $\dfrac{2x}{5} = \dfrac{3}{8}$

32. $\dfrac{4x}{3} = \dfrac{2}{5}$

33. $\dfrac{3x}{8} = \dfrac{-1}{6}$

34. $\dfrac{5x}{12} = \dfrac{-4}{9}$

Solve for x and check your result.

35. $\dfrac{x}{6} = \dfrac{5}{3}$

36. $\dfrac{x}{4} = \dfrac{7}{8}$

37. $\dfrac{x}{2} = \dfrac{-7}{3}$

38. $\dfrac{x}{4} = \dfrac{-5}{12}$

39. $\dfrac{x}{5} = \dfrac{2}{3}$

40. $\dfrac{x}{7} = \dfrac{3}{4}$

41. $\dfrac{x}{6} = \dfrac{-9}{4}$

42. $\dfrac{x}{8} = \dfrac{-7}{12}$

43. $\dfrac{x}{12} = \dfrac{5}{9}$

44. $\dfrac{x}{15} = \dfrac{3}{10}$

45. $\dfrac{2x}{5} = \dfrac{3}{8}$

46. $\dfrac{4x}{3} = \dfrac{2}{5}$

47. $\dfrac{3x}{8} = \dfrac{-1}{6}$

48. $\dfrac{5x}{12} = \dfrac{-4}{9}$

For exercises 49 to 56, decide whether you are given an equation or an expression. Then state whether you could look for a solution.

49. $\dfrac{x}{6} = \dfrac{7}{3}$

50. $\dfrac{x}{5} + \dfrac{3}{8}$

51. $\dfrac{2x}{5} - \dfrac{7}{8}$

52. $\dfrac{3x}{4} = \dfrac{12}{5}$

53. $\dfrac{x}{2} - 3$

54. $\dfrac{x}{8} + 4 = 0$

55. $\dfrac{3}{5}x - 2 = 6$

56. $\dfrac{2}{3}x + 4 - \dfrac{5}{6}$

Solve for x and check your result.

57. $\dfrac{x}{5} - \dfrac{1}{3} = \dfrac{x-7}{3}$

58. $\dfrac{x}{6} + \dfrac{3}{4} = \dfrac{x-1}{4}$

59. $\dfrac{x}{4} - \dfrac{1}{5} = \dfrac{4x+3}{20}$

60. $\dfrac{x}{12} - \dfrac{1}{6} = \dfrac{2x-7}{12}$

61. $\dfrac{x}{3} + 4 = \dfrac{5}{2}$

62. $\dfrac{x}{5} + 2 = \dfrac{7}{10}$

63. $\dfrac{x}{6} - 5 = \dfrac{3}{4}$

64. $\dfrac{x}{4} - 3 = \dfrac{5}{3}$

In exercises 65 to 67, indicate whether the given statement is true or false.

65. An expression can be an equation.

66. An equation can be an expression.

67. An equation is really two expressions separated by an equals sign.

68. In your own words, describe the steps used in solving an equation involving fractions. Include an example of your own.

Answers

1. $x = 32$ **3.** $x = 40$ **5.** $x = -18$ **7.** $x = -20$ **9.** $x = 20$
11. $x = 24$ **13.** $x = -18$ **15.** $x = 25$ **17.** $x = 16$ **19.** $x = -9$
21. 6 **23.** 6 **25.** 15 **27.** 12 **29.** 36 **31.** 40 **33.** 24
35. $x = 10$ **37.** $x = \dfrac{-14}{3}$ **39.** $x = \dfrac{10}{3}$ **41.** $x = \dfrac{-27}{2}$
43. $x = \dfrac{20}{3}$ **45.** $x = \dfrac{15}{16}$ **47.** $x = \dfrac{-4}{9}$ **49.** Equation; yes
51. Expression; no **53.** Expression; no **55.** Equation; yes **57.** $x = 15$
59. $x = 7$ **61.** $x = \dfrac{-9}{2}$ **63.** $x = \dfrac{69}{2}$ **65.** False **67.** True

4.5 Applications of Linear Equations in One Variable

4.5 OBJECTIVES

1. Solve applications involving linear equations
2. Solve applications involving a perimeter

The main reason for learning how to set up and solve algebraic equations is so that we can use them to solve word problems. In fact, algebraic equations were *invented* to make solving word problems much easier. The first word problems that we know about are over 4000 years old. They were literally "written in stone," on Babylonian tablets, about 500 years before the first algebraic equation made its appearance.

Before algebra, people solved word problems primarily by **substitution,** which is a method of finding unknown numbers by using trial and error in a logical way. Example 1 shows how to solve a word problem using substitution.

Example 1 Solving a Word Problem by Substitution

The sum of two consecutive integers is 37. Find the two integers.

If the two integers were 20 and 21, their sum would be 41. Because that's more than 37, the integers must be smaller. If the integers were 15 and 16, the sum would be 31. More trials yield that the sum of 18 and 19 is 37.

 CHECK YOURSELF 1

The sum of two consecutive integers is 91. Find the two integers.

Most word problems are not so easily solved by substitution. For more complicated word problems, a five-step procedure is used. Using this step-by-step approach will, with practice, allow you to organize your work. Organization is the key to solving word problems. Here are the five steps.

Step by Step: Using Equations to Solve Word Problems

Step 1 Read the problem carefully. Then reread it to decide what you are asked to find.

Step 2 Choose a letter to represent one of the unknowns in the problem. Then represent all other unknowns of the problem with expressions that use the same letter.

Step 3 Translate the problem to the language of algebra to form an equation.

Step 4 Solve the equation and answer the question of the original problem.

Step 5 Check your solution by returning to the original problem.

RECALL We discussed these translations in Section 2.6. You might find it helpful to review that section before going on.

The third step is usually the hardest part. We must translate words to the language of algebra. Before we look at a complete example, the table may help you review that translation step.

Translating Words to Algebra

Words	Algebra
The sum of x and y	$x + y$
3 plus a	$3 + a$ or $a + 3$
5 more than m	$m + 5$
b increased by 7	$b + 7$
The difference of x and y	$x - y$
4 less than a	$a - 4$
s decreased by 8	$s - 8$
The product of x and y	$x \cdot y$ or xy
5 times a	$5 \cdot a$ or $5a$
Twice m	$2m$
The quotient of x and y	$\dfrac{x}{y}$
a divided by 6	$\dfrac{a}{6}$
One-half of b	$\dfrac{b}{2}$ or $\dfrac{1}{2}b$

Now we will look at some typical examples of translating phrases to algebra.

Example 2   Translating Statements

Translate each statement to an algebraic expression.

(a) The sum of a and 2 times b $a + 2b$

Sum   2 times b

(b) 5 times m, increased by 1 $5m + 1$

5 times m Increased by 1

(c) 5 less than 3 times x $3x - 5$

3 times x 5 less than

(d) The product of x and y, divided by 3 $\dfrac{xy}{3}$

The product of x and y

Divided by 3

CHECK YOURSELF 2

Translate to algebra.

(a) 2 more than twice x **(b)** 4 less than 5 times n
(c) The product of twice a and b **(d)** The sum of s and t, divided by 5

Now we will work through a complete example. Although this problem could be solved by substitution, it is presented here to help you practice the five-step approach.

OBJECTIVE 1   **Example 3   Solving an Application**

The sum of a number and 5 is 17. What is the number?

Step 1   *Read carefully.* You must find the unknown number.

Step 2   *Choose letters or variables.* Let x represent the unknown number. There are no other unknowns.

Step 3   *Translate.*

The sum of

$x + 5 = 17$

is

Step 4   *Solve.*

$$\begin{array}{r} x + 5 = 17 \\ \underline{-5 \quad -5} \\ x = 12 \end{array} \quad \text{Add } -5.$$

So the number is 12.

NOTE Always return to the *original problem* to check your result and *not* to the equation of step 3. This will prevent possible errors!

Step 5   *Check.* Is the sum of 12 and 5 equal to 17? Yes ($12 + 5 = 17$). We have checked our solution.

CHECK YOURSELF 3

The sum of a number and 8 is 35. What is the number?

Definition: Representing Consecutive Integers

Consecutive integers are integers that follow one another. To represent them in algebra:

If x is an integer, then $x + 1$ is the next consecutive integer, $x + 2$ is the next, and so on.

We need this idea in Example 4.

Example 4   Solving an Application

The sum of two consecutive integers is 41. What are the two integers?

Step 1   We want to find the two consecutive integers.

Step 2   Let x be the first integer. Then $x + 1$ must be the next.

Step 3

The first integer ⟶ The second integer

$$x + \overbrace{x + 1} = 41$$

The sum   Is

Solve the equation.

Step 4

$$x + x + 1 = 41$$
$$2x + 1 = 41$$
$$2x = 40$$
$$x = 20$$

The first integer (x) is 20, and the next integer ($x + 1$) is 21.

Check.

Step 5   The sum of the two integers 20 and 21 is 41.

CHECK YOURSELF 4

The sum of three consecutive integers is 51. What are the three integers?

Sometimes algebra is used to reconstruct missing information. Example 5 does just that with some election information.

Example 5   Solving an Application

There were 55 more yes votes than no votes on an election measure. If 735 votes were cast in all, how many yes votes were there? How many no votes?

Step 1   We want to find the number of yes votes and the number of no votes.

Step 2   Let x be the number of no votes. Then

$$\underline{x + 55}$$

55 more than x

is the number of yes votes.

Step 3

$$x + x + 55 = 735$$

No votes   Yes votes

Step 4

$$x + x + 55 = 735$$
$$2x + 55 = 735$$
$$2x = 680$$
$$x = 340$$

No votes $(x) = 340$

Yes votes $(x + 55) = 395$

Step 5   Thus, 340 no votes plus 395 yes votes equals 735 total votes. The solution checks.

 CHECK YOURSELF 5

Francine earns $120 per month more than Rob. If they earn a total of $2,680 per month, what are their monthly salaries?

Similar methods will allow you to solve a variety of word problems. Example 6 includes three unknown quantities but uses the same basic solution steps.

Example 6   **Solving an Application**

Juan worked twice as many hours as Jerry. Marcia worked 3 more hours than Jerry. If they worked a total of 31 h, find out how many hours each worked.

Step 1   We want to find the hours each worked, so there are three unknowns.

Step 2   Let x be the hours that Jerry worked.

Twice Jerry's hours

NOTE There are other choices for x, but choosing the smallest quantity will usually give the easiest equation to write and solve.

Then $2x$ is Juan's hours worked

3 more hours than Jerry worked

and $x + 3$ is Marcia's hours.

Step 3

Jerry   Juan   Marcia

$$x + 2x + x + 3 = 31$$

Sum of their hours

Step 4

$$x + 2x + x + 3 = 31$$
$$4x + 3 = 31$$
$$4x = 28$$
$$x = 7$$

Jerry's hours $(x) = 7$

Juan's hours $(2x) = 14$

Marcia's hours $(x + 3) = 10$

Step 5   The sum of their hours $(7 + 14 + 10)$ is 31, and the solution is verified.

CHECK YOURSELF 6

Lucy jogged twice as many miles as Paul but 3 less than Isaac. If the three ran a total of 23 mi, how far did each person run?

The solutions for many problems from geometry will also yield linear equations. Consider Example 7.

OBJECTIVE 2

Example 7   Solving a Geometry Application

NOTE Whenever you are working on an application involving geometric figures, you should draw a sketch of the problem, including the labels assigned in step 2.

The length of a rectangle is 1 cm less than 3 times the width. If the perimeter is 54 cm, find the dimensions of the rectangle.

Step 1   You want to find the dimensions (the width and length).

Step 2   Let x be the width.

Then $3x - 1$ is the length.

3 times   1 less than
the width

Length $3x - 1$

Width
x

Step 3   To write an equation, we use this formula for the perimeter of a rectangle:

$$P = 2W + 2L$$

So

$$54 = 2x + 2(3x - 1)$$

Perimeter   Twice the   Twice the
            width   length

Step 4 Solve the equation.

$$2x + 2(3x - 1) = 54$$
$$2x + 6x - 2 = 54$$
$$8x = 56$$
$$x = 7$$

NOTE Be sure to return to the original statement of the problem when checking your result.

The width x is 7 cm, and the length, $3x - 1$, is 20 cm. We leave step 5, the check, to you.

CHECK YOURSELF 7

The length of a rectangle is 5 in. more than twice the width. If the perimeter of the rectangle is 76 in., what are the dimensions of the rectangle?

READING YOUR TEXT

The following fill-in-the-blank exercises are designed to assure that you understand the key vocabulary used in this section. Each sentence usually comes directly from the section. You will find the correct answers in Appendix C.

Section 4.5

(a) The main reason for learning how to set up and solve equations is to use them to solve _____ problems.

(b) _____ is a method of finding unknown numbers by using trial and error in a logical way.

(c) _____ integers are integers that follow one another.

(d) For a rectangle, twice the width plus twice the length equals the _____.

CHECK YOURSELF ANSWERS

1. 45 and 46   **2. (a)** $2x + 2$; **(b)** $5n - 4$; **(c)** $2ab$; **(d)** $\dfrac{s + t}{5}$

3. The equation is $x + 8 = 35$. The number is 27.

4. The equation is $x + x + 1 + x + 2 = 51$. The integers are 16, 17, and 18.

5. The equation is $x + x + 120 = 2,680$. Rob's salary is $1,280, and Francine's is $1,400.

6. Paul: 4 mi; Lucy: 8 mi; Isaac: 11 mi

7. The width is 11 in.; the length is 27 in.

4.5 Exercises

Name _____

Section _____ Date _____

Translate each statement to an algebraic equation. Let *x* represent the number in each case. Do not solve.

1. 3 more than a number is 7. **ALEKS**

2. 5 less than a number is 12. **ALEKS**

3. 7 less than 3 times a number is twice that same number.

4. 4 more than 5 times a number is 6 times that same number.

5. 2 times the sum of a number and 5 is 18 more than that same number.

6. 3 times the sum of a number and 7 is 4 times that same number.

7. 3 more than twice a number is 7.

8. 5 less than 3 times a number is 25.

9. 7 less than 4 times a number is 41.

10. 10 more than twice a number is 44.

11. 3 times a number is 12 more than that number. **ALEKS**

12. 5 times a number is 8 less than that number. **ALEKS**

Solve the word problems. Be sure to label the unknowns and to show the equation you use for the solution.

13. Number Problem The sum of a number and 7 is 33. What is the number?

14. Number Problem The sum of a number and 15 is 22. What is the number?

15. Number Problem The sum of a number and −15 is 7. What is the number?

16. Number Problem The sum of a number and −8 is 17. What is the number?

17. Social Science In an election, the winning candidate has 1,840 votes. If the total number of votes cast was 3,260, how many votes did the losing candidate receive?

18. Business and Finance Mike and Stefanie work at the same company and make a total of $4,760 per month. If Stefanie makes $2,400 per month, how much does Mike earn every month?

ANSWERS

1. _____
2. _____
3. _____
4. _____
5. _____
6. _____
7. _____
8. _____
9. _____
10. _____
11. _____
12. _____
13. _____
14. _____
15. _____
16. _____
17. _____
18. _____

19. Number Problem   The sum of twice a number and 7 is 33. What is the number?

20. Number Problem   3 times a number, increased by 8, is 50. Find the number.

21. Number Problem   5 times a number, minus 12, is 78. Find the number.

22. Number Problem   4 times a number, decreased by 20, is 44. What is the number?

23. Number Problem   The sum of two consecutive integers is 71. Find the two integers.

24. Number Problem   The sum of two consecutive integers is 145. Find the two integers.

25. Number Problem   The sum of three consecutive integers is 63. What are the three integers?

26. Number Problem   If the sum of three consecutive integers is 93, find the three integers.

27. Number Problem   The sum of two consecutive even integers is 66. What are the two integers? (*Hint:* Consecutive even integers such as 10, 12, and 14 can be represented by $x, x + 2, x + 4$, and so on.)

28. Number Problem   If the sum of two consecutive even integers is 86, find the two integers.

29. Number Problem   If the sum of two consecutive odd integers is 52, what are the two integers? (*Hint:* Consecutive odd integers such as 21, 23, and 25 can be represented by $x, x + 2, x + 4$, and so on.)

30. Number Problem   The sum of two consecutive odd integers is 88. Find the two integers.

31. Number Problem   The sum of three consecutive odd integers is 105. What are the three integers?   VIDEO

32. Number Problem   The sum of three consecutive even integers is 126. What are the three integers?

33. Number Problem   The sum of four consecutive integers is 86. What are the four integers?

34. Number Problem   The sum of four consecutive integers is 62. What are the four integers?

35. Number Problem   4 times an integer is 9 more than 3 times the next consecutive integer. What are the two integers?

36. **Number Problem** 4 times an even integer is 30 less than 5 times the next consecutive even integer. Find the two integers.

37. **Social Science** In an election, the winning candidate had 160 more votes than the loser. If the total number of votes cast was 3,260, how many votes did each candidate receive? **ALEKS**

38. **Business and Finance** Jody earns $140 more per month than Frank. If their monthly salaries total $2,760, what amount does each earn? **ALEKS**

39. **Business and Finance** A washer-dryer combination costs $650. If the washer costs $70 more than the dryer, what does each appliance cost? **ALEKS**

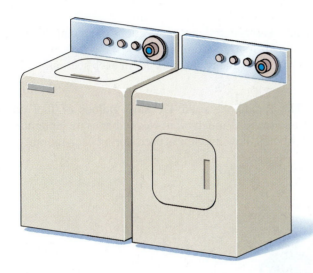

40. **Construction** Yuri has a board that is 98 in. long. He wishes to cut the board into two pieces so that one piece will be 10 in. longer than the other. What should be the length of each piece?

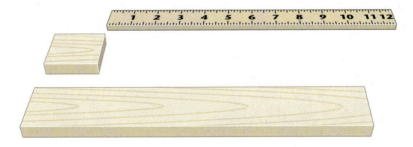

41. **Number Problem** Yan Ling is 1 year less than twice as old as his sister. If the sum of their ages is 14 years, how old is Yan Ling?

42. **Number Problem** Diane is twice as old as her brother Dan. If the sum of their ages is 27 years, how old are Diane and her brother?

43. _____

44. _____

45. _____

46. _____

47. _____

48. _____

49. _____

50. _____

51. _____

52. _____

43. Number Problem   Maritza is 3 years less than 4 times as old as her daughter. If the sum of their ages is 37, how old is Maritza?

44. Number Problem   Mrs. Jackson is 2 years more than 3 times as old as her son. If the difference between their ages is 22 years, how old is Mrs. Jackson?

45. Business and Finance   On her vacation in Europe, Jovita's expenses for food and lodging were $60 less than twice as much as her airfare. If she spent $2,400 in all, what was her airfare?

46. Business and Finance   Rachel earns $6,000 less than twice as much as Tom. If their two incomes total $48,000, how much does each earn?

47. Statistics   There are 99 students registered in three sections of algebra. There are twice as many students in the 10 A.M. section as the 8 A.M. section and 7 more students at 12 P.M. than at 8 A.M. How many students are in each section?

48. Business and Finance   The Randolphs used 12 more gallons of fuel oil in October than in September and twice as much oil in November as in September. If they used 132 gal for the 3 months, how much was used during each month?

In exercises 49 to 52, find the length of each side of the figure for the given perimeter.

49.

$2x - 2$
x
$x + 2$

$P = 24$ in.

50.
$3x - 4$
x

$P = 32$ cm

51.

$3x - 1$
$3x$
$2x - 1$
$x + 2$

$P = 90$ in.

52.
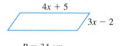
$4x + 5$
$3x - 2$

$P = 34$ cm

53. "I make \$2.50 an hour more in my new job." If x = the amount I used to make per hour and y = the amount I now make, which equations say the same thing as the given statement? Explain your choices by translating the equation into English and comparing with the original statement.

    (a) $x + y = 2.50$ **(b)** $x - y = 2.50$

    (c) $x + 2.50 = y$ **(d)** $2.50 + y = x$

    (e) $y - x = 2.50$ **(f)** $2.50 - x = y$

54. "The river rose 4 ft above flood stage last night." If a = the river's height at flood stage, b = the river's height last night, which equations say the same thing as the given statement? Explain your choices by translating the equations into English and comparing the meaning with the original statement.

    (a) $a + b = 4$ **(b)** $b - 4 = a$

    (c) $a - 4 = b$ **(d)** $a + 4 = b$

    (e) $b + 4 = b$ **(f)** $b - a = 4$

55. Maxine lives in Pittsburgh, Pennsylvania, and pays $8\frac{33}{100}$ cents per kilowatt hour (kWh) for electricity. During the 6 months of cold winter weather, her household uses about 1,500 kWh of electric power per month. During the two hottest summer months, the usage is also high because the family uses electricity to run an air conditioner. During these summer months, the usage is 1,200 kWh per month; the rest of the year, usage averages 900 kWh per month.

    (a) Write an expression for the total yearly electric bill.

    (b) Maxine is considering spending \$2,000 for more insulation for her home so that it is less expensive to heat and to cool her home. The insulation company claims that "with proper installation the insulation will reduce your heating and cooling bills by 25%." If Maxine invests the money in insulation, how long will it take her to get her money back in savings on her electric bill? Write to her about what information she needs to answer this question. Give her your opinion about how long it will take to save \$2,000 on heating and cooling bills and explain your reasoning. What is your advice to Maxine?

56. **Number Problem** If one-third of a number is subtracted from three-fourths of that number, the difference is 15. What is the number?

57. **Number Problem** If one-fourth of a number is subtracted from two-fifths of the number, the difference is 3. Find the number.

58. **Number Problem** If five-sixths of a number is added to one-fifth of the number, the sum is 31. What is the number?

53. _____

54. _____

55. _____

56. _____

57. _____

58. _____

In exercises 59 to 61, indicate whether the given statement is always true, sometimes true, or never true.

59. The sum of two consecutive odd integers is even.

60. The sum of three consecutive integers is odd.

61. The sum of two consecutive integers is even.

Answers

1. $x + 3 = 7$ **3.** $3x - 7 = 2x$ **5.** $2(x + 5) = x + 18$ **7.** $2x + 3 = 7$
9. $4x - 7 = 41$ **11.** $3x = x + 12$ **13.** 26 **15.** 22 **17.** 1,420
19. 13 **21.** 18 **23.** 35, 36 **25.** 20, 21, 22 **27.** 32, 34
29. 25, 27 **31.** 33, 35, 37 **33.** 20, 21, 22, 23 **35.** 12, 13
37. 1,710; 1,550 **39.** Washer, $360; dryer, $290 **41.** 9 years old
43. 29 years old **45.** $820 **47.** 8 A.M.: 23; 10 A.M.: 46; 12 P.M.: 30
49. 6 in., 8 in., 10 in. **51.** 12 in., 19 in., 29 in., 30 in. **53.**
55. **57.** 20 **59.** Always **61.** Never

4.6 Complex Fractions

NOTE TO THE INSTRUCTOR
This section is considered
optional. Its coverage may be
omitted without loss of
continuity.

4.6 OBJECTIVE

1. Simplify complex fractions

In Section 3.4, we learned to write a division of fractions such as $\dfrac{2}{3} \div \dfrac{4}{5}$ as the complex

fraction $\dfrac{\frac{2}{3}}{\frac{4}{5}}$. A **complex fraction** has a fraction in its numerator, in its denominator, or in

both. We then learned to rewrite the original division as the first fraction multiplied by the

reciprocal of the second, so $\dfrac{2}{3} \div \dfrac{4}{5} = \dfrac{2}{3} \cdot \dfrac{5}{4}$. To simplify a complex fraction, then, we can

multiply the numerator by the reciprocal of the denominator.

OBJECTIVE 1

Example 1 Simplifying Complex Fractions

NOTE Unless we are solving an
application problem, we write
improper fractions rather than
mixed numbers.

$$\frac{\frac{3}{4}}{\frac{5}{8}} = \frac{3}{4} \div \frac{5}{8} = \frac{3}{\underset{1}{\cancel{4}}} \cdot \frac{\overset{2}{\cancel{8}}}{5} = \frac{6}{5}$$

CHECK YOURSELF 1

Simplify.

(a) $\dfrac{\frac{4}{7}}{\frac{3}{7}}$

(b) $\dfrac{\frac{3}{8}}{\frac{5}{6}}$

Like parentheses, the fraction bar acts as a grouping symbol. Operations in the numerator or denominator should be applied before the fraction can be simplified.

Example 2 Simplifying a Complex Fraction

Simplify $\dfrac{2 - \frac{3}{5}}{4 - \frac{1}{3}}$.

To apply the operations in the numerator and denominator, we first find common denominators.

$$\frac{2 - \frac{3}{5}}{4 - \frac{1}{3}} = \frac{\frac{10}{5} - \frac{3}{5}}{\frac{12}{3} - \frac{1}{3}}$$

Now we continue with the subtractions.

$$\dfrac{\dfrac{10}{5} - \dfrac{3}{5}}{\dfrac{12}{3} - \dfrac{1}{3}} = \dfrac{\dfrac{7}{5}}{\dfrac{11}{3}}$$

$$\dfrac{\dfrac{7}{5}}{\dfrac{11}{3}} = \dfrac{7}{5} \div \dfrac{11}{3} = \dfrac{7}{5} \cdot \dfrac{3}{11} = \dfrac{21}{55}$$

 CHECK YOURSELF 2 _____

Simplify.

$$\dfrac{5 + \dfrac{3}{7}}{7 - \dfrac{2}{3}}$$

Consider the complex fraction we just studied in Example 2. It may be easier to simplify *without* applying the operations first. We simply must remember to correctly apply the distributive property. This approach is shown in Example 3.

Example 3   Simplifying a Complex Fraction

Simplify $\dfrac{2 - \dfrac{3}{5}}{4 - \dfrac{1}{3}}$.

First we note that the LCD of all the fractions that appear is 15. Then we multiply the main numerator and main denominator by 15.

$$\dfrac{15 \cdot \left(2 - \dfrac{3}{5}\right)}{15 \cdot \left(4 - \dfrac{1}{3}\right)} = \dfrac{15 \cdot 2 - 15 \cdot \dfrac{3}{5}}{15 \cdot 4 - 15 \cdot \dfrac{1}{3}} = \dfrac{30 - 9}{60 - 5} = \dfrac{21}{55}$$

 CHECK YOURSELF 3 _____

Simplify $\dfrac{5 + \dfrac{3}{7}}{7 - \dfrac{2}{3}}$ *using the approach shown in Example 3.*

READING YOUR TEXT

The following fill-in-the-blank exercises are designed to assure that you understand the key vocabulary used in this section. Each sentence usually comes directly from the section. You will find the correct answers in Appendix C.

Section 4.6

(a) A _____ fraction has a fraction in its numerator, in its denominator, or in both.

(b) To simplify a complex fraction, multiply the numerator by the _____ of the denominator.

(c) Like parentheses, the fraction bar acts as a _____ symbol.

(d) Another way to simplify a complex fraction is to multiply the main numerator and the main denominator by the _____ of all the fractions that appear.

CHECK YOURSELF ANSWERS

1. (a) $\dfrac{4}{3}$; (b) $\dfrac{9}{20}$ **2.** $\dfrac{6}{7}$ **3.** $\dfrac{6}{7}$

Name _____

Section _____ Date _____

ANSWERS

1. _____
2. _____
3. _____
4. _____
5. _____
6. _____
7. _____
8. _____
9. _____
10. _____
11. _____
12. _____
13. _____
14. _____
15. _____
16. _____
17. _____
18. _____
19. _____
20. _____
21. _____

4.6   Exercises

Simplify each complex fraction.

1. $\dfrac{\frac{2}{3}}{\frac{6}{8}}$

2. $\dfrac{\frac{5}{6}}{\frac{10}{15}}$

3. $\dfrac{\frac{1}{2}}{\frac{1}{4}}$

4. $\dfrac{\frac{3}{4}}{\frac{1}{8}}$

5. $\dfrac{\frac{4}{5}}{\frac{9}{10}}$

6. $\dfrac{\frac{5}{8}}{\frac{6}{10}}$

7. $\dfrac{\frac{6}{19}}{\frac{5}{38}}$

8. $\dfrac{\frac{8}{5}}{\frac{23}{25}}$

9. $\dfrac{6\frac{1}{2}}{\frac{2}{3}}$

10. $\dfrac{5\frac{1}{3}}{\frac{7}{6}}$

11. $\dfrac{4\frac{1}{2}}{5\frac{1}{4}}$

12. $\dfrac{8\frac{1}{3}}{9\frac{1}{6}}$

13. $\dfrac{2 - \frac{1}{2}}{2 + \frac{1}{4}}$

14. $\dfrac{3 + \frac{5}{6}}{7 - \frac{2}{3}}$

15. $\dfrac{3 + \frac{1}{8}}{4 - \frac{1}{4}}$

16. $\dfrac{5 - \frac{1}{3}}{5 + \frac{2}{3}}$

17. $\dfrac{2 + \frac{1}{6}}{5 - \frac{1}{3}}$

18. $\dfrac{4 + \frac{1}{5}}{6 + \frac{3}{20}}$

19. $\dfrac{8 + \frac{1}{2}}{4 - \frac{1}{4}}$

20. $\dfrac{5 + \frac{1}{2}}{2 - \frac{1}{10}}$

21. $\dfrac{4 - \frac{1}{3}}{3 + \frac{1}{2}}$

22. $\dfrac{2 - \dfrac{1}{4}}{5 + \dfrac{2}{3}}$

23. $\dfrac{5 + \dfrac{3}{4}}{7 - \dfrac{1}{6}}$

24. $\dfrac{3 - \dfrac{2}{3}}{5 + \dfrac{1}{5}}$

25. Care must be taken when writing a complex fraction. Show how the following fraction may be interpreted in two different ways. Simplify each.

$$\dfrac{\dfrac{2}{3}}{5}$$

26. Care must be taken when writing a complex fraction. Show how the following fraction may be interpreted in two different ways. Simplify each.

$$\dfrac{\dfrac{4}{7}}{5}$$

Answers

1. $\dfrac{8}{9}$ **3.** 2 **5.** $\dfrac{8}{9}$ **7.** $\dfrac{12}{5}$ **9.** $\dfrac{39}{4}$ **11.** $\dfrac{6}{7}$ **13.** $\dfrac{2}{3}$ **15.** $\dfrac{5}{6}$

17. $\dfrac{13}{28}$ **19.** $\dfrac{34}{15}$ **21.** $\dfrac{22}{21}$ **23.** $\dfrac{69}{82}$ **25.** $\dfrac{2}{15}; \dfrac{10}{3}$

ACTIVITY 4: HOME REMODELING

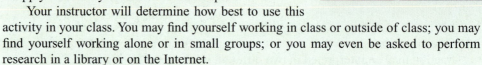

Each chapter in this text includes an activity. The activity is related to the vignette you encountered in the chapter opening. The activity provides you with the opportunity to apply the math you studied in the chapter.

Your instructor will determine how best to use this activity in your class. You may find yourself working in class or outside of class; you may find yourself working alone or in small groups; or you may even be asked to perform research in a library or on the Internet.

Benjamin and Olivia are putting a new floor in their kitchen. To get the floor up to the desired height, they need to add $1\frac{1}{8}$ in. of subfloor. They can do this in one of two ways. They can put $\frac{1}{2}$-in. sheet on top of $\frac{5}{8}$-in. board (note that the total would be $\frac{9}{8}$ in. or $1\frac{1}{8}$ in.). They could also put $\frac{3}{8}$-in. board on top of $\frac{3}{4}$-in. sheet.

The table below gives the price for each sheet of plywood from Home Depot.

Thickness	Cost for a 4 ft × 8 ft Sheet
$\frac{1}{8}$ in.	$9.15
$\frac{1}{4}$ in.	13.05
$\frac{3}{8}$ in.	14.99
$\frac{1}{2}$ in.	17.88
$\frac{5}{8}$ in.	19.13
$\frac{3}{4}$ in.	21.36
$\frac{7}{8}$ in.	25.23
1 in.	28.49

1. What is the combined price for a $\frac{1}{2}$-in. sheet and a $\frac{5}{8}$-in. sheet?

2. What is the combined price for a $\frac{3}{8}$-in. sheet and a $\frac{3}{4}$-in. sheet?

3. What other combinations of two sheets of plywood yields the needed $1\frac{1}{8}$-in. thickness?

4. Of the four combinations, which is most economical?

5. The kitchen is to be 12 ft × 12 ft. Find the total cost of the plywood you have suggested using in question 4.

4 Summary

DEFINITION/PROCEDURE	EXAMPLE	REFERENCE
Addition and Subtraction of Fractions		**Section 4.1**
To Add (Subtract) Like Fractions 1. Add (subtract) the numerators. 2. Place the sum (difference) over the common denominator. 3. Simplify the resulting fraction if necessary.	$$\frac{5}{18} + \frac{7}{18} = \frac{12}{18} = \frac{2}{3}$$	*p. 293, 295*
Least Common Multiple (LCM) The LCM is the *smallest* number that is a multiple of each of a group of numbers.		*p. 298*
To Find the LCD of a Group of Fractions 1. Write the prime factorization for each of the denominators. 2. Find all the prime factors that appear in any one of the prime factorizations. 3. Form the product of those prime factors, using each factor the greatest number of times it occurs in any one factorization.	To find the LCD of fractions with denominators 4, 6, and 15: $$4 = 2 \cdot 2$$ $$6 = 2 \quad \cdot 3$$ $$15 = \qquad 3 \cdot 5$$ $$\overline{2 \cdot 2 \cdot 3 \cdot 5}$$ The LCD $= 2 \cdot 2 \cdot 3 \cdot 5$, or 60.	*p. 302*
To Add or Subtract Unlike Fractions 1. Find the LCD of the fractions. 2. Change each fraction to an equivalent fraction with the LCD as a common denominator. 3. Add (subtract) the resulting like fractions as before.	$$\frac{3}{4} + \frac{7}{10} = \frac{15}{20} + \frac{14}{20}$$ $$= \frac{29}{20}$$	*p. 303, 305*
Operations on Mixed Numbers		**Section 4.2**
Mixed Number The sum of a whole number and a proper fraction.	$2\frac{1}{3}$ and $5\frac{7}{8}$ are mixed numbers. Note that $2\frac{1}{3}$ means $2 + \frac{1}{3}$.	*p. 313*
To Change an Improper Fraction into a Mixed Number 1. Divide the numerator by the denominator. The quotient is the whole-number portion of the mixed number. 2. If there is a remainder, write the remainder over the original denominator. This gives the fractional portion of the mixed number.	$$\frac{22}{5} = 4\frac{2}{5}$$ $$\begin{array}{r} 4 \\ 5\overline{)22} \\ \underline{20} \\ 2 \end{array}$$ ← Quotient ← Remainder	*p. 314*
To Change a Mixed Number to an Improper Fraction 1. Multiply the denominator of the fraction by the whole-number portion of the mixed number. 2. Add the numerator of the fraction to that product. 3. Write that sum over the original denominator to form the improper fraction.	Denominator Whole number Numerator $$5\frac{3}{4} = \frac{(4 \cdot 5) + 3}{4} = \frac{23}{4}$$ Denominator	*p. 315*

© 2010 McGraw-Hill Companies

DEFINITION/PROCEDURE	EXAMPLE	REFERENCE
Multiplying or Dividing Mixed Numbers Convert any mixed or whole numbers to improper fractions. Then multiply or divide the fractions as before.	$$6\frac{2}{3} \cdot 3\frac{1}{5} = \frac{\overset{4}{\cancel{20}}}{3} \cdot \frac{16}{\underset{1}{\cancel{5}}}$$ $$= \frac{64}{3} = 21\frac{1}{3}$$	*p. 316, 318*
To Add or Subtract Mixed Numbers 1. Rewrite as improper fractions. 2. Add or subtract the fractions. 3. Rewrite the results as a mixed number if required.	$$5\frac{1}{2} - 3\frac{3}{4} = \frac{11}{2} - \frac{15}{4}$$ $$= \frac{22}{4} - \frac{15}{4}$$ $$= \frac{7}{4}$$ $$= 1\frac{3}{4}$$	*p. 320, 321*
Equations Containing Fractions		**Section 4.4**
To solve an equation containing one or more fractions: 1. Find the LCD of the denominators. 2. Multiply *every term* by the LCD. 3. Solve the resulting equation as before.	To solve $\dfrac{x}{5} + 1 = \dfrac{1}{2}$: 1. The LCD is 10. 2. $10\left(\dfrac{x}{5} + 1\right) = 10\left(\dfrac{1}{2}\right)$ 3. $2x + 10 = 5$ $\quad\quad 2x = -5$ $\quad\quad\ x = \dfrac{-5}{2}$	*p. 351*
Applications of Linear Equations in One Variable		**Section 4.5**
To use an equation to solve a word problem: 1. Read the problem carefully to decide what you are asked to find. 2. Choose a letter to represent one of the unknowns. Then represent all other unknowns with expressions using that same letter. 3. Translate the problem to algebra to form an equation. 4. Solve the equation and answer the original question. 5. Check your solution by returning to the original problem.		*p. 361*
Consecutive integers If x is an integer, then $x + 1$ is the next consecutive integer, $x + 2$ is the next, and so on.	If 20 is an integer, $20 + 1 = 21$ is the next consecutive integer.	*p. 362*
Complex Fractions		**Section 4.6**
A complex fraction has a fraction in its numerator or denominator (or both).	$\dfrac{\frac{2}{3}}{\frac{5}{6}}$ is a complex fraction.	*p. 373*
To simplify a complex fraction, multiply the numerator and denominator by the LCD of the fractions within the complex fraction.	$\dfrac{\frac{2}{3}}{\frac{5}{6}} = \dfrac{\frac{2}{3} \cdot 6}{\frac{5}{6} \cdot 6} = \dfrac{4}{5}$	*p. 374*

Summary Exercises

This summary exercise set is provided to give you practice with each of the objectives of this chapter. Each exercise is keyed to the appropriate chapter section. When you are finished, you can check your answers to the odd-numbered exercises against those presented in the back of the text. If you have difficulty with any of these questions, go back and reread the examples from that section. The answers to the even-numbered exercises appear in the *Instructor's Manual*. Your instructor will give you guidelines on how to best use these exercises in your class.

[4.1] Add. Simplify when possible.

1. $\dfrac{8}{15} + \dfrac{2}{15}$

2. $\dfrac{-4}{7} + \dfrac{-3}{7}$

3. $\dfrac{2}{9} + \dfrac{5}{9} + \dfrac{4}{9}$

4. $\dfrac{4}{15} + \dfrac{7}{15} + \dfrac{7}{15}$

Find the LCM for each group of numbers.

5. 9, 12, and 24

6. 14, 21, and 28

Arrange the fractions in order from smallest to largest.

7. $\dfrac{5}{8}, \dfrac{7}{12}$

8. $\dfrac{5}{6}, \dfrac{4}{5}, \dfrac{7}{10}$

Complete the statements using the symbol $<$, $=$, or $>$.

9. $\dfrac{5}{12}$ ____ $\dfrac{3}{8}$

10. $\dfrac{3}{7}$ ____ $\dfrac{9}{21}$

11. $\dfrac{9}{16}$ ____ $\dfrac{7}{12}$

Write as equivalent fractions with the LCD as a common denominator.

12. $\dfrac{1}{6}, \dfrac{7}{8}$

13. $\dfrac{3}{10}, \dfrac{5}{8}, \dfrac{7}{12}$

Find the LCD for fractions with the given denominators.

14. 6 and 24

15. 12 and 18

16. 2, 5, and 8

17. 3, 6, and 8

Add or subtract as indicated.

18. $\dfrac{-3}{10} + \dfrac{-7}{12}$

19. $\dfrac{3}{8} + \dfrac{5}{12}$

20. $\dfrac{5}{36} + \dfrac{7}{24}$

21. $\dfrac{-2}{15} + \dfrac{-9}{20}$

22. $\dfrac{3}{8} + \dfrac{5}{12} + \dfrac{7}{18}$

23. $\dfrac{5}{6} + \dfrac{8}{15} + \dfrac{9}{20}$

24. $\dfrac{8}{9} + \dfrac{-3}{9}$

25. $\dfrac{9}{10} - \dfrac{6}{10}$

26. $\dfrac{5}{8} - \dfrac{1}{8}$

27. $\dfrac{11}{12} + \dfrac{-7}{12}$

28. $\dfrac{7}{8} - \dfrac{2}{3}$

29. $\dfrac{5}{6} - \dfrac{3}{5}$

30. $\dfrac{11}{18} + \dfrac{-2}{9}$

31. $\dfrac{5}{6} + \dfrac{-1}{4}$

32. $\dfrac{11}{12} - \dfrac{1}{4} - \dfrac{1}{3}$

33. $\dfrac{13}{15} + \dfrac{2}{3} - \dfrac{3}{5}$

[4.2]   Convert to mixed or whole numbers.

34. $\dfrac{41}{6}$

35. $\dfrac{-32}{8}$

36. $\dfrac{23}{3}$

37. $\dfrac{47}{4}$

Convert to improper fractions.

38. $7\dfrac{5}{8}$

39. $-4\dfrac{3}{10}$

40. $5\dfrac{2}{7}$

41. $-12\dfrac{8}{13}$

Multiply.

42. $-5\dfrac{1}{3} \cdot 1\dfrac{4}{5}$

43. $1\dfrac{5}{12} \cdot 8$

44. $3\dfrac{1}{5} \cdot \dfrac{7}{8} \cdot 2\dfrac{6}{7}$

Divide.

45. $3\dfrac{3}{8} \div 2\dfrac{1}{4}$

46. $3\dfrac{3}{7} \div 8$

Perform the indicated operations.

47. $6\dfrac{5}{7} + 3\dfrac{4}{7}$

48. $-5\dfrac{7}{10} + \left(-3\dfrac{11}{12}\right)$

49. $-7\dfrac{7}{9} + 3\dfrac{4}{9}$

50. $2\dfrac{1}{3} + 5\dfrac{1}{6} - 2\dfrac{4}{5}$

[4.3]   Solve the applications.

51. Social Science   The scale on a map is 1 in. = 80 mi. If two cities are $2\dfrac{3}{4}$ in. apart on the map, what is the actual distance between the cities?

52. Construction   A kitchen measures $5\dfrac{1}{3}$ yd by $4\dfrac{1}{4}$ yd. If you purchase linoleum costing $9 per square yard, what will it cost to cover the floor?

53. Construction   Your living room measures $6\dfrac{2}{3}$ yd by $4\dfrac{1}{2}$ yd. If you purchase carpeting at $18 per square yard, what will it cost to carpet the room?

54. Construction   A living room has dimensions $5\dfrac{2}{3}$ yd by $4\dfrac{1}{2}$ yd. How much carpeting must be purchased to cover the room?

55. Science and Medicine   If you drive 126 mi in $2\dfrac{1}{4}$ h, what is your average speed?

56. Science and Medicine   If you drive 117 mi in $2\dfrac{1}{4}$ h, what is your average speed?

57. Construction   An 18-acre piece of land is to be subdivided into home lots that are each $\dfrac{3}{8}$ acre. How many lots can be formed?

58. Crafts   A recipe calls for $\dfrac{1}{3}$ cup of milk. You have $\dfrac{3}{4}$ cup. How much milk will be left over?

59. Construction   Bradley needs two shelves, one $32\dfrac{3}{8}$ in. long and the other $36\dfrac{11}{16}$ in. long. What is the total length of shelving that is needed?

60. Geometry   Find the perimeter of the triangle.

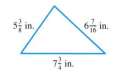

61. Construction   A sheet of plywood consists of two outer sections that are $\dfrac{3}{16}$ in. thick and a center section that is $\dfrac{3}{8}$ in. thick. How thick is the plywood overall?

[4.4] Solve the equations.

62. $\dfrac{x}{4} = 8$

63. $-\dfrac{x}{5} = -3$

64. $\dfrac{2}{3}x = 18$

65. $\dfrac{3}{4}x = 24$

66. $\dfrac{x}{3} - 5 = 1$

67. $\dfrac{3}{4}x - 2 = 7$

68. $\dfrac{x}{11} = \dfrac{12}{33}$

69. $\dfrac{x}{10} = \dfrac{9}{30}$

70. $\dfrac{x + 1}{5} = \dfrac{20}{25}$

71. $\dfrac{2}{5} = \dfrac{x - 2}{20}$

[4.5] Solve the word problems. Be sure to label the unknowns and to show the equation you used.

72. The sum of 3 times a number and 7 is 25. What is the number?

73. 5 times a number, decreased by 8, is 32. Find the number?

74. If the sum of two consecutive integers is 85, find the two integers?

75. The sum of three consecutive odd integers is 57. What are the three integers?

76. Rafael earns $35 more per week than Andrew. If their weekly salaries total $715, what amount does each earn?

77. Larry is 2 years older than Susan, and Nathan is twice as old as Susan. If the sum of their ages is 30 years, find each of their ages.

[4.6] Simplify.

78. $\dfrac{\frac{2}{3}}{\frac{3}{5}}$

79. $\dfrac{\frac{5}{8}}{\frac{3}{4}}$

80. $\dfrac{6 - \frac{2}{3}}{3 + \frac{1}{6}}$

81. $\dfrac{5 + \frac{1}{5}}{17 - \frac{3}{10}}$

Self-Test for Chapter 4

The purpose of this self-test is to help you check your progress so that you can find sections and concepts that you need to review before the next in-class exam. Allow yourself about an hour to take this test. At the end of that hour, check your answers against those given in the back of this text. If you missed a question, notice the section reference that accompanies the question. Go back to that section and reread the examples until you have mastered that particular concept.

Name _____

Section _____ Date _____

ANSWERS

In exercises 1 and 2, add.

1. $\dfrac{3}{10} + \dfrac{6}{10}$

2. $\dfrac{5}{12} + \dfrac{3}{12}$

1. _____

2. _____

3. Find the least common multiple of 18, 24, and 36.

3. _____

4. _____

In exercises 4 and 5, find the least common denominator for fractions with the given denominators.

4. 12 and 15

5. 3, 4, and 18

5. _____

6. _____

In exercises 6 to 8, add.

6. $\dfrac{2}{5} + \dfrac{4}{10}$

7. $\dfrac{-1}{6} + \dfrac{-3}{7}$

8. $\dfrac{1}{4} + \dfrac{5}{8} + \dfrac{7}{10}$

7. _____

8. _____

9. _____

In exercises 9 to 11, subtract.

9. $\dfrac{7}{9} - \dfrac{-4}{9}$

10. $\dfrac{-7}{18} - \dfrac{5}{18}$

11. $\dfrac{11}{12} - \dfrac{3}{20}$

10. _____

11. _____

12. _____

12. Convert $\dfrac{17}{4}$ to a mixed number.

13. _____

13. Convert $8\dfrac{2}{9}$ to an improper fraction.

14. _____

15. _____

Perform the indicated operation.

14. $2\dfrac{2}{3} \cdot 1\dfrac{2}{7}$

15. $5\dfrac{1}{3} \cdot \dfrac{3}{4}$

16. $5\dfrac{3}{5} \div 2\dfrac{1}{10}$

16. _____

17. _____

17. $4\dfrac{1}{6} + 3\dfrac{3}{4}$

18. $7\dfrac{3}{8} - 5\dfrac{5}{8}$

19. $7 - 5\dfrac{7}{15}$

18. _____

19. _____

20. _____

21. _____

22. _____

23. _____

24. _____

25. _____

26. _____

27. _____

28. _____

29. _____

30. _____

Solve the applications.

20. A $31\frac{1}{3}$-acre piece of land is subdivided into home lots. Each home lot is to be $\frac{2}{3}$ acre. How many homes can be built?

21. A bookshelf is 66 in. long. If the thickness of each book on the shelf is $1\frac{3}{8}$ in., how many books can be placed on the shelf?

22. The average person drinks about $3\frac{1}{5}$ cups of coffee per workday. If a person works 5 days a week for 50 weeks every year, estimate how many cups of coffee that person will drink in a working lifetime of $51\frac{3}{4}$ years.

Solve for x and check your result.

23. $\dfrac{x}{4} = -8$

24. $\dfrac{2x}{3} = \dfrac{-8}{9}$

25. $\dfrac{2}{5}x - 3 = 17$

26. $\dfrac{x}{2} - 3 = \dfrac{5}{3}$

Solve the applications.

27. 3 times an integer is 41 less than 5 times the next consecutive integer. Find the two integers.

28. On a shopping trip, Caitlin spent $10 less than twice the amount that Ben spent. Together they spent $95. How much did Ben spend?

Simplify each complex fraction.

29. $\dfrac{\dfrac{2}{3}}{\dfrac{5}{6}}$

30. $\dfrac{2 - \dfrac{1}{5}}{4 + \dfrac{3}{10}}$

Cumulative Review for Chapters 1 to 4

Name _____

Section _____ Date _____

ANSWERS

The following exercises are presented to help you review concepts from earlier chapters that you may have forgotten. This section is meant as review material and not as a comprehensive exam. The answers are presented in the back of the text. If you have difficulty with any of these exercises, be certain to at least read through the summary related to that section.

In exercises 1 to 14, perform the indicated operations.

1. $8 + (-4)$

2. $-7 + (-5)$

3. $\dfrac{7}{3} + \dfrac{11}{3}$

4. $\dfrac{4}{5} - \dfrac{2}{3}$

5. $(-6)(3)$

6. $\dfrac{1}{3} \cdot \dfrac{3}{5}$

7. $\dfrac{3}{7} \cdot \dfrac{-2}{3}$

8. $(-50) \div (-5)$

9. $\dfrac{-2}{5} \div \dfrac{3}{10}$

10. $15 \div 0$

11. $0 \div \dfrac{1}{2}$

12. $\dfrac{1}{5} + \dfrac{3}{4} - \dfrac{1}{3}$

13. $(-3)(-9)$

14. $\dfrac{3}{5} \cdot \dfrac{1}{3} \cdot \dfrac{5}{7}$

Convert to mixed numbers.

15. $\dfrac{16}{9}$

16. $\dfrac{36}{5}$

Convert to improper fractions.

17. $5\dfrac{3}{4}$

18. $6\dfrac{1}{9}$

Identify each as an expression or an equation.

19. $2x + 1$

20. $3x - 4 = 2$

1. _____

2. _____

3. _____

4. _____

5. _____

6. _____

7. _____

8. _____

9. _____

10. _____

11. _____

12. _____

13. _____

14. _____

15. _____

16. _____

17. _____

18. _____

19. _____

20. _____

21.

22.

23.

24.

25.

26.

27.

28.

29.

30.

31.

32.

33.

34.

35.

36.

37.

38.

39.

40.

21. $\dfrac{2}{3}x + \dfrac{1}{5}$

22. $\dfrac{1}{4}x - 3 = 5$

In exercises 23 to 30, perform the indicated operations.

23. $3\dfrac{2}{5} \cdot \dfrac{-5}{8}$

24. $1\dfrac{5}{12} \cdot 8$

25. $3\dfrac{5}{7} + 2\dfrac{4}{7}$

26. $-8\dfrac{1}{9} + 3\dfrac{5}{9}$

27. $9 - 5\dfrac{3}{8}$

28. $3\dfrac{1}{6} + 3\dfrac{1}{4} - 2\dfrac{7}{8}$

29. $2\dfrac{1}{4} \div (-3)$

30. $-5 \div 3\dfrac{1}{3}$

Solve each application.

31. A $6\dfrac{1}{2}$-in. bolt is placed through a wall that is $5\dfrac{7}{8}$ in. thick. How far does the bolt extend beyond the wall?

32. On a 6-h trip, Carlos drove $1\dfrac{3}{4}$ h. Then Maria drove for another $2\dfrac{1}{3}$ h. How many hours remained on the trip?

Solve for x and check your result.

33. $3x - 2 = 5x + 4$

34. $\dfrac{3}{4}x = \dfrac{x + 4}{2}$

Solve the word problems. Be sure to show the equation used for the solution.

35. If 4 times a number decreased by 7 is 45, find that number.

36. The sum of two consecutive integers is 93. What are those two integers?

37. If 3 times an odd integer is 12 more than the next consecutive odd integer, what is that integer?

38. Michelle earns $120 more per week than Dimitri. If their weekly salaries total $720, how much does Michelle earn?

39. The length of a rectangle is 2 cm more than 3 times its width. If the perimeter of the rectangle is 44 cm, what are the dimensions of the rectangle?

40. One side of a triangle is 5 in. longer than the shortest side. The third side is twice the length of the shortest side. If the triangle perimeter is 37 in., find the length of each leg.

chapter

5

DECIMALS

CHAPTER 5 OUTLINE

Section 5.1   Introduction to Decimals, Place Value,
and Rounding   page 391

Section 5.2   Addition and Subtraction of Decimals   page 403

Section 5.3   Multiplication of Decimals   page 414

Section 5.4   Division of Decimals   page 423

Section 5.5   Fractions and Decimals   page 434

Section 5.6   Equations Containing Decimals   page 446

Section 5.7   Square Roots and the Pythagorean
Theorem   page 451

Section 5.8   Applications   page 463

INTRODUCTION

Whether looked at as entertainment, competition, or business, the world of sports is dominated by statistics. Here are only a few of the most interesting sports-related statistics.

As entertainment:

- The largest attendance for a basketball game occurred on December 13, 2003, between Michigan State and Kentucky; 78,129 people attended.
- 26 of the 100 most-watched shows in the history of television were Super Bowls.

As a competition:

- In his 20-year career, Wayne Gretzky scored 2,857 points. The second most points scored in a career was Mark Messier's 1,887.
- In 1879, Will White pitched 75 complete games for the Cincinnati Reds. In 2003, no major league pitcher had more than 9 complete games.
- On March 2, 1962, Wilt Chamberlain scored 100 points in a professional basketball game. The second highest score in a game was 78, also by Chamberlain. The third highest score on the list was—perhaps you guessed it—73 by Wilt Chamberlain.

As a business:

- In 1995, Malcolm Glazer purchased the NFL's Tampa Bay Buccaneers for $195,000,000. In 2004, the franchise was valued at over $700,000,000.
- One-fourth of all big-screen televisions sold in the United States are sold the week before the World Series or the week before the Super Bowl.

We will look at mathematical applications of sports statistics throughout this chapter. You will find related exercises in Sections 5.2 to 5.6.

Pretest Chapter 5

This pretest will provide a preview of the types of exercises you will encounter in each section of this chapter. The answers for these exercises can be found in the back of the text. If you are working on your own or ahead of the class, this pretest can help you identify the sections in which you should focus more of your time.

[5.1] **1.** Give the place value of 5 in the decimal 13.4658.

2. Write $2\dfrac{371}{1,000}$ in decimal form and in words.

[5.2] **3.** **(a)** Add $56 + (-5.16) + 1.8 + (-0.33)$ **(b)** Subtract $4.6 - 2.225$

[5.8] **4. Business and Finance** You have $20 in cash and make purchases of $6.89 and $10.75. How much cash do you have left?

[5.3] **5.** **(a)** Multiply $(0.357)(-2.41)$ **(b)** Multiply $0.5362 \cdot 1,000$

[5.1] **6.** Round 2.35878 to the nearest hundredth.

[5.8] **7. Business and Finance** You fill up your car with 9.2 gal of fuel at $4.299 per gallon. What is the cost of the fill-up (to the nearest cent)?

[5.6] **8.** Solve $4.2 + 1.5x = -3.3$.

9. Solve $2.38 - (6.8 + 4.5x) = -3.25x$.

[5.4] **10.** Divide $57\overline{)242.25}$.

11. Divide $1.6\overline{)3.896}$.

[5.8] **12. Business and Finance** Manny worked 27.5 h in a week and earned $319. What was his hourly rate of pay?

[5.4] **13.** Divide $53.4 \div 1,000$.

[5.5] **14.** Find the decimal equivalent of each of the following.

(a) $\dfrac{3}{8}$ **(b)** $\dfrac{7}{24}$ (to the nearest hundredth)

[5.7] **15.** Use the Pythagorean theorem to find the length of the hypotenuse for $\triangle ABC$.

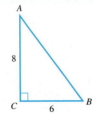

5.1 Introduction to Decimals, Place Value, and Rounding

5.1 OBJECTIVES

1. Identify place value in a decimal fraction
2. Write a decimal in words
3. Write a decimal as a fraction or mixed number
4. Compare the size of several decimals
5. Round a decimal to any specified decimal place

In Chapter 4, we looked at common fractions. We will turn now to a special kind of fraction, a **decimal fraction,** which is a fraction whose denominator is a *power of 10.* Some examples of decimal fractions are $\frac{3}{10}$, $\frac{45}{100}$, and $\frac{123}{1,000}$.

In Chapter 1, we talked about the idea of place value. Recall that in our decimal place-value system, each place has *one-tenth* the value of the place to its left.

Example 1   Identifying Place Values

RECALL The powers of 10 are 1, 10, 100, 1,000, and so on. You might want to review Section 1.7 before going on.

Label the place values for the number 538.

5	3	8
↑	↑	↑
Hundreds	Tens	Ones

The ones place value is one-tenth of the tens place value; the tens place value is one-tenth of the hundreds place value; and so on.

 CHECK YOURSELF 1 _____

Label the place values for the number 2,793.

We now want to extend this idea *to the right* of the ones place. Write a period to the *right* of the ones place. This is called the **decimal point.** Each digit to the right of that decimal point will represent a fraction whose denominator is a power of 10. The first place to the right of the decimal point is the tenths place:

NOTE The decimal point separates the whole-number part and the fractional part of a decimal fraction.

$$0.1 = \frac{1}{10}$$

Example 2   Writing a Number in Decimal Form

Write the mixed number $3\frac{2}{10}$ in decimal form.

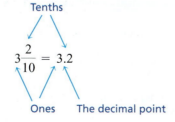

CHECK YOURSELF 2 _____

Write $5\dfrac{3}{10}$ *in decimal form.*

As you move farther to the *right*, each place value must be one-tenth of the value before it. The second place value is hundredths $\left(0.01 = \dfrac{1}{100}\right)$. The next place is thousandths, the fourth position is the ten thousandths place, and so on. The figure illustrates the value of each position as we move to the right of the decimal point.

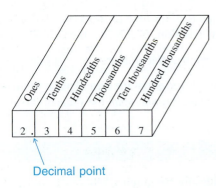

Decimal point

OBJECTIVE 1

NOTE For convenience we will shorten the term "decimal fraction" to "decimal" from this point on.

| **Example 3**   **Identifying Place Values** |

What are the place values for 4 and 6 in the decimal 2.34567? The place value of 4 is hundredths, and the place value of 6 is ten thousandths.

CHECK YOURSELF 3 _____

What is the place value of 5 in the decimal of Example 3?

Understanding place values will allow you to read and write decimals by using these steps.

| **Step by Step:**   Reading or Writing Decimals in Words |

NOTE If there are *no* nonzero digits to the left of the decimal point, start directly with Step 3.

Step 1	Read the digits *to the left* of the decimal point as a whole number.
Step 2	Read the decimal point as the word *and*.
Step 3	Read the digits *to the right* of the decimal point as a whole number followed by the place value of the rightmost digit.

OBJECTIVE 2

| **Example 4**   **Writing a Decimal Number in Words** |

Write each decimal number in words.

5.03 is read as "five and three hundredths."

Hundredths   The rightmost digit, 3, is in
                    the hundredths position.

NOTE An informal way of reading decimals is to simply read the digits in order and use the word *point* to indicate the decimal point. 2.58 can be read "two point five eight." 0.689 can be read "zero point six eight nine."

12.057 is read as "twelve and fifty-seven thousandths."

Thousandths   The rightmost digit, 7, is in the thousandths position.

0.5321 is read as "five thousand three hundred twenty-one ten thousandths."

When the decimal has no whole-number part, we have chosen to write a 0 to the left of the decimal point. This simply makes sure that you don't miss the decimal point. However, both 0.5321 and .5321 are correct.

CHECK YOURSELF 4 _____

Write 2.58 in words.

NOTE The number of digits to the right of the decimal point is called the number of **decimal places** in a decimal number. So, 0.35 has two decimal places.

One quick way to write a decimal as a common fraction is to remember that the number of decimal places must be the same as the number of zeros in the denominator of the common fraction.

OBJECTIVE 3

| Example 5 | **Writing a Decimal Number as a Mixed Number** |

Write each decimal as a common fraction or mixed number.

$$0.35 = \frac{35}{100}$$ This fraction can then be simplified to $\frac{7}{20}$.

Two places   Two zeros

The same method can be used with decimals that are greater than 1. Here the result will be a mixed number.

NOTE The 0 to the right of the decimal point is a **placeholder** that is not needed in the common-fraction form.

$$2.058 = 2\frac{58}{1,000}$$ This mixed number can be simplified to $2\frac{29}{500}$.

Three places   Three zeros

CHECK YOURSELF 5 _____

Write as common fractions or mixed numbers. Do not simplify.

(a) 0.528   **(b)** 5.08

RECALL By the Fundamental Principle of Fractions, multiplying the numerator and denominator of a fraction by the same nonzero number does not change the value of the fraction.

It is often useful to compare the sizes of two decimal fractions. One approach to comparing decimals uses this fact: Writing zeros to the right of the rightmost digit *does not change* the value of a decimal. 0.53 is the same as 0.530. Look at the fractional form:

$$\frac{53}{100} = \frac{530}{1,000}$$

The fractions are equivalent. We have multiplied the numerator and denominator by 10. We will see how this is used to compare decimals in Example 6.

OBJECTIVE 4   **Example 6**   **Comparing the Sizes of Two Decimal Numbers**

Which is larger?

0.84   or   0.842

Write 0.84 as 0.840. Then we see that 0.842 (or 842 thousandths) is greater than 0.840 (or 840 thousandths), and we can write

0.842 > 0.84

 CHECK YOURSELF 6 _____

Complete the statement, using the symbol < or >.

0.588 _____ 0.59

When working with a decimal, it may be helpful to picture the location of the decimal on a number line.

Example 7   **Plotting Decimals on a Number Line**

Plot the number 4.6 on the given number line. Then estimate the location for 4.68.

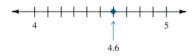

The number 4.6 is located six-tenths of the distance from 4 to 5. Since each tick mark represents one-tenth, we count to the sixth tick mark and draw a dot.

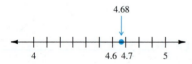

The number 4.68 is eight-tenths of the distance from 4.6 to 4.7. We might estimate its location as:

 CHECK YOURSELF 7 _____

Plot the number 8.3 on the number line. Then estimate the location for 8.51.

Whenever a decimal represents a measurement made by some instrument (a rule or a scale), the decimals are not exact. They are accurate only to a certain number of places and are called **approximate numbers.** Usually, we want to make all decimals in a particular problem accurate to a specified decimal place or tolerance. This will require **rounding** the decimals. We can picture the process on a number line.

Example 8   Rounding to the Nearest Tenth

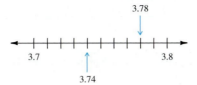

NOTE 3.74 is closer to 3.7 than it is to 3.8. 3.78 is closer to 3.8 than it is to 3.7.

3.74 is rounded down to the nearest tenth, 3.7. 3.78 is rounded up to 3.8.

 CHECK YOURSELF 8

Use the number line in Example 8 to round 3.77 to the nearest tenth.

Rather than using the number line, this rule can be applied.

Step by Step: To Round a Decimal

Step 1   Find the place to which the decimal is to be rounded.

Step 2   If the next digit to the right is 5 or more, increase the digit in the place you are rounding by 1. Discard the remaining digits to the right.

Step 3   If the next digit to the right is less than 5, just discard that digit and any remaining digits to the right.

OBJECTIVE 5

Example 9   Rounding to the Nearest Tenth

Round 34.58 to the nearest tenth.

NOTE Many students find it easiest to mark this digit with an arrow.

34.58   Locate the digit you are rounding to. The 5 is in the tenths place.

Because the next digit to the right, 8, is 5 or more, increase the tenths digit by 1. Then discard the remaining digits.

34.58 is rounded to 34.6.

 CHECK YOURSELF 9

Round 48.82 to the nearest tenth.

Example 10   Rounding to the Nearest Hundredth

Round 5.673 to the nearest hundredth.

5.673   The 7 is in the hundredths place.

The next digit to the right, 3, is less than 5. Leave the hundredths digit as it is and discard the remaining digits to the right.

5.673 is rounded to 5.67.

CHECK YOURSELF 10 _____

Round 29.247 to the nearest hundredth.

Example 11   Rounding to a Specified Decimal Place

Round 3.14159 to four decimal places.

NOTE The fourth place to the *right* of the decimal point is the ten thousandths place.

3.14159   The 5 is in the ten thousandths place.

The next digit to the right, 9, is 5 or more, so increase the digit you are rounding to by 1. Discard the remaining digits to the right.

3.14159 is rounded to 3.1416.

 CHECK YOURSELF 11 _____

Round 0.8235 to three decimal places.

READING YOUR TEXT _____

The following fill-in-the-blank exercises are designed to assure that you understand the key vocabulary used in this section. Each sentence comes directly from the section. You will find the correct answers in Appendix C.

Section 5.1

(a) A _____ fraction is a fraction whose denominator is a power of 10.

(b) The period to the right of the ones place is called the _____ point.

(c) The number of digits to the right of the decimal point is called the number of decimal _____.

(d) When a decimal represents a measurement made by some instrument, it is called an _____ number.

CHECK YOURSELF ANSWERS _____

1.   2   7   9   3   **2.** $5\dfrac{3}{10} = 5.3$ **3.** Thousandths

          Thousands / | Ones
               Hundreds   Tens

4. Two and fifty-eight hundredths   **5. (a)** $\dfrac{528}{1,000}$; **(b)** $5\dfrac{8}{100}$ **6.** $0.588 < 0.59$

7.   8.3   8.51   **8.** 3.8   **9.** 48.8   **10.** 29.25   **11.** 0.824

          8   9

5.1 Exercises

For the decimal 8.57932:

1. What is the place value of 7?

2. What is the place value of 5?

3. What is the place value of 3?

4. What is the place value of 2?

Write in decimal form.

5. $\dfrac{23}{100}$

6. $\dfrac{371}{1,000}$

7. $\dfrac{209}{10,000}$

8. $3\dfrac{5}{10}$

9. $23\dfrac{56}{1,000}$

10. $7\dfrac{431}{10,000}$

Write in words.

11. 0.23

12. 0.371

13. 0.071

14. 0.0251

15. 12.07

16. 23.056

Write in decimal form.

17. Fifty-one thousandths

18. Two hundred fifty-three ten thousandths

19. Seven and three tenths

20. Twelve and two hundred forty-five thousandths

Write each as a common fraction or mixed number.

21. 0.65

22. 0.00765

23. 5.231

24. 4.0171

© 2010 McGraw-Hill Companies

25. _____

26. _____

27. _____

28. _____

29. _____

30. _____

31. _____

32. _____

33. _____

34. _____

35. _____

36. _____

37. _____

38. _____

39. _____

40. _____

41. _____

42. _____

43. _____

44. _____

45. _____

46. _____

Complete each statement, using the symbol $<$, $=$, or $>$.

25. 0.69 _____ 0.689 **ALEKS**   **26.** 0.75 _____ 0.752 **ALEKS**

27. 1.23 _____ 1.230 **ALEKS**   **28.** 2.451 _____ 2.45 **ALEKS**

29. 10 _____ 9.9 **ALEKS**   **30.** 4.98 _____ 5 **ALEKS**

31. 1.459 _____ 1.46 **ALEKS**   **32.** 0.235 _____ 0.2350 **ALEKS**

33. Arrange in order from smallest to largest. **ALEKS**

0.71, 0.072, $\dfrac{7}{10}$, 0.007, 0.0069

$\dfrac{7}{100}$, 0.0701, 0.0619, 0.0712

34. Arrange in order from smallest to largest. **ALEKS**

2.05, $\dfrac{25}{10}$, 2.0513, 2.059

$\dfrac{251}{100}$, 2.0515, 2.052, 2.051

Round to the indicated place.

35. 53.48   tenths   **ALEKS**   **36.** 6.785   hundredths   **ALEKS**

37. 21.534   hundredths   **ALEKS**   **38.** 5.842   tenths   **ALEKS**

39. 0.342   hundredths   **ALEKS**   **40.** 2.3576   thousandths   **ALEKS**

41. 2.71828   thousandths   **ALEKS**   **42.** 1.543   tenths   **ALEKS**

43. 0.0475   tenths   **ALEKS**   **44.** 0.85356   ten thousandths   **ALEKS**

45. 4.85344   ten thousandths   **ALEKS**   **46.** 52.8728   thousandths   **ALEKS**

47. 6.734   two decimal places

48. 12.5467   three decimal places

49. 6.58739   four decimal places

50. 503.824   two decimal places

Round 56.35829 to the nearest:

51. Tenth

52. Ten thousandth

53. Thousandth

54. Hundredth

In exercises 55 to 60, determine the decimal that corresponds to the shaded portion of each "decimal square." Note that the total value of a decimal square is 1.

55.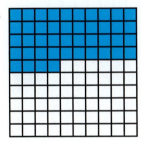

56.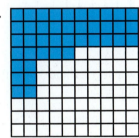

57.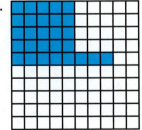

58.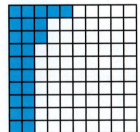

59.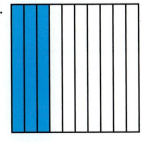

60.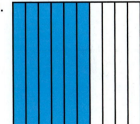

© 2010 McGraw-Hill Companies

ANSWERS

47. _____

48. _____

49. _____

50. _____

51. _____

52. _____

53. _____

54. _____

55. _____

56. _____

57. _____

58. _____

59. _____

60. _____

In exercises 61 to 64, shade the portion of the square that is indicated by the given decimal.

61. 0.23

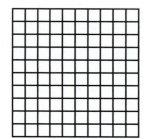

62. 0.89

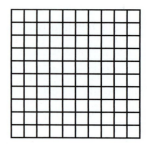

63. 0.3

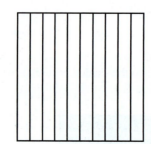

64. 0.30

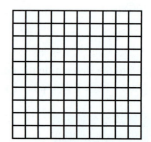

65. Plot (draw a dot) 3.2 and 3.7 on the number line. Then estimate the location for 3.62.

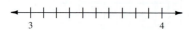

66. Plot 12.51 and 12.58 on the number line. Then estimate the location for 12.537.

67. Plot 7.124 and 7.127 on the number line. Then estimate the location of 7.1253.

68. Plot 5.73 and 5.74 on the number line. Then estimate the location for 5.782.

69. Estimate, to the tenth of a degree, the reading of the Fahrenheit thermometer shown.

ANSWERS

70. _____

71. _____

72. _____

73. _____

74. _____

75. _____

76. _____

77. _____

70. Estimate, to the tenth of a centimeter, the length of the pencil shown.

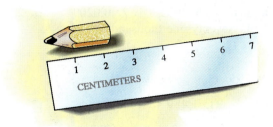

71. (a) What is the difference in these values: 0.120, 0.1200, and 0.12000?

(b) Explain in your own words why placing zeros to the right of a decimal point does not change the value of the number.

72. Lula wants to round 76.24491 to the nearest hundredth. She first rounds 76.24491 to 76.245 and then rounds 76.245 to 76.25 and claims that this is the final answer. What is wrong with this approach?

73. Allied Health A nurse calculates a child's dose of Reglan to be 1.53 milligrams (mg). Round this dose to the nearest tenth of a milligram.

74. Allied Health A nurse calculates a young boy's dose of Dilantin to be 23.375 mg every 5 min. Round this dose to the nearest hundredth of a milligram.

In exercises 75 to 77, indicate whether the given statement is always true, sometimes true, or never true.

75. A decimal can be written as a fraction or a mixed number.

76. A decimal written to the thousandth is greater than a decimal written to the hundredth.

77. Zeros can be written to the right of the rightmost decimal place without changing the size of the number.

Answers

1. Hundredths   **3.** Ten thousandths   **5.** 0.23   **7.** 0.0209   **9.** 23.056
11. Twenty-three hundredths   **13.** Seventy-one thousandths

15. Twelve and seven hundredths   **17.** 0.051   **19.** 7.3

21. $\dfrac{65}{100}$ $\left(\text{or } \dfrac{13}{20}\right)$ **23.** $5\dfrac{231}{1000}$ **25.** $0.69 > 0.689$

27. $1.23 = 1.230$ **29.** $10 > 9.9$ **31.** $1.459 < 1.46$

33. 0.0069, 0.007, 0.0619, $\dfrac{7}{100}$, 0.0701, 0.0712, 0.072, $\dfrac{7}{10}$, 0.71

35. 53.5   **37.** 21.53   **39.** 0.34   **41.** 2.718   **43.** 0.0   **45.** 4.8534
47. 6.73   **49.** 6.5874   **51.** 56.4   **53.** 56.358   **55.** 0.44   **57.** 0.28
59. 0.3   **61.** 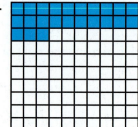   **63.**

65.
```
◄──┼──┼──┼──┼──┼──╳──┼──┼──┼──►
   3              4
```
67.
```
◄──┼──┼──┼──┼──┼──╳──┼──┼──┼──►
  7.12             7.13
```
69. 98.6°F   **71.**   **73.** 1.5 mg   **75.** Always   **77.** Always

5.2 Addition and Subtraction of Decimals

5.2 OBJECTIVES

1. Add two or more decimals
2. Subtract one decimal from another

Working with decimals rather than common fractions makes the basic operations much easier. We will start by looking at addition. One method for adding decimals is to write the decimals as common fractions, add, and then change the sum back to a decimal.

$$0.34 + 0.52 = \frac{34}{100} + \frac{52}{100} = \frac{86}{100} = 0.86$$

It is much more efficient to leave the numbers in decimal form and perform the addition in the same way as we did with whole numbers. You can use this rule.

Step by Step: To Add Decimals

Step 1 Write the numbers being added in column form *with their decimal points aligned vertically.*

Step 2 Add just as you would with whole numbers.

Step 3 Place the decimal point of the sum in line with the decimal points of the addends.

Since the decimal point in a number determines the place value of each digit, aligning the decimal points vertically ensures that tenths will be added to tenths, hundredths will be added to hundredths, and so on.

Example 1 illustrates the use of this rule.

OBJECTIVE 1

Example 1 Adding Decimals

Add 0.13, 0.42, and 0.31.

NOTE Placing the decimal points in a vertical line ensures that we are adding digits of the same place value.

$$
\begin{array}{r}
0.13 \\
0.42 \\
+\ 0.31 \\
\hline
0.86
\end{array}
$$

 CHECK YOURSELF 1

Add 0.23, 0.15, *and* 0.41.

In adding decimals, you can use the *carrying process* just as you did in adding whole numbers. Consider Example 2.

Example 2 **Adding Decimals Involving Carrying**

Add 0.35, 1.58, and 0.67.

```
  1 2  ←——— Carries        In the hundredths column:
  0.35                     5 + 8 + 7 = 20
  1.58                     Write 0 and carry 2 to the tenths column.
+ 0.67                     In the tenths column:
  ————                     2 + 3 + 5 + 6 = 16
  2.60                     Write 6 and carry 1 to the ones column.
```

Note: The carrying process works with decimals, just as it did with whole numbers, because each place value is again *one-tenth* the value of the place to its left.

CHECK YOURSELF 2 _____

Add 23.546, 0.489, 2.312, and 6.135.

In adding decimals, the numbers may not have the same number of decimal places. Just fill in as many zeros as needed so that all of the numbers added have the same number of decimal places.

Recall that adding zeros to the right *does not change* the value of a decimal. 0.53 is the same as 0.530.

We see how this is used in Example 3.

Example 3 **Adding Decimals**

Add 0.53, 4, 2.7, and 3.234.

NOTE Be sure that the decimal points are aligned.

```
  0.53
  4.        Note that for a whole number, the decimal
  2.7       is understood to be to its right. So 4 = 4.
+ 3.234
```

Now fill in the missing zeros and add as before.

```
  0.530
  4.000     Now all the numbers being added
  2.700     have three decimal places.
+ 3.234
  ——————
  10.464
```

CHECK YOURSELF 3 _____

Add 6, 2.583, 4.7, and 2.54.

The addition of negative decimals follows the same rules as the addition of negative integers or fractions.

1. If the signs are the same, find the sum of the magnitudes (absolute values) of the numbers and give the result the sign of the numbers.

2. If the signs are opposite, find the difference of the absolute values of the numbers and give the result the sign of the number with greater magnitude (absolute value).

Example 4   Adding Signed Decimals

Find each sum.

(a) $-2.34 + (-15.7)$

The signs are the same, so we find the sum of the absolute values and give the result a negative sign.

$$-2.34 + (-15.7) = -18.04$$

(b) $-3.56 + 2.14$

The signs are opposite, so we find the difference of the absolute values and give the result the sign of the number with greater magnitude. The -3.56 has greater magnitude, so the result will be negative.

$$-3.56 + 2.14 = -1.42$$

CHECK YOURSELF 4 _____

Find each sum.

(a) $-9.4 + (-19.26)$ **(b)** $-12.3 + 7.2$

Much of what we have said about adding decimals is also true of subtraction. To subtract decimals, we use this rule:

Step by Step: To Subtract Decimals

Step 1 Write the numbers being subtracted in column form *with their decimal points aligned vertically.*

Step 2 Subtract just as you would with whole numbers.

Step 3 Place the decimal point of the difference in line with the decimal points of the numbers being subtracted.

As noted before, vertically aligning the decimal points guarantees that digits of equal place value line up. This means we will be subtracting tenths from tenths, hundredths from hundredths, and so on.

Example 5 illustrates the use of this rule.

OBJECTIVE 2

Example 5   Subtracting a Decimal

Subtract 1.23 from 3.58.

$$
\begin{array}{r}
3.58 \\
-\ 1.23 \\
\hline
2.35
\end{array}
$$

Subtract in the hundredths, the tenths, and then the ones columns.

CHECK YOURSELF 5 _____

Subtract 9.87 − 5.45.

Because each place value is one-tenth the value of the place to its left, borrowing, when you are subtracting decimals, works just as it did in subtracting whole numbers.

Example 6   Subtraction of a Decimal that Involves Borrowing

Subtract 1.86 from 6.54.

$$\begin{array}{r} {}^{5}{}^{14}{}_1 \\ \cancel{6}.\cancel{5}4 \\ -\ 1.86 \\ \hline 4.68 \end{array}$$ Here, borrow from the tenths and ones places to do the subtraction.

CHECK YOURSELF 6 _____

Subtract 35.35 − 13.89.

In subtracting decimals, as in adding, we can write zeros to the right of the decimal point so that both decimals have the same number of decimal places.

Example 7   Subtracting a Decimal

(a) Subtract 2.36 from 7.5.

NOTE When you are subtracting, align the decimal points, then write zeros to the right to align the digits.

$$\begin{array}{r} {}^{4}\ {}_1 \\ 7.\cancel{5}\,0 \\ -\ 2.36 \\ \hline 5.14 \end{array}$$ We have written a 0 at the end of 7.5. Next, borrow 1 tenth from the 5 tenths in the minuend.

(b) Subtract 3.657 from 9.

NOTE 9 has been rewritten as 9.000.

$$\begin{array}{r} 8\ 99 \\ \cancel{9}.\cancel{0}\cancel{0}0 \\ -\ 3.657 \\ \hline 5.343 \end{array}$$ In this case, move left to the ones place to begin the borrowing process.

CHECK YOURSELF 7 _____

Subtract 5 − 2.345.

When subtracting with negative decimals, change the exercise to an addition problem and follow the rules stated just before Example 5.

Example 8   Subtracting with Negative Decimals

Perform each subtraction.

(a) Subtract 4.31 from -6.55.

First, translate the statement into an expression.

$$-6.55 - 4.31$$

Rewrite as an addition problem.

$$-6.55 + (-4.31)$$

The signs are the same, so add the absolute values and make the result negative.

$$-6.55 + (-4.31) = -10.86$$

(b) Subtract -12.4 from -7.2.

Translate the statement into an expression.

$$-7.2 - (-12.4)$$

Rewrite as an addition problem.

$$-7.2 + 12.4$$

The signs are opposite, so find the difference of the absolute values and make the answer positive (12.4 has greater magnitude).

$$-7.2 + 12.4 = 5.2$$

CHECK YOURSELF 8 _____

(a) Subtract 6.39 from -5.12.   **(b)** Subtract -12.34 from 1.5.

(c) Subtract -8.1 from -15.62.

Using Your Calculator to Add and Subtract Decimals

RECALL The reason for this book is to help you review the basic skills of arithmetic. We are using these calculator sections to show you how the calculator can be helpful as a tool. Unless your instructor says otherwise, you should be using your calculator *only on the problems in these special sections.*

Entering decimals in your calculator is similar to entering whole numbers. There is just one difference: The decimal point key ⊡ is used to place the decimal point as you enter the number.

Example 9   Entering a Decimal Number into a Calculator

To enter 12.345, press

[1] [2] [•] [3] [4] [5]

Display   12.345

CHECK YOURSELF 9 _____

Enter 14.367 on your calculator.

NOTE You don't have to press the 0 key for the digit 0 to the left of the decimal point.

Example 10   Entering a Decimal Number into a Calculator

To enter 0.678, press

 . 6 7 8

Display   0.678

 CHECK YOURSELF 10 _____

Enter 0.398 on your calculator.

The process of adding and subtracting with negative decimals on your calculator is the same as we saw in Chapter 1 when we were adding and subtracting integers.

Example 11   Adding Decimals

To add 2.567 + (−0.89), enter

2.567 + 0.89 +/− =

or, on a graphing calculator,

2.567 + (−) 0.89 Enter

Display   1.677

 CHECK YOURSELF 11 _____

Add on your calculator.

5.39 + (−9.7)

Subtraction of decimals on the calculator is similar.

Example 12   Subtracting with Negative Decimals

To subtract −4.2 − (−2.875), enter

4.2 +/− − 2.875 +/− =   or   (−) 4.2 − (−) 2.875 Enter

Display   −1.325

 CHECK YOURSELF 12 _____

Subtract on your calculator.

−16.3 − (−7.895)

Often both addition and subtraction are involved in a calculation. In this case, just enter the decimals and the operation signs, + or −, as they appear in the problem.

Example 13   Adding and Subtracting Decimals

NOTE Again there are differences in the operation of various calculators. Try this problem on yours to check that its operation sequence is correct.

To find $23.7 - 5.2 + 3.87 - 2.341$, enter

$23.7 \boxed{-} 5.2 \boxed{+} 3.87 \boxed{-} 2.341 \boxed{=}$

Display   20.029

CHECK YOURSELF 13 ___

Use your calculator to find

$52.8 - 36.9 + 15.87 - 9.36$

READING YOUR TEXT ___

The following fill-in-the-blank exercises are designed to assure that you understand the key vocabulary used in this section. Each sentence comes directly from the section. You will find the correct answers in Appendix C.

Section 5.2

(a) The first step in adding decimals is to write the numbers so that their decimal _____ align vertically.

(b) In adding decimals, the numbers may not have the same number of decimal _____ .

(c) Because each place value is one-tenth of the value of the place to its left, _____ , when subtracting decimals, works just as it did in subtracting whole numbers.

(d) We use the calculator sections in this text to show you how the calculator can be helpful as a _____ .

CHECK YOURSELF ANSWERS ___

1. 0.79   **2.** 32.482   **3.**   6.000   **4. (a)** −28.66; **(b)** −5.1   **5.** 4.42
                                     2.583
                                     4.700
                                   + 2.540
                                   ‾‾‾‾‾‾‾
                                    15.823

6. 21.46   **7.** 2.655   **8. (a)** −11.51; **(b)** 13.84; **(c)** −7.52   **9.** 14.367
10. 0.398   **11.** −4.31   **12.** −8.405   **13.** 22.41

Name _____

Section _____ Date _____

ANSWERS

1. _____
2. _____
3. _____
4. _____
5. _____
6. _____
7. _____
8. _____
9. _____
10. _____
11. _____
12. _____
13. _____
14. _____
15. _____
16. _____
17. _____
18. _____
19. _____
20. _____

5.2 Exercises

Add.

1.   0.28   ALEKS
   + 0.79

2.   2.59   ALEKS
   + 0.63

3. −1.045 + (−0.23)

4. −2.485 + (−1.25)

5.   0.62
    4.23
   + 12.5

6.   0.50
    2.99
   + 24.8

7.   5.28   ALEKS
   + 19.455

8.   23.845   ALEKS
   + 7.29

9.   13.58
    7.239
   + 1.5

10. −8.625 + (−2.45) + (−12.6)

11. 25.3582
     6.5
     1.898
   + 0.69

12.   1.336
     15.6857
      7.9
   +   0.85

13. 0.43 + 0.8 + 0.561

14. 1.25 + 0.7 + 0.259

15. 5 + 23.7 + 8.7 + 9.85

16. 28.3 + 6 + 8.76 + 3.8

17. −25.83 + (−5.62)

18. −32.59 + (−9.56)

19. 42.731 + 1.058 + 103.24

20. 27.4 + 213.321 + 39.38

In exercises 21 to 24, use decimal square shading to represent the addition process. Shade each square and the total.

21.

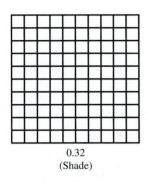 + 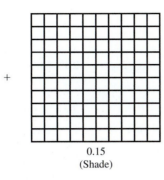 =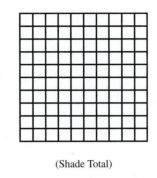

0.32
(Shade)

0.15
(Shade)

(Shade Total)

22.

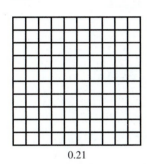 + 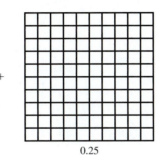 =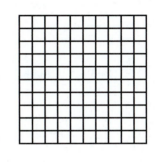

0.21

0.25

23.

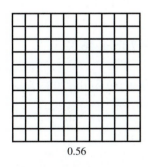 + 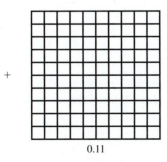 =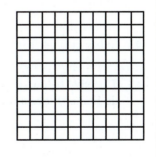

0.56

0.11

24.

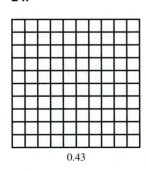 + 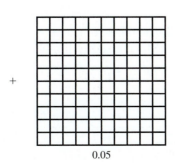 =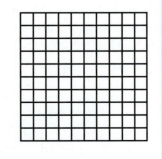

0.43

0.05

Add or subtract as indicated.

25. 0.85
  − 0.59

26. 5.68
  − 2.65

27. 23.81 + (−6.57)   **ALEKS**

28. 48.03 + (−19.95) **ALEKS**

29. 17.134
  − 3.502

30. 40.092
  − 21.595

31. −35.8 + 7.45 **ALEKS**

32. −7.83 + 5.2 **ALEKS**

33. 3.82 **ALEKS**
  − 1.565

34. 8.59 **ALEKS**
  − 5.6

35. 7.02 **ALEKS**
  − 4.7

36. 45.6 **ALEKS**
  − 8.75

37. −12 − (−5.35)   **ALEKS**

38. −15 − (−8.85) **ALEKS**

39. Subtract 2.87 from 6.84.

40. Subtract 3.69 from 10.57.

41. Subtract −7.75 from 9.4.

42. Subtract 5.82 from 12.

43. Subtract 0.24 from 5.

44. Subtract −8.7 from 16.32.

45. The table below shows the lengths, in miles, of the first five stages of the 2005 Tour de France bicycle race. Find the total distance for these five stages.

Stage number	1	2	3	4	5
Length in miles	11.8	112.8	132	41.9	113.7

46. This table shows the lengths, in miles, of stages 6 through 10 of the 2005 Tour de France bicycle race. Find the total distance for these five stages.

Stage number	6	7	8	9	10
Length in miles	123.7	142	143.8	106.3	110.9

47. Manufacturing Technology A pin is defined to have a clearance fit in a 0.618-in.-diameter hole. A clearance of 0.013 in. is called out. What is the diameter of the pin?

48. Manufacturing Technology A dimension on a computer-aided design (CAD) plan is given as 3.084 in. ± 0.125 in. What are the minimum and maximum lengths of this feature?

Calculator Exercises

Solve the exercises using your calculator.

49. $5.87 + 3.6 + 9.25$

50. $3.456 + 10 + 2.8 + 5.62$

51. $-28.21 + (-387.6) + (-3,935.21)$

52. $-10,345.2 + (-2,308.35) + (-153.58)$

53. $-4.59 - (-2.389)$ **54.** $-19.375 - (-14.2)$

55. $27.85 - 3.45 - 2.8$ **56.** $8.8 - 4.59 - 2.325 + 8.5$

57. $14 + 3.2 - 9.35 - 3.375$ **58.** $8.7675 + 2.8 - 3.375 - 6$

Answers

1. 1.07   **3.** −1.275   **5.** 17.35   **7.** 24.735   **9.** 22.319

11. 34.4462   **13.** 1.791   **15.**
$$\begin{array}{r} \overset{22}{5.00} \\ 23.70 \\ 8.70 \\ +\ 9.85 \\ \hline 47.25 \end{array}$$
   17. −31.45   **19.** 147.029

21. 0.47   **23.** 0.67   **25.** 0.26   **27.**
$$\begin{array}{r} {}^{11\ 71} \\ 23.\overset{}{8}1 \\ -\ 6.57 \\ \hline 17.24 \end{array}$$
   29. 13.632

31. −28.35   **33.** 2.255   **35.** 2.32   **37.** −6.65   **39.** 3.97
41. 17.15   **43.** 4.76   **45.** 499.38 miles   **47.** 0.605 in.
49. 18.72   **51.** −4,351.02   **53.** −2.201   **55.** 21.6   **57.** 4.475

ANSWERS

47.

48.

49.

50.

51.

52.

53.

54.

55.

56.

57.

58.

5.3 Multiplication of Decimals

5.3 OBJECTIVES

1. Multiply two or more decimals
2. Estimate the product of decimals
3. Multiply a decimal by a power of ten

To start our discussion of the multiplication of decimals, we will write the decimals in common-fraction form and then multiply.

OBJECTIVE 1

Example 1 Multiplying Two Decimals

$$0.32 \cdot 0.2 = \frac{32}{100} \cdot \frac{2}{10} = \frac{64}{1000} = 0.064$$

Here 0.32 has *two* decimal places, and 0.2 has *one* decimal place. The product 0.064 has *three* decimal places.

Note:

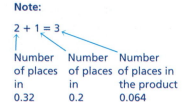

| Number of places in 0.32 | Number of places in 0.2 | Number of places in the product 0.064 |

 CHECK YOURSELF 1

Find the product and the number of decimal places.

$0.14 \cdot 0.054$

You do not need to write decimals as common fractions to multiply. Our work suggests this rule.

Step by Step: To Multiply Decimals

Step 1 Multiply the decimals as though they were whole numbers.
Step 2 Count the number of decimal places in each of the numbers being multiplied.
Step 3 Place the decimal point in the product so that the number of decimal places in the product is the sum of the number of decimal places in the factors.

Example 2 illustrates this rule.

Example 2 Multiplying Two Decimals

Multiply 0.23 by 0.7.

$$
\begin{array}{r}
0.23 \\
\times\ 0.7 \\
\hline
0.161
\end{array}
$$

← Two places
← One place
← Three places

© 2010 McGraw-Hill Companies

CHECK YOURSELF 2 _____

Multiply 0.36 · 1.52.

You may have to affix zeros to the left in the product to place the decimal point. Consider Example 3.

Example 3   Multiplying Two Decimals

Multiply.

$$
\begin{array}{r}
0.136 \\
\times\ 0.28 \\
\hline
1088 \\
272\ \ \\
\hline
0.03808
\end{array}
$$

0.136 ← Three places
× 0.28 ← Two places   3 + 2 = 5

0.03808 ← Five places   Insert a 0 to mark off five decimal places.

↑
Insert 0

 CHECK YOURSELF 3 _____

Multiply 0.234 · 0.24.

Estimation is also helpful in multiplying decimals.

OBJECTIVE 2   ### Example 4   Estimating the Product of Two Decimals

Estimate the product 24.3 · 5.8.

Round

$$
\begin{array}{r}
24.3 \\
\times\ 5.8
\end{array}
\quad \longrightarrow \quad
\begin{array}{r}
24 \\
\times\ 6 \\
\hline
144
\end{array}
$$

Multiply for the estimate.

Note that the exact product is 140.94. This is reasonably close to our estimate. If we had made a mistake such as misplacing the decimal point in the exact calculation, we would surely notice!

 CHECK YOURSELF 4 _____

Estimate the product.

17.95 · 8.17

When multiplying negative decimals, recall these rules:

1. The product of two numbers with the same sign will be positive.

2. The product of two numbers with opposite signs will be negative.

Example 5 Multiplying Signed Decimals

Find each product.

(a) $(2.5)(-1.4)$

The signs are opposite, so the product will be negative.

$$\begin{array}{r} 1.4 \\ \times\ 2.5 \\ \hline 70 \\ 28 \\ \hline 3.50 \end{array}$$

So, $(2.5)(-1.4) = -3.5$.

(b) $(-4.6)(-1.3)$

The signs are the same, so the product will be positive.

$$\begin{array}{r} 1.3 \\ \times\ 4.6 \\ \hline 78 \\ 52 \\ \hline 5.98 \end{array}$$

So, $(-4.6)(-1.3) = 5.98$.

 CHECK YOURSELF 5

Find each product.

(a) $-1.7 \cdot 4.2$ **(b)** $(-2.6)(-2.3)$ **(c)** $(6.8)(-1.6)$

We will study applications involving operations with decimals in Section 5.8. For now, we will note that there are enough applications involving multiplication by the powers of 10 to make it worthwhile to develop a special rule so you can do such operations quickly and easily. Look at the patterns in some of these special multiplications.

$$\begin{array}{r} 0.679 \\ \times\quad 10 \\ \hline 6.790 \end{array} \quad \text{or} \quad 6.79 \qquad\qquad \begin{array}{r} 23.58 \\ \times\quad 10 \\ \hline 235.80 \end{array} \quad \text{or} \quad 235.8$$

Do you see that multiplying by 10 has moved the decimal point *one place to the right?* Now we will look at what happens when we multiply by 100.

NOTE The rule will be used to multiply by 10, 100, 1,000, and so on.

$$\begin{array}{r} 0.892 \\ \times\quad 100 \\ \hline 89.200 \end{array} \quad \text{or} \quad 89.2 \qquad\qquad \begin{array}{r} 5.74 \\ \times\quad 100 \\ \hline 574.00 \end{array} \quad \text{or} \quad 574$$

NOTE The digits remain the same. Only the *position* of the decimal point is changed.

Multiplying by 100 shifts the decimal point *two places to the right.* The pattern of these examples gives us this rule:

NOTE Multiplying by 10, 100, or any other larger power of 10 makes the number *larger.* Move the decimal point *to the right.*

Property: To Multiply by a Power of 10

Move the decimal point to the right the same number of places as there are zeros in the power-of-10 term.

OBJECTIVE 3

Example 6   Multiplying by Powers of Ten

$2.356 \cdot 10 = 23.56$

One zero

The decimal point has moved one place to the right.

$3.67 \cdot 1,000 = 3,670.$

Three zeros

The decimal point has moved three places to the right. Note that we inserted a 0 to place the decimal point correctly.

RECALL 10^5 is just a 1 followed by five zeros.

$0.005672 \cdot 10^5 = 567.2$

Five zeros

The decimal point has moved five places to the right.

CHECK YOURSELF 6

Multiply.

(a) $43.875 \cdot 100$

(b) $0.0083 \cdot 10^3$

The steps for finding the product of decimals on a calculator are similar to the ones we used for multiplying whole numbers. To find the product of a group of decimals, just extend the process.

Example 7   Multiplying a Group of Decimals

To multiply $2.8 \cdot 3.45 \cdot 3.725$, enter

2.8 ☒ 3.45 ☒ 3.725 =

Display   35.9835   Again, look at the original expression to see if this answer is reasonable.

CHECK YOURSELF 7

Multiply $3.1 \cdot 5.72 \cdot 6.475$.

You can also easily find powers of decimals with your calculator by using a procedure similar to that in Example 7.

Example 8   Finding the Power of a Decimal Number

RECALL

$(2.35)^3 = 2.35 \cdot 2.35 \cdot 2.35$

Find $(2.35)^3$.

Enter

2.35 ☒ 2.35 ☒ 2.35 =

Display   12.977875

CHECK YOURSELF 8 _____

Evaluate $(6.2)^4$.

As we stated earlier, some calculators have keys that will find powers more quickly. Look for keys marked $\boxed{x^2}$ or $\boxed{y^x}$. Other calculators have a power key marked $\boxed{\wedge}$.

> **Example 9** **Finding the Power of a Decimal Number Using Power Keys**

Find $(2.35)^3$.
  Enter

2.35 $\boxed{\wedge}$ 3 $\boxed{\text{Enter}}$ or   2.35 $\boxed{y^x}$ 3 $\boxed{=}$

Display   12.977875.

CHECK YOURSELF 9 _____

Find $(6.2)^4$ *using the power key.*

How many places can your calculator display? Most calculators can display either 8, 9, or 10 digits. To find the display capability of your calculator, just enter digits until the calculator can accept no more numbers. For example, try entering

1 $\boxed{-}$ 0.226592266 $\boxed{=}$

Does your calculator display 10 digits? Now turn the calculator upside down. What does it say? (It may take a little imagination to see it.)

What happens when your calculator wants to display an answer that is too big to fit in the display? We will try an experiment to see. Enter

10 $\boxed{\times}$ 10 $\boxed{=}$

Now continue to multiply this answer by 10. Many calculators will let you do this by simply pressing $\boxed{=}$. Others require you to enter "$\boxed{\times}$ 10" for each calculation. Multiply by 10 until the display is no longer a 1 followed by a series of zeros. The new display represents the power of 10 of the answer. It will be displayed as either

1^{10}

(which looks like 1 to the tenth power, but means 1 times 10^{10}) or

1 E 10

(which also means 1 times 10^{10}).

Answers that are displayed in this way are said to be in **scientific notation.** This is a topic that you will study in your next math course. In this text, we will avoid exercises with answers that are too large to display in the decimal notation that you already know. If you do get such an answer, you should go back and check your work. Do not be afraid to try experimenting with your calculator. It is amazing how much math you can (accidently) learn while playing!

> ### Example 10   Multiplying by a Power of Ten Using the Power Key on a Calculator

Find the product $3.485 \cdot 10^4$.
Use your calculator to enter

$3.485 \boxed{\times} 10 \boxed{y^x} 4 \boxed{=}$ or $3.485 \boxed{\times} 10 \boxed{\wedge} 4 \boxed{\text{Enter}}$

The result will be 34,850. Note that the decimal point has moved four places (the power of 10) to the right.

CHECK YOURSELF 10

Find the product $8.755 \cdot 10^6$.

READING YOUR TEXT

The following fill-in-the-blank exercises are designed to assure that you understand the key vocabulary used in this section. Each sentence comes directly from the section. You will find the correct answers in Appendix C.

Section 5.3

(a) To determine the number of decimal places in a product, _____ the number of decimal places in the numbers being multiplied.

(b) When multiplying, you may have to affix _____ to the left in the product to place the decimal point.

(c) The product of two numbers with opposite signs is _____.

(d) To multiply by a power of ten, move the decimal point to the _____ the same number of places as there are zeros in the power-of-10 term.

CHECK YOURSELF ANSWERS

1. 0.00756; 5 decimal places   **2.** 0.5472   **3.** 0.05616   **4.** 144
5. **(a)** −7.14; **(b)** 5.98; **(c)** −10.88   **6.** **(a)** 4,387.5; **(b)** 8.3   **7.** 114.8147
8. 1,477.6336   **9.** 1,477.6336   **10.** 8,755,000

5.3   Exercises

Multiply.

1. 2.3
      × 3.4   **ALEKS**

2. 6.5
      × 4.3   **ALEKS**

3. 8.4
      × 5.2
                 ALEKS

4. 9.2
      × 4.6   **ALEKS**

5. 2.56
      × 72   **ALEKS**

6. 56.7
      × 35   **ALEKS**

7. 0.78
      × 2.3   **ALEKS**

8. 9.5
      × 0.45   **ALEKS**

9. 15.7
      × 2.35

10. 28.3
       × 0.59   **ALEKS**

11. 0.354
       × 0.8   **ALEKS**

12. 0.624
       × 0.85   **ALEKS**

13. $(3.28)(-5.07)$

14. $(0.582)(-6.3)$

15. $5.238 \cdot 0.48$

16. $0.372 \cdot 58$

17. $(-1.053)(-0.552)$

18. $-2.375 \cdot 0.28$

19. $(0.0056)(-0.082)$

20. $(-1.008)(-0.046)$

21. $0.8 \cdot 2.376$

22. $3.52 \cdot 58$

23. $0.3085 \cdot 4.5$

24. $0.028 \cdot 0.685$

Multiply.

25. $5.89 \cdot 10$ **ALEKS**

26. $-0.895 \cdot 100$

27. $-23.79 \cdot 100$

28. $2.41 \cdot 10$ **ALEKS**

29. $0.045 \cdot 10$ **ALEKS**

30. $5.8 \cdot 100$ **ALEKS**

31. $(0.431)(-100)$

32. $(0.025)(-10)$

33. $0.471 \cdot 100$
        ALEKS

34. $0.95 \cdot 10,000$ **ALEKS**

35. $(-0.7125)(-1,000)$

36. $(-23.42)(-1,000)$

37. $4.25 \cdot 10^2$ **38.** $0.36 \cdot 10^3$

39. $3.45 \cdot 10^4$ **40.** $0.058 \cdot 10^5$

In exercises 41 and 42, fill in the blank with the correct decimal place.

41. A number written to the tenths multiplied by a number written to the hundredths will usually give a product to the _____.

42. A number written to the hundredths multiplied by a number written to the thousandths will usually give a product to the _____.

43. Each lap around a running track is 0.25 miles. How many miles are there in 13.5 laps?

44. Each lap around a running track is 0.25 miles. How many miles are there in 43 laps?

45. Each lap around Daytona International Speedway is 2.5 miles. How many miles are there in 13.5 laps?

46. Each lap around Daytona International Speedway is 2.5 miles. How many miles are there in 199 laps?

 # Calculator Exercises

Compute.

47. $(-127.85)(0.055)(15.84)$ **48.** $(-18.28)(143.45)(-0.075)$

49. $(2.65)^2$ **50.** $(0.08)^3$

51. $(3.95)^3$ **52.** $(0.521)^2$

Find the products using your calculator.

53. $3.365 \cdot 10^3$ **54.** $4.128 \cdot 10^3$

55. $4.316 \cdot 10^5$ **56.** $8.163 \cdot 10^6$

ANSWERS

37. _____

38. _____

39. _____

40. _____

41. _____

42. _____

43. _____

44. _____

45. _____

46. _____

47. _____

48. _____

49. _____

50. _____

51. _____

52. _____

53. _____

54. _____

55. _____

56. _____

ANSWERS

57. _____

58. _____

59. _____

60. _____

61. _____

62. _____

57. $7.236 \cdot 10^8$

58. $5.234 \cdot 10^7$

59. $32.136 \cdot 10^5$

60. $41.234 \cdot 10^4$

61. $31.789 \cdot 10^4$

62. $61.356 \cdot 10^3$

Answers

1. 7.82   **3.** 43.68   **5.** 184.32   **7.** 1.794   **9.** 36.895   **11.** 0.2832
13. -16.6296 **15.** 2.51424   **17.** 0.581256   **19.** -0.0004592
21. 1.9008   **23.** 1.38825   **25.** 58.9   **27.** $-2,379$ **29.** 0.45
31. -43.1 **33.** 47.1   **35.** 712.5   **37.** 425   **39.** 34,500
41. thousandths   **43.** 3.375 miles   **45.** 33.75 miles   **47.** -111.38292
49. 7.0225   **51.** 61.629875   **53.** 3,365   **55.** 431,600
57. 723,600,000   **59.** 3,213,600   **61.** 317,890

5.4 Division of Decimals

5.4 OBJECTIVES

1. Divide a decimal by a whole number
2. Divide a decimal by a decimal
3. Divide a decimal by a power of ten
4. Use order of operations with decimals

The division of decimals is very similar to our earlier work with dividing whole numbers. The only difference is in learning to place the decimal point in the quotient. We will start with the case of dividing a decimal by a whole number. Here, placing the decimal point is easy. You can apply this rule.

Step by Step: To Divide a Decimal by a Whole Number

Step 1 Place the decimal point in the quotient *directly above* the decimal point of the dividend.

Step 2 Divide as you would with whole numbers.

OBJECTIVE 1

Example 1 Dividing a Decimal by a Whole Number

Divide 29.21 by 23.

NOTE Do the division just as if you were dealing with whole numbers. Just remember to place the decimal point in the quotient *directly above* the one in the dividend.

$$
\begin{array}{r}
1.27 \\
23\overline{)29.21} \\
\underline{23} \\
6\,2 \\
\underline{4\,6} \\
1\,61 \\
\underline{1\,61} \\
0
\end{array}
$$

The quotient is 1.27. You can check by multiplying $1.27 \cdot 23$.

 CHECK YOURSELF 1 _____

Divide 80.24 by 34.

We will look at another example of dividing a decimal by a whole number.

Example 2 Dividing a Decimal by a Whole Number

Divide 122.2 by 52.

NOTE Again place the decimal point of the quotient above that of the dividend.

$$
\begin{array}{r}
2.3 \\
52\overline{)122.2} \\
\underline{104} \\
18\,2 \\
\underline{15\,6} \\
2\,6
\end{array}
$$

We normally do not use a remainder when dealing with decimals. Instead, we write an extra 0 in the dividend and continue.

RECALL Writing a 0 in the rightmost position does not change the value of the dividend. It simply allows us to complete the division process in this case.

$$
\begin{array}{r}
2.35 \\
52\overline{)122.20} \quad \longleftarrow \text{ Add a 0.} \\
\underline{104} \\
18\ 2 \\
\underline{15\ 6} \\
2\ 60 \\
\underline{2\ 60} \\
0
\end{array}
$$

So $122.2 \div 52 = 2.35$. The quotient is 2.35.

 CHECK YOURSELF 2

Divide 234.6 by 68.

Often you will be asked to give a quotient that is rounded to a certain place value. In this case, continue the division process to *one digit past* the indicated place value. Then round the result back to the desired accuracy.

When working with money, for instance, we normally give the quotient to the nearest hundredth of a dollar (the nearest cent). This means carrying the division out to the thousandths place and then rounding back.

> **Example 3** **Dividing a Decimal by a Whole Number and Rounding the Result**

NOTE Find the quotient to *one place past* the desired place and then round the result.

Find the quotient of $25.75 \div 15$ to the nearest hundredth.

$$
\begin{array}{r}
1.716 \\
15\overline{)25.750} \\
\underline{15} \\
10\ 7 \\
\underline{10\ 5} \\
25 \\
\underline{15} \\
100 \\
\underline{90} \\
10
\end{array}
$$

Write a 0 to carry the division to the thousandths place.

So $25.75 \div 15 = 1.72$ (to the nearest hundredth).

 CHECK YOURSELF 3

Find $99.26 \div 35$ to the nearest hundredth.

We want now to look at division *by* decimals. Here is an example using a fractional form.

Example 4   Rewriting an Exercise that Requires Dividing by a Decimal

Rewrite the division exercise so that the divisor is a whole number.

$$2.57 \div 3.4 = \frac{2.57}{3.4}$$ Write the division as a fraction.

$$= \frac{2.57 \cdot 10}{3.4 \cdot 10}$$ We multiply the numerator and denominator by 10 so the divisor is a whole number. This *does not change* the value of the fraction.

$$= \frac{25.7}{34}$$ Multiplying by 10, shift the decimal point in the numerator and denominator *one place to the right*.

$$= 25.7 \div 34$$ Our division exercise is rewritten so that the divisor is a whole number.

So

$$2.57 \div 3.4 = 25.7 \div 34$$ After we multiply the numerator and denominator by 10, we see that 2.57 ÷ 3.4 is the same as 25.7 ÷ 34.

NOTE It's always easier to rewrite a division exercise so that you're dividing by a whole number. Dividing by a whole number makes it easy to place the decimal point in the quotient.

CHECK YOURSELF 4 _____

Rewrite the division exercise so that the divisor is a whole number.

3.42 ÷ 2.5

NOTE Of course, multiplying by any whole-number power of 10 greater than 1 is just a matter of shifting the decimal point to the right.

Do you see the rule suggested by Example 4? We multiplied the numerator and the denominator (the dividend and the divisor) by 10. We made the divisor a whole number without altering the actual digits involved. All we did was shift the decimal point in the divisor and dividend the same number of places. This leads us to the next rule.

Step by Step: To Divide by a Decimal

Step 1 Move the decimal point in the divisor *to the right,* making the divisor a whole number.

Step 2 Move the decimal point in the dividend to the right *the same number of places.* Add zeros if necessary.

Step 3 Place the decimal point in the quotient directly above the decimal point of the new dividend.

Step 4 Divide as you would with whole numbers.

We will now look at an example of the use of our division rule.

OBJECTIVE 2

Example 5   Dividing by a Decimal

Divide 1.573 by 0.48 and give the quotient to the nearest tenth.

Write

$$0.48\overline{)1.57\,3}$$ Shift the decimal points two places to the right to make the divisor a whole number.

Now divide:

NOTE Once the division statement is rewritten, place the decimal point in the quotient above that in the dividend.

$$
\begin{array}{r}
3.27 \\
48\overline{)157.30} \\
144 \\
\hline
13\ 3 \\
9\ 6 \\
\hline
3\ 70 \\
3\ 36 \\
\hline
34
\end{array}
$$

Note that we add a 0 to carry the division to the hundredths place. In this case, we want to find the quotient to the nearest tenth.

Round 3.27 to 3.3. So

$1.573 \div 0.48 = 3.3$ (to the nearest tenth)

CHECK YOURSELF 5

Divide, rounding the quotient to the nearest tenth.

$3.4 \div 1.24$

Recall that you can multiply decimals by powers of 10 by simply shifting the decimal point to the right. A similar approach will work for division by powers of 10.

OBJECTIVE 3   **Example 6   Dividing a Decimal by a Power of 10**

(a) Divide 35.3 by 10.

$$
\begin{array}{r}
3.53 \\
10\overline{)35.30} \\
30 \\
\hline
5\ 3 \\
5\ 0 \\
\hline
30 \\
30 \\
\hline
0
\end{array}
$$

The dividend is 35.3. The quotient is 3.53. The decimal point has been shifted *one place to the left.* Note also that the divisor, 10, has *one* zero.

(b) Divide 378.5 by 100.

$$
\begin{array}{r}
3.785 \\
100\overline{)378.500} \\
300 \\
\hline
78\ 5 \\
70\ 0 \\
\hline
8\ 50 \\
8\ 00 \\
\hline
500 \\
500 \\
\hline
0
\end{array}
$$

Here the dividend is 378.5, whereas the quotient is 3.785. The decimal point is now shifted *two places to the left.* In this case the divisor, 100, has *two* zeros.

✓ CHECK YOURSELF 6 _____

Perform each of the divisions.

(a) $52.6 \div 10$ **(b)** $267.9 \div 100$

Example 6 suggests this rule.

> **Property:** To Divide a Decimal by a Power of 10
>
> Move the decimal point *to the left* the same number of places as there are zeros in the power-of-10 term.

Example 7   Dividing a Decimal by a Power of 10

Divide.

(a) $27.3 \div 10 = 2\,7.3$ Shift one place to the left.

                  $= 2.73$

(b) $57.53 \div 100 = 0\,57.53$ Shift two places to the left.

                  $= 0.5753$

NOTE As you can see, we may have to insert zeros to correctly place the decimal point.

(c) $39.75 \div 1,000 = 0\,039.75$ Shift three places to the left.

                  $= 0.03975$

(d) $85 \div 1,000 = 0\,085.$ The decimal after the whole number 85 is implied.

                  $= 0.085$

RECALL 10^4 is a 1 followed by four zeros.

(e) $235.72 \div 10^4 = 0\,0235.72$ Shift four places to the left.

                  $= 0.023572$

✓ CHECK YOURSELF 7 _____

Divide.

(a) $3.84 \div 10$ **(b)** $27.3 \div 1,000$

Recall the rules associated with the division of real numbers:

1. The quotient of two numbers with the same sign is positive.

2. The quotient of two numbers with opposite signs is negative.

| Example 8 | **Dividing with Negative Decimals** |

Find each quotient.

(a) $-1.61 \div 2.3$

The signs are opposite, so the quotient will be negative. Dividing 1.61 by 2.3, we find

$$2.3\overline{)1.61}$$

$$\begin{array}{r} .7 \\ 23\overline{)16.1} \\ \underline{16\ 1} \\ 0 \end{array}$$

So, $-1.61 \div 2.3 = -0.7$.

(b) $-5.13 \div (-6.84)$

The signs are the same, so the result will be positive. Dividing 5.13 by 6.84, we find

$$6.84\overline{)5.13}$$

$$\begin{array}{r} .75 \\ 684\overline{)513.00} \\ \underline{478\ 8} \\ 34\ 20 \\ \underline{34\ 20} \\ 0 \end{array}$$

So, $-5.13 \div (-6.84) = 0.75$.

CHECK YOURSELF 8 _____

Find each quotient.

(a) $-3.612 \div 2.58$ **(b)** $-2.224 \div (-2.78)$

Recall that the order of operations is always used to simplify a mathematical expression with several operations. You should recall the order of operations as given here.

| Property: | The Order of Operations |

1. Perform any operations enclosed in *parentheses.*
2. Apply any *exponents.*
3. Do any *multiplication* and *division,* moving from left to right.
4. Do any *addition* and *subtraction,* moving from left to right.

OBJECTIVE 4

Example 9   Applying the Order of Operations

Simplify each expression.

(a) $4.6 + (0.5 \cdot 4.4)^2 - 3.93$ parentheses
$= 4.6 + (2.2)^2 - 3.93$ exponent
$= 4.6 + 4.84 - 3.93$ add (left of the subtraction)
$= 9.44 - 3.93$ subtract
$= 5.51$

(b) $16.5 - (2.8 + 0.2)^2 + 4.1 \cdot 2$ parentheses
$= 16.5 - (3)^2 + 4.1 \cdot 2$ exponent
$= 16.5 - 9 + 4.1 \cdot 2$ multiply
$= 16.5 - 9 + 8.2$ subtraction (left of the addition)
$= 7.5 + 8.2$ add
$= 15.7$

(c) $4.8 + 6(8.9 - 10.9)^2 - 8.5 \cdot 4$ parentheses
$= 4.8 + 6(-2)^2 - 8.5 \cdot 4$ exponent
$= 4.8 + 6 \cdot 4 - 8.5 \cdot 4$ multiply
$= 4.8 + 24 - 34$ add (left of the subtraction)
$= 28.8 - 34$ subtract
$= -5.2$

CHECK YOURSELF 9 _____

Simplify each expression.

(a) $6.35 + (0.2 \cdot 8.5)^2 - 3.7$ **(b)** $2.5^2 - (3.57 - 2.14) + 3.2 \cdot 1.5$
(c) $2.7 + 5(4.75 - 9.75)^2 - 5^3$

Using Your Calculator to Divide Decimals

It would be most surprising if you had reached this point without using your calculator to divide decimals. It is a good way to check your work and a reasonable way to solve applications.

Example 10   Dividing Decimals

Use your scientific calculator to find the quotient.

$211.56 \div (-82)$

Enter the problem in the calculator.

211.56 ⟦÷⟧ 82 ⟦+/−⟧ ⟦=⟧

We find that the answer is -2.58.

NOTE On a graphing calculator, we type:

211.56 ⟦÷⟧ ⟦(−)⟧ 82 ⟦ENTER⟧

CHECK YOURSELF 10 _____

Use your calculator to find the quotient.

$-304.32 \div 9.6$

READING YOUR TEXT _____

The following fill-in-the-blank exercises are designed to assure that you understand the key vocabulary used in this section. Each sentence comes directly from the section. You will find the correct answers in Appendix C.

Section 5.4

(a) When dividing decimals, place the decimal in the _____ directly above the decimal point of the dividend.

(b) We normally do not use a _____ when dealing with decimal division. Instead we write an extra zero in the dividend and continue dividing.

(c) When dividing by a power of ten, move the decimal point to the _____ the same number of digits as there are zeros in the power of ten.

(d) The rules for order of operations are always used to simplify a mathematical _____.

CHECK YOURSELF ANSWERS _____

1. 2.36 **2.** 3.45 **3.** 2.84 **4.** $34.2 \div 25$ **5.** 2.7
6. (a) 5.26; **(b)** 2.679 **7. (a)** 0.384; **(b)** 0.0273 **8. (a)** -1.4; **(b)** 0.8
9. (a) 5.54; **(b)** 9.62; **(c)** 2.7 **10.** -31.7

5.4 Exercises

Divide.

1. $16.68 \div 6$

2. $43.92 \div 8$ ALEKS

3. $1.92 \div (-4)$

4. $-5.52 \div (-6)$

5. $-5.48 \div (-8)$

6. $-2.76 \div 8$

7. $13.89 \div 6$ VIDEO ALEKS

8. $21.92 \div 5$ ALEKS

9. $185.6 \div (-32)$

10. $-165.6 \div (-36)$

11. $-79.9 \div (-34)$

12. $179.3 \div (-55)$

13. $52 \overline{)13.78}$ VIDEO

14. $76 \overline{)26.22}$

15. $0.6 \overline{)11.07}$ VIDEO

16. $0.8 \overline{)10.84}$

17. $3.8 \overline{)7.22}$ ALEKS

18. $2.9 \overline{)13.34}$ ALEKS

19. $5.2 \overline{)11.622}$

20. $6.4 \overline{)3.616}$

21. $0.27 \overline{)1.8495}$ ALEKS

22. $0.038 \overline{)0.8132}$ ALEKS

23. $0.046 \overline{)1.587}$

24. $0.52 \overline{)3.2318}$

25. $-0.658 \div 2.8$ VIDEO

26. $0.882 \div (-0.36)$

Divide by moving the decimal point.

27. $5.8 \div 10$ ALEKS

28. $5.1 \div 10$ ALEKS

29. $-4.568 \div 100$

30. $-3.817 \div 100$

31. $24.39 \div 1,000$ VIDEO ALEKS

32. $8.41 \div 100$ ALEKS

33. $-6.9 \div 1,000$

34. $-7.2 \div 1,000$

ANSWERS

1. _____   2. _____
3. _____   4. _____
5. _____   6. _____
7. _____   8. _____
9. _____   10. _____
11. _____   12. _____
13. _____   14. _____
15. _____   16. _____
17. _____   18. _____
19. _____   20. _____
21. _____   22. _____
23. _____   24. _____
25. _____
26. _____
27. _____
28. _____
29. _____
30. _____
31. _____
32. _____
33. _____
34. _____

Name _____

Section _____ Date _____

35.

36.

37.

38.

39.

40.

41.

42.

43.

44.

45.

46.

47.

48.

49.

50.

51.

52.

53.

54.

55.

56.

35. $7.8 \div 10^2$

36. $3.6 \div 10^3$

37. $-45.2 \div 10^5$

38. $-57.3 \div 10^4$

Divide and round the quotient to the indicated decimal place.

39. $23.8 \div 9$ tenths

40. $5.27 \div 8$ hundredths

41. $38.48 \div 46$ hundredths

42. $3.36 \div 36$ thousandths

43. $0.7\overline{)1.642}$ hundredths

44. $0.6\overline{)7.695}$ tenths

45. $4.5\overline{)8.415}$ tenths

46. $5.8\overline{)16}$ hundredths

47. $3.12\overline{)4.75}$ hundredths

48. $64.2\overline{)16.3}$ thousandths

Simplify each expression.

49. $4.2 - 3.1 \cdot 1.5 + (3.1 + 0.4)^2$

50. $150 + 4.1 \cdot 1.5 - (2.5 \cdot 1.6)^3 \cdot 2.4$

51. $17.9 \cdot 1.1 - (2.3 \cdot 1.1)^2 + (13.4 - 2.1 \cdot 4.6)$

52. $6.89^2 - 3.14 \cdot 2.5 + (4.1 - 3.2 \cdot 1.6)^2$

In exercises 53 to 56, fill in the blank.

53. If a number is multiplied by 1,000, the decimal point should be moved _____ places to the _____ .

54. If a number is divided by 10,000, the decimal point should be moved _____ places to the _____ .

55. If a number is divided by 100, the decimal point should be moved _____ places to the _____ .

56. If a number is multiplied by 1,000,000, the decimal point should be moved _____ places to the _____ .

In baseball, a pitcher's earned run average (ERA) is computed by dividing the number of earned runs allowed by the number of *nine-inning segments* pitched. To make the computation, we use the formula:

$$\text{ERA} = \frac{9 \cdot (\text{earned runs})}{(\text{innings pitched})}$$

Use this formula to compute the earned run average in exercises 57 and 58. Round the answers to the nearest hundredth.

57. Pedro gave up 57 earned runs in 161 innings pitched.

58. Randy gave up 71 earned runs in 197 innings pitched.

Calculator Exercises

Divide and check.

59. $-99.705 \div (-34.5)$ **60.** $-171.25 \div 2.74$

61. $-0.01372 \div 0.056$ **62.** $0.200754 \div (-0.00855)$

Divide and round to the indicated place.

63. $0.5782 \div 1.236$ thousandths

64. $1.25 \div 0.785$ hundredths

65. $1.34 \div 2.63$ two decimal places

66. $12.364 \div 4.361$ three decimal places

Answers

1. 2.78   **3.** -0.48 **5.** 0.685   **7.** 2.315   **9.** -5.8 **11.** 2.35
13. 0.265   **15.** 18.45   **17.** 1.9   **19.** 2.235   **21.** 6.85   **23.** 34.5
25. -0.235 **27.** 0.58   **29.** -0.04568 **31.** 0.02439   **33.** -0.0069
35. 0.078   **37.** -0.000452 **39.** 2.6   **41.** 0.84   **43.** 2.35
45. 1.9   **47.** 1.52   **49.** 11.8   **51.** 17.0291   **53.** 3, right
55. 2, left   **57.** 3.19   **59.** 2.89   **61.** -0.245 **63.** 0.468
65. 0.51

5.5 Fractions and Decimals

1. Convert a common fraction to a decimal
2. Convert a common fraction to a repeating decimal
3. Convert a mixed number to a decimal
4. Convert a decimal to a common fraction or mixed number
5. Compare the sizes of fractions and decimals

Because a common fraction can be interpreted as division, you can divide the numerator of the common fraction by its denominator to convert a common fraction to a decimal. The result is called a **decimal equivalent.**

OBJECTIVE 1

Example 1 Converting a Fraction to a Decimal Equivalent

Write $\dfrac{5}{8}$ as a decimal.

RECALL 5 can be written as 5.0, 5.00, 5.000, and so on. In this case, we continue the division by adding zeros to the dividend until a 0 remainder is reached.

$$
\begin{array}{r}
0.625 \\
8\overline{)5.000} \\
\underline{4\,8} \\
20 \\
\underline{16} \\
40 \\
\underline{40} \\
0
\end{array}
$$

Because $\dfrac{5}{8}$ means $5 \div 8$, divide 8 into 5.

We see that $\dfrac{5}{8} = 0.625$; 0.625 is the decimal equivalent of $\dfrac{5}{8}$.

CHECK YOURSELF 1

Find the decimal equivalent of $\dfrac{7}{8}$.

Some fractions are used so often that we have listed their decimal equivalents for your reference.

NOTE The division used to find these decimal equivalents stops when a 0 remainder is reached. The equivalents are called **terminating decimals.**

Some Common Decimal Equivalents

$\dfrac{1}{2} = 0.5$	$\dfrac{1}{4} = 0.25$	$\dfrac{1}{5} = 0.2$	$\dfrac{1}{8} = 0.125$
	$\dfrac{3}{4} = 0.75$	$\dfrac{2}{5} = 0.4$	$\dfrac{3}{8} = 0.375$
		$\dfrac{3}{5} = 0.6$	$\dfrac{5}{8} = 0.625$
		$\dfrac{4}{5} = 0.8$	$\dfrac{7}{8} = 0.875$

If a decimal equivalent does not terminate, you can round the result to approximate the fraction to some specified number of decimal places. Consider Example 2.

Example 2   Converting a Fraction to a Decimal Equivalent

Write $\dfrac{3}{7}$ as a decimal. Round the answer to the nearest thousandth.

$$
\begin{array}{r}
0.4285 \\
7\overline{)3.0000} \\
2\,8 \\
\hline
20 \\
14 \\
\hline
60 \\
56 \\
\hline
40 \\
35 \\
\hline
5
\end{array}
$$

In this example, we are choosing to round to three decimal places, so we must add enough zeros to carry the division to four decimal places.

NOTE Multiply 0.429 times 7 and then multiply 0.428 times 7. Which product is closer to 3?

So $\dfrac{3}{7} = 0.429$ (to the nearest thousandth).

CHECK YOURSELF 2

Find the decimal equivalent of $\dfrac{5}{11}$ to the nearest thousandth.

If a fraction's decimal equivalent does *not* terminate, it will *repeat* a sequence of digits. These decimals are called **repeating decimals.**

OBJECTIVE 2

Example 3   Converting a Fraction to a Repeating Decimal

(a) Write $\dfrac{1}{3}$ as a decimal.

$$
\begin{array}{r}
0.333 \\
3\overline{)1.000} \\
9 \\
\hline
10 \\
9 \\
\hline
10 \\
9 \\
\hline
\end{array}
$$

The digit 3 will just repeat itself indefinitely because each new remainder will be 1.

Adding more zeros and going on will simply lead to more threes in the quotient.

We can say $\dfrac{1}{3} = 0.333 \cdots$ The three dots mean "and so on" and tell us that 3 will repeat itself indefinitely.

(b) Write $\dfrac{5}{12}$ as a decimal.

$$
\begin{array}{r}
0.4166\cdots \\
12\overline{)5.0000} \\
\underline{4\,8} \\
20 \\
\underline{12} \\
80 \\
\underline{72} \\
80 \\
\underline{72} \\
8
\end{array}
$$

In this example, the digit 6 will just repeat itself because the remainder, 8, will keep occurring if we add more zeros and continue the division.

CHECK YOURSELF 3

Find the decimal equivalent of each fraction.

(a) $\dfrac{2}{3}$ **(b)** $\dfrac{7}{12}$

Some important decimal equivalents (rounded to the nearest thousandth) are given here for reference.

$$\dfrac{1}{6} = 0.167 \qquad \dfrac{1}{3} = 0.333 \qquad \dfrac{2}{3} = 0.667 \qquad \dfrac{5}{6} = 0.833$$

Another way to write a repeating decimal is with a bar placed over the digit or digits that repeat. For example, we can write

$0.37373737\cdots$

as

$0.\overline{37}$

The bar placed over the digits indicates that "37" repeats indefinitely.

$$\dfrac{1}{6} = 0.1\overline{6}$$

The bar placed over the digit 6 indicates that only the 6 repeats.

Example 4   Converting a Fraction to a Repeating Decimal

Write $\dfrac{5}{11}$ as a decimal.

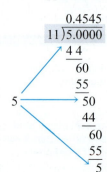

As soon as a remainder repeats itself, as 5 does here, the pattern of digits will repeat in the quotient.

$$\dfrac{5}{11} = 0.\overline{45}$$
$$= 0.4545\cdots$$

CHECK YOURSELF 4 _____

Use the bar notation to write the decimal equivalent of $\frac{5}{7}$. (Be patient. You'll have to divide for a while to find the repeating pattern.)

You can find the decimal equivalents for mixed numbers in a similar way. Find the decimal equivalent of the fractional part of the mixed number and then combine that with the whole-number part. Example 5 illustrates this approach.

OBJECTIVE 3

Example 5   Converting a Mixed Number to a Decimal Equivalent

RECALL

$3\frac{5}{16} = 3 + \frac{5}{16}$

Find the decimal equivalent of $3\frac{5}{16}$.

$\frac{5}{16} = 0.3125$ First find the equivalent of $\frac{5}{16}$ by division.

$3\frac{5}{16} = 3.3125$ Add 3 to the result.

CHECK YOURSELF 5 _____

Find the decimal equivalent of $2\frac{5}{8}$.

To this point in the section, we have seen that to find the decimal equivalent of a fraction, we use long division. Because the remainder must be less than the divisor, the remainder must *either repeat or become 0.* Thus *every common fraction* will have a *repeating* or a *terminating* decimal as its decimal equivalent.

Using what we have learned about place values, you can easily write decimals as common fractions. Use this rule.

Step by Step: To Convert a Terminating Decimal Less Than 1 to a Common Fraction

Step 1   Write the digits of the decimal without the decimal point. This will be the numerator of the common fraction.

Step 2   The denominator of the fraction is a 1 followed by as many zeros as there are places in the decimal.

Example 6   Converting a Decimal to a Common Fraction

$$0.7 = \frac{7}{10} \qquad 0.09 = \frac{9}{100} \qquad 0.257 = \frac{257}{1,000}$$

One place   One zero   Two places   Two zeros   Three places   Three zeros

CHECK YOURSELF 6

Write as common fractions.

(a) 0.3 **(b)** 0.311

When a decimal is converted to a common fraction, the common fraction that results should be written in lowest terms.

OBJECTIVE 4

Example 7   Converting a Decimal to a Common Fraction

Convert 0.395 to a fraction and write the result in lowest terms.

NOTE To simplify, divide the numerator and denominator by 5.

$$0.395 = \frac{395}{1,000} = \frac{79}{200}$$

CHECK YOURSELF 7

Write 0.275 as a common fraction.

If the decimal has a whole-number portion, write the digits to the right of the decimal point as a proper fraction and then form a mixed number for your result.

Example 8   Converting a Decimal to a Mixed Number

Write 12.277 as a mixed number.

NOTE Repeating decimals can also be written as common fractions, although the process is more complicated. We limit ourselves to the conversion of terminating decimals in this textbook.

$$0.277 = \frac{277}{1,000} \qquad \text{so} \qquad 12.277 = 12\frac{277}{1,000}$$

CHECK YOURSELF 8

Write 32.433 as a mixed number.

Comparing the sizes of common fractions and decimals requires finding the decimal equivalent of the common fraction and then comparing the resulting decimals.

OBJECTIVE 5

Example 9   Comparing the Sizes of Common Fractions and Decimals

Which is larger, $\frac{3}{8}$ or 0.38?

Write the decimal equivalent of $\frac{3}{8}$. That decimal is 0.375. Now comparing 0.375 and 0.38, we see that 0.38 is the larger of the numbers:

$$0.38 > \frac{3}{8}$$

CHECK YOURSELF 9 _____

Which is larger, $\dfrac{3}{4}$ or 0.8?

A calculator is very useful in converting common fractions to decimals. Just divide the numerator by the denominator, and the decimal equivalent will be in the display.

Example 10   Converting Fractions to Decimals

Find the decimal equivalent of $\dfrac{7}{16}$.

7 ÷ 16 =

Display 0.4375   0.4375 is the decimal equivalent of $\dfrac{7}{16}$.

CHECK YOURSELF 10 _____

Find the decimal equivalent of $\dfrac{5}{16}$.

Often, you will want to round the result in the display.

Example 11   Converting Fractions to Decimals

NOTE Some calculators show the 0 to the left of the decimal point and seven digits to the right. Others omit the 0 and show eight digits to the right. Check yours with this example.

Find the decimal equivalent of $\dfrac{5}{24}$ to the nearest hundredth.

5 ÷ 24 =

Display 0.2083333

$$\dfrac{5}{24} = 0.21 \quad \text{(nearest hundredth)}$$

CHECK YOURSELF 11 _____

Find the decimal equivalent of $\dfrac{7}{29}$ to the nearest hundredth.

To find the decimal equivalent of a mixed number, use the sequence given in Example 12.

NOTE There are several ways to do this, depending on the calculator you are using. For example,

7 $\boxed{+}$ 5 $\boxed{\div}$ 8 $\boxed{=}$

will work on most scientific calculators. Try it on yours.

Example 12   Converting Mixed Numbers to Decimals

Change $7\dfrac{5}{8}$ to a decimal.

5 $\boxed{\div}$ 8 $\boxed{+}$ 7 $\boxed{=}$

Display   7.625

CHECK YOURSELF 12 _____

Find the decimal equivalent of $3\dfrac{3}{8}$.

Depending on the calculator you are using, the result may be rounded at its last displayed digit.

Example 13   Converting Fractions to Decimals

For $\dfrac{5}{9}$:

5 $\boxed{\div}$ 9 $\boxed{=}$

Display   0.555555555 or 0.555555556   ◄—Rounded display; actually $\dfrac{5}{9} = 0.\overline{5}$

CHECK YOURSELF 13 _____

Find the decimal equivalent of $\dfrac{7}{11}$.

READING YOUR TEXT _____

The following fill-in-the-blank exercises are designed to assure that you understand the key vocabulary used in this section. Each sentence comes directly from the section. You will find the correct answers in Appendix C.

Section 5.5

(a)   When we divide the numerator of a fraction by its denominator, we get the decimal _____.

(b)   If a fraction's decimal equivalent does not terminate, the result is called a _____ decimal.

(c)   To convert a decimal to a fraction, we can write the denominator of the fraction as a one followed by as many zeros as there are _____ in the decimal.

(d)   When a decimal is converted to a common fraction, the common fraction that results should be written in _____ terms.

CHECK YOURSELF ANSWERS

1. 0.875   **2.** $\dfrac{5}{11} = 0.455$ (to the nearest thousandth)

3. **(a)** $0.666 \cdots$; **(b)** $\dfrac{7}{12} = 0.583 \cdots$ The digit 3 will continue indefinitely.

4. $\dfrac{5}{7} = 0.\overline{714285}$ **5.** 2.625   **6.** **(a)** $\dfrac{3}{10}$; **(b)** $\dfrac{311}{1,000}$ **7.** $0.275 = \dfrac{11}{40}$

8. $32\dfrac{433}{1,000}$ **9.** $0.8 > \dfrac{3}{4}$ **10.** 0.3125   **11.** 0.24   **12.** 3.375

13. 0.63636363 or 0.63636364

Name _____

Section _____ Date _____

ANSWERS

1. _____	2. _____
3. _____	4. _____
5. _____	6. _____
7. _____	
8. _____	
9. _____	
10. _____	
11. _____	
12. _____	
13. _____	
14. _____	
15. _____	
16. _____	
17. _____	
18. _____	
19. _____	
20. _____	
21. _____	
22. _____	
23. _____	
24. _____	
25. _____	

5.5 Exercises

Find the decimal equivalent for each fraction.

1. $\dfrac{3}{4}$ **ALEKS**

2. $\dfrac{4}{5}$ **ALEKS**

3. $\dfrac{9}{20}$ **ALEKS**

4. $\dfrac{3}{10}$ **ALEKS**

5. $\dfrac{1}{5}$ **ALEKS**

6. $\dfrac{1}{8}$ **ALEKS**

7. $\dfrac{5}{16}$ **ALEKS**

8. $\dfrac{11}{20}$ **ALEKS**

9. $\dfrac{7}{10}$ **ALEKS**

10. $\dfrac{7}{16}$ **ALEKS**

11. $\dfrac{27}{40}$ **VIDEO ALEKS**

12. $\dfrac{17}{32}$ **ALEKS**

Find the decimal equivalent rounded to the indicated place.

13. $\dfrac{5}{6}$ thousandths

14. $\dfrac{7}{12}$ hundredths

15. $\dfrac{4}{15}$ thousandths

Write the decimal equivalent, using the bar notation.

16. $\dfrac{1}{18}$ **ALEKS**

17. $\dfrac{4}{9}$ **VIDEO ALEKS**

18. $\dfrac{3}{11}$ **ALEKS**

Find the decimal equivalent for each of the mixed numbers.

19. $5\dfrac{3}{5}$

20. $7\dfrac{3}{4}$

21. $4\dfrac{7}{16}$

Find the decimal equivalent for each fraction. Use the bar notation.

22. $\dfrac{1}{11}$ **ALEKS**

23. $\dfrac{1}{111}$

24. $\dfrac{1}{1,111}$

25. From the pattern of exercises 22 to 24, can you guess the decimal representation for $\dfrac{1}{11,111}$?

Insert > or < to form a true statement.

26. $\dfrac{18}{21}$ ___ 0.863

27. $\dfrac{31}{34}$ ___ 0.9118

28. $\dfrac{21}{37}$ ___ 0.5664

29. $\dfrac{13}{17}$ ___ 0.7657

Write each decimal as a common fraction or mixed number. Write your answer in lowest terms.

30. 0.3

31. 0.8

32. 0.6

33. 0.37

34. 0.97

35. 0.587

36. 0.379

37. 0.48

38. 0.75

39. 0.58

40. 0.65

41. 0.425

42. 0.116

43. 0.375

44. 0.225

45. 0.136

46. 0.575

47. 0.059

48. 0.067

49. 0.0625

50. 0.0425

51. 6.3

52. 5.7

53. 2.17

54. 3.31

55. 5.28

56. 15.35

Complete each statement, using the symbol < or >.

57. $\dfrac{7}{8}$ ___ 0.87

58. $\dfrac{5}{16}$ ___ 0.313

59. $\dfrac{9}{25}$ ___ 0.4

60. $\dfrac{11}{17}$ ___ 0.638

ANSWERS

26. _____
27. _____
28. _____
29. _____
30. _____
31. _____
32. _____
33. _____
34. _____
35. _____
36. _____
37. _____ 38. _____
39. _____ 40. _____
41. _____ 42. _____
43. _____ 44. _____
45. _____ 46. _____
47. _____ 48. _____
49. _____ 50. _____
51. _____ 52. _____
53. _____ 54. _____
55. _____ 56. _____
57. _____ 58. _____
59. _____ 60. _____

61. _____

62. _____

63. _____

64. _____

65. _____

66. _____

67. _____

68. _____

69. _____

70. _____

71. _____

72. _____

73. _____

74. _____

75. _____

76. _____

In exercises 61 and 62, indicate whether the given statement is always true, sometimes true, or never true.

61. A decimal can be written as a common fraction or mixed number.

62. A fraction can be written as a terminating decimal.

63. Every fraction has a decimal equivalent that either terminates (for example, $\frac{1}{4} = 0.25$), or repeats (for example, $\frac{2}{9} = 0.\overline{2}$). Work with a group to discover which fractions have terminating decimals and which have repeating decimals. You may assume that the numerator of each fraction you consider is 1 and focus your attention on the denominator. (*Hint:* Study the prime factorization of each denominator.)

64. Find the decimal equivalent, using bar notation, for $\frac{1}{7}, \frac{2}{7}, \frac{3}{7}$, and $\frac{4}{7}$. Describe any patterns that you see. Based on these results, predict the decimal equivalent for $\frac{5}{7}$ and $\frac{6}{7}$.

65. Each lap around a running track is 0.25 miles. Write the distance as a fraction.

66. Each lap around Daytona International Speedway is 2.5 miles. Write the distance as a fraction.

Calculator Exercises

Find the decimal equivalent rounded as indicated.

67. $\frac{9}{16}$

68. $\frac{7}{24}$ hundredths

69. $\frac{5}{32}$ thousandths

70. $\frac{11}{75}$ thousandths

71. $\frac{3}{11}$ use bar notation

72. $\frac{7}{11}$ use bar notation

73. $\frac{16}{33}$ use bar notation

74. $3\frac{4}{5}$

75. $3\frac{7}{8}$

76. $8\frac{3}{16}$

Convert each decimal to a fraction.

77. 0.3

78. 0.55

79. 0.305

80. 0.1

81. 0.875

82. 0.125

Answers

1. 0.75 **3.** 0.45 **5.** 0.2 **7.** 0.3125 **9.** 0.7 **11.** 0.675
13. 0.833 **15.** 0.267 **17.** $0.\overline{4}$ **19.** 5.6 **21.** 4.4375 **23.** $0.\overline{009}$

25. $0.\overline{00009}$ **27.** $<$ **29.** $<$ **31.** $\dfrac{4}{5}$ **33.** $\dfrac{37}{100}$ **35.** $\dfrac{587}{1,000}$

37. $\dfrac{12}{25}$ **39.** $\dfrac{29}{50}$ **41.** $\dfrac{17}{40}$ **43.** $\dfrac{3}{8}$ **45.** $\dfrac{17}{125}$ **47.** $\dfrac{59}{1,000}$

49. $\dfrac{1}{16}$ **51.** $6\dfrac{3}{10}$ **53.** $2\dfrac{17}{100}$ **55.** $5\dfrac{7}{25}$ **57.** $>$ **59.** $<$

61. Always **63.** **65.** $\dfrac{1}{4}$ mile **67.** 0.5625 **69.** 0.156

71. $0.\overline{27}$ **73.** $0.\overline{48}$ **75.** 3.875 **77.** $\dfrac{3}{10}$ **79.** $\dfrac{61}{200}$ **81.** $\dfrac{7}{8}$

ANSWERS

77. _____

78. _____

79. _____

80. _____

81. _____

82. _____

5.6 Equations Containing Decimals

5.6 OBJECTIVE

1. Solve equations containing decimals

Equations involving decimals can be solved by the methods we studied in Chapter 3. For instance, to solve $2.3x = 6.9$, we simply use the multiplication property of equality to divide both sides of the equation by 2.3. This will isolate the variable on the left as desired. This is illustrated in Example 1.

OBJECTIVE 1

Example 1 Solving an Equation Involving Decimals

Solve the equation.

$$2.3x = 6.9$$

Dividing both sides by 2.3, we get

$$\frac{2.3x}{2.3} = \frac{6.9}{2.3} = \frac{69}{23}$$

RECALL Always check your solution.

or $x = 3$.

CHECK YOURSELF 1

Solve the equation.

$4.1x = 20.5$

The addition property of equality is also used to solve equations containing decimals.

Example 2 Solving an Equation Involving Decimals

Solve the equation.

$$1.7x - 1.68 = 2.5x + 5.8$$

First, we can use the addition property to collect variable terms on the left and constant terms on the right.

$$
\begin{array}{rcl}
1.7x - 1.68 = & 2.5x + 5.8 \\
+ 1.68 & + 1.68 \\
\hline
1.7x = & 2.5x + 7.48 \\
-2.5x & -2.5x \\
\hline
-0.8x = & 7.48
\end{array}
$$

We can now divide each side of the equation by -0.8.

$$\frac{-0.8x}{-0.8} = \frac{7.48}{-0.8}$$ We used the multiplication property of equality to eliminate the coefficient.

$$x = -9.35$$

Note that we could have collected variable terms on the right and constant terms on the left. Try this. You should get the same solution.

The key is this: The addition property of equality is used to collect variable terms on one side and constant terms on the other.

 CHECK YOURSELF 2 _____

Solve the equation.

$5.3x - 3.46 = 7.1x + 5.09$

Care must be taken when an equation contains parentheses.

Example 3   Solving an Equation Containing Parentheses

Solve the equation.

$3.1x - (x - 4.3) = 1.3x - 3.2$

First, we rewrite the equation so that subtraction is changed to the addition of the opposite. To accomplish this, we distribute the negative over the expression $x - 4.3$.

$3.1x + (-x + 4.3) = 1.3x + (-3.2)$

or

$3.1x + (-x) + 4.3 = 1.3x + (-3.2)$

Collecting variable terms on the left and constant terms on the right

$$
\begin{array}{rcl}
2.1x + 4.3 &=& 1.3x + (-3.2) \\
-\ 4.3 && -4.3 \\
\hline
2.1x &=& 1.3x + (-7.5) \\
-1.3x && -1.3x \\
\hline
0.8x &=& -7.5
\end{array}
$$ Subtract 4.3 from each side.

Dividing both sides by 0.8, we find

$x = -9.375$

 CHECK YOURSELF 3 _____

Solve the equation.

$2.3x + 13.2 = 1.52 - (x - 4.75)$

READING YOUR TEXT

The following fill-in-the-blank exercises are designed to assure that you understand the key vocabulary used in this section. Each sentence comes directly from the section. You will find the correct answers in Appendix C.

Section 5.6

(a) The _____ property of equality is used to collect variable terms on one side and constant terms on the other.

(b) We use the _____ property of equality to eliminate the coefficient in the variable term.

(c) Care must be taken when an equation contains _____.

(d) We may rewrite an equation so that subtraction is changed to the addition of the _____.

CHECK YOURSELF ANSWERS

1. $x = 5$ **2.** $x = -4.75$ **3.** $x = -2.1$

5.6 Exercises

Name _____

Section _____ Date _____

Solve each equation for x.

1. $3.2x = 12.8$

2. $5.1x = -15.3$

3. $-4.5x = 13.5$

4. $-8.2x = -32.8$

5. $1.3x + 2.8x = 12.3$

6. $2.7x + 5.4x = -16.2$

7. $9.3x + (-6.2x) = 12.4$

8. $12.5x + (-7.2x) = -21.2$

9. $5.3x + (-7) = 2.3x + 5$

10. $9.8x + 2 = 3.8x + 20$

11. $3x - 0.54 = 2(x - 0.15)$

12. $7x + 0.125 = 6x - 0.289$

13. $6x + 3(x - 0.2789) = 4(2x + 0.3912)$

14. $9x - 2(3x - 0.124) = 2x + 0.965$

15. $5x - (0.345 - x) = 5x + 0.8713$

16. $-3(0.234 - x) = 2(x + 0.974)$

17. $2.3x - 4.25 = 3.3x + 2.15$

18. $5.7x + 3.84 = 6.7x + 5.26$

19. $7.1x - 14 = 4.3x - 8.54$

20. $5.6x + 7.5 = 2.9x + 0.885$

21. $3.2x + 8.36 = 5x + 13.94$

22. $4.8x - 2.35 = 7x - 8.51$

23. $7.4(x - 1.2) = 8.4(x - 0.5)$

24. $4.8(x + 3.5) = 5.8(x - 2.5)$

25.

26.

27.

28.

29.

30.

31.

32.

33.

34.

25. $3.5(x - 1.4) = 1.3x - 3.14$

26. $2.6x - 12.6 = 5.9(x - 1.8)$

27. $5 - 1.6(x + 2) = 2(x - 1.3) - 0.1$

28. $3 - 2.4(x - 5) = 4(x + 1.5) - 7.64$

In exercises 29 and 30, write and solve an equation that models the given situation.

29. Lance Armstrong won the fifteenth stage of the 2003 Tour de France, cycling 99 miles in 4.5 hours (in the mountains!). Find his average speed. (*Hint:* Distance, rate, and time are related by the formula $d = rt$.)

30. Julia earned $1,200.50 over a three-week period. Her hourly wage is $12.25. How many hours did she work?

In exercises 31 to 34, match the given equation with an equivalent equation at the right.

31. $8.3x = -4.15$

(a) $x = -4.15 - 8.3$

32. $\dfrac{x}{8.3} = -4.15$

(b) $x = 8.3 + 4.15$

33. $8.3 + x = -4.15$

(c) $x = \dfrac{-4.15}{8.3}$

34. $x - 8.3 = 4.15$

(d) $x = (8.3)(-4.15)$

Answers

1. 4   **3.** −3   **5.** 3   **7.** 4   **9.** 4   **11.** 0.24   **13.** 2.4015
15. 1.2163   **17.** −6.4   **19.** 1.95   **21.** −3.1   **23.** −4.68
25. 0.8   **27.** 1.25   **29.** $22\,\dfrac{\text{mi}}{\text{h}}$ **31.** c   **33.** a

5.7 Square Roots and the Pythagorean Theorem

5.7 OBJECTIVES

1. Find the square root of a perfect square
2. Apply the Pythagorean theorem
3. Approximate the square root of a number
4. Use a calculator to find a square root

Some numbers can be written as the product of two identical factors, for example,

$$9 = 3 \cdot 3$$

Either factor is called a **square root** of the number. The symbol $\sqrt{}$ (called a **radical sign**) is used to indicate a square root. Thus $\sqrt{9} = 3$ because $3 \cdot 3 = 9$. The term *square root* is used because three *squared* is nine.

$$3^2 = 9 \qquad \text{and} \qquad \sqrt{9} = 3$$

OBJECTIVE 1

Example 1 Finding the Square Root

Find the square root of (a) 49 and of (b) 16.

(a) $\sqrt{49} = 7$ Because $7 \cdot 7 = 7^2 = 49$

(b) $\sqrt{16} = 4$ Because $4 \cdot 4 = 4^2 = 16$

CHECK YOURSELF 1 _____

Find each square root.

(a) $\sqrt{121}$ **(b)** $\sqrt{36}$

When the square root of a whole number is itself a whole number, the original number is called a **perfect square.** In Example 1, we see that 49 and 16 are perfect squares.

The most frequently used theorem in geometry is undoubtedly the Pythagorean theorem. In this section, you will use that theorem. You will also learn a little about the history of the theorem. It is a theorem that applies only to right triangles.

The side opposite the right angle of a right triangle is called the **hypotenuse.** The hypotenuse will always be the largest side of a right triangle. The two shorter sides are called the **legs** of the right triangle.

NOTE A **right** triangle is a triangle that contains a 90° angle.

Example 2 Identifying the Hypotenuse

In the right triangle shown, the side labeled c is the hypotenuse.

NOTE The sides labeled a and b are the legs.

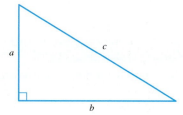

451

CHECK YOURSELF 2 _____

Which side represents the hypotenuse of the given right triangle?

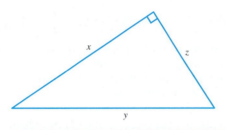

NOTE Such a triple is also called a **Pythagorean triple.**

The numbers 3, 4, and 5 have a special relationship. Together they are called a **perfect triple,** which means that when you square all three numbers, the sum of the squares of the smaller numbers equals the squared value of the largest number.

Example 3   Identifying Perfect Triples

Show that each group of numbers is a perfect triple.

(a) 3, 4, and 5

$3^2 = 9$, $4^2 = 16$, $5^2 = 25$

and $9 + 16 = 25$, so we can say that $3^2 + 4^2 = 5^2$.

(b) 7, 24, and 25

$7^2 = 49$, $24^2 = 576$, $25^2 = 625$

and $49 + 576 = 625$, so we can say that $7^2 + 24^2 = 25^2$.

CHECK YOURSELF 3 _____

Show that each group of numbers is a perfect triple.

(a) 5, 12, and 13   **(b)** 6, 8, and 10

All the triples that you have seen, and many more, were known by the Babylonians more than 4,000 years ago. Stone tablets that had dozens of perfect triples carved into them have been found. The basis of the Pythagorean theorem was understood long before the time of Pythagoras (ca. 540 B.C.). The Babylonians not only understood perfect triples but also knew how triples related to a right triangle.

Property: The Pythagorean Theorem (Version 1)

If the lengths of the three sides of a right triangle are all integers, they will form a perfect triple.

There are two other forms in which the Pythagorean theorem is regularly presented. It is important that you see the connection between the three forms.

> **Property:** The Pythagorean Theorem (Version 2)
>
> The square of the hypotenuse of a right triangle is equal to the sum of the squares of the other two sides.

NOTE This is the version that you will refer to in your algebra classes.

**CAUTION**

This equation *only* applies to right triangles.

> **Property:** The Pythagorean Theorem (Version 3)
>
> Given a right triangle with sides *a* and *b* and hypotenuse *c*, it is always true that
>
> $c^2 = a^2 + b^2$

OBJECTIVE 2   **Example 4   Finding the Length of a Side of a Right Triangle**

Find the missing integer length for each right triangle.

(a)

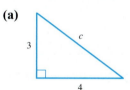

$$c^2 = 3^2 + 4^2$$
$$= 9 + 16$$
$$= 25$$

We need the square root of 25.

$$c = 5$$

(b)

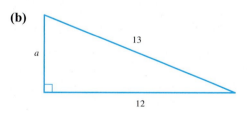

$$13^2 = a^2 + 12^2$$

$$169 = a^2 + 144$$
$$-144 \quad\quad -144$$
$$\overline{25 = a^2}$$
$$5 = a$$

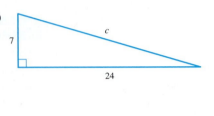

CHECK YOURSELF 4 _____

Find the missing integer length for each right triangle.

(a)

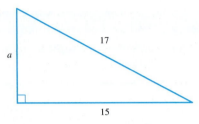

17

a

15

(b)

7

c

24

| Example 5 | Using the Pythagorean Theorem |

If the lengths of the two legs of a right triangle are 6 and 8, find the length of the hypotenuse.

$$c^2 = a^2 + b^2$$ The value of the hypotenuse is found from the Pythagorean theorem with $a = 6$ and $b = 8$.

$$c^2 = (6)^2 + (8)^2 = 36 + 64 = 100$$

NOTE The triangle has sides 6, 8, and 10.

6

10

8

$$c = \sqrt{100} = 10$$ The length of the hypotenuse is 10 (because $10^2 = 100$).

CHECK YOURSELF 5 _____

Find the hypotenuse of a right triangle whose legs measure 9 and 12.

In some right triangles, the lengths of the hypotenuse and one leg are given and we are asked to find the length of the missing leg.

| Example 6 | Using the Pythagorean Theorem |

Find the missing length.

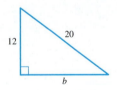

12

20

b

$$a^2 + b^2 = c^2$$ Use the Pythagorean theorem with $a = 12$ and $c = 20$.

$$(12)^2 + b^2 = (20)^2$$

$$144 + b^2 = 400$$

$$b^2 = 400 - 144 = 256$$

$$b = \sqrt{256} = 16$$ The missing side is 16.

CHECK YOURSELF 6

Find the missing length for a right triangle with one leg measuring 8 cm and the hypotenuse measuring 10 cm.

Not every square root is a whole number. In fact, there are only 10 whole-number square roots for the numbers from 1 to 100. They are the square roots of 1, 4, 9, 16, 25, 36, 49, 64, 81, and 100. However, we can approximate square roots that are not whole numbers. For example, we know that the square root of 12 is not a whole number. We also know that its value must lie somewhere between the square root of 9 ($\sqrt{9} = 3$) and the square root of 16 ($\sqrt{16} = 4$). That is, $\sqrt{12}$ is between 3 and 4.

OBJECTIVE 3   **Example 7   Approximating Square Roots**

Approximate $\sqrt{29}$.
$\sqrt{25} = 5$ and $\sqrt{36} = 6$, so $\sqrt{29}$ must be between 5 and 6.

CHECK YOURSELF 7

$\sqrt{19}$ *is between which two numbers?*

(a) 4 and 5   **(b)** 5 and 6   **(c)** 6 and 7

To find a square root on your scientific calculator, you use the square root key. On some calculators, you simply enter the number and then press the square root key. With others, you must use the second function on the $\boxed{x^2}$ (or $\boxed{y^x}$) key and specify the root you wish to find.

OBJECTIVE 4   **Example 8   Finding a Square Root Using a Calculator**

Find the square root of 256.

RECALL To use the radical key ($\boxed{\sqrt{}}$) with a scientific calculator, first enter 256 and then press the radical key. With a graphing calculator, press the radical key first, and then enter 256 and a closing parenthesis, $\boxed{)}$, followed by $\boxed{\text{ENTER}}$.

256 $\boxed{\sqrt{}}$

Display   16

or

256 $\boxed{\text{2nd}}$ $\boxed{\sqrt[x]{y} \atop y^x}$ 2 $\boxed{=}$

Display   16   The "2" is entered for the 2nd (square) root.

CHECK YOURSELF 8

Find the square root of 361.

As we saw in Example 7, not every square root is a whole number. Your calculator can help give you the *approximate* square root of any number.

> ### Example 9 Finding an Approximate Square Root Using a Calculator
>
> Approximate the square root of 29. Round your answer to the nearest tenth.
>
> Enter
>
> 29

Your calculator display will read something like this:

Display 5.385164807

C A U T I O N

If we square 5.4, we will not get exactly 29. Try this.

This is an *approximation* of the square root. It is rounded to the nearest billionth. The calculator cannot display the exact answer because there is no end to the sequence of digits (and also no pattern.) If the square root of a whole number is not another whole number, then the answer has an infinite number of digits.

To find the approximate square root, we round to the nearest tenth. Our approximation for the square root of 29 is 5.4.

CHECK YOURSELF 9 _____

Approximate the square root of 19. Round your answer to the nearest tenth.

> ### Example 10 Evaluating Expressions Using a Calculator
>
> Use a scientific calculator to approximate the value of $\sqrt{177}$.
>
> $\sqrt{177}$ Using the calculator, you find $\sqrt{177} = 13.3041\cdots$. To the nearest hundredth, $\sqrt{177} = 13.30$.

CHECK YOURSELF 10 _____

Use a scientific calculator to approximate the value of each expression. Round your answer to the nearest hundredth.

(a) $\sqrt{357}$ **(b)** $7(\sqrt{71})$

READING YOUR TEXT _____

The following fill-in-the-blank exercises are designed to assure that you understand the key vocabulary used in this section. Each sentence comes directly from the section. You will find the correct answers in Appendix C.

Section 5.7

(a) The symbol $\sqrt{}$ is called a _____ sign.

(b) Three _____ is equal to nine.

(c) The side opposite the right angle of a right triangle is called the _____.

(d) If the lengths of three sides of a right triangle are all integers, they will form a _____ triple.

CHECK YOURSELF ANSWERS

1. **(a)** 11; **(b)** 6   **2.** Side y **3.** **(a)** $5^2 + 12^2 = 25 + 144 = 169$, $13^2 = 169$, so $5^2 + 12^2 = 13^2$; **(b)** $6^2 + 8^2 = 36 + 64 = 100$, $10^2 = 100$, so $6^2 + 8^2 = 10^2$

4. **(a)** 8; **(b)** 25   **5.** 15   **6.** 6 cm   **7.** **(a)** 4 and 5   **8.** 19   **9.** 4.4

10. **(a)** 18.89; **(b)** 58.98

5.7 **Exercises**

In exercises 1 to 4, find the square root.

1. $\sqrt{64}$ **ALEKS**

2. $\sqrt{121}$ **ALEKS**

3. $\sqrt{169}$ **VIDEO**   **ALEKS**

4. $\sqrt{196}$ **ALEKS**

Identify the hypotenuse of each triangle by giving its letter.

5.

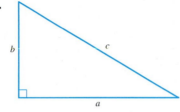

6.

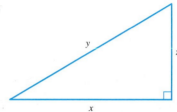

For exercises 7 to 12, identify which numbers are perfect triples.

7. 3, 4, 5

8. 4, 5, 6

9. 7, 12, 13   **VIDEO**

10. 5, 12, 13

11. 8, 15, 17

12. 9, 12, 15

For exercises 13 to 16, find the missing length for each right triangle.

13.   **VIDEO**   **ALEKS**

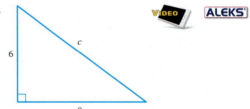

14.   **ALEKS**

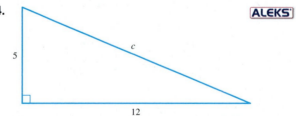

15.   **VIDEO**   **ALEKS**

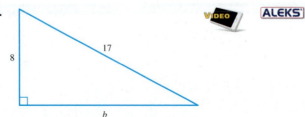

16.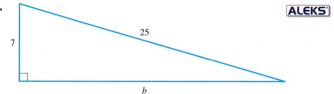

ALEKS

Select the correct approximation.

17. Is $\sqrt{23}$ between **(a)** 3 and 4, **(b)** 4 and 5, or **(c)** 5 and 6?   ALEKS

18. Is $\sqrt{15}$ between **(a)** 1 and 2, **(b)** 2 and 3, or **(c)** 3 and 4?   ALEKS

19. Is $\sqrt{44}$ between **(a)** 6 and 7, **(b)** 7 and 8, or **(c)** 8 and 9?   VIDEO   ALEKS

20. Is $\sqrt{31}$ between **(a)** 3 and 4, **(b)** 4 and 5, or **(c)** 5 and 6?   ALEKS

In exercises 21 to 24, find the perimeter of each triangle shown. (*Hint:* First find the missing length.)

21.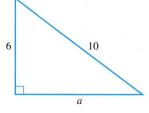

22.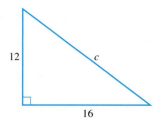

23.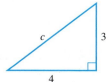

24.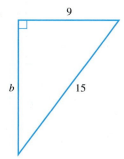

25. Find the altitude, h, of the triangle shown.

VIDEO

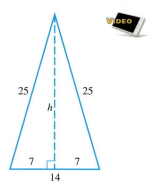

ANSWERS

16. _____

17. _____

18. _____

19. _____

20. _____

21. _____

22. _____

23. _____

24. _____

25. _____

26. Find the altitude of the triangle shown.

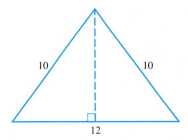

10   10

12

In exercises 27 and 28, find the length of the diagonal of each rectangle.

27.

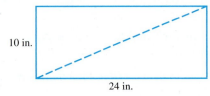

10 in.

24 in.

28.

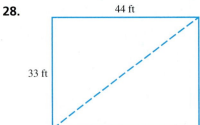

44 ft

33 ft

29. A castle wall, 24 ft high, is surrounded by a moat 7 ft across. Will a 26-ft ladder, placed at the edge of the moat, be long enough to reach the top of the wall?

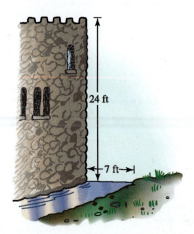

24 ft

7 ft

30. A baseball diamond is the shape of a square that has sides of length 90 ft. Find the distance from home plate to second base. Round your answer to the nearest hundredth.

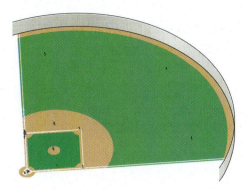

Calculator Exercises

Use your calculator to find the square root of each number.

31. 64

32. 144

33. 289

34. 1,024

35. 1,849

36. 784

37. 8,649

38. 5,329

39. 3,844

40. 3,364

Use your calculator to approximate each square root. Round to the nearest tenth.

41. $\sqrt{23}$

42. $\sqrt{31}$

43. $\sqrt{51}$

44. $\sqrt{42}$

45. $\sqrt{134}$

46. $\sqrt{251}$

ANSWERS

30. _____

31. _____

32. _____

33. _____

34. _____

35. _____

36. _____

37. _____

38. _____

39. _____

40. _____

41. _____

42. _____

43. _____

44. _____

45. _____

46. _____

In exercises 47 and 48, fill in the blank with always, sometimes, or never.

47. In a right triangle, the hypotenuse is ——————— the longest side.

48. If the lengths of the two legs in a right triangle are whole numbers, then the length of the hypotenuse is ——————— a whole number.

49. Investigate the following: Suppose you have a collection of three numbers that form a perfect triple. If you double every number in the collection, will the resulting numbers form a perfect triple? What happens if you triple the original numbers in the collection? Provide examples.

Answers

1. 8   **3.** 13   **5.** c **7.** Yes   **9.** No   **11.** Yes   **13.** 10
15. 15   **17.** b   **19.** a   **21.** 24   **23.** 12   **25.** 24   **27.** 26 in.
29. Yes   **31.** 8   **33.** 17   **35.** 43   **37.** 93   **39.** 62   **41.** 4.8
43. 7.1   **45.** 11.6   **47.** Always   **49.**

5.8 Applications

5.8 OBJECTIVES

1. Use addition of decimals to solve application problems
2. Use subtraction of decimals to solve application problems
3. Use multiplication of decimals to solve application problems
4. Use multiplication of decimals by a power of 10 to solve application problems
5. Use division of decimals to solve application problems
6. Use division of decimals by a power of 10 to solve application problems
7. Use the Pythagorean theorem to solve application problems

Many applied problems require working with decimals. For instance, filling up at a gas station means reading decimal amounts.

OBJECTIVE 1

| Example 1 | An Application of the Addition of Decimals |

On a trip, the Chang family kept track of their gas purchases. If they bought 12.3, 14.2, 10.7, and 13.8 gal, how much gas did they use on the trip?

NOTE Because we want a total amount, we use addition for the solution.

$$
\begin{array}{r}
12.3 \\
14.2 \\
10.7 \\
+\ 13.8 \\
\hline
51.0 \text{ gal}
\end{array}
$$

CHECK YOURSELF 1 _____

The Higueras kept track of the gasoline they purchased on a recent trip. If they bought 12.4, 13.6, 9.7, 11.8, and 8.3 gal, how much gas did they buy on the trip?

Every day you deal with amounts of money. Because our system of money is a decimal system, most problems involving money also involve operations with decimals.

| Example 2 | An Application of the Addition of Decimals |

Andre makes deposits of $3.24, $15.73, $50, $28.79, and $124.38 during May. What is the total of his deposits for the month?

$$
\begin{array}{r}
\$\ \ \ 3.24 \\
15.73 \\
50.00 \\
28.79 \\
+\ \ 124.38 \\
\hline
\$222.14
\end{array}
$$

Simply add the amounts of money deposited as decimals. Note that we write $50 as $50.00.

$222.14 ⟵ The total of deposits for May

CHECK YOURSELF 2 _____

Your textbooks for the fall term cost $63.50, $78.95, $43.15, $82, and $85.85. What was the total cost of textbooks for the term?

In Chapter 1, we defined *perimeter* as the distance around the outside of a straight-edged shape. Finding the perimeter often requires that we add decimal numbers.

Example 3 An Application Involving the Addition of Decimals

Rachel is going to put a fence around the perimeter of her farm. The figure shows a picture of the land, measured in kilometers. How much fence does she need to buy?

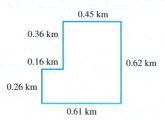

The perimeter is the sum of the lengths of the sides, so we add those lengths to find the total fencing needed.

$$0.16 + 0.36 + 0.45 + 0.62 + 0.61 + 0.26 = 2.46$$

Rachel needs 2.46 km of fence for the perimeter of her farm.

 CHECK YOURSELF 3 _____

Manuel intends to build a walkway around the perimeter of his garden (shown in the figure). What will the total length of the walkway be?

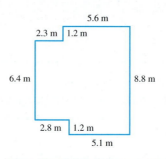

We can apply the subtraction methods of Section 5.2 in solving applications involving decimals.

OBJECTIVE 2 | ### Example 4 An Application of the Subtraction of Decimal Numbers

Jonathan was 98.3 cm tall on his sixth birthday. On his seventh birthday he was 104.2 cm. How much did he grow during the year?

NOTE We want to find the difference between the two measurements, so we subtract.

$$\begin{array}{r} 104.2 \text{ cm} \\ -\ \ 98.3 \text{ cm} \\ \hline 5.9 \text{ cm} \end{array}$$

Jonathan grew 5.9 cm during the year.

 CHECK YOURSELF 4 _____

A car's highway mileage before a tune-up was 28.8 miles per gallon $\left(\dfrac{mi}{gal}\right)$. After the tune-up, it measured 30.1 $\dfrac{mi}{gal}$. What was the increase in mileage?

The same method can be used in working with money.

> ### Example 5   An Application of the Subtraction of Decimal Numbers
>
> At the grocery store, Sally buys a roast that is marked $12.37. She pays for her purchase with a $20 bill. How much change does she get?

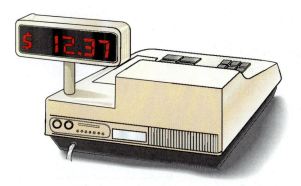

NOTE Sally's change will be the *difference* between the price of the roast and the $20 paid. We must use subtraction for the solution.

$$
\begin{array}{r}
\$20.00 \\
-\ \ 12.37 \\
\hline
\$\ \ 7.63
\end{array}
$$

Add zeros to write $20 as $20.00. Then subtract as before.

Sally will receive $7.63 in change after her purchase.

CHECK YOURSELF 5 _____

A stereo system that normally sells for $549.50 *is discounted (or marked down) to* $499.95 *for a sale. What is the savings?*

Keeping your checkbook balanced requires addition and subtraction of decimal numbers.

> ### Example 6   An Application Involving the Addition and Subtraction of Decimals
>
> For the check register, find the running balance.

Beginning balance	$234.15
Check # 301	23.88
Balance	_____
Check # 302	38.98
Balance	_____
Check # 303	114.66
Balance	_____
Deposit	175.75
Balance	_____
Check # 304	212.55
Ending balance	_____

To keep a running balance, we add the deposits and subtract the checks.

Beginning balance	$234.15	
Check # 301	23.88	Subtract
Balance	210.27	
Check # 302	38.98	Subtract
Balance	171.29	
Check # 303	114.66	Subtract
Balance	56.63	
Deposit	175.75	Add
Balance	232.38	
Check # 304	212.55	Subtract
Ending balance	19.83	

CHECK YOURSELF 6

For the check register, add the deposit amounts and subtract the check amounts to find the balance.

	Beginning balance	$398.00
	Check # 401	19.75
(a)	Balance	_____
	Check # 402	56.88
(b)	Balance	_____
	Check # 403	117.59
(c)	Balance	_____
	Deposit	224.67
(d)	Balance	_____
	Check # 404	411.48
(e)	Ending balance	_____

Now we will look at some applications of our work in multiplying decimals.

OBJECTIVE 3

Example 7   An Application Involving the Multiplication of Two Decimals

A sheet of paper has dimensions 27.5 cm by 21.5 cm. What is its area?

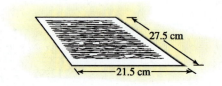

27.5 cm

21.5 cm

We multiply to find the required area.

RECALL Area is length times width, so multiplication is the necessary operation.

```
   27.5 cm
×  21.5 cm
  137 5
  275
  550
  591.25 cm²
```

The area of the paper is 591.25 cm².

CHECK YOURSELF 7 _____

If 1 kg is 2.2 lb, how many pounds equal 5.3 kg?

Example 8	An Application Involving the Multiplication of Two Decimals

Jack buys 8.7 gal of kerosene at 98.9 cents per gallon. Find the cost of the kerosene.

NOTE Usually in problems dealing with money, we round the result to the nearest cent (hundredth of a dollar).

We multiply the cost per gallon by the number of gallons. Then we round the result to the nearest cent. Note that the units of the answer will be cents.

```
   98.9
×   8.7
  69 23
 791 2
 860.43
```
The product 860.43 (cents) is rounded to 860 (cents), or $8.60.

The cost of Jack's kerosene will be $8.60.

CHECK YOURSELF 8 _____

One liter (L) is approximately 0.265 gal. On a trip to Europe, the Bernards purchased 88.4 L of gas for their rental car. How many gallons of gas did they purchase, to the nearest tenth of a gallon?

Sometimes we will have to use more than one operation for a solution, as Example 9 shows.

Example 9 An Application Involving Two Operations

Steve purchased a television set for $299.50. He agreed to pay for the set by making payments of $27.70 for 12 months. How much extra did he pay on the installment plan?

First we multiply to find the amount actually paid.

$$
\begin{array}{r}
\$\ 27.70 \\
\times\quad 12 \\
\hline
55\ 40 \\
277\ 0 \\
\hline
\$332.40 \\
\end{array}
$$
$\longleftarrow$ Amount paid

Now subtract the listed price. The difference will give the extra amount Steve paid.

$$
\begin{array}{r}
\$332.40 \\
-\quad 299.50 \\
\hline
\$\ 32.90 \\
\end{array}
$$
$\longleftarrow$ Extra amount

Steve will pay an additional $32.90 on the installment plan.

 CHECK YOURSELF 9 _____

Sandy's new car had a list price of $20,985. She paid $3,000 down and will pay $664.45 per month for 36 months on the balance. How much extra will she pay with this loan arrangement?

Example 10 is just one of many applications that require multiplying by a power of 10.

OBJECTIVE 4 ### Example 10 An Application Involving Multiplication by a Power of 10

NOTE There are 1,000 meters in a kilometer.

To convert from kilometers to meters, multiply by 1,000. Find the number of meters in 2.45 km.

NOTE If the result is a whole number, there is no need to write the decimal point.

2.45 km = 2450. m   Just move the decimal point three places to the right to make the conversion. Note that we added a zero to place the decimal point correctly.

 CHECK YOURSELF 10 _____

To convert from kilograms to grams, multiply by 1,000. Find the number of grams in 5.23 kg.

OBJECTIVE 5

Example 11   An Application Involving the Division of a Decimal by a Whole Number

RECALL You might want to review the rules for rounding decimals in Section 5.1.

A carton of 144 items costs $56.10. What is the price per item to the nearest cent?

To find the price per item, divide the total price by 144.

$$
\begin{array}{r}
0.389 \\
144\overline{)56.100} \\
43\ 2 \\
\overline{12\ 90} \\
11\ 52 \\
\overline{1\ 380} \\
1\ 296 \\
\overline{84}
\end{array}
$$

Carry the division to the thousandths place and then round back.

The cost per item is rounded to $0.39, or 39¢.

CHECK YOURSELF 11 _____

An office paid $26.55 for 72 pens. What was the cost per pen to the nearest cent?

Example 12   Solving an Application Involving the Division of Decimals

Andrea worked 37.5 h in a week and earned $291.75. What was her hourly rate of pay?

To find the hourly rate of pay we must use division. We divide the number of hours worked into the total pay.

NOTE We must add a zero to the dividend to complete the division process.

$$
\begin{array}{r}
7.78 \\
37.5\,\overline{)291.7\ 50} \\
262\ 5 \\
\overline{29\ 2\ 5} \\
26\ 2\ 5 \\
\overline{3\ 0\ 00} \\
3\ 0\ 00 \\
\overline{0}
\end{array}
$$

Andrea's hourly rate of pay was $7.78.

CHECK YOURSELF 12 _____

A developer wants to subdivide a 12.6-acre piece of land into 0.45-acre lots. How many lots are possible?

Example 13   Solving an Application Involving the Division of Decimals

At the start of a trip, the odometer read 34,563. At the end of the trip, it read 36,235. If 86.7 gal of gas were used, find the number of miles per gallon (to the nearest tenth).

First, find the number of miles traveled by subtracting the initial reading from the final reading.

$$
\begin{array}{r}
36{,}235 \\
-\ 34{,}563 \\
\hline
1{,}672
\end{array}
$$

Final reading
Initial reading
Miles covered

Next, divide the miles traveled by the number of gallons used. This will give us the miles per gallon.

$$
\begin{array}{r}
1\,9.\,28 \\
86.7\ \overline{)1672.0\ 00} \\
\underline{867} \\
805\ 0 \\
\underline{780\ 3} \\
24\ 7\ 0 \\
\underline{17\ 3\ 4} \\
7\ 3\ 60 \\
\underline{6\ 9\ 36} \\
4\ 24
\end{array}
$$

Round 19.28 to $19.3 \dfrac{\text{mi}}{\text{gal}}$.

CHECK YOURSELF 13

John starts his trip with an odometer reading of 15,436 and ends with a reading of 16,238. If he used 45.9 gal of gas, find the number of miles per gallon (to the nearest tenth).

Now we will look at an application of our work in dividing by powers of 10.

OBJECTIVE 6 | **Example 14   Solving an Application Involving a Power of 10**

To convert from millimeters (mm) to meters, we divide by 1,000. How many meters does 3,450 mm equal?

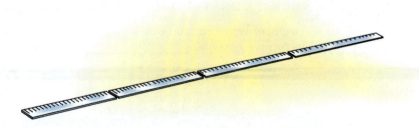

3,450 mm = 3 450. m   Shift three places to the left to divide by 1,000.

= 3.450 m

CHECK YOURSELF 14 _____

A shipment of 1,000 notebooks cost a stationery store $658. What was the cost per notebook to the nearest cent?

> **Example 15   An Application Involving the Division of Decimals Using the Calculator**

Omar drove 256.3 mi on a tank of gas. When he filled up the tank, it took 9.1 gal. What was his gas mileage?

Here's where students get into trouble when they use a calculator. Entering these values, you may be tempted to answer "$28.16483516 \dfrac{\text{mi}}{\text{gal}}$." The difficulty is that there is no way you can compute gas mileage to the nearest hundred-millionth mile. How do you decide where to round off the answer that the calculator gives you? A good rule of thumb when multiplication or division is involved is to report in your answer the least number of digits given in the problem. In this case, you were given a number with four digits and another with two digits. Your answer should not have more than two digits. Instead of 28.16483516, the answer could be $28 \dfrac{\text{mi}}{\text{gal}}$. Think about the question. If you were asked for gas mileage, how precise an answer would you give? The best answer to this question would be to give the nearest whole number of miles per gallon: $28 \dfrac{\text{mi}}{\text{gal}}$.

CHECK YOURSELF 15 _____

Emmet gained a total of 857 yards in 209 times that he carried the football. How many yards did he average for each time he carried the ball?

The Pythagorean theorem can be applied to solve a variety of geometric problems.

OBJECTIVE 7

> **Example 16   Solving for the Length of the Diagonal**

Find, to the nearest tenth, the length of the diagonal of a rectangle that is 8 cm long and 5 cm wide. Let x be the unknown length of the diagonal:

NOTE Always draw and label a sketch showing the information in a problem where geometric figures are involved.

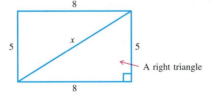
A right triangle

So

$$x^2 = 5^2 + 8^2$$
$$= 25 + 64$$
$$= 89$$
$$x = \sqrt{89}$$

Thus,

$$x \approx 9.4 \text{ cm}$$

RECALL The symbol $\approx$ means "is approximately equal to."

CHECK YOURSELF 16

The diagonal of a rectangle is 12 *in. and its width is* 6 *in. Find its length to the nearest tenth.*

The next application also makes use of the Pythagorean theorem.

Example 17   Solving an Application

How long must a guywire be to reach from the top of a 30-ft pole to a point on the ground 20 ft from the base of the pole? Round your answer to the nearest foot.

Again be sure to draw a sketch of the problem.

NOTE Always check to see if your final answer is reasonable.

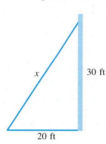

$$x^2 = 20^2 + 30^2$$
$$= 400 + 900$$
$$= 1{,}300$$
$$x = \sqrt{1{,}300}$$
$$\approx 36 \text{ ft}$$

 CHECK YOURSELF 17

A 16.0-ft ladder leans against a wall with its base 4.00 ft from the wall. How far off the floor is the top of the ladder?

 Example 18   Approximating Length with a Calculator

Approximate the length of the diagonal of a rectangle. The diagonal forms the hypotenuse of a triangle with legs 12.2 in. and 15.7 in. The length of the diagonal would be $\sqrt{12.2^2 + 15.7^2} = \sqrt{395.33} \approx 19.9$ in.

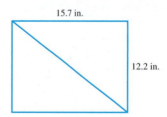

15.7 in.

12.2 in.

CHECK YOURSELF 18

Approximate the length of the diagonal of the rectangle to the nearest tenth.

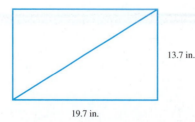

13.7 in.

19.7 in.

READING YOUR TEXT

The following fill-in-the-blank exercises are designed to assure that you understand the key vocabulary used in this section. Each sentence comes directly from the section. You will find the correct answers in Appendix C.

Section 5.8

(a) Our system of money is an example of a _____ system.

(b) The _____ is the distance around the outside of a straight-edged shape.

(c) The area of a rectangle is its _____ times its width.

(d) There are 1,000 meters in a _____.

CHECK YOURSELF ANSWERS

1. 55.8 gal   **2.** $353.45   **3.** 33.4 m   **4.** $1.3 \dfrac{\text{mi}}{\text{gal}}$ **5.** $49.55

6. (a) $378.25; **(b)** $321.37; **(c)** $203.78; **(d)** $428.45; **(e)** $16.97   **7.** 11.66 lb

8. 23.4 gal   **9.** $5,935.20   **10.** 5,230 g   **11.** $0.37, or 37¢   **12.** 28 lots

13. $17.5 \dfrac{\text{mi}}{\text{gal}}$ **14.** 66¢   **15.** 4.10 yd   **16.** Length is approximately 10.4 in.

17. Height is approximately 15.5 ft.   **18.** Length is approximately 24.0 in.

Name _____

Section _____ Date _____

ANSWERS

1. _____

2. _____

3. _____

4. _____

5. _____

6. _____

7. _____

8. _____

9. _____

5.8 Exercises

Perform the additions.

1. Add twenty-three hundredths, five tenths, and two hundred sixty-eight thousandths.

2. Add seven tenths, four hundred fifty-eight thousandths, and fifty-six hundredths.

3. Add five and three tenths, seventy-five hundredths, twenty and thirteen hundredths, and twelve and seven tenths. **VIDEO**

4. Add thirty-eight and nine tenths, five and fifty-eight hundredths, seven, and fifteen and eight tenths.

Solve the applications.

5. Business and Finance On a 3-day trip, Dien bought 12.7, 15.9, and 13.8 gal of gas. How many gallons of gas did he buy? **ALEKS**

6. Statistics Felix ran 2.7 mi on Monday, 1.9 mi on Wednesday, and 3.6 mi on Friday. How far did he run during the week? **ALEKS**

7. Statistics Rainfall was recorded in centimeters during the winter months as indicated on the bar graph.

(a) How much rain fell during those months?

(b) How much more rain fell in December than in February?

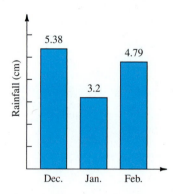

8. Construction A metal fitting has three sections, with lengths 2.5, 1.775, and 1.45 in. What is the total length of the fitting? **ALEKS**

9. Business and Finance Nicole had the following expenses on a business trip: gas, $45.69; food, $123; lodging, $95.60; and parking and tolls, $8.65. What were her total expenses during the trip? **VIDEO** **ALEKS**

ANSWERS

10. _____

11. _____

12. _____

13. _____

14. _____

15. _____

16. _____

10. Business and Finance Hok Sum's textbooks for one term cost $29.95, $47, $52.85, $33.35, and $10. What was his total cost for textbooks?

11. Business and Finance Jordan wrote checks of $50, $11.38, $112.57, and $9.73 during a single week. What was the total amount of the checks he wrote?

12. Business and Finance The deposit slip shown indicates the amounts that made up a deposit Peter Rabbit made. What was the total amount of his deposit?

DEPOSIT TICKET

☑ ☐ CASH ▶ 75.35

Peter Rabbit
123 East Derbunny St.

3–50/310

▶ 58.00

▶ 7.89

DATE _____

DEPOSITS MAY NOT BE AVAILABLE FOR IMMEDIATE WITHDRAWAL

▶ 100.00

(OR TOTAL FROM OTHER SIDE)

SUB TOTAL ▶ .

_____

SIGN HERE FOR CASH RECEIVED (IF REQUIRED) *

* LESS CASH ▶ .
RECEIVED

Briarpatch National Bank

$.

13. Construction Lupe is putting a fence around her yard. Her yard is rectangular and measures 8.16 yd long and 12.68 yd wide. How much fence should Lupe purchase? **ALEKS**

14. Geometry Find the perimeter of the given figure. **ALEKS**

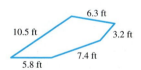

6.3 ft

10.5 ft

3.2 ft

7.4 ft

5.8 ft

15. Construction The figure gives the distance in miles of the boundary sections around a ranch. How much fencing is needed for the property?

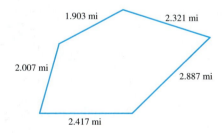

1.903 mi

2.321 mi

2.007 mi

2.887 mi

2.417 mi

16. Business and Finance A television set selling for $399.50 is discounted (or marked down) to $365.75. What is the savings? **ALEKS**

17. _____

18. _____

19. _____

20. _____

21. _____

22. _____

17. Business and Finance   For the check register, find the running balance.

Beginning balance	$456.00
Check # 601	$199.29
Balance	_____
Service charge	$ 18.00
Balance	_____
Check # 602	$ 85.78
Balance	_____
Deposit	$250.45
Balance	_____
Check # 603	$201.24
Ending balance	_____

18. Business and Finance   For the check register, find the running balance.

Beginning balance	$589.21
Check # 678	$175.63
Balance	_____
Check # 679	$ 56.92
Balance	_____
Deposit	$121.12
Balance	_____
Check # 680	$345.99
Ending balance	_____

19. Business and Finance   Your bill for a car tune-up includes $7.80 for oil, $5.90 for a filter, $3.40 for spark plugs, $4.10 for points, and $28.70 for labor. Estimate your total cost by rounding each amount to the nearest dollar.   **VIDEO**

20. Business and Finance   The payroll at a car repair shop for 1 week was $456.73, utilities were $123.89, advertising was $212.05, and payments to distributors were $415.78. Estimate the amount spent in 1 week by rounding each amount to the nearest dollar.

21. Business and Finance   On a recent business trip your expenses were $343.78 for airfare, $412.78 for lodging, $148.89 for food, and $102.15 for other items. Estimate your total expenses by rounding each amount to the nearest dollar.

VIDEO

22. Here are charges on a credit card account:
$8.97, $32.75, $15.95, $67.32, $215.78, $74.95, $83.90, and $257.28

  (a) Estimate the total bill for the charges by rounding each number to the nearest dollar and adding the results.

  (b) Estimate the total bill by adding the charges and then rounding to the nearest dollar.

  (c) What are the advantages and disadvantages of the methods in **(a)** and **(b)**?

23. Find the next number in the sequence: 3.125, 3.375, 3.625, . . .

Recall that a magic square is one in which the sum of every row, column, and diagonal is the same. Complete the magic squares in exercises 24 and 25.

24.

1.6		1.2
	1	
0.8		

25.

2.4		7.2
10.8		
4.8		

26. Find the next two numbers in each of the sequences:

   (a) 0.75   0.62   0.5   0.39

   (b) 1.0   1.5   0.9   3.5   0.8

27. Business and Finance Kurt bought four shirts on sale as pictured. What was the total cost of the purchase?

28. Business and Finance A light plane uses $5.8 \frac{\text{gal}}{\text{h}}$ of fuel. How much fuel is used on a flight of 3.2 h? Give your answer to the nearest tenth of a gallon.

29. Construction The Hallstons select a carpet costing $15.49 per square yard. If they need 7.8 yd² of carpet, what is the cost to the nearest cent?

ANSWERS

23. _____

24. _____

25. _____

26. _____

27. _____

28. _____

29. _____

30. _____

31. _____

32. _____

33. _____

34. _____

35. _____

36. _____

37. _____

38. _____

39. _____

40. _____

41. _____

42. _____

43. _____

30. Business and Finance Maureen's car payment is $242.38 per month for 4 years. How much will she pay altogether?

31. Geometry A classroom is 7.9 m wide and 11.2 m long. Estimate its area.

32. Crafts You buy a roast that weighs 6.2 lb and costs $3.89 per pound. Estimate the cost of the roast.

33. Business and Finance A store purchases 100 items at a cost of $1.38 each. Find the total cost of the order.

34. Science and Medicine To convert from meters to centimeters, multiply by 100. How many centimeters are there in 5.3 m? **ALEKS**

35. Science and Medicine How many grams are there in 2.2 kg? Multiply by 1,000 to make the conversion. **VIDEO**

36. Business and Finance An office purchases 1,000 pens at a cost of 17.8 cents each. What is the cost of the purchase in dollars?

Item	Quantity	Item Price	Total
Pens	1000	$0.178	

37. Crafts We have 91.25 in. of plastic labeling tape and wish to make labels that are 1.25 in. long. How many labels can be made?

38. Business and Finance Alberto worked 32.5 h, earning $306.15. How much did he make per hour?

39. Crafts A roast weighing 5.3 lb sold for $14.89. Find the cost per pound to the nearest cent.

40. Construction One nail weighs 0.025 oz. How many nails are there in 1 lb? (1 lb is 16 oz.)

41. Statistics A family drove 1,390 mi, stopping for gas three times. If they purchased 15.5, 16.2, and 10.8 gal of gas, find the number of miles per gallon (the mileage) to the nearest tenth.

42. Statistics On a trip an odometer changed from 36,213 to 38,319. If 136 gal of gas were used, find the number of miles per gallon (to the nearest tenth).

43. Science and Medicine To convert from millimeters to inches, we can divide by 25.4. If film is 35 mm wide, find the width to the nearest hundredth of an inch.

ALEKS

44.

45.

46.

47.

48.

49.

50.

44. Statistics To convert from centimeters to inches, we can divide by 2.54. The rainfall in Paris was 11.8 cm during 1 week. What was that rainfall to the nearest hundredth of an inch? **ALEKS**

45. Science and Medicine The blood alcohol content (BAC) of a person who has been drinking is determined by Widmark's formula:

$$BAC = \frac{total\ oz \cdot percent\ alcohol \cdot 1.055}{body\ weight \cdot 0.68} - (hours\ drinking \cdot 0.017)$$

A 125-lb person is driving and is stopped by a policewoman on suspicion of driving under the influence (DUI). The driver claims that in the past 2 h he consumed only six 12-oz bottles of 3.9% beer. If he undergoes a breathalyzer test, what will his BAC be? Will this amount be under the legal limit for your state?

46. Four brands of soap are available in a local store.

Brand	Ounces	Total Price	Unit Price
Squeaky Clean	5.5	$0.36	
Smell Fresh	7.5	0.41	
Feel Nice	4.5	0.31	
Look Bright	6.5	0.44	

Compute the unit price and decide which brand is the best buy.

47. Sophie is a quality control expert. She inspects boxes of #2 pencils. Each pencil weighs 4.4 g. The contents of a box of pencils weigh 66.6 g. If a box is labeled CONTENTS: 16 PENCILS, should Sophie approve the box as meeting specifications? Explain your answer.

48. Write a plan to determine the number of miles per gallon your car (or your family car) gets. Use this plan to determine your car's actual miles per gallon.

49. Express the width and length of a $1 bill in centimeters. Then express the same dimensions in millimeters.

50. Geometry If the perimeter of a square is 19.2 cm, how long is each side?

$P = 19.2$ cm

51. _____

52. _____

53. _____

54. _____

55. _____

56. _____

57. _____

58. _____

51. Geometry   If the perimeter of an equilateral triangle (all sides have equal length) is 16.8 cm, how long is each side?

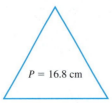

$P = 16.8$ cm

52. Geometry   If the perimeter of a regular pentagon (all sides have equal length) is 23.5 in., how long is each side?

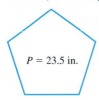

$P = 23.5$ in.

In exercises 53 to 58, express your answer to the nearest thousandth.

53. Geometry   Find the length of the diagonal of a rectangle with a length of 10 cm and a width of 7 cm.

54. Geometry   Find the length of the diagonal of a rectangle with 5 in. width and 7 in. length.

55. Geometry   Find the width of a rectangle whose diagonal is 12 ft and whose length is 10 ft.

56. Geometry   Find the length of a rectangle whose diagonal is 9 in. and whose width is 6 in.

57. Construction   How long must a guywire be to run from the top of a 20-ft pole to a point on the ground 8 ft from the base of the pole?

58. Construction   The base of a 15-ft ladder is 5 ft away from a wall. How high from the floor is the top of the ladder?

Calculator Exercises

Solve the applications using your calculator.

59. Business and Finance Your checking account has a balance of $532.89. You write checks of $50, $27.54, and $134.75 and make a deposit of $50. What is your ending balance?

60. Business and Finance Your checking account has a balance of $278.45. You make deposits of $200 and $135.46. You write checks for $389.34, $249, and $53.21. What is your ending balance? Be careful with this problem. A negative balance means that your account is overdrawn.

61. Business and Finance Dr. Rogers is concerned with the increasing cost of making photocopies. She wants to examine alternatives to the current financing plan. The office currently leases a copy machine for $110 per month and $0.025 per copy. A 3-year payment plan is available that costs $125 per month and $0.015 per copy.

(a) If the office expects to run 100,000 copies per year, which is the better plan?

(b) How much money will the better plan save over the other plan?

62. Business and Finance In a bottling company, a machine can fill a 2-L bottle in 0.5 s and move the next bottle into place in 0.1 s. How many 2-L bottles can be filled by the machine in 2 h?

63. The owner of a bakery sells a finished cake for $8.99. The cost of baking 16 cakes is $75.63. Write a plan to find out how much profit the baker can make on each cake sold.

64. Construction An 80.5-acre piece of land is being subdivided into 0.35-acre lots. How many lots are possible in the subdivision?

65. Business and Finance In 1 week, Tom earned $395.85 by working 36.25 h. What was his hourly rate of pay to the nearest cent?

In exercises 66 to 68, indicate whether the given statement is true or false.

66. The Pythagorean theorem can be used on any triangle.

67. The perimeter of a figure is the distance around its edge.

68. The perimeter of a rectangle can be found by multiplying the length times the width.

ANSWERS

59. _____

60. _____

61. _____

62. _____

63. _____

64. _____

65. _____

66. _____

67. _____

68. _____

Answers

1. 0.998

3.
$$\begin{array}{r} \overset{1}{5.30} \\ 0.75 \\ 20.13 \\ +\ 12.70 \\ \hline 38.88 \end{array}$$

5. 42.4 gal

7. **(a)** 13.37 cm; **(b)** 0.59 cm

9. $272.94   **11.** $183.68   **13.** 41.68 yd   **15.** 11.535 mi

17. End balance: $202.14   **19.** $50

21. $1,008   **23.**   **25.**

2.4	8.4	7.2
10.8	6	1.2
4.8	3.6	9.6

27. $39.92

29. $120.82   **31.** 88 m²   **33.** $138   **35.** 2,200 g   **37.** 73 labels

39. $2.81   **41.** $32.7\,\dfrac{\text{mi}}{\text{gal}}$ **43.** 1.38 in.   **45.**   **47.**

49.   **51.** 5.6 cm   **53.** Approximately 12.207 cm

55. Approximately 6.633 ft   **57.** Approximately 21.541 ft   **59.** $370.60

61. **(a)** Current plan: $11,460; 3-year lease: $9,000; **(b)** Savings: $2,460

63.   **65.** $10.92   **67.** True

ACTIVITY 5: STATISTICS IN SPORTS

Each activity in this text is designed to either enhance your understanding of the topics of the preceding chapter, provide you with a mathematical extension of those topics, or both. The activities can be undertaken by one student, but they are better suited for a small group project. Occasionally, it is only through discussion that different facets of the activity become apparent. For material related to this activity, visit the text website at www.mhhe.com/baratto.

Part 1: Baseball

There are many statistics in the sport of baseball that are expressed in decimal form. Two of these are the batting average and the earned run average. Both are actually examples of rates.

A batting average is a rate where the units are *hits per at-bat*. To compute the batting average for a hitter, divide the number of hits (H) by the number of times at-bat (AB). The result will be a decimal less than 1 (unless the batter always gets a hit!) and is always expressed to the nearest thousandth. For example, if a hitter has 2 hits in 7 at-bats, we divide 2 by 7, getting 0.285714. . . . The batting average is then rounded to 0.286.

Compute the batting average for each of the following major league players.

	Player	Hits	At-Bats	Average
1	Pujols	139	373	
2	Helton	134	384	
3	Guillen	99	290	
4	Suzuki	142	419	
5	Mora	99	293	

The earned run average (ERA) for a pitcher is also a rate; its units are earned runs per nine innings. It represents the number of earned runs the pitcher gives up in nine innings. To compute the ERA for a pitcher, multiply the number of earned runs allowed by the pitcher by nine, and then divide by the number of innings pitched. The result is always rounded to the nearest hundredth.

Compute the earned run average for each of the following major league players.

	Player	Earned Runs	Innings	ERA
6	Loaiza	33	$137\frac{1}{3}$	
7	Martinez	27	110	
8	Brown	31	$123\frac{1}{3}$	

Challenge: Suppose a hitter has 54 hits in 200 times at bat. How many hits in a row must the hitter get in order to raise his average to at least 0.300?

Part 2: Cycling in the Tour de France

The Tour de France is perhaps the most grueling of all sporting events. It is a bicycle race that spans 22 days (including 2 rest days) and involves 20 stages of riding in a huge circuit around the country of France. This includes several stages that take the riders through two mountainous regions, the Alps and the Pyrenees. Below are the winners of each stage in the 2003 race, along with the winning time in hours and the length of the stage expressed in miles. For each stage, compute the winner's average speed by dividing the miles traveled by the winning time. Round your answers to the nearest tenth of a mile per hour.

Stage	Winner	Miles	Time	Speed
Prologue	McGee	4.03	0.12	
1	Petacchi	104	3.74	
2	Cooke	127	5.11	
3	Petacchi	103.5	3.46	
4	US Postal	42.85	1.31	
5	Petacchi	122.03	4.16	
6	Petacchi	142.8	5.14	
7	Virenque	143.2	6.10	
8	Mayo	135.78	5.96	
9	Vinokourov	114.4	5.03	
10	Piil	136.1	5.16	
11	Flecha	95.2	3.49	
12	Ullrich	29.19	0.98	
13	Sastre	122.5	5.27	
14	Simoni	118.92	5.53	
15	Armstrong	98.9	4.49	
16	Hamilton	122.5	4.99	
17	Knaven	112	3.91	
18	Lastras	124	4.06	
19	Millar	30.4	0.90	
20	Nazon	94.4	3.65	

Find the total number of miles traveled.

By examining the speeds of the stages, can you identify which stages occurred in the mountains?

The overall tour winner was Lance Armstrong with a total winning time of 83.69 hours. Compute his average speed for the entire race, again rounding to the nearest tenth of a mile per hour.

5 Summary

DEFINITION/PROCEDURE	EXAMPLE	REFERENCE
Introduction to Decimals, Place Value, and Rounding		**Section 5.1**
Decimal Fraction A fraction whose denominator is a power of 10. We call decimal fractions **decimals.**	$\dfrac{7}{10}$ and $\dfrac{47}{100}$ are decimal fractions.	*p. 391*
Decimal Place Each position for a digit to the right of the decimal point. Each decimal place has a place value that is one-tenth the value of the place to its left.	2.3456 ↑↑↑↑ ⎯ Ten thousandths ⎯ Thousandths ⎯ Hundredths ⎯ Tenths	*p. 392*
Reading and Writing Decimals in Words 1. Read the digits *to the left* of the decimal point as a whole number. 2. Read the decimal point as the word *and.* 3. Read the digits *to the right* of the decimal point as a whole number followed by the place value of the rightmost digit.	Hundredths ↙ 8.15 is read "eight and fifteen hundredths."	*p. 392*
Rounding Decimals 1. Find the place to which the decimal is to be rounded. 2. If the next digit to the right is 5 or more, increase the digit in the place you are rounding to by 1. Discard any remaining digits to the right. 3. If the next digit to the right is less than 5, just discard that digit and any remaining digits to the right.	To round 5.87 to the nearest tenth: ↓ 5.87 is rounded to 5.9 To round 12.3454 to the nearest thousandth: ↓ 12.3454 is rounded to 12.345.	*p. 395*
Addition and Subtraction of Decimals		**Section 5.2**
To Add or Subtract Decimals 1. Write the numbers being added (or subtracted) in column form with their decimal points in a vertical line. You may have to place zeros to the right of the existing digits. 2. Add (or subtract) just as you would with whole numbers. 3. Place the decimal point of the sum (or difference) in line with the decimal points.	To subtract 5.875 from 8.5: $\begin{array}{r} 8.500 \\ -\ 5.875 \\ \hline 2.625 \end{array}$	*p. 403, 405*
Multiplication of Decimals		**Section 5.3**
To Multiply Decimals 1. Multiply the decimals as though they were whole numbers. 2. Add the number of decimal places in the factors. 3. Place the decimal point in the product so that the number of decimal places in the product is the sum of the number of decimal places in the factors.	To multiply $2.85 \cdot 0.045$: $\begin{array}{r} 2.85 \leftarrow\text{Two places} \\ \times\ 0.045 \leftarrow\text{Three places} \\ \hline 1425 \\ 1140\ \ \\ \hline 0.12825 \leftarrow\text{Five places} \end{array}$	*p. 414*
Multiplying by Powers of 10 Move the decimal point to the right the same number of places as there are zeros in the power of 10.	$2.37 \cdot 10 = 23.7$ $0.567 \cdot 1{,}000 = 567$	*p. 416*

Continued

DEFINITION/PROCEDURE	EXAMPLE	REFERENCE
Division of Decimals		Section 5.4
To Divide by a Decimal 1. Move the decimal point to the right, making the divisor a whole number. 2. Move the decimal point in the dividend to the right the same number of places. Add zeros if necessary. 3. Place the decimal point in the quotient directly above the decimal point of the dividend. 4. Divide as you would with whole numbers.	To divide 16.5 by 5.5, move the decimal points: $$\begin{array}{r} 3 \\ 5.5\overline{)16.5} \\ \underline{16\ 5} \\ 0 \end{array}$$	p. 425
To Divide by a Power of 10 Move the decimal point to the left the same number of places as there are zeros in the power of 10.	$25.8 \div 10 = 2{\scriptstyle\wedge}5.8 = 2.58$	p. 427
Fractions and Decimals		Section 5.5
To Convert a Common Fraction to a Decimal 1. Divide the numerator of the common fraction by its denominator. 2. The quotient is the decimal equivalent of the common fraction.	To convert $\frac{1}{2}$ to a decimal: $$\begin{array}{r} 0.5 \\ 2\overline{)1.0} \\ \underline{1\ 0} \\ 0 \end{array}$$	p. 434
To Convert a Terminating Decimal Less Than 1 to a Common Fraction 1. Write the digits of the decimal without the decimal point. This will be the numerator of the common fraction. 2. The denominator of the fraction is a 1 followed by as many zeros as there are places in the decimal.	To convert 0.275 to a common fraction: $$0.275 = \frac{275}{1000} = \frac{11}{40}$$	p. 437
Equations Containing Decimals		Section 5.6
To solve an equation that contains decimals, use the same procedure used for solving other linear equations.	$$\begin{aligned} 1.2x + 3.9x &= 2.7x - 3.2 \\ 5.1x &= 2.7x - 3.2 \\ 2.4x &= -3.2 \\ x &= \frac{-3.2}{2.4} \\ x &= -1.\overline{33} \end{aligned}$$	p. 446
Square Roots and the Pythagorean Theorem		Section 5.7
The square root of a number is a value that, when squared, gives us that number. The length of the three sides of a right triangle will form a perfect triple.	$3^2 + 4^2 = 5^2$	p. 451
The Pythagorean theorem is usually written as $$c^2 = a^2 + b^2$$		p. 453

Summary Exercises

This summary exercise set is provided to give you practice with each of the objectives of this chapter. Each exercise is keyed to the appropriate chapter section. When you are finished, you can check your answers to the odd-numbered exercises against those presented in the back of the text. If you have difficulty with any of these questions, go back and reread the examples from that section. The answers to the even-numbered exercises appear in the *Instructor's Manual*. Your instructor will give you guidelines on how to best use these exercises in your class.

[5.1] In exercises 1 and 2, find the indicated place values.

1. 7 in 3.5742

2. 3 in 0.5273

In exercises 3 and 4, write the fractions in decimal form.

3. $\dfrac{37}{100}$

4. $\dfrac{307}{10,000}$

In exercises 5 and 6, write the decimals in words.

5. 0.071

6. 12.39

In exercises 7 and 8, write the fractions in decimal form.

7. Four and five tenths

8. Four hundred and thirty-seven thousandths

In exercises 9 to 12, complete each statement using the symbol <, =, or >.

9. 0.79 _____ 0.785

10. 1.25 _____ 1.250

11. 12.8 _____ 13

12. 0.832 _____ 0.83

In exercises 13 to 15, round to the indicated place.

13. 5.837   hundredths

14. 9.5723   thousandths

15. 4.87625   three decimal places

[5.2] In exercises 16 to 19, add.

16.
$$
\begin{array}{r}
2.58 \\
+\ 0.89 \\
\hline
\end{array}
$$

17.
$$
\begin{array}{r}
3.14 \\
0.8 \\
2.912 \\
+\ 12 \\
\hline
\end{array}
$$

18. -1.3, 25, -5.27, and 6.158

19. Add eight, forty-three thousandths, five and nineteen hundredths, and seven and three tenths.

In exercises 20 to 23, subtract.

20.   29.21
        − 5.89

21.   6.73
        − 2.485

22. 1.735 from −2.81

23. −12.38 from 19

[5.3]   In exercises 24 to 29, multiply.

24.   22.8
        × 0.72

25.   0.0045
        × 0.058

26. −1.24 · 56

27. (−0.0025)(−0.491)

28. 0.052 · 1,000

29. 0.045 · 10^4

[5.4]   In exercises 30 to 32, divide. Round answers to the nearest hundredth.

30. 8)3.08

31. 58)269.7

32. 55)17.69

In exercises 33 to 36, divide. Round answers to the nearest thousandth.

33. 0.7)1.865

34. −3.042 ÷ (−0.37)

35. 5.3)6.748

36. 0.2549 ÷ (−2.87)

In exercises 37 to 39, divide.

37. 7.6 ÷ 10

38. 80.7 ÷ 1,000

39. 457 ÷ 10^4

[5.5]   In exercises 40 to 43, find the decimal equivalents.

40. $\dfrac{7}{16}$

41. $\dfrac{3}{7}$ (round to the thousandths)

42. $\dfrac{4}{15}$ (use bar notation)

43. $3\dfrac{3}{4}$

In exercises 44 to 48, write as common fractions or mixed numbers. Simplify your answers.

44. 0.21

45. 0.084

46. 5.28

47. 0.0067

48. 21.857

[5.6] Solve the equations and check your results.

49. $3.7x + 8 = 1.7x + 16$

50. $5.4x + (-3) = 8.4x + 9$

51. $2.9x = 4.9x - 3.3$

52. $1.4x = 4.4x + 9.75$

53. $2(x - 1.8) = 4.2(x + 0.9) + 3.18$

54. $3(x + 5.8) = 6.5(x - 1.2) + 12.25$

[5.7] In exercises 55 and 56, find the square root.

55. $\sqrt{324}$

56. $\sqrt{784}$

57. Find the hypotenuse of the triangle whose sides are 33 and 44.

[5.8] In exercises 58 to 71, solve the applications.

58. Geometry Find the perimeter (to the nearest hundredth of a centimeter) of a rectangle that has dimensions 5.37 cm by 8.64 cm.

59. Statistics Janice ran 4.8 mi on Sunday, 5.3 mi on Tuesday, 3.9 mi on Thursday, and 8.2 mi on Saturday. How far did she run during the week?

60. Geometry Find dimension a in the figure.

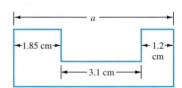

61. Business and Finance A stereo system that normally sells for $499.50 is discounted (or marked down) to $437.75 for a sale. Find the savings.

62. Business and Finance If you cash a $50 check and make purchases of $8.71, $12.53, and $9.83, how much money do you have left?

63. Business and Finance Neal worked for 37.4 h during a week. If his hourly rate of pay was $9.25, how much did he earn?

64. Business and Finance   To find the simple interest on a loan at $11\frac{1}{2}$% for 1 year, we must multiply the amount of the loan by 0.115. Find the simple interest on a $2,500 loan at $11\frac{1}{2}$% for 1 year.

65. Business and Finance   A television set has an advertised price of $499.50. You buy the set and agree to make payments of $27.15 per month for 2 years. How much extra are you paying by buying on this installment plan?

66. Business and Finance   A stereo dealer buys 100 portable radios for a promotion sale. If she pays $57.42 per radio, what is her total cost?

67. Business and Finance   During a charity fund-raising drive 37 employees of a company donated a total of $867.65. What was the average donation per employee?

68. Statistics   In six readings, Faith's gas mileage was 38.9, 35.3, 39.0, 41.2, 40.5, and 40.8 $\frac{\text{mi}}{\text{gal}}$. What was the average mileage to the nearest tenth of a mile per gallon? (*Hint:* First find the sum of the mileages. Then divide the sum by 6, because there are 6 mileages.)

69. Construction   A developer is planning to subdivide an 18.5-acre piece of land. She estimates that 5 acres will be used for roads and wants individual lots of 0.25 acre. How many lots are possible?

70. Statistics   Paul drives 949 mi using 31.8 gal of gas. What is his mileage for the trip (to the nearest tenth of a mile per gallon)?

71. Business and Finance   A shipment of 1,000 videotapes cost a dealer $7,090. What was the cost per tape to the dealer?

Solve each of the applications. Approximate your answer to one decimal place where necessary.

72. Geometry   Find the length of the diagonal of a rectangle whose length is 12 in. and whose width is 9 in.

73. Geometry   Find the length of a rectangle whose diagonal has a length of 10 cm and whose width is 5 cm.

74. Construction   How long must a guywire be to run from the top of an 18-ft pole to a point on level ground 16 ft away from the base of the pole?

Self-Test for Chapter 5

The purpose of this self-test is to help you check your progress so that you can find sections and concepts that you need to review before the next in-class exam. Allow yourself about an hour to take this test. At the end of that hour, check your answers against those given in the back of this text. If you missed a question, notice the section reference that accompanies the question. Go back to that section and reread the examples until you have mastered that particular concept.

Name _____

Section _____ Date _____

ANSWERS

1. Find the place value of 8 in 0.5248.

2. Write $\dfrac{49}{1,000}$ in decimal form.

1. _____

2. _____

3. Write 2.53 in words.

4. Write twelve and seventeen thousandths in decimal form.

3. _____

4. _____

In exercises 5 and 6, complete the statement, using the symbol $<$ or $>$.

5. _____

5. 0.889 _____ 0.89

6. 0.531 _____ 0.53

6. _____

7. _____

In exercises 7 to 9, add.

8. _____

7.
```
    3.45
    0.6
 + 12.59
```

8. 2.4, 35, -4.73, and -5.123.

9. _____

10. _____

9. Seven, seventy-nine hundredths, and five and thirteen thousandths.

11. _____

12. _____

13. _____

In exercises 10 and 11, round to the indicated place.

14. _____

10. 0.5977   thousandths

11. 23.5724   two decimal places

In exercises 12 to 14, subtract.

12.
```
   18.32
 −  7.78
```

13.
```
    40
 − 15.625
```

14. -1.742 from -5.63

15. _____

16. _____

17. _____

18. _____

19. _____

20. _____

21. _____

22. _____

23. _____

24. _____

25. _____

26. _____

27. _____

28. _____

29. _____

30. _____

31. _____

32. _____

33. _____

In exercises 15 to 19, multiply.

15.   32.9
    $\times\ 0.53$

16.   0.049
    $\times\ \ 0.57$

17. $(2.75)(-0.53)$

18. $0.735 \cdot 1{,}000$

19. $1.257 \cdot 10^4$

In exercises 20 to 26, divide. When indicated, round to the given place value.

20. $8\overline{)3.72}$

21. $27\overline{)63.45}$

22. $2.72 \div 53$ thousandths

23. $4.1\overline{)10.455}$

24. $0.6\overline{)1.431}$

25. $-3.969 \div 0.54$

26. $0.263 \div 3.91$ three decimal places

In exercises 27 and 28, divide.

27. $4.983 \div 1{,}000$

28. $523 \div 10^5$

In exercises 29 to 31, find the decimal equivalents of the common fractions. When indicated, round to the given place value.

29. $\dfrac{9}{16}$

30. $\dfrac{4}{7}$ (thousandths)

31. $\dfrac{7}{11}$ (use bar notation)

In exercises 32 and 33, write the decimals as common fractions or mixed numbers. Simplify your answer.

32. 0.072

33. 4.44

34. Insert $<$ or $>$ to form a true statement.

$$0.168 \underline{\hspace{1cm}} \frac{3}{25}$$

35. A baseball team has a winning percentage of 0.458. Write this as a fraction in simplest form.

In exercises 36 to 40, solve each equation.

36. $5.2x = 7.54$

37. $2.3x = -8.28$

38. $1.83 + 2x = 5.04 + 5x$

39. $6.5(x - 1.4) = 4.4(x - 3.2) + 9.495$

40. $1.3 + 2(x + 6.8) = 3x + 20.2$

41. Find the square root of 441.

42. The legs of a right triangle are 39 m and 52 m in length. Find the length of the hypotenuse.

43. Find the perimeter of the triangle shown.

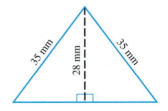

44. Business and Finance On a business trip, Martin bought the following amounts of gasoline: 14.4, 12, 13.8, and 10 gal. How much gasoline did he purchase on the trip?

45. Business and Finance You pay for purchases of $13.99, $18.75, $9.20, and $5 with a $50 bill. How much cash will you have left?

34. _____

35. _____

36. _____

37. _____

38. _____

39. _____

40. _____

41. _____

42. _____

43. _____

44. _____

45. _____

46. **Geometry**   Find the area of a rectangle with length 3.5 in. and width 2.15 in.

47. **Business and Finance**   A college bookstore purchases 1,000 pens at a cost of 54.3 cents per pen. Find the total cost of the order in dollars.

48. **Construction**   A 14-acre piece of land is being developed into home lots. If 2.8 acres of land will be used for roads and each home site is to be 0.35 acre, how many lots can be formed?

49. **Business and Finance**   A street improvement project will cost $57,340 and that cost is to be divided among the 100 families in the area. What will be the cost to each individual family?

© 2010 McGraw-Hill Companies

Cumulative Review
for Chapters 1 to 5

Name _____

Section _____ Date _____

The following exercises are presented to help you review concepts from earlier chapters that you may have forgotten. This section is meant as review material and not as a comprehensive exam. The answers are presented in the back of the text. If you have difficulty with any of these exercises, be certain to at least read through the summary related to that section.

ANSWERS

Perform the indicated operations.

1. $9 - (-3)$

2. $(-4)(-5)$

3. $0 \div (-7)$

4. $27 - 3 \cdot 2^2$

5. $8(-9 + 7)$

6. $\dfrac{2}{5} \cdot \dfrac{15}{8}$

7. $\dfrac{-3}{4} + \dfrac{-1}{3}$

8. $\dfrac{2}{5} \div \dfrac{-3}{10}$

9. $12.3 + 8.52$

10. $19.3 - 8.47$

11. $0.03 \cdot 425$

12. $0.0042 \cdot 1{,}000$

13. $53.26 \div 100$

Round to the indicated place.

14. 6.82148   thousandths

15. 5.982   tenths

Complete each statement using the symbol $<$, $=$, or $>$.

16. 15.6295 _____ 15.631

17. 7.04 _____ 7.040

Write as a common fraction.

18. 0.15

19. 0.08

Write as a decimal.

20. $\dfrac{5}{8}$

21. $\dfrac{6}{7}$ round to the hundredths

22. Write the prime factorization of 210.

23. Find the greatest common factor of 30 and 48.

1. _____   2. _____

3. _____   4. _____

5. _____

6. _____

7. _____

8. _____

9. _____

10. _____

11. _____

12. _____

13. _____

14. _____

15. _____

16. _____

17. _____

18. _____

19. _____

20. _____

21. _____

22. _____

23. _____

24. _____

25. _____

26. _____

27. _____

28. _____

29. _____

30. _____

31. _____

32. _____

33. _____

34. _____

35. _____

36. _____

37. _____

38. _____

39. _____

40. _____

41. _____

42. _____

43. _____

44. _____

24. Find the least common multiple of 15 and 18.

Identify each as an expression or an equation.

25. $5x - 3 = 8$

26. $\dfrac{3}{4}x - \dfrac{1}{4}$

Evaluate each expression if $x = -3$, $y = 6$, $z = -4$, and $w = 2$.

27. $x + 3y - 3z$

28. $\dfrac{3x - y}{w - x}$

Combine like terms.

29. $7x + 5y - 4x - 6y$

30. $\dfrac{13}{5}x + 2 - \dfrac{3}{5}x + 5$

Solve each equation and check your result.

31. $4x + 3 = 5x + 7$

32. $2x - 6 = 5x + 9$

33. $\dfrac{x}{4} = 12$

34. $\dfrac{2}{5}x = -10$

35. $-1.2x = 3$

36. $4.2x - 5 = 1.2x - 29$

In exercises 37 and 38, use the figure:

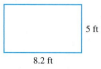

5 ft

8.2 ft

37. Find the perimeter of the figure shown.

38. Find the area of the figure shown.

Solve each application.

39. If a number increased by 8 is 23, find that number.

40. A dresser costs $200 more than a bed frame. Together the two pieces cost $900. How much does the dresser cost alone?

41. Francisco drove 273 mi using 15 gal of gas. What was his mileage (miles per gallon) for the trip? Round to the nearest tenths.

42. The length of a rectangle is 4 cm less than 3 times its width. The perimeter of the rectangle is 88 cm. What are the dimensions of the rectangle?

43. Find the square root of 289.

44. The legs of a right triangle are 13 in. and 18 in. in length. Find the length of the hypotenuse. Round to the nearest tenth.

RATIO, RATE, AND PROPORTION

CHAPTER 6 OUTLINE

Section 6.1   Ratios   page 499

Section 6.2   Rates   page 507

Section 6.3   Proportions   page 517

Section 6.4   Similar Triangles and Proportions   page 537

Section 6.5   Linear Measurement and Conversion   page 545

chapter

6

INTRODUCTION

How long is a yard? Has it always been the same length? The history of measurement is an interesting one. In addition to the more familiar units in the English system (inch, pound, cup, etc.), history contains numerous examples of other units. Length was measured in cubits and rods, weight was measured by comparing an object with grains of barley, and cordwood was measured in "steres." Here is some measurement history:

- A yard was defined by King Henry I, early in the twelfth century, as the distance from the tip of his nose to the tip of his outstretched finger.
- A cubit was defined as the distance from the elbow to the tip of the middle finger.
- A rod was a traditional Saxon land measure that was originally defined as the total length of the left feet of the first sixteen men to leave church on Sunday morning.

People have always felt the need to measure, and many units of measure were first related to parts of the body. As units became more standardized, many tools were invented to aid in the process of measuring. At the same time, people applied mathematics to measurement problems involving construction, determination of inaccessible distances or heights, and the conversion from one type of unit to another.

The mathematics needed here often involves the ideas of ratio, rate, and proportion. In Sections 6.1, 6.4, and 6.5, you will find exercises that relate to these topics and units of measure. At the end of the chapter, but before the chapter summary, you will find a more in-depth activity on measurement and its history.

1. _____

2. _____

3. _____

4. _____

5. _____

6. _____

7. _____

8. _____

9. _____

10. _____

11. _____

12. _____

13. _____

14. _____

Pretest Chapter 6

This pretest will provide a preview of the types of exercises you will encounter in each section of this chapter. The answers for these exercises can be found in the back of the text. If you are working on your own or ahead of the class, this pretest can help you identify the sections in which you should focus more of your time.

[6.1] **1.** Write the ratio of 7 to 10.

2. Write the ratio of 20 to 15 in lowest terms.

[6.2] **3.** Find the rate equivalent to $\dfrac{551 \text{ mi}}{19 \text{ gal}}$.

4. Find the unit price given that a dozen cans of cat food cost $9.48.

[6.3] **5.** Is $\dfrac{4}{7} = \dfrac{12}{21}$ a true proportion?

6. Is $\dfrac{5}{9} = \dfrac{9}{16}$ a true proportion?

7. Solve for x: $\dfrac{x}{4} = \dfrac{5}{2}$

8. Solve for a: $\dfrac{5}{a} = \dfrac{7}{21}$

9. Solve for n: $\dfrac{\frac{1}{2}}{2} = \dfrac{3}{n}$

[6.4] **10.** **Geometry** A 6-ft-tall man casts a 10-ft shadow. Find the height of a building casting a 220-ft shadow.

11. **Business and Finance** Cans of tomato juice are marked 2 for $1.05. At this price, what will 12 cans cost?

12. **Business and Finance** The ratio of compact cars to larger model cars sold during a month was 9 to 4. If 72 compact cars were sold during that period, how many larger cars were sold?

13. **Business and Finance** If 2 gal of paint will cover 450 ft², how many gallons will be needed to paint a room with 2,475 ft² of wall surface?

[6.4] **14.** **Geometry** Use a proportion to find the unknown side, labeled x, given the following pair of similar triangles.

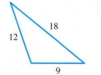

6.1 Ratios

6.1 OBJECTIVES

1. Write the ratio of two numbers in simplest form
2. Write the ratio of two quantities in simplest form

In Chapter 3, you saw two meanings for a fraction:

1. A fraction can name a certain number of parts of a whole. For example, $\frac{3}{5}$ names 3 parts of a whole that has been divided into 5 equal parts.

2. A fraction can indicate division. The same $\frac{3}{5}$ can be thought of as $3 \div 5$.

We now want to turn to a third meaning for a fraction:

3. A fraction can be a ratio. A **ratio** is a means of comparing two numbers or quantities that have the same units.

OBJECTIVE 1

Example 1 Writing a Ratio as a Fraction

NOTE Another way of writing the ratio of 3 to 5 is 3:5. We have chosen to use fraction notation for ratios in this textbook.

Write the ratio 3 to 5 as a fraction.

To compare 3 to 5, we write the ratio of 3 to 5 as $\frac{3}{5}$. So $\frac{3}{5}$ also means "the ratio of 3 to 5."

 CHECK YOURSELF 1

Write the ratio of 7 to 12 as a fraction.

Example 2 illustrates the use of a ratio in comparing *like quantities,* which means we're comparing inches to inches, cm to cm, apples to apples, etc.

OBJECTIVE 2

Example 2 Applying the Concept of Ratio

The width of a rectangle is 7 cm and its length is 19 cm. Write the ratio of its width to its length as a fraction.

$$\frac{7 \text{ cm}}{19 \text{ cm}} = \frac{7}{19}$$ We are comparing centimeters to centimeters, so the units "cancel."

NOTE A ratio fraction can be greater than 1.

NOTE In this case, the ratio is *never* written as a mixed number. It is left as an improper fraction.

The ratio of its length to its width is

$$\frac{19 \text{ cm}}{7 \text{ cm}} = \frac{19}{7}$$

 CHECK YOURSELF 2

A basketball team wins 17 of its 29 games in a season.

(a) Write the ratio of wins to games played.   **(b)** Write the ratio of wins to losses.

Because a ratio is a fraction, we can reduce it to simplest form. Consider Example 3.

NOTE When simplifying a fraction, you are actually multiplying by one.

$$\frac{20 \div 10}{30 \div 10} \text{ is the same as}$$

$$\frac{20}{30} \cdot \frac{\overset{1}{\cancel{10}}}{\underset{1}{\cancel{10}}}$$

This is another application of the fundamental rule of fractions.

Example 3 Writing a Ratio in Simplest Form

Write the ratio of 20 to 30 in lowest terms.

$$\frac{20}{30} = \frac{2}{3}$$ Divide the numerator and denominator by the common factor of 10.

 CHECK YOURSELF 3 _____

Write the ratio of 24 to 32 in lowest terms.

Example 4 is an application using ratios.

Example 4 Simplifying the Ratio of Two Dimensions

A common size for a movie screen is 32 ft by 18 ft. Write this as a ratio in simplest form.

$$\frac{32 \text{ ft}}{18 \text{ ft}} = \frac{32}{18} = \frac{16}{9}$$

 CHECK YOURSELF 4 _____

A common computer display mode is 640 pixels (picture elements) by 480 pixels. Write this as a ratio in simplest form.

Some ratios include fractions or decimals, as in Examples 5 and 6.

Example 5 Simplifying a Ratio Involving a Fraction

Loren sank a $22\frac{1}{2}$-ft putt, and Carrie sank a 30-ft putt. Express the ratio of the two distances as a ratio of whole numbers.

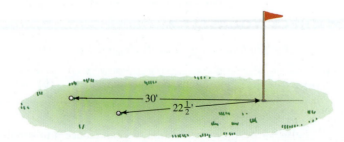

© 2010 McGraw-Hill Companies

$$\frac{22\frac{1}{2}}{30} = \frac{\frac{45}{2}}{30} = \frac{\frac{45}{2}}{\frac{30}{1}}$$

Because we are dividing a fraction by a fraction, we invert and multiply.

$$\frac{45}{2} \div \frac{30}{1} = \frac{45}{2} \cdot \frac{1}{30} = \frac{3}{4}$$

The ratio $22\frac{1}{2}$ to 30 is equivalent to the ratio 3 to 4.

 CHECK YOURSELF 5

Rita jogged $3\frac{1}{2}$ mi this morning, and Yi jogged $4\frac{1}{4}$ mi. Express the ratio of the two distances as a ratio of whole numbers.

Example 6 simplifies a ratio involving decimals.

Example 6   Simplifying a Ratio Involving Decimals

The diameter of a 20-oz bottle is 2.8 in., and the diameter of a 2-L bottle is 5.25 in. Express the ratio of the two diameters as a ratio of whole numbers.

$$\frac{2.8}{5.25} = \frac{2.8 \cdot 100}{5.25 \cdot 100} = \frac{280}{525}$$

$$\frac{280}{525} = \frac{8}{15}$$

The ratio of the diameters 2.8 to 5.25 is equivalent to the ratio 8 to 15.

 CHECK YOURSELF 6

The width of a standard newspaper column is 2.625 in., and the length of a standard column is 19.5 in. Express the ratio of the two measurements as a ratio of whole numbers.

In Example 7, we see that sometimes, to find a ratio, we must rewrite two denominate numbers so that they have the same units.

Example 7   Rewriting Denominate Numbers to Find a Ratio

Joe took 2 full hours to complete his final, but Jaymie finished hers in 75 min. Find the ratio of the two times.

To find a ratio, both numbers must have the same units. If we first convert the 2 h to 120 min, both units are minutes.

NOTE

$$\frac{2\ \cancel{h}}{1} \cdot \frac{60\ \text{min}}{1\ \cancel{h}} = \frac{120\ \text{min}}{1}$$

$$\frac{2\ \text{h}}{75\ \text{min}} = \frac{120\ \text{min}}{75\ \text{min}} = \frac{8}{5}$$

CHECK YOURSELF 7

Find the ratio of whole numbers that is equivalent to the ratio of 16 ft to 10 yd.

READING YOUR TEXT

The following fill-in-the-blank exercises are designed to assure that you understand the key vocabulary used in this section. Each sentence usually comes directly from the section. You will find the correct answers in Appendix C.

Section 6.1

(a) A _____ is a means of comparing two numbers or quantities.

(b) When simplifying a ratio of two like quantities, the ratio is left as an _____ fraction.

(c) When simplifying a fraction, you are actually multiplying by _____.

(d) Sometimes, to find a ratio, we must rewrite two denominate numbers so that they have the same _____.

CHECK YOURSELF ANSWERS

1. $\dfrac{7}{12}$ **2. (a)** $\dfrac{17}{29}$; **(b)** $\dfrac{17}{12}$ (the team lost 12 games)   **3.** $\dfrac{3}{4}$ **4.** $\dfrac{4}{3}$ **5.** $\dfrac{14}{17}$

6. $\dfrac{7}{52}$ **7.** $\dfrac{8}{15}$

6.1 Exercises

Name _____

Section _____ Date _____

ANSWERS

Write each of the ratios in simplest form.

1. The ratio of 9 to 13

2. The ratio of 5 to 4

3. The ratio of 9 to 4

4. The ratio of 5 to 12

5. The ratio of 10 to 15

6. The ratio of 16 to 12

7. The ratio of $3\frac{1}{2}$ to 14

8. The ratio of $5\frac{3}{5}$ to $2\frac{1}{10}$

9. The ratio of 10.5 to 2.7

10. The ratio of 2.2 to 0.6

11. The ratio of 12 mi to 18 mi

12. The ratio of 100 cm to 90 cm

13. The ratio of 40 ft to 65 ft

14. The ratio of 12 oz to 18 oz

15. The ratio of $48 to $42

16. The ratio of 20 ft to 24 ft

17. The ratio of 75 s to 3 min

18. The ratio of 7 oz to 3 lb

19. The ratio of 4 nickels to 5 dimes

20. The ratio of 8 in. to 3 ft

21. The ratio of 2 days to 10 h

22. The ratio of 4 ft to 4 yd

23. The ratio of 5 gal to 12 quarts (qt)

24. The ratio of 7 dimes to 3 quarters

1. _____
2. _____
3. _____
4. _____
5. _____
6. _____
7. _____
8. _____
9. _____
10. _____
11. _____ 12. _____
13. _____ 14. _____
15. _____ 16. _____
17. _____ 18. _____
19. _____ 20. _____
21. _____ 22. _____
23. _____ 24. _____

Solve the applications.

25. **Social Science**   An algebra class has 7 men and 13 women. Write the ratio of men to women. Write the ratio of women to men.   **ALEKS**

26. **Statistics**   A football team wins 9 of its 16 games with no ties. Write the ratio of wins to games played. Write the ratio of wins to losses.   **ALEKS**

27. **Social Science**   In a school election 4,500 yes votes were cast, and 3,000 no votes were cast. Write the ratio of yes to no votes.   **VIDEO**

28. **Statistics**   A basketball player made 42 of the 70 shots taken in a tournament. Write the ratio of shots made to shots taken.

29. **Number Problem**   Carla walked $2\frac{1}{4}$ mi this afternoon, and Mario walked $5\frac{3}{8}$ mi this afternoon. Express the ratio of the two distances as a ratio of whole numbers.

30. **Technology**   One car has an $11\frac{1}{2}$-gal tank and another has a $17\frac{3}{4}$-gal tank. Express the ratio of the capacities as a ratio of whole numbers.

31. **Technology**   One refrigerator holds $2\frac{2}{3}$ ft^3 of food and another holds $5\frac{3}{4}$ ft^3 of food. Express the ratio of the capacities as a ratio of whole numbers.   **VIDEO**

32. **Business and Finance**   The price of an antibiotic in one drugstore is \$12.50. The price of the same antibiotic in another drugstore is \$8.75. Write the ratio of the prices as a ratio of whole numbers.

33. **Geometry**   The width of a notebook is 3.5 in., and the length is 6.75 in. Write the ratio of length to width as a ratio of whole numbers.

34. Statistics Marc took 3 h to mow a lawn. Angelina took 150 min to mow the same lawn a week earlier. Write the ratio of Marc's time to Angelina's time as a ratio of whole numbers.

35. Business and Finance A company employs 24 women and 18 men. Write the ratio of men to women employed by the company.

36. Geometry If a room is 30 ft long and 6 yd wide, write the ratio of the length to the width of the room.

37. (a) Buy a 1.69-oz (medium size) bag of M&Ms. Determine the ratio of the number of M&Ms of each color (yellow, red, blue, orange, brown, and green) to the total number of M&Ms in the bag.

(b) Compare your ratios to those your classmates obtain.

(c) Use the information from parts **(a)** and **(b)** to determine a ratio for all the different colors in a bag of M&Ms.

(d) E-mail the manufacturer of M&Ms (Mars, Inc.) at www.m™-ms.com and see if they use fixed ratios to determine the distribution of the colors in a bag. If they do, compare these ratios to yours.

34. _____

35. _____

36. _____

37. _____

38. Geometry Two pencils are shown. Write the ratio of the length of the smaller pencil to the larger pencil.

Answers

1. $\dfrac{9}{13}$ **3.** $\dfrac{9}{4}$ **5.** $\dfrac{2}{3}$ **7.** $\dfrac{1}{4}$ **9.** $\dfrac{35}{9}$ **11.** $\dfrac{2}{3}$ **13.** $\dfrac{8}{13}$ **15.** $\dfrac{8}{7}$

17. $\dfrac{5}{12}$ **19.** $\dfrac{2}{5}$ **21.** $\dfrac{24}{5}$ **23.** $\dfrac{5}{3}$ **25.** $\dfrac{7}{13}, \dfrac{13}{7}$ **27.** $\dfrac{3}{2}$ **29.** $\dfrac{18}{43}$

31. $\dfrac{32}{69}$ **33.** $\dfrac{27}{14}$ **35.** $\dfrac{3}{4}$ **37.**

6.2 Rates

6.2 OBJECTIVES

1. Find a unit rate
2. Find a unit price

A ratio compares two denominate numbers. Examples of a ratio include

$$\frac{9 \text{ s}}{12 \text{ s}} = \frac{9}{12} = \frac{3}{4}$$

or

$$\frac{5 \text{ mi}}{7 \text{ mi}} = \frac{5}{7}$$

In these examples, the units of the numerator and denominator are identical. When we compare denominate numbers with different units, we get a **rate.**

OBJECTIVE 1

| Example 1   Finding a Rate

Find each rate.

(a) $\dfrac{12 \text{ ft}}{16 \text{ s}} = \dfrac{12 \text{ ft}}{16 \text{ s}} = \dfrac{3 \text{ ft}}{4 \text{ s}}$

(b) $\dfrac{200 \text{ mi}}{10 \text{ gal}} = \dfrac{200}{10} \dfrac{\text{mi}}{\text{gal}} = 20 \dfrac{\text{mi}}{\text{gal}}$

(c) $\dfrac{10 \text{ gal}}{200 \text{ mi}} = \dfrac{10}{200} \dfrac{\text{gal}}{\text{mi}} = \dfrac{1}{20} \dfrac{\text{gal}}{\text{mi}}$

✔ **CHECK YOURSELF 1**

Find each rate.

(a) $\dfrac{250 \text{ mi}}{10 \text{ h}}$ (b) $\dfrac{\$60{,}000}{2 \text{ yr}}$ (c) $\dfrac{2 \text{ yr}}{\$60{,}000}$

In each part of Example 1, notice that we start with a comparison of two denominate numbers such as 12 ft and 16 s. When the rate is formed, $\dfrac{12 \text{ ft}}{16 \text{ s}}$, we may read this as "12 feet per 16 seconds." Our next move is to separate the numerical portion of the rate from the units portion, writing $\dfrac{12}{16}\dfrac{\text{ft}}{\text{s}}$. This may be read as "$\dfrac{12}{16}$ of a foot per second." Reducing the fraction, we arrive at $\dfrac{3}{4}\dfrac{\text{ft}}{\text{s}}$, reading this as "$\dfrac{3}{4}$ of a foot per second."

When the numerical portion of the rate is separated from the units portion, we call the resulting rate a **unit rate.** A unit rate can be read as "a certain number of numerator units per *one* denominator unit." In part (b) above, we have "20 miles per 1 gallon," and in part (c) above we have "$\dfrac{1}{20}$ gallon per 1 mile."

When creating a unit rate, we often express the numerical portion as a decimal, as in Example 2.

Example 2   Finding a Unit Rate

Randy Johnson had 320 strikeouts in 280 innings. What was his strikeout per inning rate?

RECALL In Section 6.1, we stated that mixed numbers were inappropriate for ratios. When we write a unit rate, a mixed number, usually rewritten as a decimal, is preferred.

$$\frac{320 \text{ strikeouts}}{280 \text{ innings}} = \frac{320}{280}\frac{\text{strikeouts}}{\text{inning}}$$

$$= 1\frac{1}{7}\frac{\text{strikeouts}}{\text{inning}}$$

To rewrite this as a decimal, we approximate. The approximation is indicated by the symbol $\approx$.

$$1\frac{1}{7}\frac{\text{strikeouts}}{\text{inning}} \approx 1.14\frac{\text{strikeouts}}{\text{inning}}$$

CHECK YOURSELF 2

Chamique Holdsclaw scored 450 points in 20 games. What was her $\dfrac{points}{game}$ rate?

One purpose for computing a unit rate is for comparison. Example 3 shows this.

Example 3   Comparing Unit Rates

Player A scores 50 points in 9 games, and player B scores 260 points in 36 games. Which player scored at a higher rate?

Player A's rate was $\dfrac{50 \text{ points}}{9 \text{ games}} = \dfrac{50}{9} \dfrac{\text{points}}{\text{game}}$

$\approx 5.56 \dfrac{\text{points}}{\text{game}}$

Player B's rate was $\dfrac{260 \text{ points}}{36 \text{ games}} = \dfrac{260}{36} \dfrac{\text{points}}{\text{game}}$

$= \dfrac{65}{9} \dfrac{\text{points}}{\text{game}}$

$\approx 7.22 \dfrac{\text{points}}{\text{game}}$

Player B scored at a higher rate.

  CHECK YOURSELF 3 _____

Hassan scored 25 goals in 8 games, and Lee scored 52 goals in 18 games. Which player scored at a higher rate?

One type of unit rate, seen commonly in supermarkets, is unit price.

> **Definition: Unit Price**
>
> **NOTE** A unit price is a price *per unit.* The unit used may be ounces, pints, pounds, or some other unit.
>
> The **unit price** relates a price to some common unit.

OBJECTIVE 2   **Example 4   Finding a Unit Price**

Find the unit price for each item.

(a) 8 oz of cream cost $1.53

$\dfrac{\$1.53}{8 \text{ oz}} = \dfrac{153 \text{ cents}}{8 \text{ oz}} = \dfrac{153}{8} \dfrac{\text{cents}}{\text{oz}}$

$\approx 19 \dfrac{\text{cents}}{\text{oz}}$

(b) 20 lb of potatoes cost $3.98

POTATOES
20lb $3.98

$$\frac{\$3.98}{20 \text{ lb}} = \frac{398 \text{ cents}}{20 \text{ lb}} = \frac{398}{20} \frac{\text{cents}}{\text{lb}}$$

$$\approx 20 \frac{\text{cents}}{\text{lb}}$$

 CHECK YOURSELF 4

Find the unit price for each item.

(a) 12 soda cans cost $2.98   **(b)** 25 lb of dog food cost $9.99

Example 5   Finding a Unit Rate and a Unit Price

In England, several centuries ago, when nails were made by hand, you could purchase 100 one-inch nails for 2 pennies (or twopence). Because the abbreviation for pennies was *d*, these nails became known as 2d nails. This system for describing nails is still in use today.

(a) How many nails could one get for 1 penny?

Here we are finding a unit rate. We write: $\dfrac{100 \text{ nails}}{2 \text{ pennies}} = \dfrac{100}{2} \dfrac{\text{nails}}{\text{penny}} = 50 \dfrac{\text{nails}}{\text{penny}}$

So, one could get 50 nails for 1 penny.

(b) What would the unit price have been?

We want to find the price per nail.

NOTE A penny in England is different from a penny in the U.S.

$$\frac{2 \text{ pennies}}{100 \text{ nails}} = \frac{1}{50} \frac{\text{pennies}}{\text{nail}} \quad \text{or} \quad 0.02 \text{ pennies per nail}$$

 CHECK YOURSELF 5

6-inch nails are classified as 60d nails. As in example 5, this meant that 100 six-inch nails cost 60 pennies (sixtypence).

(a) How many 60d nails could one get for 1 penny?
(b) What would have been the unit price of 60d nails?

READING YOUR TEXT

The following fill-in-the-blank exercises are designed to assure that you understand the key vocabulary used in this section. Each sentence usually comes directly from the section. You will find the correct answers in Appendix C.

Section 6.2

(a) When we compare denominate numbers with different units, we get a _____.

(b) A _____ rate can be read as "a certain number of numerator units per one denominator unit."

(c) When we write a unit rate, a _____ number, usually rewritten as a decimal, is preferred.

(d) The _____ relates a price to some common unit.

CHECK YOURSELF ANSWERS

1. (a) $25\ \dfrac{\text{mi}}{\text{h}}$; **(b)** $30{,}000\ \dfrac{\text{dollars}}{\text{yr}}$; **(c)** $\dfrac{1}{30{,}000}\ \dfrac{\text{yr}}{\text{dollar}}$ **2.** $22.5\ \dfrac{\text{points}}{\text{game}}$

3. Hassan had a higher rate.   **4. (a)** $\approx 25\ \dfrac{\text{cents}}{\text{can}}$; **(b)** $\approx 40\ \dfrac{\text{cents}}{\text{lb}}$

5. (a) $\dfrac{5}{3}\ \dfrac{\text{nails}}{\text{penny}} = 1\dfrac{2}{3}\ \dfrac{\text{nails}}{\text{penny}}$, which means that for 1 penny, one could only get 1 nail;

(b) $\dfrac{3}{5}\ \dfrac{\text{pennies}}{\text{nail}}$ or 0.6 pennies per nail

Name _____

Section _____ Date _____

ANSWERS

1. _____
2. _____
3. _____
4. _____
5. _____
6. _____
7. _____
8. _____
9. _____
10. _____
11. _____
12. _____
13. _____
14. _____
15. _____
16. _____
17. _____
18. _____

6.2 **Exercises**

Find each unit rate.

1. $\dfrac{300 \text{ mi}}{4 \text{ h}}$

2. $\dfrac{95 \text{ cents}}{5 \text{ pencils}}$

3. $\dfrac{69 \text{ ft}}{3 \text{ s}}$

4. $\dfrac{3 \text{ s}}{69 \text{ ft}}$

5. $\dfrac{\$10,000}{5 \text{ yr}}$

6. $\dfrac{5 \text{ yr}}{\$10,000}$

7. $\dfrac{680 \text{ ft}}{17 \text{ s}}$

8. $\dfrac{480 \text{ mi}}{15 \text{ gal}}$

9. $\dfrac{15 \text{ gal}}{480 \text{ mi}}$

10. $\dfrac{7,200 \text{ revolutions}}{16 \text{ mi}}$

11. $\dfrac{57 \text{ oz}}{3 \text{ cans}}$

12. $\dfrac{\$2,000,000}{4 \text{ yr}}$

13. $\dfrac{150 \text{ calories}}{3 \text{ oz}}$

14. $\dfrac{240 \text{ lb of fertilizer}}{6 \text{ lawns}}$

15. $\dfrac{192 \text{ diapers}}{32 \text{ babies}}$

16. $\dfrac{657,200 \text{ library books}}{5,200 \text{ students}}$

Solve the applications.

17. **Technology** Trac drives 256 mi using 8 gal of gasoline. How many miles per gallon does his car get? **ALEKS**

18. **Technology** Seven pounds of fertilizer cover fourteen hundred square feet. How many square feet are covered by one pound of fertilizer? **ALEKS**

19. **Statistics** A local college has 6,000 registered vehicles for 2,400 campus parking spaces. How many vehicles are there for each parking space? **VIDEO** **ALEKS**

19. _____

20. _____

21. _____

22. _____

23. _____

24. _____

25. _____

26. _____

20. **Statistics** Curt Schilling has 141 strikeouts in 163 innings. What was his strike-out per inning rate? **ALEKS**

21. **Construction** A water pump can produce 280 gal in 24 h. How many gallons per hour is this? **ALEKS**

22. **Business and Finance** If 214 shares of stock cost $5,992, what was the cost per share? **ALEKS**

23. **Technology** A printer produces 4 pages in 6 s. How many pages are produced per second? **VIDEO** **ALEKS**

24. **Statistics** Augie eats 12 hamburgers in 48 min. How many minutes does it take Augie to eat one hamburger? **ALEKS**

25. **Business and Finance** A 12-oz can of tuna costs $4.80. What is the cost of tuna per ounce? **ALEKS**

26. **Business and Finance** The fabric for a dress costs $76.45 for 9 yd. What is the cost per yard? **ALEKS**

27. Business and Finance   Gerry laid 634 bricks in 35 min, and his friend Matt laid 515 bricks in 27 min. Who is the faster bricklayer?

28. Technology   Mike drove 135 mi in 2.5 h. Sam drove 91 mi in 1.75 h. Who drives at the faster speed?

29. Statistics   Luis Gonzalez has 137 hits in 387 at bats. Larry Walker has 119 hits in 324 at bats. Who has the higher batting average?

30. Business and Finance   What is the better buy—5 lb of sugar for $4.75 or 20 lb of sugar for $19.92?

Find the unit price for each item.

31. $57.50 for 5 shirts   **ALEKS**   **32.** $104.93 for 7 CDs   **ALEKS**

33. $5.16 for a dozen oranges **ALEKS**   **34.** $10.44 for 18 bottles of water
                                                            ALEKS

Find the best buy in each of the exercises.

35. Dishwashing liquid: **36.** Canned corn:

    (a) 12 oz for 79¢   **(a)** 10 oz for 21¢

    (b) 22 oz for $1.29   **(b)** 17 oz for 39¢

37. Syrup:

(a) 12 oz for 99¢

(b) 24 oz for $1.59

(c) 36 oz for $2.19

38. Shampoo:

(a) 4 oz for $1.16

(b) 7 oz for $1.52

(c) 15 oz for $3.39

37. _____

38. _____

39. _____

40. _____

41. _____

42. _____

39. Salad oil (1 qt is 32 oz):

(a) 18 oz for 89¢

(b) 1 qt for $1.39

(c) 1 qt 16 oz for $2.19

40. Tomato juice [1 pint (pt) is 16 oz]:

(a) 8 oz for 37¢

(b) 1 pt 10 oz for $1.19

(c) 1 qt 14 oz for $1.99

41. Peanut butter (1 lb is 16 oz):

(a) 12 oz for $1.25

(b) 18 oz for $1.72

(c) 1 lb 12 oz for $2.59

(d) 2 lb 8 oz for $3.76

42. Laundry detergent:

(a) 1 lb 2 oz for $1.99

(b) 1 lb 12 oz for $2.89

(c) 2 lb 8 oz for $4.19

(d) 5 lb for $7.99

43. Construction 6d nails were so called because 100 of them cost 6 pennies (see Example 5). Find the unit price of 6d nails.

44. Construction 10d nails cost 10 pennies per 100. Find the unit price of 10d nails.

45. Construction Today, bulk nails are generally sold by weight.

    (a) Go to your local hardware store and find the weight of 100 standard 3-inch nails (these historically were called 10d nails).

    (b) Determine the cost of the nails.

    (c) Find the unit cost of the nails.

    (d) Using the historical price (10 pennies per 100 nails) as a benchmark, write a brief paragraph discussing how prices have changed since the nails cost 10 pennies. Use another product as an example as well.

    (e) Compare the British penny and the U.S. penny. How do your findings affect what you wrote for part (d)?

Answers

1. $75 \dfrac{\text{mi}}{\text{h}}$ **3.** $23 \dfrac{\text{ft}}{\text{s}}$ **5.** $2{,}000 \dfrac{\text{dollars}}{\text{yr}}$ **7.** $40 \dfrac{\text{ft}}{\text{s}}$ **9.** $\approx 0.03 \dfrac{\text{gal}}{\text{mi}}$

11. $19 \dfrac{\text{oz}}{\text{can}}$ **13.** $50 \dfrac{\text{calories}}{\text{oz}}$ **15.** $6 \dfrac{\text{diapers}}{\text{baby}}$ **17.** $32 \dfrac{\text{mi}}{\text{gal}}$

19. 2.5 vehicles per space   **21.** $\approx 11.67 \dfrac{\text{gal}}{\text{h}}$ **23.** $\dfrac{2}{3} \dfrac{\text{page}}{\text{s}}$ **25.** $40 \dfrac{\text{cents}}{\text{oz}}$

27. Matt   **29.** Larry Walker   **31.** $\dfrac{\$11.50}{\text{shirt}}$ **33.** $\dfrac{\$0.43}{\text{orange}}$ **35.** b

37. c   **39.** b   **41.** c   **43.** $\dfrac{3}{50} \dfrac{\text{pennies}}{\text{nail}}$ or 0.06 pennies per nail

45.

6.3 Proportions

6.3 OBJECTIVES

1. Write a proportion
2. Determine whether two rates are proportional
3. Solve a proportion for an unknown value
4. Solve an application involving a proportion
5. Solve an application involving unit pricing

> **Definition:** Proportion
>
> A **proportion** is an equation that compares two equal fractions (or rates).

NOTE This is the same as saying the fractions are equivalent. They name the same number.

RECALL A letter representing an unknown value is called a **variable**. Here a, b, c, and d are variables. We could have chosen any letters.

Because the ratio of 1 to 3 is equal to the ratio of 2 to 6, we can write the proportion

$$\frac{1}{3} = \frac{2}{6}$$

The proportion $\dfrac{a}{b} = \dfrac{c}{d}$ is read "a is to b as c is to d." We read the proportion $\dfrac{1}{3} = \dfrac{2}{6}$ as "one is to three as two is to six."

OBJECTIVE 1

Example 1 Writing a Proportion

Write the proportion 3 is to 7 as 9 is to 21.

$$\frac{3}{7} = \frac{9}{21}$$

CHECK YOURSELF 1 _____

Write the proportion 4 is to 12 as 6 is to 18.

When you write a proportion for two rates, placement is important.

Example 2 Writing a Proportion with Two Rates

Write a proportion that is equivalent to the statement: If it takes 3 h to mow 4 acres of grass, it will take 6 h to mow 8 acres.

$$\frac{3 \text{ h}}{4 \text{ acres}} = \frac{6 \text{ h}}{8 \text{ acres}}$$

Note that, in both fractions, the hours units are in the numerator and the acres units are in the denominator.

CHECK YOURSELF 2 _____

Write a proportion that is equivalent to the statement: If it takes 5 rolls of wallpaper to cover 400 ft², it will take 7 rolls to cover 560 ft².

If two fractions are equal, they form a true proportion.

> **Property:** **The Proportion Rule**
>
> If $\dfrac{a}{b} = \dfrac{c}{d}$, then $a \cdot d = b \cdot c$.
>
> We say that the fractions $\dfrac{a}{b}$ and $\dfrac{c}{d}$ are equivalent. The fractions form a true proportion.

Example 3 Determining Whether Two Fractions Are Equivalent

Determine whether each pair of fractions is equivalent.

(a) $\dfrac{5}{6} \overset{?}{=} \dfrac{10}{12}$

NOTE This process is sometimes called **cross multiplication.** We have

$$\frac{a}{b} \overset{?}{\underset{\nearrow}{\searrow}} \frac{c}{d}$$

$$a \cdot d \overset{?}{=} b \cdot c$$

Multiply:

$$\left. \begin{array}{l} 6 \cdot 10 = 60 \\ 5 \cdot 12 = 60 \end{array} \right) \text{Equal}$$

Because $b \cdot c = a \cdot d$, $\dfrac{5}{6} = \dfrac{10}{12}$ is a proportion. It is a *true proportion* or a *true equation.*

(b) $\dfrac{3}{7} \overset{?}{=} \dfrac{4}{9}$

Multiply:

$$\left. \begin{array}{l} 7 \cdot 4 = 28 \\ 3 \cdot 9 = 27 \end{array} \right) \text{Not equal}$$

The products are not equal, so $\dfrac{3}{7} = \dfrac{4}{9}$ is not a proportion. It is a *false proportion.*

✔ CHECK YOURSELF 3

Determine whether each pair of fractions is equivalent.

(a) $\dfrac{5}{8} \overset{?}{=} \dfrac{20}{32}$

(b) $\dfrac{7}{9} \overset{?}{=} \dfrac{3}{4}$

If two fractions are equivalent, we say that they are **proportional.**

OBJECTIVE 2 ### Example 4 Verifying a Proportion

Determine whether each pair of fractions is proportional.

(a) $\dfrac{3}{1} \overset{?}{=} \dfrac{30}{5}$
$$\dfrac{3}{\frac{1}{2}}$$

$$\frac{1}{2} \cdot 30 = 15$$

$$3 \cdot 5 = 15$$

Because the products are equal, the fractions are proportional.

(b) $\dfrac{0.4}{20} \overset{?}{=} \dfrac{3}{100}$

$$20 \cdot 3 = 60$$

$$0.4 \cdot 100 = 40$$

Because the products are *not* equal, the fractions are not proportional.

 CHECK YOURSELF 4 _____

Determine whether each pair of fractions is proportional.

(a) $\dfrac{0.5}{8} \overset{?}{=} \dfrac{3}{48}$

(b) $\dfrac{\frac{1}{4}}{6} \overset{?}{=} \dfrac{3}{80}$

The proportion rule can also be used to verify that rates are proportional.

Example 5   Verifying a Proportion Involving Rates

Is the rate $\dfrac{5 \text{ U.S. dollars}}{2{,}500 \text{ colones}}$ equivalent to the rate $\dfrac{27 \text{ U.S. dollars}}{13{,}500 \text{ colones}}$?

US Dollars	Colones
$1.00	500
$0.002	1.0

We want to know if the equation

$$\frac{5}{2{,}500} \overset{?}{=} \frac{27}{13{,}500}$$

is true.

$$5 \cdot 13{,}500 = 67{,}500$$

$$27 \cdot 2{,}500 = 67{,}500$$

The rates are equivalent.

CHECK YOURSELF 5

Is the rate $\dfrac{50 \text{ pages}}{45 \text{ min}}$ *equivalent to the rate* $\dfrac{30 \text{ pages}}{25 \text{ min}}$?

We can use what we have learned so far about proportions and equations in order to *solve a proportion.* A proportion consists of four values. If three of the four values of a proportion are known, you can always find the missing or unknown value.

In the proportion $\dfrac{a}{3} = \dfrac{10}{15}$, the first value is unknown. We have chosen to represent the unknown value with the letter a. Using the proportion rule, we can proceed.

NOTE $\dfrac{?}{3} = \dfrac{10}{15}$ is a proportion in which the first value is unknown. In this section, we learn how to find that unknown value.

$$\frac{a}{3} = \frac{10}{15}$$

$$15 \cdot a = 3 \cdot 10 \qquad \text{or} \qquad 15 \cdot a = 30$$

The equal sign tells us that $15 \cdot a$ and 30 are just different names for the same number. In Section 2.9, we learned that this type of statement is called an **equation.**

Recall that we can divide both sides of an equation by the same nonzero number.

$$15 \cdot a = 30 \qquad \text{Divide both sides by 15.}$$

$$\frac{15 \cdot a}{15} = \frac{30}{15}$$

$$\frac{\overset{1}{\cancel{15}} \cdot a}{\underset{1}{\cancel{15}}} = \frac{\overset{2}{\cancel{30}}}{\underset{1}{\cancel{15}}} \qquad \text{Divide by the coefficient of the variable. Do you see why we divided by 15? It leaves our unknown } a \text{ by itself in the left term.}$$

$$a = 2$$

You should always check your result. It is easy in this case. We found a value of 2 for a. Replace the unknown a with that value. Then verify that the fractions are proportional. We started with $\dfrac{a}{3} = \dfrac{10}{15}$ and found a value of 2 for a. So we write

NOTE Replace a with 2 and multiply.

$$\frac{(2)}{3} \overset{?}{=} \frac{10}{15}$$

$$3 \cdot 10 \overset{?}{=} 2 \cdot 15$$

$$30 = 30$$

The value of 2 for a is correct.

The procedure for solving a proportion is summarized here.

NOTE This gives us the unknown value. Now check the result.

> **Step by Step: To Solve a Proportion**
>
> **Step 1** Use the proportion rule to write the equivalent equation $a \cdot d = b \cdot c$.
> **Step 2** Divide both terms of the equation by the coefficient of the variable.
> **Step 3** Use the value found to replace the unknown in the original proportion. Check that the ratios or the rates are proportional.

OBJECTIVE 3   **Example 6   Solving Proportions for Unknown Values**

Find the unknown value.

$$\frac{8}{x} = \frac{6}{9}$$

Step 1   Using the proportion rule, we have

$$6 \cdot x = 8 \cdot 9$$

or $6x = 72$

Step 2   Note that the coefficient of the variable is 6 and so divide both sides of the equation by that coefficient.

$$\frac{\overset{1}{\cancel{6}}x}{\underset{1}{\cancel{6}}} = \frac{\overset{12}{\cancel{72}}}{\underset{1}{\cancel{6}}}$$

$$x = 12$$

Step 3   To check, replace x with 12 in the original proportion.

$$\frac{8}{(12)} \overset{?}{=} \frac{6}{9}$$

Multiply:

$$12 \cdot 6 \overset{?}{=} 8 \cdot 9$$

$$72 = 72 \qquad \text{The value of 12 for } x \text{ checks.}$$

 CHECK YOURSELF 6 ⎯⎯⎯⎯⎯⎯⎯⎯⎯⎯⎯⎯⎯

Solve the proportions for n. Check your result.

(a) $\dfrac{4}{5} = \dfrac{n}{25}$ **(b)** $\dfrac{7}{9} = \dfrac{42}{n}$

In solving for a missing term in a proportion, we may find an equation involving fractions or decimals. Example 7 involves finding the unknown value in such cases.

Example 7   Solving Proportions for Unknown Values

(a) Solve the proportion for x.

$$\frac{\frac{1}{4}}{3} = \frac{4}{x}$$

$$\frac{1}{4}x = 12$$

$$4 \cdot \frac{1}{4}x = 4 \cdot 12$$

$$x = \boxed{48}$$

To check, replace x with 48 in the original proportion.

$$\frac{\frac{1}{4}}{3} \overset{?}{=} \frac{4}{(48)}$$

$$3 \cdot 4 \overset{?}{=} \frac{1}{4} \cdot 48$$

$$12 = 12$$

(b) Solve the proportion for a.

NOTE Here we must divide 6 by 0.5 to find the unknown value. The steps of that division are shown here for review.

$$\frac{0.5}{2} = \frac{3}{a}$$

$$0.5a = 6$$

$$\frac{0.5a}{0.5} = \frac{6}{0.5} \qquad \text{Divide by the coefficient, 0.5.}$$

$$a = 12$$

We will leave it to you to confirm that $0.5 \cdot 12 = 2 \cdot 3$.

CHECK YOURSELF 7 _____

(a) Solve for a.

$$\frac{\frac{1}{2}}{5} = \frac{3}{a}$$

(b) Solve for x.

$$\frac{0.4}{x} = \frac{2}{30}$$

Now that you have learned how to find an unknown value in a proportion, this can be used in solving other types of applications.

Step by Step: Solving Applications of Proportions

Step 1 Read the problem carefully to determine the given information.

Step 2 Write the proportion necessary to solve the problem. Use a letter to represent the unknown quantity. Be sure to include the units in writing the proportion.

Step 3 Solve, answer the question of the original problem, and check the proportion as before.

OBJECTIVE 4   **Example 8   Using a Proportion to Find an Unknown Value**

In a shipment of 400 parts, 14 are found to be defective. How many defective parts should be expected in a shipment of 1,000 parts?

Assume that the ratio of defective parts to the total number remains the same.

$$\frac{14 \text{ defective}}{400 \text{ total}} = \frac{x \text{ defective}}{1{,}000 \text{ total}} \qquad \text{We have decided to let } x \text{ be the unknown number of defective parts.}$$

Multiply:

$400x = 14,000$

Divide by the coefficient, 400.

$x = 35$

35 defective parts should be expected in the shipment.
    Checking the original proportion, we get

$14 \cdot 1,000 \overset{?}{=} 400 \cdot 35$
$14,000 = 14,000$

CHECK YOURSELF 8

An investment of $3,000 earned $330 for 1 year. How much will an investment of $10,000 earn at the same rate for 1 year?

Now we look at an application involving fractions in the proportion.

Example 9   Using Proportions to Find an Unknown Value

The scale on a map is given as $\frac{1}{4}$ in. = 3 mi. The distance between two towns is 4 in. on the
map. How far apart are the towns in miles?

NOTE We could divide both
sides by $\frac{1}{4}$:

$\dfrac{\frac{1}{4} \cdot x}{\frac{1}{4}} = \dfrac{3 \cdot 4}{\frac{1}{4}}$

$x = \dfrac{3 \cdot 4}{\frac{1}{4}}$

$x = \dfrac{12}{\frac{1}{4}}$

and then invert and multiply

$x = \dfrac{12}{1} \cdot \dfrac{4}{1}$

$= 48$

For this solution we use the fact that the ratio of inches (on the map) to miles remains
the same.

$\dfrac{\frac{1}{4}\ \text{in.}}{3\ \text{mi}} = \dfrac{4\ \text{in.}}{x\ \text{mi}}$

$\dfrac{1}{4} \cdot x = 3 \cdot 4 = 12$

$4\left(\dfrac{1}{4}\right) \cdot x = 4 \cdot 12$

$1 \cdot x = 4 \cdot 12$

$x = 48\ \text{(mi)}$

CHECK YOURSELF 9 _____

Jack drives 125 mi in 2$\frac{1}{2}$ h. At the same rate, how far will he be able to travel in

4 h? (Hint: Write 2$\frac{1}{2}$ as an improper fraction.)

We may also find decimals in the solution of an application.

Example 10   Using Proportions to Find an Unknown Value

Jill works 4.2 h and receives $21. How much will she get if she works 10 h?
   The ratio of hours worked to the amount of pay remains the same.

$$\frac{4.2 \text{ h}}{\$21} = \frac{10 \text{ h}}{\$a} \qquad \text{Let } a \text{ be the unknown amount of pay.}$$

$$4.2a = 210$$

$$\frac{4.2a}{4.2} = \frac{210}{4.2} \qquad \text{Divide both sides by 4.2.}$$

$$a = \$50$$

CHECK YOURSELF 10 _____

A piece of cable 8.5 cm long weighs 68 g. What will a 10-cm length of the same cable weigh?

In Example 11, we must convert the units stated in the problem.

Example 11   Using Proportions to Find an Unknown Value

A machine produces 15 cans in 2 min. At this rate, how many cans can it make in an 8-h period?
   In writing a proportion for this problem, we must write the times involved in terms of the same units.

$$\frac{15 \text{ cans}}{2 \text{ min}} = \frac{x \text{ cans}}{480 \text{ min}} \qquad \begin{array}{l}\text{Because 1 h is 60 min, convert 8 h} \\ \text{to 480 min.}\end{array}$$

$$2x = 15 \cdot 480$$

$$\text{or} \quad 2x = 7{,}200$$

$$x = 3{,}600 \text{ cans}$$

CHECK YOURSELF 11 _____

Instructions on a can of film developer call for 2 oz of concentrate to 1 qt of water. How much of the concentrate is needed to mix with 1 gal of water? (4 qt = 1 gal)

In practical applications, you may have to round the result after using your calculator in the solution of a proportion. Example 12 shows such a situation.

Example 12   Using Proportions to Find an Unknown Value

Micki drives 278 mi on 13.6 gal of gas. If the gas tank of her car holds 21 gal, how far can she travel on a full tank of gas?

We can write the proportion

$$\frac{278 \text{ mi}}{13.6 \text{ gal}} = \frac{x \text{ mi}}{21 \text{ gal}}$$

Multiply.

$$13.6\, x = 278 \cdot 21$$

Now divide both terms by 13.6.

$$\frac{\overset{1}{\cancel{13.6}}x}{\underset{1}{\cancel{13.6}}} = \frac{278 \cdot 21}{13.6}$$

Now to find x, we must multiply 278 by 21 and then divide by 13.6. On the calculator,

$$278 \;\boxed{\times}\; 21 \;\boxed{\div}\; 13.6 \;\boxed{=}\; 429.26471$$

We round the result to the nearest mile; Micki can drive about 429 mi on a full tank of gas.

CHECK YOURSELF 12

Life insurance costs $4.37 for each $1,000 of insurance. How much does a $25,000 policy cost?

Your calculator can be very handy for comparing prices at the grocery store. As we said in Section 6.2, to find the unit price, just divide the cost of the item by the number of units.

OBJECTIVE 5   **Example 13   Finding Unit Prices**

A dishwashing liquid comes in three sizes:

(a) 12 oz for 77¢
(b) 22 oz for $1.33
(c) 32 oz for $1.85

Which is the best buy?

For each size, find the unit price in cents per ounce.

(a) For the first size (77¢ for 12 oz), using your calculator, divide.

$$77 \;\boxed{\div}\; 12 \;\boxed{=}\; 6.4166667$$

$$\frac{77¢}{12 \text{ oz}} \approx 6.4¢ \text{ per ounce}$$ We chose to round to the nearest tenth of a cent.

(b) To find the unit price for the second size, divide again.

NOTE We treat $1.33 as 133¢ to find the rate "cents per ounce."

133 $\boxed{\div}$ 22 $\boxed{=}$ 6.0454545

$$\frac{\$1.33}{22 \text{ oz}} \approx 6.0¢ \text{ per ounce}$$ Again, round to the nearest tenth of a cent.

(c) For the third size,

185 $\boxed{\div}$ 32 $\boxed{=}$ 5.78125

$$\frac{\$1.85}{32 \text{ oz}} \approx 5.8¢ \text{ per ounce}$$

Comparing the three unit prices, we see that the 32-oz size of dishwashing liquid, at 5.8¢ per ounce, is the best buy.

CHECK YOURSELF 13

A floor cleaner comes in three sizes:

32 oz for $2.89
48 oz for $3.43
70 oz for $4.96

Which is the best buy?

All rates used must be in terms of the same units. If quantities involve different units, they must be converted.

Example 14 Finding Unit Prices

Vegetable oil is sold in the following quantities:

(a) 16 oz for $1.27
(b) 1 pt 8 oz for $1.79
(c) 1 qt 6 oz for $2.89

Which is the best buy?

(a) $\dfrac{\$1.27}{16 \text{ oz}} \approx 7.9¢$ per ounce

(b) Because 1 pt is 16 oz, 1 pt 8 oz is (16 + 8) oz, or 24 oz. So we write 1 pt 8 oz as 24 oz.

$$\frac{\$1.79}{24 \text{ oz}} \approx 7.5¢ \text{ per ounce}$$

(c) Because 1 qt is 32 oz, 1 qt 6 oz is (32 + 6) oz, or 38 oz. Write 1 qt 6 oz as 38 oz.

$$\frac{\$2.89}{38 \text{ oz}} \approx 7.6¢ \text{ per ounce}$$

In this case, by comparing the unit prices we see that the 1-pt 8-oz size is the best buy.

CHECK YOURSELF 14

Ketchup is sold in the following quantities:

12 oz for $0.68
1 pt 5 oz for $1.05
1 qt 7 oz for $1.89

Which is the best buy?

READING YOUR TEXT

The following fill-in-the-blank exercises are designed to assure that you understand the key vocabulary used in this section. Each sentence usually comes directly from the section. You will find the correct answers in Appendix C.

Section 6.3

(a) A _____ is an equation that compares two equal fractions or rates.

(b) If $\dfrac{a}{b} = \dfrac{c}{d}$, the fractions form a _____ proportion.

(c) If two fractions are equivalent, we say that they are _____.

(d) When solving a proportion, the second step involves dividing both sides of the equation by the _____ of the variable.

CHECK YOURSELF ANSWERS

1. $\dfrac{4}{12} = \dfrac{6}{18}$ **2.** $\dfrac{5 \text{ rolls}}{400 \text{ ft}^2} = \dfrac{7 \text{ rolls}}{560 \text{ ft}^2}$ **3. (a)** Yes; **(b)** no   **4. (a)** Yes; **(b)** no

5. No

6. (a) $5n = 100$ To check: **(b)** $7n = 42 \cdot 9$ To check:

$\dfrac{5n}{5} = \dfrac{100}{5}$ $\dfrac{4}{5} \stackrel{?}{=} \dfrac{20}{25}$ $\dfrac{7n}{7} = \dfrac{42 \cdot 9}{7}$ $\dfrac{7}{9} \stackrel{?}{=} \dfrac{42}{54}$

$n = 20$ $5 \cdot 20 \stackrel{?}{=} 4 \cdot 25$ $n = 6 \cdot 9$ $7 \cdot 54 \stackrel{?}{=} 9 \cdot 42$

$100 = 100$ $n = 54$ $378 = 378$

7. (a) 30; **(b)** 6   **8.** $1,100   **9.** $\dfrac{125 \text{ mi}}{\dfrac{5}{2} \text{ h}} = \dfrac{x \text{ mi}}{4 \text{ h}}$

$\dfrac{5}{2} x = 500$ Divide both sides by $\dfrac{5}{2}$.

$x = 200 \text{ mi}$

10. 80 g   **11.** 8 oz   **12.** $109.25   **13.** 70-oz size at 7.1¢ per oz

14. 1 qt 7 oz for 4.8¢ per oz

Name _____

Section _____ Date _____

ANSWERS

1. _____

2. _____

3. _____

4. _____

5. _____

6. _____

7. _____

8. _____

9. _____

10. _____

11. _____

12. _____

13. _____

14. _____

6.3 **Exercises**

Write each as a proportion.

1. 4 is to 9 as 8 is to 18.

2. 6 is to 11 as 18 is to 33.

3. 2 is to 9 as 8 is to 36.

4. 10 is to 15 as 20 is to 30.

5. 3 is to 5 as 15 is to 25.

6. 8 is to 11 as 16 is to 22.

7. 9 is to 13 as 27 is to 39.

8. 15 is to 21 as 60 is to 84.

In exercises 9 to 14, write the proportion that is equivalent to the given statement.

9. **Business and Finance** If 15 lb of string beans cost $4, then 45 lb will cost $12.

10. **Statistics** If Maria hit 8 home runs in 15 softball games, then she should hit 24 home runs in 45 games.

11. **Business and Finance** If 3 credit hours at Bucks County Community College cost $216, then 12 credits will cost $864.

12. **Technology** If 16 lb of fertilizer cover 1,520 ft^2, then 21 lb should cover 1,995 ft^2.

13. **Number Problem** If Audrey travels 180 mi on interstate I-95 in 3 h, then he should travel 300 mi in 5 h.

14. **Number Problem** If 2 vans can transport 18 people, then 5 vans can transport 45 people.

Determine whether each pair of fractions is proportional.

15. $\dfrac{3}{4} \overset{?}{=} \dfrac{9}{12}$

16. $\dfrac{6}{7} \overset{?}{=} \dfrac{18}{21}$

17. $\dfrac{3}{4} \overset{?}{=} \dfrac{15}{20}$

18. $\dfrac{3}{5} \overset{?}{=} \dfrac{6}{10}$

19. $\dfrac{11}{15} \overset{?}{=} \dfrac{9}{13}$

20. $\dfrac{9}{10} \overset{?}{=} \dfrac{2}{7}$

21. $\dfrac{8}{3} \overset{?}{=} \dfrac{24}{9}$

22. $\dfrac{5}{8} \overset{?}{=} \dfrac{15}{24}$

23. $\dfrac{6}{17} \overset{?}{=} \dfrac{9}{11}$

24. $\dfrac{5}{12} \overset{?}{=} \dfrac{8}{20}$

25. $\dfrac{7}{16} \overset{?}{=} \dfrac{21}{48}$

26. $\dfrac{2}{5} \overset{?}{=} \dfrac{7}{9}$

27. $\dfrac{10}{3} \overset{?}{=} \dfrac{150}{50}$

28. $\dfrac{5}{8} \overset{?}{=} \dfrac{75}{120}$

29. $\dfrac{3}{7} \overset{?}{=} \dfrac{18}{42}$

30. $\dfrac{12}{7} \overset{?}{=} \dfrac{96}{50}$

31. $\dfrac{7}{15} \overset{?}{=} \dfrac{84}{180}$

32. $\dfrac{76}{24} \overset{?}{=} \dfrac{19}{6}$

33. $\dfrac{60}{36} \overset{?}{=} \dfrac{25}{15}$

34. $\dfrac{\frac{1}{2}}{4} \overset{?}{=} \dfrac{5}{40}$

35. $\dfrac{3}{1\frac{1}{5}} \overset{?}{=} \dfrac{30}{6}$

36. $\dfrac{\frac{2}{3}}{6} \overset{?}{=} \dfrac{1}{12}$

ANSWERS

15. _____

16. _____

17. _____

18. _____

19. _____

20. _____

21. _____

22. _____

23. _____

24. _____

25. _____

26. _____

27. _____

28. _____

29. _____

30. _____

31. _____

32. _____

33. _____

34. _____

35. _____

36. _____

37. $\dfrac{\frac{3}{4}}{12} \overset{?}{=} \dfrac{1}{16}$

38. $\dfrac{0.3}{4} \overset{?}{=} \dfrac{1}{20}$

39. $\dfrac{3}{60} \overset{?}{=} \dfrac{0.3}{6}$

40. $\dfrac{0.6}{0.12} \overset{?}{=} \dfrac{2}{0.4}$

41. $\dfrac{0.6}{15} \overset{?}{=} \dfrac{2}{75}$

In exercises 42 to 50, determine if the given rates are equivalent.

42. $\dfrac{7 \text{ cups of flour}}{4 \text{ loaves of bread}} \overset{?}{=} \dfrac{4 \text{ cups of flour}}{3 \text{ loaves of bread}}$

43. $\dfrac{6 \text{ American dollars}}{50 \text{ krone}} \overset{?}{=} \dfrac{15 \text{ American dollars}}{125 \text{ krone}}$

44. $\dfrac{22 \text{ mi}}{15 \text{ gal}} \overset{?}{=} \dfrac{55 \text{ mi}}{35 \text{ gal}}$

45. $\dfrac{46 \text{ pages}}{30 \text{ min}} \overset{?}{=} \dfrac{18 \text{ pages}}{8 \text{ min}}$

46. $\dfrac{9 \text{ in.}}{57 \text{ mi}} \overset{?}{=} \dfrac{6 \text{ in.}}{38 \text{ mi}}$

47. $\dfrac{12 \text{ yen}}{5 \text{ pesos}} \overset{?}{=} \dfrac{108 \text{ yen}}{45 \text{ pesos}}$

48. $\dfrac{12 \text{ gal of paint}}{8,329 \text{ ft}^2} \overset{?}{=} \dfrac{9 \text{ gal of paint}}{1,240 \text{ ft}^2}$

49. $\dfrac{12 \text{ in. of snow}}{1.4 \text{ in. of rain}} \overset{?}{=} \dfrac{36 \text{ in. of snow}}{7 \text{ in. of rain}}$

50. $\dfrac{9 \text{ people}}{2 \text{ cars}} \overset{?}{=} \dfrac{11 \text{ people}}{3 \text{ cars}}$

Solve for the unknown in each of the proportions.

51. $\dfrac{x}{3} = \dfrac{6}{9}$ **ALEKS**

52. $\dfrac{x}{6} = \dfrac{3}{9}$ **ALEKS**

53. $\dfrac{10}{n} = \dfrac{15}{6}$ **ALEKS**

54. $\dfrac{x}{8} = \dfrac{15}{24}$ **ALEKS**

55. $\dfrac{a}{42} = \dfrac{5}{7}$ **ALEKS**

56. $\dfrac{7}{12} = \dfrac{m}{24}$ **ALEKS**

57. $\dfrac{18}{12} = \dfrac{12}{p}$ **ALEKS**

58. $\dfrac{x}{32} = \dfrac{7}{8}$ **ALEKS**

59. $\dfrac{x}{18} = \dfrac{64}{72}$ **ALEKS**

60. $\dfrac{20}{15} = \dfrac{100}{a}$ **ALEKS**

61. $\dfrac{6}{n} = \dfrac{75}{100}$ **ALEKS**

62. $\dfrac{36}{x} = \dfrac{8}{6}$ **ALEKS**

63. $\dfrac{5}{35} = \dfrac{a}{28}$ **ALEKS**

64. $\dfrac{20}{24} = \dfrac{p}{18}$ **ALEKS**

65. $\dfrac{12}{100} = \dfrac{3}{x}$ **ALEKS**

66. $\dfrac{b}{7} = \dfrac{21}{49}$ **ALEKS**

67. $\dfrac{p}{24} = \dfrac{25}{120}$ **ALEKS**

68. $\dfrac{5}{x} = \dfrac{20}{88}$ **ALEKS**

69. $\dfrac{\frac{1}{2}}{2} = \dfrac{3}{a}$

70. $\dfrac{x}{5} = \dfrac{2}{\frac{1}{3}}$

71. $\dfrac{\frac{1}{4}}{12} = \dfrac{m}{96}$

72. $\dfrac{12}{\frac{1}{3}} = \dfrac{108}{y}$

73. $\dfrac{\frac{2}{5}}{8} = \dfrac{1.2}{n}$

74. $\dfrac{4}{a} = \dfrac{\frac{2}{5}}{10}$

75. $\dfrac{0.2}{2} = \dfrac{1.2}{a}$

76. $\dfrac{n}{3} = \dfrac{6}{0.5}$

77. $\dfrac{p}{7} = \dfrac{8}{0.7}$

78. $\dfrac{y}{12} = \dfrac{5}{0.6}$

79. $\dfrac{x}{3.3} = \dfrac{1.1}{6.6}$

80. $\dfrac{0.5}{a} = \dfrac{1.25}{5}$

ANSWERS

57. _____

58. _____

59. _____

60. _____

61. _____

62. _____

63. _____

64. _____

65. _____

66. _____

67. _____

68. _____

69. _____

70. _____

71. _____

72. _____

73. _____

74. _____

75. _____

76. _____

77. _____

78. _____

79. _____

80. _____

81. _____

82. _____

83. _____

84. _____

85. _____

86. _____

87. _____

88. _____

89. _____

90. _____

91. _____

Solve each application.

81. **Business and Finance**   If 12 books are purchased for $96, how much will you pay for 18 books at the same rate?   **ALEKS**

82. **Construction**   If an 8-ft two-by-four costs $2.80, what should a 12-ft two-by-four cost?

83. **Business and Finance**   A box of 18 tea bags is marked 90¢. At that price, what should a box of 48 tea bags cost?   **VIDEO**   **ALEKS**

84. **Business and Finance**   Cans of orange juice are marked 2 for 93¢. What would the price of a case of 24 cans be?   **ALEKS**

85. **Business and Finance**   A worker can complete the assembly of 15 tape players in 6 h. At this rate, how many can the worker complete in a 40-h workweek?   **ALEKS**

86. **Business and Finance**   If 3 lb of apples cost 90¢, what will 10 lb cost?   **ALEKS**

87. **Social Science**   The ratio of yes to no votes in an election was 3 to 2. How many no votes were cast if there were 2,880 yes votes?   **ALEKS**

88. **Social Science**   The ratio of men to women at a college is 7 to 5. How many women students are there if there are 3,500 men?   **ALEKS**

89. **Technology**   A photograph 5 in. wide by 6 in. high is to be enlarged so that the new width is 15 in. What will be the height of the enlargement?

90. **Business and Finance**   Meg's job is assembling lawn chairs. She can put together 55 chairs in 4 h. At this rate, how many chairs can she assemble in an 8-h shift?   **ALEKS**

91. **Statistics**   Christy can travel 110 mi in her new car on 5 gal of gas. How far can she travel on a full tank, which has 12 usable gallons?   **ALEKS**

ANSWERS

92. _____

93. _____

94. _____

95. _____

96. _____

97. _____

92. **Construction** The Changs purchased an $280,000 home, and the property taxes were $2,200. If they make improvements and the house is now valued at $350,000, what will the new property tax be? **ALEKS**

93. **Statistics** A car travels 165 mi in 3 h. How far will it travel in 8 h if it continues at the same speed? **ALEKS**

94. **Business and Finance** A battery pack is on sale at two for $5. At this rate, how much will seven packs cost? **ALEKS**

95. **Technology** The ratio of teeth on a smaller gear to those on a larger gear is 3 to 7. If the smaller gear has 15 teeth, how many teeth does the larger gear have?

96. **Business and Finance** A store has T-shirts on sale at two for $15.50. At this rate, what will five shirts cost? **ALEKS**

Statistics Using the given map, estimate the distances between the cities named in exercises 97 to 100.

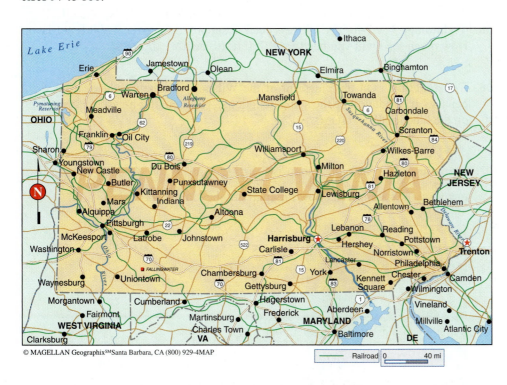

© MAGELLAN Geographix℠Santa Barbara, CA (800) 929-4MAP

Railroad 0   40 mi

97. Find the distance from Harrisburg to Philadelphia.

ANSWERS

98. _____

99. _____

100. _____

101. _____

102. _____

103. _____

104. _____

105. _____

106. _____

107. _____

98. Find the distance from Punxsutawney (home of the groundhog) to State College (home of the Nittany Lions).

99. Find the distance from Gettysburg to Meadville.

100. Find the distance from Scranton to Waynesburg.

101. Technology An inspection reveals 30 defective parts in a shipment of 500. How many defective parts should be expected in a shipment of 1,200? **ALEKS**

102. Business and Finance You invest $4,000 in a stock that pays a $180 dividend in 1 year. At the same rate, how much will you need to invest to earn $270? **ALEKS**

103. Statistics A football back ran 212 yd in the first two games of the season. If he continues at the same pace, how many yards should he gain in the 11-game season? **ALEKS**

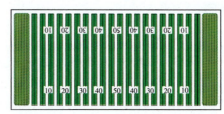

104. Crafts A 6-lb roast will serve 14 people. What size roast is needed to serve 21 people? **ALEKS**

105. Technology A 2-lb box of grass seed is supposed to cover 2,500 ft^2 of lawn. How much seed will you need for 8,750 ft^2 of lawn? **ALEKS**

106. Construction On the blueprint of the Wilsons' new home, the scale is 5 in. equals 7 ft. What will be the actual length of a bedroom if it measures 10 in. long on the blueprint? **ALEKS**

107. Statistics The scale on a map is $\frac{1}{2}$ in. = 50 mi. If the distance between two towns on the map is 6 in., how far apart are they in miles?

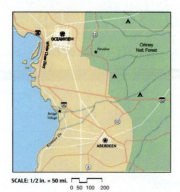

SCALE: 1/2 in. = 50 mi.
0 50 100 200

ANSWERS

108.

109.

110.

111.

112.

113.

108. Construction Construction-grade lumber costs $384.50 per 1,000 board-feet. What will be the cost of 686 board-feet?

109. Technology A shipment of 75 parts is inspected, and 6 are found to be faulty. At the same rate, how many defective parts should be found in a shipment of 139? Round your result to the nearest whole number.

110. Technology A machine produces 158 items in 12 min. At the same rate, how many items will it produce in 8 h?

111. Business and Finance Tom and Jerry operate a food concession stand at a local amusement park. They sell the most food when the attendance at the park is at maximum capacity. When this happens, they sell an average of 450 pork roll sandwiches and 550 cheese steak sandwiches. The company that owns the park is going to expand; they will increase the capacity of the park from 6,000 to 9,000 people next season. Tom and Jerry plan to expand their concession stand so they can sell more sandwiches.

(a) Using the same ratio of attendance to sandwiches, how many additional sandwiches of each kind would Tom and Jerry expect to sell?

(b) The following costs are associated with the anticipated expansion:

Item	Cost per Unit
Construction	$70/square foot
Supplies	$0.65/pork roll sandwich
	$1.25/steak sandwich
Employee costs	$8/hour

Currently, pork roll sandwiches sell for $1.50, and steak sandwiches sell for $2.75. Based on these prices and the information in part **(a)**, how would you plan the expansion, and what would you charge for a sandwich?

Find the best buy in each of the exercises.

112. Dishwashing liquid:

(a) 12 oz for 79¢

(b) 22 oz for $1.29

113. Canned corn:

(a) 10 oz for 21¢

(b) 17 oz for 39¢

114. Syrup:

   (a) 12 oz for 99¢

   (b) 24 oz for $1.59

   (c) 36 oz for $2.19

116. Salad oil (1 qt is 32 oz):

   (a) 18 oz for 89¢

   (b) 1 qt for $1.39

   (c) 1 qt 16 oz for $2.19

118. Peanut butter (1 lb is 16 oz):

   (a) 12 oz for $1.25

   (b) 18 oz for $1.72

   (c) 1 lb 12 oz for $2.59

   (d) 2 lb 8 oz for $3.76

115. Shampoo:

   (a) 4 oz for $1.16

   (b) 7 oz for $1.52

   (c) 15 oz for $3.39

117. Tomato juice (1 pt is 16 oz):

   (a) 8 oz for 37¢

   (b) 1 pt 10 oz for $1.19

   (c) 1 qt 14 oz for $1.99

Answers

1. $\dfrac{4}{9} = \dfrac{8}{18}$ **3.** $\dfrac{2}{9} = \dfrac{8}{36}$ **5.** $\dfrac{3}{5} = \dfrac{15}{25}$ **7.** $\dfrac{9}{13} = \dfrac{27}{39}$ **9.** $\dfrac{15\text{ lb}}{\$4} = \dfrac{45\text{ lb}}{\$12}$

11. $\dfrac{3\text{ credits}}{\$216} = \dfrac{12\text{ credits}}{\$864}$ **13.** $\dfrac{180\text{ mi}}{3\text{ h}} = \dfrac{300\text{ mi}}{5\text{ h}}$ **15.** Yes   **17.** Yes

19. No   **21.** Yes   **23.** No   **25.** Yes   **27.** No   **29.** Yes

31. Yes   **33.** Yes   **35.** No   **37.** Yes   **39.** Yes   **41.** No

43. Yes   **45.** No   **47.** Yes   **49.** No   **51.** $9x = 18; x = 2$ **53.** 4

55. 30   **57.** $18p = 144; p = 8$ **59.** 16   **61.** 8

63. $35a = 140; a = 4$ **65.** 25   **67.** 5   **69.** 12   **71.** 2   **73.** 24

75. 12   **77.** 80   **79.** 0.55   **81.** $144   **83.** $2.40   **85.** 100 players

87. 1,920 no votes   **89.** 18 in.   **91.** 264 mi   **93.** 440 mi

95. 35 teeth   **97.** 110 mi   **99.** 215 mi   **101.** 72 defective parts

103. $\dfrac{212\text{ yd}}{2\text{ games}} = \dfrac{x\text{ yd}}{11\text{ games}}; x = 1{,}166\text{ yd}$ **105.** 7 lb   **107.** 600 mi

109. 11 parts   **111.**   **113.** a   **115.** b   **117.** c

6.4 Similar Triangles and Proportions

6.4 OBJECTIVES

1. Use proportions to find a missing measurement from similar triangles
2. Solve an application involving similar triangles

Have you ever wondered how tall a certain tree was? In this section, you will learn how to find the height of a tree without climbing it! First we must introduce the idea of similar right triangles. Recall that a triangle in which two of the sides are perpendicular is called a **right triangle.**

NOTE The square always indicates a right triangle.

Right triangles

Other triangles

Definition: Similar Right Triangles

Two right triangles are **similar** if the ratios of corresponding sides are equivalent.

Example 1 **Confirming Similar Triangles**

Show that the two right triangles are similar.

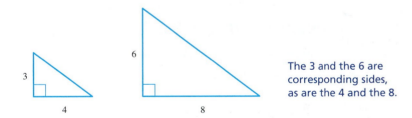

The 3 and the 6 are corresponding sides, as are the 4 and the 8.

We need to use two different pairs of sides to show that two triangles are similar. In this case,

$$\frac{3}{6} \stackrel{?}{=} \frac{4}{8}$$

is a true proportion since $3 \cdot 8 = 6 \cdot 4$, so the triangles are similar.

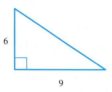

CHECK YOURSELF 1 _____

Show that the two triangles are similar.

If we know that two triangles (even if they do not happen to be right triangles) are similar, we can find the length of a missing side.

OBJECTIVE 1   **Example 2   Using Similar Triangles to Find a Missing Length**

Given the two similar triangles, find the length of the side marked x.

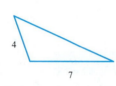 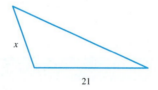

The triangles are similar, so the sides must be proportional.

$$\frac{4}{x} = \frac{7}{21}$$
$$4 \cdot 21 = x \cdot 7$$
$$84 = 7x$$
$$12 = x$$

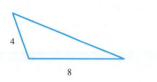

CHECK YOURSELF 2 _____

Find the length of the side marked x if the triangles are similar.

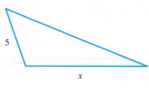

Proportions can also be used to find the length of the long side (called the **hypotenuse**) of a right triangle.

Example 3   Finding a Missing Length

Find the length of the side marked with an *x*, given that the triangles are similar.

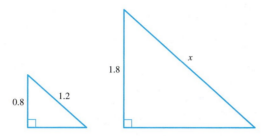

Again, we use proportions

$$\frac{0.8}{1.8} = \frac{1.2}{x}$$

$$0.8 \cdot x = 1.8 \cdot 1.2$$

$$0.8x = 2.16$$

$$x = \frac{2.16}{0.8} = 2.7$$

CHECK YOURSELF 3

Find the length of the side marked with an x, given that the triangles are similar.

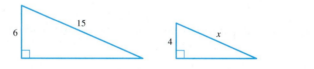

A common application of similar triangles occurs when one is looking for the height of a tall object. Example 4 illustrates this application of similar triangles.

OBJECTIVE 2

Example 4   Solving an Application Using Similar Triangles

If a 6-ft-tall man casts a shadow that is 10 ft long, how tall is a tree that casts a shadow that is 70 ft long?

We can look at a picture of the two triangles involved.

NOTE Connect the top of the tree to the end of the shadow to create a triangle. Connecting the top of the man to the end of his shadow creates a similar triangle.

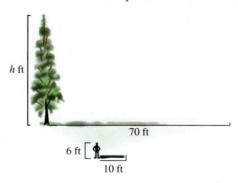

From the similar triangles, we have the proportion

$$\frac{h}{6} = \frac{70}{10}$$

Using the proportion rule, we have $6 \cdot 70 = 10 \cdot h$

$$h \cdot 10 = 420$$

$$\frac{10h}{10} = \frac{420}{10}$$

$$h = 42$$

The tree is 42 ft tall.

CHECK YOURSELF 4

If a woman who is $5\frac{1}{2}$ ft tall casts a shadow that is 3 ft long, how tall is a building that casts a shadow that is 90 ft long?

READING YOUR TEXT

The following fill-in-the-blank exercises are designed to assure that you understand the key vocabulary used in this section. Each sentence usually comes directly from the section. You will find the correct answers in Appendix C.

Section 6.4

(a) A triangle in which two of the sides are perpendicular is called a _____ triangle.

(b) Two right triangles are _____ if the ratios of corresponding sides are equivalent.

(c) If we know that two triangles are _____, we can find the length of a missing side.

(d) Proportions can be used to find the length of the long side, called the _____, of a right trangle.

CHECK YOURSELF ANSWERS

1. $\frac{4}{6} = \frac{6}{9}$ **2.** 10 **3.** 10 **4.** 165 ft tall

6.4　Exercises

Use a proportion to find the unknown side, labeled x, in each of the pairs of similar figures.

1.

x

2

6

4

ALEKS

2.

5

2

x

6

ALEKS

Name _____

Section _____ Date _____

ANSWERS

1. _____

2. _____

3. _____

4. _____

5. _____

6. _____

3.

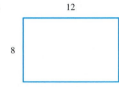

12

x

8

4

ALEKS

4.

3

x

4

12

ALEKS

In exercises 5 to 10, the two triangles shown are similar. Find the indicated side.

5. Find v.

S

3

R　5　T

V

12

U　v　W

VIDEO **ALEKS**

6. Find f.

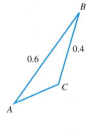

B

0.4

0.6

C

A

ALEKS

E

2

f

F

D

7. Find g.

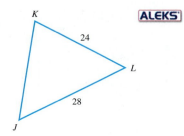

8. Find m.

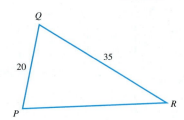

9. Find t.

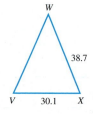

10. Find e.

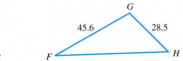

Solve the applications.

11. **Geometry** A 9-ft light pole casts a 15-ft shadow. Find the height of a nearby tree that is casting a 40-ft shadow. 6 **ALEKS**

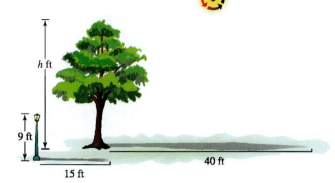

12. **Construction** A 6-ft fence post casts a 9-ft shadow. How tall is a nearby pole that casts a 15-ft shadow? 6 **ALEKS**

13. **Geometry** A light pole casts a shadow that measures 4 ft. At the same time, a yardstick casts a shadow that is 9 in. long. How tall is the pole? 6 **ALEKS**

14. **Geometry** A tree casts a shadow that measures 5 m. At the same time, a meter-stick casts a shadow that is 0.4 m long. How tall is the tree? 6 **ALEKS**

15. Use the ideas of similar triangles to determine the height of a pole or tree on your campus. Work with one or two partners. (6)

Answers

1. 3   **3.** 6   **5.** 20   **7.** 12   **9.** 39.2   **11.** 24 ft   **13.** 16 ft
15.

6.5 Linear Measurement and Conversion

6.5 OBJECTIVES

1. Convert between English system units of length
2. Simplify and perform operations involving English system units of length
3. Identify metric system prefixes and units of length
4. Convert between metric system units of length

Many arithmetic problems involve **units of measure.** When we measure an object, we give it a number and some unit. For instance, we might say a board is 6 feet long, a container holds 4 quarts, or a package weighs 5 pounds. Feet, quarts, and pounds are the units of measure.

The system you are probably most familiar with is called the **English system of measurement.** This system is used in the United States and a few other countries. The table lists the units of measurement you should be familiar with.

English Units of Measure and Equivalents

Length	Weight
1 foot (ft) = 12 inches (in.) 1 yard (yd) = 3 ft 1 mile (mi) = 5,280 ft	1 pound (lb) = 16 ounces (oz) 1 ton = 2,000 lb

Capacity	
1 pint (pt) = 16 fluid ounces (fl oz) 1 quart (qt) = 2 pt 1 gallon (gal) = 4 qt	

In this section, we focus on units of length. From the table, you can see that the units we focus on first include inches, feet, yards, and miles.

You may want to use the equivalencies shown in the table to change from one unit to another. We will look at one approach.

Property: Converting Units in the English System

To change from one unit to another, replace the unit of measure with the appropriate equivalent measure and multiply.

OBJECTIVE 1

Example 1   Converting within the English System

NOTE We write 5 ft as 5(1 ft) and then change 1 ft to 12 in.

$5 \text{ ft} = 5(1 \text{ ft}) = 5(12 \text{ in.}) = 60 \text{ in.}$ Replace 1 ft with 12 in.

$48 \text{ in.} = 48(1 \text{ in.}) = 48\left(\dfrac{1}{12}\text{ ft}\right) = 4 \text{ ft}$ Because 12 in. = 1 ft, 1 in. = $\dfrac{1}{12}$ ft.

$4 \text{ yd} = 4(1 \text{ yd}) = 4(3 \text{ ft}) = 12 \text{ ft}$ Replace 1 yd with 3 ft.

CHECK YOURSELF 1 _____

Complete each of the statements.

(a) 4 ft = _____ in.   **(b)** 12 yd = _____ ft

(c) 144 in. = _____ ft   **(d)** 3 ft = _____ in.

Here is another idea that may help you convert units. You can use a *units fraction* to convert from one unit to another. A units fraction always has a value of 1.

> **Property:** Using a Units Fraction
>
> To decide which units fraction to use, just choose one with the unit you *want* in the numerator (inches in Example 2) and the unit you *want to remove* in the denominator (feet in Example 2).

NOTE This is a variation on the method of units analysis discussed throughout this text.

> **Example 2** Using a Units Fraction to Convert

Convert 5 ft to inches.

To convert from feet to inches, you can multiply by $\dfrac{12\ \text{in.}}{1\ \text{ft}}$. So, to convert 5 ft to inches, write

NOTE $\dfrac{12\ \text{in.}}{1\ \text{ft}}$ is a *units fraction.* It can be reduced to 1.

$5\ \text{ft} = 5\ \cancel{\text{ft}} \left(\dfrac{12\ \text{in.}}{1\ \cancel{\text{ft}}} \right)$ We are multiplying by 1, and so the value of the expression is not changed.

$= 60\ \text{in.}$ Note that we can divide out units just as we do numbers.

CHECK YOURSELF 2 _____

Use a units fraction to complete each of the statements.

(a) 240 in. = _____ ft   **(b)** 7 ft = _____ in.

You have now had a chance to use two different methods for converting from one unit of measurement to another. Use whichever approach seems easier for you.

From our work so far, it should be clear that one big disadvantage of the English system is that the relationships between units are all different. One foot is 12 in., 1 lb is 16 oz, and so on. We will see that this problem does not exist in the metric system.

Example 3 shows the steps used to simplify English system units of length with multiple units.

OBJECTIVE 2

> **Example 3** Simplifying English System Units of Length

Simplify 4 ft 18 in.

NOTE 18 in. is larger than 1 ft and can be simplified.

$\underset{\text{18 in.}}{4\ \text{ft}\ 18\ \text{in.}} = 4\ \text{ft} + \underbrace{1\ \text{ft} + 6\ \text{in.}}$ Write 18 in. as 1 ft 6 in. because 12 in. is 1 ft.

$= 5\ \text{ft}\ 6\ \text{in.}$

CHECK YOURSELF 3 _____

(a) Simplify 2 yd 8 ft   **(b)** Simplify 7 ft 20 in.

We can always add or subtract units of length according to this rule.

Step by Step: Adding Like Units of Length

Step 1 Arrange the numbers so that the like units are in the same vertical column.

Step 2 Add in each column.

Step 3 Simplify if necessary.

Example 4 illustrates this rule for adding English system units of length.

Example 4 **Adding English System Units of Length**

Add 5 ft 4 in., 6 ft 7 in., and 7 ft 9 in.

NOTE The columns here represent inches and feet.

$$\begin{array}{r} 5 \text{ ft} \quad 4 \text{ in.} \\ 6 \text{ ft} \quad 7 \text{ in.} \\ +\ \ 7 \text{ ft} \quad 9 \text{ in.} \\ \hline 18 \text{ ft} \ 20 \text{ in.} \end{array}$$ Arrange in a vertical column.

Add in each column.

NOTE Be sure to simplify the results.

$= 19 \text{ ft} \quad 8 \text{ in.}$ Simplify as before.

 CHECK YOURSELF 4 _____

Add 4 ft 6 in., 2 ft 9 in., and 5 ft 11 in.

To subtract units of length, we have a similar rule.

Step by Step: Subtracting Like Units of Length

Step 1 Arrange the numbers so that the like units are in the same vertical column.

Step 2 Subtract in each column. You may have to borrow from the larger unit at this point.

Step 3 Simplify if necessary.

In Example 5, we subtract English system units of length.

Example 5 **Subtracting English System Units of Length**

(a) Subtract 3 ft 6 in. from 8 ft 10 in.

$$\begin{array}{r} 8 \text{ ft} \ 10 \text{ in.} \\ -\ 3 \text{ ft} \quad 6 \text{ in.} \\ \hline 5 \text{ ft} \quad 4 \text{ in.} \end{array}$$ Arrange vertically.

Subtract in each column.

(b) Subtract 5 ft 8 in. from 9 ft 3 in.

NOTE Borrowing with units of length is not the same as in the place-value system, in which we always borrowed 10.

$$\begin{array}{r} 9 \text{ ft } 3 \text{ in.} \\ -\ 5 \text{ ft } 8 \text{ in.} \\ \hline \end{array}$$ Do you see the problem? We cannot subtract in the inches column.

To complete the subtraction, we borrow 1 ft and rename. The "borrowed" number will depend on the units involved.

$$\begin{array}{r} \cancel{9} \text{ ft } \cancel{3} \text{ in.} \\ 8 \text{ ft } 15 \text{ in.} \\ -\ 5 \text{ ft }\ \ 8 \text{ in.} \\ \hline 3 \text{ ft }\ \ 7 \text{ in.} \end{array}$$ 9 ft becomes 8 ft 12 in. Combine the 12 in. with the original 3 in.

We can now subtract.

CHECK YOURSELF 5

(a) Subtract 5 ft 9 in. from 10 ft 11 in.

(b) Find the difference between 7 ft 2 in. and 4 ft 5 in.

Certain types of problems involve multiplying or dividing units of length by abstract numbers, that is, numbers without a unit of measure attached. This rule is used.

Step by Step: Multiplying or Dividing by Abstract Numbers

Step 1 Multiply or divide each part of the unit of length by the abstract number.

Step 2 Simplify if necessary.

Example 6 illustrates this procedure.

Example 6 **Multiplying English System Units of Length by Abstract Numbers**

(a) Multiply $4 \cdot 5$ in.

$4 \cdot 5$ in. $= 20$ in. or 1 ft 8 in.

(b) Multiply 3(2 ft 7 in.).

NOTE Multiply each part of the unit of length by 3.

$$\begin{array}{r} 2 \text{ ft }\ \ 7 \text{ in.} \\ \times\ \ \ \ \ \ \ \ \ 3 \\ \hline 6 \text{ ft } 21 \text{ in.} \end{array}$$

Simplify. The product is 7 ft 9 in.

(c) Divide 8 ft 4 in. by 4.

$$\frac{8 \text{ ft } 4 \text{ in.}}{4} = 2 \text{ ft } 1 \text{ in.}$$

CHECK YOURSELF 6

(a) Multiply 2 yd 2ft 2 in. by 6.

(b) Divide 9 ft 6 in. by 3.

NOTE The basic unit of length in the metric system is also spelled *metre* (the British spelling).

Thus far, we have studied the English system of measurement, which is used in the United States and a few other countries. Our work will now concentrate on the **metric system,** which is used throughout the rest of the world.

The metric system is based on one unit of length, the **meter (m).** In the eighteenth century, the meter was defined to be one ten-millionth of the distance from the north pole to the equator. Today the meter is scientifically defined in terms of a wavelength in the spectrum of krypton-86 gas.

One big advantage of the metric system is that you can convert from one unit to another by simply multiplying or dividing by powers of 10. This advantage and the need for uniformity throughout the world have led to legislation that will promote the use of the metric system in the United States.

NOTE Even in the United States, the metric system is used in science, medicine, the automotive industry, the food industry, and many other areas.

We will see how the metric system works by starting with measures of length and comparing a basic English unit, the yard, with the meter.

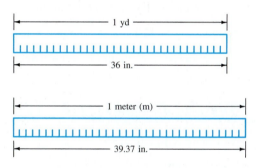

NOTE The meter is one of the basic units of the International System of Units (abbreviated SI). This is a standardization of the metric system agreed to by scientists in 1960.

As you can see, the meter is just slightly longer than the yard. It is used for measuring the same things you might measure in feet or yards. Look at Example 7 to get a feel for the size of the meter.

OBJECTIVE 3

Example 7   Estimating Metric Length

NOTE There is a standard pattern of abbreviation in the metric system. We will introduce the abbreviation for each term as we go along. The abbreviation for meter is m (no period!).

A room might be 6 m long.

A building lot could be 30 m wide.

A fence is 2 m tall.

CHECK YOURSELF 7

Try to estimate each length in meters.

(a) A traffic lane is _____ m wide.

(b) A small car is _____ m long.

(c) You are _____ m tall.

For other units of length, the meter is multiplied or divided by powers of 10. One commonly used unit is the **centimeter (cm),** which is smaller than a meter.

> **Definition:** Comparing Centimeters (cm) to Meters (m)
>
> 1 centimeter (cm) = $\frac{1}{100}$ meter (m)

NOTE The prefix *centi* means one hundredth. This should be no surprise. What is our cent? It is one hundredth of a dollar.

The drawing relates the centimeter and the meter:

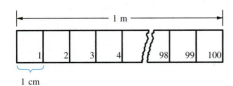

There are 100 cm in 1 m.

Just to give you an idea of the size of the centimeter, it is about the width of your little finger. There are about $2\frac{1}{2}$ cm to 1 in., and the unit is used to measure small objects. Look at Example 8 to get a feel for the length of a centimeter.

> **Example 8** **Estimating Metric Length**
>
> A small paperback book is 10 cm wide.
>
> A playing card is 8 cm long.
>
> A ballpoint pen is 16 cm long.

CHECK YOURSELF 8

Try to estimate each value. Then use a metric ruler to check your guess.

(a) This page is _____ cm long.

(b) A dollar bill is _____ cm long.

(c) The seat of the chair you are on is _____ cm from the floor.

To measure *very* small things, the **millimeter (mm)** is used. To give you an idea of its size, the millimeter is about the thickness of a new dime.

> **Definition:** Comparing Millimeters (mm) to Meters (m)
>
> 1 millimeter (mm) = $\frac{1}{1,000}$ m

NOTE The prefix *milli* means one thousandth.

The diagram will help you see the relationships of the three units we have looked at.

NOTE Notice that there are 10 mm to 1 cm.

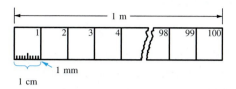

To get used to the millimeter, consider Example 9.

Example 9   Estimating Metric Length

Standard camera film is 35 mm wide.

A small paper clip is 5 mm wide.

A new pencil is about 200 mm long.

 CHECK YOURSELF 9 _____

Try to estimate each value. Then use a metric ruler to check your guess.

(a) Your pencil is _____ mm wide.
(b) The tabletop you are working on is _____ mm thick.

NOTE The prefix *kilo* means 1,000. You are already familiar with this. For instance, 1 kilowatt (kW) = 1,000 watts (W).

The **kilometer (km)** is used to measure long distances. The kilometer is about six-tenths of a mile.

Definition: Comparing Kilometers (km) to Meters (m)

1 kilometer (km) = 1,000 m

Example 10 shows how to get used to the kilometer.

Example 10   Estimating Metric Length

The distance from New York to Boston is 325 km.

A popular distance for road races is 10 km.

Now that you have seen the four commonly used units of length in the metric system, you can review with the following Check Yourself exercise.

CHECK YOURSELF 10

Choose the most reasonable measure in each of the statements.

(a) The width of a doorway: 50 mm, 1 m, or 50 cm.
(b) The length of your pencil: 20 m, 20 mm, or 20 cm.
(c) The distance from your house to school: 500 km, 5 km, or 50 m.
(d) The height of a basketball center: 2.2 m, 22 m, or 22 cm.
(e) The width of a matchbook: 30 cm, 30 mm, or 3 mm.

NOTE Of course, this is easy. All we need to do is move the decimal point to the right or left the required number of places. Again, that's the big advantage of the metric system.

NOTE The *smaller* the unit, the *more* units it takes, so *multiply*.

As we said earlier, to convert units of measure within the metric system, all we have to do is multiply or divide by the appropriate power of 10.

Property: Converting Metric Measurements to Smaller Units

To convert to a *smaller* unit of measure, we *multiply* by a power of 10, moving the decimal point *to the right.*

OBJECTIVE 4

For example, the first solution should be $5.2 \, \cancel{m} \left(\dfrac{100 \text{ cm}}{1 \, \cancel{m}} \right) =$ 520 cm since $1 \text{ cm} = \dfrac{1}{100} \text{ m}$, 100 cm $= 1 \text{ m}$.

Example 11   Converting Metric Length

5.2 m = 520 cm	Multiply by 100 to convert from meters to centimeters.
8 km = 8,000 m	Multiply by 1,000.
6.5 m = 6,500 mm	Multiply by 1,000.
2.5 cm = 25 mm	Multiply by 10.

CHECK YOURSELF 11

Complete the statements. Remember, just move the decimal point the appropriate number of places and write the answer.

(a) 3 km = _____ m   **(b)** 4.5 m = _____ cm
(c) 1.2 m = _____ mm   **(d)** 6.5 cm = _____ mm

Property: Converting Metric Measurements to Larger Units

To convert to a *larger* unit of measure, we *divide* by a power of 10, moving the decimal point *to the left.*

NOTE The *larger* the unit, the *fewer* units it takes, so *divide*.

Example 12   Converting Metric Length

43 mm = 4.3 cm   Divide by 10.

3,000 m = 3 km   Divide by 1,000.

450 cm = 4.5 m   Divide by 100.

CHECK YOURSELF 12

Complete the statements.

(a) 750 cm = _____ m   **(b)** 5,000 m = _____ km

(c) 78 mm = _____ cm   **(d)** 3,500 mm = _____ m

We have introduced all the commonly used units of linear measure in the metric system. There are other prefixes that can be used to form other linear measures. The prefix *deci* means $\frac{1}{10}$, *deka* means 10, and *hecto* means 100. Their use is illustrated in the chart.

Definition: Using Metric Prefixes

1 *milli*meter (mm) $= \dfrac{1}{1,000}$ m

1 *centi*meter (cm) $= \dfrac{1}{100}$ m

1 *deci*meter (dm) $= \dfrac{1}{10}$ m

1 meter (m)

1 *deka*meter (dam) = 10 m

1 *hecto*meter (hm) = 100 m

1 *kilo*meter (km)   = 1,000 m

Example 13   Converting between Metric Lengths

(a) 800 dm = ? m

To convert from decimeters to meters, you can see from the chart that you must move the decimal point *one place to the left.*

800 dm = 80͜0 m = 80 m

(b) 500 m = ? km

To convert from meters to kilometers, move the decimal point *three places to the left.*

500 m = ͜500 km = 0.5 km

(c) 6 m = ? mm

To convert from meters to millimeters, move the decimal point *three places to the right.*

6 m = 6,000͜ mm

 CHECK YOURSELF 13

Complete each statement.

(a) 300 cm = _____ m   **(b)** 370 mm = _____ m

(c) 4,500 m = _____ km

READING YOUR TEXT

The following fill-in-the-blank exercises are designed to assure that you understand the key vocabulary used in this section. Each sentence usually comes directly from the section. You will find the correct answers in Appendix C.

Section 6.5

(a) A _____ fraction always has a value of 1.

(b) Historically, the _____ was the distance from the end of a nose to the fingertip of an outstretched arm.

(c) The prefix _____ means *one hundredth*.

(d) The metric unit of _____ is used to measure long distances.

CHECK YOURSELF ANSWERS

1. **(a)** 48 in.; **(b)** 36 ft; **(c)** 12 ft; **(d)** 36 in.   **2. (a)** 20 ft; **(b)** 84 in.
3. **(a)** 4 yd 2 ft; **(b)** 8 ft 8 in.   **4.** 13 ft 2 in.   **5. (a)** 5 ft 2 in.; **(b)** 2 ft 9 in.
6. **(a)** 16 yd 1 ft; **(b)** 3 ft 2 in.
7. **(a)** About 3 m; **(b)** perhaps 5 m; **(c)** You are probably between 1.5 and 2 m tall.
8. **(a)** About 28 cm; **(b)** almost 16 cm; **(c)** about 45 cm
9. **(a)** About 8 mm; **(b)** probably between 25 and 30 mm
10. **(a)** 1 m; **(b)** 20 cm; **(c)** 5 km; **(d)** 2.2 m; **(e)** 30 mm
11. **(a)** 3,000 m; **(b)** 450 cm; **(c)** 1,200 mm; **(d)** 65 mm
12. **(a)** 7.5 m; **(b)** 5 km; **(c)** 7.8 cm; **(d)** 3.5 m   **13. (a)** 3 m; **(b)** 0.37 m; **(c)** 4.5 km

6.5 Exercises

Name _____

Section _____ Date _____

ANSWERS

Complete the statements.

1. 8 ft = _____ in. **ALEKS**

2. 5 mi = _____ ft **ALEKS**

3. 7 yd = _____ ft   **ALEKS**

4. 39 ft = _____ yd **ALEKS**

5. 44 in. = _____ ft **ALEKS**

6. 4.72 ft = _____ in.

Solve the application.

7. Construction A unit of measurement used in surveying is the **chain.** There are 80 chains in a mile. If you measured the distance from your home to school, how many chains would you have traveled?

Simplify.

8. 4 ft 18 in. **ALEKS**

9. 7 yd 50 in.

Add.

10.   9 ft 7 in. **ALEKS**
     + 3 ft 10 in.

11.   5 yd 2 ft  **ALEKS**
      4 yd
     + 6 yd 1 ft

12. 7 ft 8 in., 8 ft 5 in., and 9 ft 7 in. **ALEKS**

Subtract.

13.   7 ft 11 in.
     − 4 ft 3 in.

14. Subtract 2 yd 2 ft from 5 yd 1 ft.

Multiply.

15. 4 · 10 in.

16. 3(4 ft 5 in.)

Divide.

17. $\dfrac{4 \text{ ft 6 in.}}{2}$

18. $\dfrac{16 \text{ mi 28 yd}}{4}$

1. _____
2. _____
3. _____
4. _____
5. _____
6. _____

7. _____
8. _____
9. _____
10. _____
11. _____
12. _____
13. _____
14. _____
15. _____
16. _____
17. _____
18. _____

19. _____

20. _____

21. _____

22. _____

23. _____

24. _____

25. _____

26. _____

27. _____

28. _____

29. _____

30. _____

31. _____

Solve each of the applications.

19. **Construction**   A railing for a deck requires pieces of cedar 4 ft 8 in., 11 ft 7 in., and 9 ft 3 in. long. What is the total length of material that is needed?   **ALEKS**

20. **Crafts**   A pattern requires a 2-ft 10-in. length of fabric. If a 2-yd length is used, what length remains?

21. **Crafts**   A picture frame is to be 2 ft 6 in. long and 1 ft 8 in. wide. A 9-ft piece of molding is available for the frame. Will this be enough for the frame?

22. **Construction**   A plumber needs two pieces of plastic pipe that are 6 ft 9 in. long and 1 piece that is 2 ft 11 in. long. He has a 16-ft piece of pipe. Is this enough for the job?

23. **Construction**   A bookshelf requires four boards 3 ft 8 in. long and two boards 2 ft 10 in. long. How much lumber will be needed for the bookshelf?

Find the following.

24.   13 yd 15 ft 10 in.
     −   9 yd 16 ft 15 in.

25.   4 mi 5 yd 3 ft 10 in.
     ×   2

Choose the most reasonable measure.

26. The height of a ceiling

   (a) 25 m
   (b) 2.5 m
   (c) 25 cm

27. The height of a kitchen counter

   (a) 9 m
   (b) 9 cm
   (c) 90 cm

28. The diagonal measure of a television screen

   (a) 50 mm
   (b) 50 cm
   (c) 5 m

29. The height of a two-story building

   (a) 7 m
   (b) 70 m
   (c) 70 cm

30. An hour's drive on a freeway

   (a) 9 km
   (b) 90 m
   (c) 90 km

31. The width of a roll of cellophane tape

   (a) 1.27 mm
   (b) 12.7 mm
   (c) 12.7 cm

32. The width of a sheet of typing paper

(a) 21.6 cm

(b) 21.6 mm

(c) 2.16 cm

33. The thickness of window glass

(a) 5 mm

(b) 5 cm

(c) 50 mm

34. The height of a refrigerator

(a) 16 m

(b) 16 cm

(c) 160 cm

35. The length of a ballpoint pen

(a) 16 mm

(b) 16 m

(c) 16 cm

36. The width of a handheld calculator key

(a) 1.2 mm

(b) 12 mm

(c) 12 cm

Complete each statement, using a metric unit of length.

37. A playing card is 6 _____ wide.

38. The diameter of a penny is 19 _____.

39. A doorway is 2 _____ high.

40. A table knife is 22 _____ long.

41. A basketball court is 28 _____ long.

42. A commercial jet flies 800 _____ per hour.

43. The width of a nail file is 12 _____.

44. The distance from New York to Washington, D.C., is 360 _____.

45. A recreation room is 6 _____ long.

46. A ruler is 22 _____ wide.

47. A long-distance run is 35 _____.

48. A paperback book is 11 _____ wide.

ANSWERS

32. _____
33. _____
34. _____
35. _____
36. _____
37. _____
38. _____
39. _____
40. _____
41. _____
42. _____
43. _____
44. _____
45. _____
46. _____
47. _____
48. _____

Complete each statement.

49. 3,000 mm = _____ m **ALEKS**　　**50.** 150 cm = _____ m **ALEKS**

51. 8 m = _____ cm **ALEKS**　　**52.** 77 mm = _____ cm **ALEKS**

53. 250 km = _____ cm 　　**54.** 500 cm = _____ m **ALEKS**

55. 25 cm = _____ mm　　**56.** 150 mm = _____ m **ALEKS**

57. 7,000 m = _____ km **ALEKS**　　**58.** 9 m = _____ cm **ALEKS**

59. 8 cm = _____ mm　　**60.** 45 cm = _____ mm

61. 5 km = _____ m **ALEKS**　　**62.** 4,000 m = _____ km
　　　　　　　　　　　　　　　　　　　　　　　ALEKS

63. 5 m = _____ mm **ALEKS**　　**64.** 7 km = _____ m **ALEKS**

Geometry Use a metric ruler to measure the necessary dimensions and complete the statements.

65. The perimeter of the parallelogram is _____ cm.

66. The perimeter of the triangle is _____ mm.

67. The perimeter of the rectangle is _____ cm.

68. The area of the rectangle in exercise 67 is _____ cm².

69. The perimeter of the square is _____ mm.

70. The area of the square in exercise 69 is _____ mm².

71. What units in the metric system would you use to measure each of the quantities?

    (a) Distance from Los Angeles to New York

    (b) Your waist measurement

    (c) Width of a hair

    (d) Your height

Answers

1. 96   **3.** 21   **5.** $3\dfrac{2}{3}$ **7.**    **9.** 8 yd 14 in. or 8 yd 1 ft 2 in.

11. 16 yd   **13.** 3 ft 8 in.   **15.** 3 ft 4 in.   **17.** 2 ft 3 in.   **19.** 25 ft 6 in.
21. Yes, 8 in. will remain   **23.** 20 ft 4 in.   **25.** 8 mi 12 yd 1 ft 8 in.
27. (c)   **29.** (a)   **31.** (b)   **33.** (a)   **35.** (c)   **37.** cm   **39.** m
41. m   **43.** mm   **45.** m   **47.** km   **49.** 3   **51.** 800
53. 25,000,000   **55.** 250   **57.** 7   **59.** 80   **61.** 5,000   **63.** 5,000
65. 10.4   **67.** 11.8   **69.** 86
71. **(a)** km; **(b)** cm; **(c)** mm; **(d)** m or cm

ACTIVITY 6: MEASUREMENTS

Each chapter in this text includes an activity. The activity is related to the vignette you encountered in the chapter opening. The activity provides you with the opportunity to apply the math you studied in the chapter to a relevant topic.

Your instructor will determine how best to use this activity in your class. You may find yourself working in class or outside of class; you may find yourself working alone or in small groups; or you may even be asked to perform research in a library or on the Internet.

According to tradition, a **cubit** was the distance between the tip of an elbow and the tip of the middle finger.

1 cubit

1. According to your own forearm, what are the dimensions of the desk or table you are working on in cubits?
2. Ask a second person to make the same measurement using their own forearm. What are their dimensions for your desk or table in cubits?
3. How long is your cubit in inches?

According to biblical tradition, the cubit was defined according to Noah's forearm. Noah was commanded to build an ark 300 cubits long, 50 cubits wide, and 30 cubits high.

4. In inches, what would be the dimensions of such an ark (using your own forearm for the length of a cubit)?
5. Convert the dimensions found in exercise 4 to feet (round to the nearest foot).

At some point in history, the length of a cubit was standardized as 18 inches.

6. How different is your cubit from the standard cubit?
7. What are the dimensions of your desk or table in standard cubits?
8. Create an algebraic equation to convert a length from your cubit to a standard cubit.
9. In inches, what are the dimensions of the above ark?
10. Convert your answer in exercise 9 to meters.

Often, common mathematical ideas have long and storied histories.

11. Research the history of the cubit (use a library, the Internet, or some other source). Write a brief paragraph concerning the history of the cubit.
12. In the chapter opener, we discuss a yard as the distance from the tip of the nose to the tip of the outstretched finger. How does *your* cubit compare to *your* yard?
13. How does the standard cubit compare to the standard yard?
14. Research the history of the yard as a unit of measure. Write a brief paragraph of this history.
15. Research the history of the meter as a unit of measure. Write two paragraphs giving this history and how the meter is used to define measurements in the English system today.

6 Summary

DEFINITION/PROCEDURE	EXAMPLE	REFERENCE
Ratios		**Section 6.1**
Ratio A means of comparing two numbers or quantities. A ratio can be written as a fraction.	$\dfrac{4}{7}$ can be thought of as "the ratio of 4 to 7."	*p. 499*
Rates		**Section 6.2**
Rate A fraction involving two denominate numbers with different units.	$\dfrac{50 \text{ home runs}}{150 \text{ games}} = \dfrac{1}{3}\dfrac{\text{home run}}{\text{game}}$	*p. 507*
Unit price The cost per unit.	$\dfrac{\$2}{5 \text{ rolls}} = \$0.40 \text{ per roll}$	*p. 509*
Proportions		**Section 6.3**
Proportion A statement that two ratios or rates are equal.	$\dfrac{3}{5} = \dfrac{6}{10}$ is a proportion that reads, "three is to five as six is to ten."	*p. 517*
The Proportion Rule If $\dfrac{a}{b} = \dfrac{c}{d}$, then $a \cdot d = b \cdot c$	If $\dfrac{3}{5} = \dfrac{6}{10}$, then $5 \cdot 6 = 3 \cdot 10$	*p. 518*
To Solve a Proportion 1. Use the proportion rule to write the equivalent equation $a \cdot d = b \cdot c$. 2. Divide both terms of the equation by the coefficient of the variable. 3. Use the value found to replace the unknown in the original proportion. Check that the ratios or rates are proportional.	To solve: $\dfrac{x}{5} = \dfrac{16}{20}$ $20x = 5 \cdot 16$ $20x = 80$ $\dfrac{\overset{1}{\cancel{20}}x}{\underset{1}{\cancel{20}}} = \dfrac{80}{20}$ $x = 4$	*p. 520*
Similar Triangles and Proportions		**Section 6.4**
A triangle in which two of the sides are perpendicular is called a **right triangle.**	is a right triangle.	*p. 537*
Two right triangles are similar if their corresponding sides are proportional.	 are proportional.	*p. 537*
If we know that two triangles are similar, we can use a proportion to find the length of a missing side.	 $\dfrac{2}{3} = \dfrac{8}{x}$ $2x = 3 \cdot 8$ $2x = 24$ $x = 12$	*p. 538*

Continued

DEFINITION/PROCEDURE	EXAMPLE	REFERENCE
Linear Measurement and Conversion		**Section 6.5**
The English system of measurement is in common use in the United States. *English Units of Measure and Equivalents* *Length* 1 foot (ft) = 12 inches (in.) 1 yard (yd) = 3 ft 1 mile (mi) = 5,280 ft *Capacity* 1 pint (pt) = 16 fluid ounces (fl oz) 1 quart (qt) = 2 pt 1 gallon (gal) = 4 qt *Weight* 1 pound (lb) = 16 ounces (oz) 1 ton = 2,000 lb		*p. 545*
Units fractions A fraction whose value is 1. Units fractions can be used to convert units.	$\dfrac{12 \text{ in.}}{1 \text{ ft}}$ and $\dfrac{1 \text{ ft}}{3 \text{ yd}}$ are units fractions.	*p. 546*
To Add Like Units of Length **1.** Arrange the numbers so that the like units are in the same column. **2.** Add in each column. **3.** Simplify if necessary.	To add 4 ft 7 in. and 5 ft 10 in.: $\quad$ 4 ft $\;$ 7 in. $\underline{+\; 5 \text{ ft } 10 \text{ in.}}$ $\quad$ 9 ft 17 in. = 10 ft 5 in.	*p. 547*
To Subtract Like Units of Length **1.** Arrange the numbers so that the like units are in the same column. **2.** Subtract in each column. You may have to borrow from the larger unit at this point. **3.** Simplify if necessary.	To subtract: $\quad$ 4 ft 7 in. $\underline{-\; 2 \text{ ft } 9 \text{ in.}}$ Borrow and rename: $\quad$ 3 ft 19 in. $\underline{-\; 2 \text{ ft } \;\; 9 \text{ in.}}$ $\quad$ 1 ft 10 in.	*p. 547*
To Multiply or Divide Units by Abstract Numbers **1.** Multiply or divide each part of the measurement by the abstract number. **2.** Simplify if necessary.	2(3 yd 2 ft) = 6 yd 4 ft, or 7 yd 1 ft	*p. 548*
Metric Units of Length The metric system of measurement is used throughout most of the world. $\quad$ **Common metric units of length** are the meter (m), centimeter (cm), millimeter (mm), and kilometer (km).		*p. 549*
Basic Metric Prefixes *milli** means $\dfrac{1}{1,000}$ $\quad$ *kilo** means 1,000 *centi** means $\dfrac{1}{100}$ $\quad$ *hecto* means 100 *deci* means $\dfrac{1}{10}$ $\quad$ *deka* means 10	*These are the most commonly used and should be memorized.	*p. 553*

Summary Exercises

This summary exercise set is provided to give you practice with each of the objectives of this chapter. Each exercise is keyed to the appropriate chapter section. When you are finished, you can check your answers to the odd-numbered exercises against those presented in the back of the text. If you have difficulty with any of these questions, go back and reread the examples from that section. The answers to the even-numbered exercises appear in the *Instructor's Solutions Manual*. Your instructor will give you guidelines on how to best use these exercises in your class.

[6.1] In exercises 1 to 8, write each ratio in simplest form.

1. The ratio of 4 to 17

2. The ratio of 28 to 42

3. Statistics For a football team that has won 10 of its 16 games, the ratio of wins to games played

4. Geometry For a rectangle of length 30 in. and width 18 in., the ratio of its length to its width

5. The ratio of $2\frac{1}{3}$ to $5\frac{1}{4}$

6. The ratio of 7.5 to 3.25

7. The ratio of 7 in. to 3 ft

8. The ratio of 72 h to 4 days

[6.2] In exercises 9 to 16, express each rate in simplest form.

9. $\dfrac{600 \text{ mi}}{6 \text{ h}}$

10. $\dfrac{270 \text{ mi}}{9 \text{ gal}}$

11. $\dfrac{350 \text{ calories}}{7 \text{ oz}}$

12. $\dfrac{36{,}000 \text{ dollars}}{9 \text{ yr}}$

13. $\dfrac{5{,}000 \text{ ft}}{25 \text{ s}}$

14. $\dfrac{10{,}500 \text{ revolutions}}{3 \text{ min}}$

15. Statistics A baseball team has had 117 hits in 18 games. Find the team's hits per game rate.

16. Statistics A basketball team has scored 216 points in 8 quarters. Find the team's points per quarter rate.

17. Statistics Taniko scored 246 points in 20 games. Marisa scored 216 points in 16 games. Which player has the higher points per game rate?

18. Business and Finance One shop will charge $306 for a job that takes $4\frac{1}{2}$ h. A second shop can do the same job in 4 h and will charge $290. Which shop has the higher cost per hour rate?

In exercises 19 to 24, find the unit price for each item.

19. A 32-oz bottle of dishwashing liquid costs $2.88.

20. A 35-oz box of breakfast cereal costs $5.60.

21. A 24-oz loaf of bread costs $2.28.

22. Five large jars of fruit cost $67.30.

23. Three CDs cost $44.85.

24. Six tickets cost $267.60.

[6.3] In exercises 25 to 28, write each proportion.

25. 4 is to 9 as 20 is to 45.

26. 7 is to 5 as 56 is to 40.

27. Statistics If Jorge can travel 110 mi in 2 h, he can travel 385 mi in 7 h.

28. Construction If 4 gal of paint will cover 1,000 ft^2, it will take 10 gal of paint to cover 2,500 ft^2.

In exercises 29 to 34, determine whether the given fractions are proportional.

29. $\dfrac{4}{13} \stackrel{?}{=} \dfrac{7}{22}$

30. $\dfrac{8}{11} \stackrel{?}{=} \dfrac{24}{33}$

31. $\dfrac{9}{24} \stackrel{?}{=} \dfrac{12}{32}$

32. $\dfrac{7}{18} \stackrel{?}{=} \dfrac{35}{80}$

33. $\dfrac{5}{\frac{1}{6}} \stackrel{?}{=} \dfrac{120}{4}$

34. $\dfrac{0.8}{4} \stackrel{?}{=} \dfrac{12}{50}$

35. Is $\dfrac{156 \text{ francs}}{30 \text{ dollars}}$ equivalent to $\dfrac{442 \text{ francs}}{85 \text{ dollars}}$?

36. Is $\dfrac{188 \text{ words}}{8 \text{ min}}$ equivalent to $\dfrac{121 \text{ words}}{5 \text{ min}}$?

In exercises 37 to 45, solve for the unknown in each proportion.

37. $\dfrac{16}{24} = \dfrac{m}{3}$

38. $\dfrac{6}{a} = \dfrac{27}{18}$

39. $\dfrac{14}{35} = \dfrac{t}{10}$

40. $\dfrac{y}{22} = \dfrac{15}{55}$

41. $\dfrac{55}{88} = \dfrac{10}{p}$

42. $\dfrac{\frac{1}{2}}{18} = \dfrac{5}{w}$

43. $\dfrac{\frac{3}{2}}{9} = \dfrac{5}{a}$

44. $\dfrac{5}{x} = \dfrac{0.6}{12}$

45. $\dfrac{s}{2.5} = \dfrac{1.5}{7.5}$

[6.3]   In exercises 46 to 52, solve each application.

46. Business and Finance   If four tickets to a civic theater performance cost $45, what will be the price for six tickets?

47. Social Science   The ratio of first-year to second-year students at a school is 8 to 7. If there are 224 second-year students, how many first-year students are there?

48. Technology   A photograph that is 5 in. wide by 7 in. tall is to be enlarged so that the new height will be 21 in. What will be the width of the enlargement?

49. Business and Finance   Marcia assembles disk drives for a computer manufacturer. If she can assemble 11 drives in 2 h, how many can she assemble in a workweek (40 h)?

50. Business and Finance   A firm finds 14 defective parts in a shipment of 400. How many defective parts can be expected in a shipment of 800 parts?

51. Geometry   The scale on a map is $\frac{1}{4}$ in. = 10 mi. How many miles apart are two towns that are 3 in. apart on the map?

52. Construction   A piece of tubing that is 16.5 cm long weighs 55 g. What is the weight of a piece of the same tubing that is 42 cm long?

[6.4]   Find each missing length.

53.

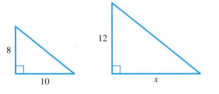

54.

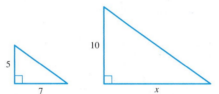

55.

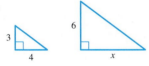

56.

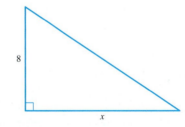

57.

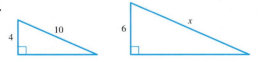

58.

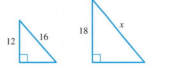

[6.5]   Complete each of the statements.

59. 11 ft = _____ in.

60. 5 mi = _____ ft

Simplify.

61. 3 ft 23 in.

62. 1 mi 6,300 ft

Add.

63.   3 ft   9 in.
     + 5 ft 10 in.

64.   2 ft   8 in.
        5 ft 10 in.
     + 4 ft 11 in.

Subtract.

65.   7 ft 11 in.
     − 2 ft   4 in.

66.   6 yd 1 ft
     − 3 yd 2 ft

Multiply.

67. 3(1 ft 5 in.)

Divide.

68. $\dfrac{5 \text{ ft } 8 \text{ in.}}{2}$

69. Construction   A room requires two pieces of floor molding 12 ft 8 in. long, one piece 6 ft 5 in. long, and one piece 10 ft long. Will 42 ft of molding be enough for the job?

Choose the most reasonable measure.

70. A marathon race

   (a) 40 km

   (b) 400 km

   (c) 400 m

71. The distance around your wrist

   (a) 15 mm

   (b) 15 cm

   (c) 1.5 m

72. The diameter of a penny

   (a) 19 cm

   (b) 1.9 mm

   (c) 19 mm

73. The width of a portable television screen

   (a) 28 mm

   (b) 28 cm

   (c) 2.8 m

Complete each statement, using a metric unit of length.

74. A matchbook is 39 _____ wide.

75. The distance from San Francisco to Los Angeles is 618 _____.

76. A 1-lb coffee can has a diameter of 10 _____.

Complete each statement.

77. 2 km = _____ m

78. 3 cm = _____ mm

79. 3,000 mm = _____ m

80. 8 m = _____ mm

81. 6 cm = _____ m

82. 8 m = _____ km

Self-Test for Chapter 6

Name _____

Section _____ Date _____

The purpose of this self-test is to help you check your progress so that you can find sections and concepts that you need to review before the next in-class exam. Allow yourself about an hour to take this test. At the end of that hour, check your answers against those given in the back of this text. If you missed a question, notice the section reference that accompanies the question. Go back to that section and reread the examples until you have mastered that particular concept.

ANSWERS

In exercises 1 to 5, write each ratio in simplest form.

1. The ratio of 7 to 19

2. The ratio of 75 to 45

3. The ratio of 8 ft to 4 yd

4. The ratio of 6 h to 3 days

5. **Statistics** A basketball team wins 26 of its 33 games during a season. What is the ratio of wins to games played? What is the ratio of wins to losses?

In exercises 6 to 8, express each rate in simplest form.

6. $\dfrac{840 \text{ mi}}{175 \text{ gal}}$

7. $\dfrac{132 \text{ dollars}}{16 \text{ h}}$

8. The unit price, if 11 gal of milk cost $28.16.

In exercises 9 to 12, determine whether the given fractions are proportional.

9. $\dfrac{3}{9} \overset{?}{=} \dfrac{27}{81}$

10. $\dfrac{6}{7} \overset{?}{=} \dfrac{9}{11}$

11. $\dfrac{9}{10} \overset{?}{=} \dfrac{27}{30}$

12. $\dfrac{\frac{1}{2}}{5} \overset{?}{=} \dfrac{2}{18}$

In exercises 13 to 18, solve for the unknown in each proportion.

13. $\dfrac{45}{75} = \dfrac{12}{x}$

14. $\dfrac{a}{24} = \dfrac{45}{60}$

15. $\dfrac{\frac{1}{2}}{p} = \dfrac{5}{30}$

16. $\dfrac{\frac{5}{6}}{8} = \dfrac{5}{a}$

17. $\dfrac{x}{0.3} = \dfrac{60}{3}$

18. $\dfrac{3}{m} = \dfrac{0.9}{4.8}$

In exercises 19 to 26, solve each application, using a proportion.

19. **Business and Finance** If ballpoint pens are marked 5 for 95¢, how much will a dozen cost?

1. _____

2. _____

3. _____

4. _____

5. _____

6. _____

7. _____

8. _____

9. _____

10. _____

11. _____

12. _____

13. _____

14. _____

15. _____

16. _____

17. _____

18. _____

19. _____

20. **Statistics**   A basketball player scores 207 points in her first 9 games. At the same rate, how many points will she score in the 28-game season?

21. **Technology**   Your new compact car travels 324 mi on 9 gal of gas. If the tank holds 16 usable gallons, how far can you drive on a tankful of gas?

22. **Social Science**   The ratio of yes to no votes in an election was 6 to 5. How many no votes were cast if 3,600 people voted yes?

23. **Business and Finance**   An assembly line can install 5 car mufflers in 4 min. At this rate, how many mufflers can be installed in an 8-h shift?

24. **Crafts**   Instructions on a package of concentrated plant food call for 2 teaspoons (tsp) to 1 qt of water. We wish to use 3 gal of water. How much of the plant food concentrate should be added to the 3 gal of water?

25. **Geometry**   A 10-ft fence casts a 16-ft shadow. How tall is a nearby tree that casts an 88-ft shadow?

26. **Geometry**   A meterstick casts a shadow that is 0.6 m. How tall is a nearby pole that casts a shadow of 12 m?

In exercises 27 and 28, do the indicated operations.

27.   7 ft 9 in.
    + 3 ft 8 in.

28.   7 lb   3 oz
    − 4 lb 10 oz

29. **Construction**   The Martins are fencing in a rectangular yard that is 110 ft long by 40 ft wide. If the fencing costs $3.50 per linear foot, what will be the total cost of the fencing?

In exercises 30 to 32, choose the reasonable measure.

30. The width of your hand

    (a) 50 cm
    (b) 10 cm
    (c) 1 m

31. The speed limit on a freeway

    (a) $9 \dfrac{km}{h}$

    (b) $90 \dfrac{km}{h}$

    (c) $90 \dfrac{m}{h}$

32. The height of a basketball player

    (a) 21 cm
    (b) 21 dm
    (c) 21 m

Cumulative Review
for Chapters 1 to 6

The following exercises are presented to help you review concepts from earlier chapters that you may have forgotten. This section is meant as review material and not as a comprehensive exam. The answers are presented in the back of the text. If you have difficulty with any of these exercises, be certain to at least read through the summary related to that section.

Name _____

Section _____ Date _____

ANSWERS

In exercises 1 to 3, name the property that is illustrated.

1. $(7 + 3) + 8 = 7 + (3 + 8)$ **2.** $6 \cdot 7 = 7 \cdot 6$

3. $5(2 + 4) = 5 \cdot 2 + 5 \cdot 4$

In exercises 4 and 5, round the numbers to the indicated place value.

4. 5,873 to the nearest hundred   **5.** 953,150 to the nearest ten thousand

6. Evaluate: $\quad 2 + 8 \cdot 3 \div 4$

7. Write the prime factorization of 264.

8. Find the least common multiple (LCM) of 6, 15, and 45.

9. Convert to a mixed number: $\quad \dfrac{22}{7}$

10. Convert to an improper fraction: $\quad 6\dfrac{5}{8}$

In exercises 11 to 14, perform the indicated operations.

11. $\dfrac{2}{3} \cdot 1\dfrac{4}{5} \cdot \dfrac{5}{8}$ **12.** $2\dfrac{2}{7} \div 1\dfrac{11}{21}$

13. $4\dfrac{7}{8} + 3\dfrac{1}{6}$ **14.** $9 + \left(-5\dfrac{3}{8}\right)$

15. Construction A $6\dfrac{1}{2}$-in. bolt is placed through a wall that is $5\dfrac{7}{8}$ in. thick. How far does the bolt extend beyond the wall?

16. Business and Finance You pay for purchases of $13.99, $18.75, $9.20, and $5 with a $50 check. How much cash will you have left?

1. _____

2. _____

3. _____

4. _____

5. _____

6. _____

7. _____

8. _____

9. _____

10. _____

11. _____

12. _____

13. _____

14. _____

15. _____

16. _____

17. _____

18. _____

19. _____

20. _____

21. _____

22. _____

23. _____

24. _____

25. _____

26. _____

27. _____

28. _____

29. _____

30. _____

31. _____

32. _____

33. _____

34. _____

35. _____

17. Geometry   Find the area of a circle whose diameter is 3.2 ft. Use 3.14 for π and round the result to the nearest hundredth.

18. Construction   A 14-acre piece of land is being developed into home lots. If 2.8 acres of land will be used for roads, and each home site is to be 0.35 acre, how many lots can be formed?

19. Write the decimal equivalent of $\dfrac{8}{11}$. Use bar notation.

20. Solve for the unknown: $\dfrac{5}{m} = \dfrac{0.4}{9}$

21. Technology   You are using a photocopy machine to reduce an advertisement that is 14 in. wide by 21 in. long. If the new width is to be 8 in., what will the new length be?

22. The opposite of 8 is _____.

23. The absolute value of -20 is _____.

In exercises 24 to 29, evaluate.

24. $-(-12)$ **25.** $|-5|$

26. $-12 + (-6)$ **27.** $-8 - (-4)$

28. $(-6)(15)$ **29.** $48 \div (-12)$

Write using symbols.

30. 3 times the sum of x and y

31. The quotient when 5 less than n is divided by 3

Solve the equations and check your results.

32. $9x - 5 = 8x$ **33.** $-\dfrac{3}{4}x = 18$

34. $2x + 3 = 7x + 5$ **35.** $\dfrac{4}{3}x - 6 = 4 - \dfrac{2}{3}x$

PERCENT

CHAPTER 7 OUTLINE

Section 7.1   Percents, Decimals, and Fractions   page 573

Section 7.2   Solving Percent Problems Using Proportions   page 588

Section 7.3   Solving Percent Applications Using Equations   page 597

Section 7.4   Applications: Simple and Compound Interest   page 607

Section 7.5   More Applications of Percent   page 612

INTRODUCTION

Every 10 years, the United States government administers a census. The U.S. census is an attempt to accurately determine the population of the nation and the characteristics of its people.

Information about the demographic trends of counties and states affect government policies and representation in Congress. Here are a few interesting results from past censuses.

- In 1990, 10.4% of students dropped out of school before the ninth grade.
- In 2000, this rate had decreased to 7.5%.
- By 2002, the number of students enrolled in postsecondary education exceeded the number of high school students.

The collection and analysis of census data is a huge undertaking involving thousands of workers.

Many of the observations that come from the data are expressed in terms of percents, the subject of this chapter. You can find exercises related to the census in Sections 7.1, 7.3, and 7.5. At the end of the chapter, before the chapter summary, you will find a more in-depth activity involving information from the 1990 and 2000 censuses.

Pretest Chapter 7

This pretest will provide a preview of the types of exercises you will encounter in each section of this chapter. The answers for these exercises can be found in the back of the text. If you are working on your own or ahead of the class, this pretest can help you identify the sections in which you should focus more of your time.

[7.1]   **1.** Write 7% as a fraction.

2. Write 23% as a decimal.

3. Write 0.035 as a percent.

4. Write $\dfrac{4}{5}$ as a percent.

[7.2]   **5.** What is 25% of 252?

6. What percent of 500 is 45?

7. 35% of a number is 210. What is the number?

[7.4]   **8. Business and Finance**   How much simple interest will you pay on a $4,000 loan for 1 year if the interest rate is 14%?

[7.3]   **9. Business and Finance**   A salesperson earns a $400 commission on sales of $8,000. What is the commission rate?

[7.5]   **10. Business and Finance**   A salary increase of 5% amounts to a $60 monthly raise. What was the monthly salary before the increase?

11. Business and Finance   A state sales tax is 6.5%. What is the sales tax on an item purchased for $174.00?

[7.4]   **12. Business and Finance**   Adriana invests $5,000 in an account that pays 6% interest per year. How much will she have in the account after 3 years?

[7.3]   **13. Business and Finance**   Julio gets a 15% discount at a local furniture store. How much will he pay for a sofa listed at $490?

7.1 Percents, Decimals, and Fractions

7.1 OBJECTIVES

1. Use percent notation
2. Convert between a percent and a fraction
3. Convert between a percent and a mixed number
4. Convert between a percent and a decimal

NOTE The first recorded use of percent occurred in Roman times when the emperor Augustus levied a $\frac{1}{100}$ tax on all goods sold at auction.

When we considered parts of a whole in earlier chapters, we used fractions and decimals. *Percent* is another useful way of naming parts of a whole. We can think of percents as ratios whose denominators are 100. In fact, the word *percent* means "for each hundred."

Definition: Percent

Percent means for each hundred. The symbol for percent, %, can be read "out of each 100."

NOTE The oldest symbol to mean percent can be found in an Italian manuscript dating back to the year 1425. The modern symbol first appeared in the eighteenth century. (*Source: NCTM, 1969.*)

Consider the sketch

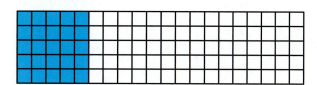

In the drawing, 25 of 100 squares are shaded. As a fraction, we write this as $\frac{25}{100}$. As a percent, we write 25%. So 25 percent of the squares are shaded.

OBJECTIVE 1

Example 1   Using Percent Notation

(a) Four out of five geography students passed their midterm exams. Write this statement, using percent notation.

NOTE The ratio of students passing to students taking the class is $\frac{4}{5}$.

$$\frac{4}{5} = \frac{80}{100} = 80\%$$

Percent means for each hundred. To obtain a denominator of 100, multiply the numerator and denominator of the original fraction by 20.

So we can say that 80% of the geography students passed.

(b) Of 50 automobiles sold by a dealer in 1 month, 35 were compact cars. Write this statement, using percent notation.

NOTE The ratio of compact cars to all cars is $\frac{35}{50}$.

$$\frac{35}{50} = \frac{70}{100} = 70\%$$

We can say that 70% of the cars sold were compact cars.

 CHECK YOURSELF 1 _____

Rewrite the statement, using percent notation: 4 *of the* 50 *parts in a shipment were defective.*

Because there are different ways of naming the parts of a whole, you need to know how to change from one of these ways to another. First we will look at changing a percent to a fraction. Because a percent is a fraction or a ratio with denominator 100, we can use this rule.

> **Property:** Changing a Percent to a Fraction
>
> To change a percent to a common fraction, divide the number before the percent symbol by 100. Note that this is equivalent to multiplying by $\dfrac{1}{100}$.

The use of this rule is shown in Example 2.

OBJECTIVE 2   **Example 2   Changing a Percent to a Fraction**

Change each percent to a fraction.

(a) $7\% = \dfrac{7}{100}$

NOTE $\dfrac{25}{100}$ is simplified to $\dfrac{1}{4}$.   **(b)** $25\% = \dfrac{25}{100} = \dfrac{1}{4}$

CHECK YOURSELF 2 _____

Write 12% *as a fraction.*

If a percent is *greater than 100,* the resulting fraction will be *greater than 1.* This is shown in Example 3.

OBJECTIVE 3   **Example 3   Changing a Percent to a Mixed Number**

Change 150% to a mixed number.

$$150\% = \frac{150}{100} = 1\frac{50}{100} = 1\frac{1}{2}$$

CHECK YOURSELF 3 _____

Write 125% *as a mixed number.*

The fractional equivalents of certain percents should be memorized.

$$33\frac{1}{3}\% = 33\frac{1}{3}\left(\frac{1}{100}\right) = \frac{100}{3} \cdot \frac{1}{100} = \frac{1}{3} \qquad 66\frac{2}{3}\% = 66\frac{2}{3}\left(\frac{1}{100}\right) = \frac{200}{3} \cdot \frac{1}{100} = \frac{2}{3}$$

It is best to try to remember these fractional equivalents.

In Example 2, we wrote percents as fractions by dividing the number before the percent symbol by 100. We use this to convert between percents and decimals. Consider 25% from Example 2(b).

$$25\% = \frac{25}{100} = 0.25$$

Because converting a percent to a fraction involves dividing by 100, we can convert a percent to a decimal by moving the decimal point two places to the left and removing the % symbol.

Similarly, we have

$$0.07 = \frac{7}{100} = 7\%$$

So, to convert from a decimal to a percent, we simply move the decimal two places to the right and include the % symbol.

> **Property: Converting Between Percent and Decimal Notation**
>
> 1. To change a percent to a decimal, divide the number by 100 and remove the % symbol. This is equivalent to shifting the decimal two places to the left.
> 2. To change a decimal to a percent, multiply the number by 100 and add the % symbol. This is equivalent to shifting the decimal two places to the right.

OBJECTIVE 4

Example 4   Changing a Percent to a Decimal

Change each percent to its decimal equivalent.

(a) $25\% = 0.25$ The decimal point in 25% is understood to be after the 5.

(b) $8\% = 0.08$ We must add a zero to move the decimal point.

Write each number as a percent.

(c) $0.18 = 18\%$ $0.18 = \frac{18}{100} = 18\%$

(d) $0.03 = 3\%$ $0.03 = \frac{3}{100} = 3\%$

CHECK YOURSELF 4

Write each number using decimal notation.

(a) 5%   **(b)** 32%

Write each number using percent notation.

(c) 0.27   **(d)** 0.02

When converting more irregular numbers, we need to be careful. Simply remember to move the decimal point two places in the proper direction. If necessary, you may need to rewrite a number in decimal form.

Example 5 Converting More Irregular Numbers

Write each number using decimal notation.

NOTE A percent greater than 100 gives a decimal greater than 1.

(a) $130\% = 1.30$

(b) $0.5\% = 0.005$

NOTE Write the common fraction as a decimal.

(c) $\dfrac{3}{4}\% = 0.75\% = 0.0075$

Write each number using percent notation.

(d) $0.045 = 4.5\%$ or $4\dfrac{1}{2}\%$ $\qquad 0.045 = \dfrac{45}{1,000} = \dfrac{45}{10}\left(\dfrac{1}{100}\right)$
$$= 4.5\%$$

(e) $0.003 = 0.3\%$ or $\dfrac{3}{10}\%$ $\qquad 0.003 = \dfrac{3}{1,000} = \dfrac{3}{10}\left(\dfrac{1}{100}\right)$
$$= 0.3\%$$

NOTE A decimal greater than 1 always gives a percent greater than 100.

(f) $1.25 = 125\%$ $\qquad 1.25 = \dfrac{125}{100} = 125\left(\dfrac{1}{100}\right) = 125\%$

CHECK YOURSELF 5

Write each number using decimal notation.

(a) 8.5% $\qquad$ **(b)** 115% $\qquad$ **(c)** $7\dfrac{1}{2}\%$

Write each number using percent notation.

(d) 0.0025 $\qquad$ **(e)** 0.125 $\qquad$ **(f)** 1.3

This next rule allows us to change fractions to percents.

Property: Changing a Fraction to a Percent

NOTE You may want to review Section 5.5 on writing decimal equivalents.

To change a fraction to a percent, write the decimal equivalent of the fraction. Then use the changing a decimal to a percent rule to obtain the percent.

Some fractions have repeating-decimal equivalents. In writing these as percents, we will either round to some indicated place or use a fractional remainder form.

OBJECTIVE 2 **Example 6   Changing a Fraction to a Percent**

(a) Write $\dfrac{3}{5}$ as a percent.

First write the decimal equivalent.

$\dfrac{3}{5} = 0.6$ To find the decimal equivalent, just divide the denominator into the numerator.

Now write the percent.

NOTE Move the decimal point two places to the right and attach the percent symbol.

$\dfrac{3}{5} = 0.6 = 60\%$

(b) Write $\dfrac{1}{3}$ as a percent.

$\dfrac{1}{3} = 0.33\overline{3} = 33\dfrac{1}{3}\%$

(c) Write $\dfrac{5}{7}$ as a percent.

NOTE In this case, we round the decimal equivalent. Then we write the percent.

$\dfrac{5}{7} = 0.714$ (to the nearest thousandth)

$= 71.4\%$ (to the nearest tenth of a percent)

 CHECK YOURSELF 6

Write as a percent.

(a) $\dfrac{3}{4}$ **(b)** $\dfrac{2}{3}$ **(c)** $\dfrac{2}{9}$ to the nearest tenth of a percent

To write a mixed number as a percent, we use exactly the same steps.

OBJECTIVE 3 **Example 7   Changing a Mixed Number to a Percent**

Write $1\dfrac{1}{4}$ as a percent.

NOTE The resulting percent must be greater than 100 because the original mixed number is greater than 1.

$1\dfrac{1}{4} = 1.25 = 125\%$

 CHECK YOURSELF 7

Write $1\dfrac{2}{5}$ as a percent.

Units Analysis

When computing a percentage, note that, in the result, the units are rarely expressed. Each time we say "percent" we are essentially saying, "numerator units per 100 denominator units."

Examples

Of 800 students, 200 were boys. What percent of the students were boys?

$$\frac{200 \text{ boys}}{800 \text{ students}} = 0.25 = 25\%$$

But what happened to our units? At the decimal, the units $\frac{\text{boys}}{\text{student}}$ (0.25 boys per student) wouldn't make much sense, but we can read the % as "25 boys per 100 students" and have a reasonable unit phrase.

Of 500 computers sold, 180 were equipped with a scanner. What percent were equipped with a scanner?

$$\frac{180}{500} = 0.36 = 36\%$$

36 computers were equipped with a scanner for each (per) 100 computers sold.

Let us briefly summarize the techniques presented in this section.

- To convert a percent to its decimal equivalent: Shift the decimal point two places to the left and remove the percent symbol.
- To convert a decimal to its percent equivalent: Shift the decimal point two places to the right and add the percent symbol.
- To convert a fraction to a percent: Use division to rewrite the fraction as a decimal and convert the decimal to a percent.
- To convert a percent to a fraction: Remove the percent symbol and divide the remaining number by 100. Simplify the result.

READING YOUR TEXT

The following fill-in-the-blank exercises are designed to assure that you understand the key vocabulary used in this section. Each sentence comes directly from the section. You will find the correct answers in Appendix C.

Section 7.1

(a) Percent means "for each _____."

(b) To change a decimal to a percent, move the decimal point two places to the _____ and attach the percent symbol.

(c) A decimal number greater than 1 always gives a percent greater than _____.

(d) When we say "percent," we are saying "numerator units per 100 _____ units."

CHECK YOURSELF ANSWERS

1. 8% were defective.   **2.** $\dfrac{3}{25}$ **3.** $1\dfrac{1}{4}$ **4. (a)** 0.05; **(b)** 0.32; **(c)** 27%; **(d)** 2%

5. (a) 0.085; **(b)** 1.15; **(c)** 0.075; **(d)** 0.25% or $\dfrac{1}{4}$%; **(e)** 12.5%; **(f)** 130%

6. (a) 75%; **(b)** $66\dfrac{2}{3}$%; **(c)** 22.2%   **7.** 140%

Name _____

Section _____ Date _____

ANSWERS

1. _____   2. _____

3. _____   4. _____

5. _____   6. _____

7. _____

8. _____

9. _____

10. _____

11. _____

12. _____

13. _____

14. _____

15. _____

16. _____

17. _____

18. _____

19. _____

20. _____

21. _____

22. _____

7.1 **Exercises**

Use percents to name the shaded portion of each drawing.

1. 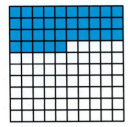 **ALEKS**

2. 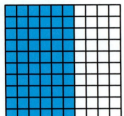 **ALEKS**

3.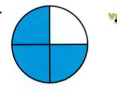

4.

Rewrite each statement, using percent notation.

5. Out of every 100 eligible people, 53 voted in a recent election.

6. You receive $5 in interest for every $100 saved for 1 year.

7. Out of every 100 entering students, 74 register for English composition.

8. 17 out of 20 college students work at part-time jobs. **ALEKS**

9. Of the 20 students in an algebra class, 5 receive a grade of A. **ALEKS**

10. Of the 50 families in a neighborhood, 31 have children in public schools. **ALEKS**

Write as fractions or mixed numbers.

11. 65% **ALEKS**   12. 48% **ALEKS**   13. 50% **ALEKS**

14. 52% **ALEKS**   15. 46% **ALEKS**   16. 35% **ALEKS**

17. 66% **ALEKS**   18. 4% **ALEKS**   19. 150%

20. 140%   21. $166\frac{2}{3}\%$ 22. $133\frac{1}{3}\%$

Write as decimals.

23. 20% ALEKS

24. 70% ALEKS

25. 5% ALEKS

26. 7%

27. 135%

28. 250% ALEKS

29. 23.6%

30. 10.5%

31. 6.4% VIDEO ALEKS

32. 3.5%

33. 0.2%

34. 0.5% ALEKS

35. $7\frac{1}{2}\%$ VIDEO

36. $8\frac{1}{4}\%$

Solve the applications.

37. Statistics Automobiles account for 85% of the travel between cities in the United States. What fraction does this percent represent?

38. Statistics Automobiles and small trucks account for 84% of the travel to and from work in the United States. What fraction does this percent represent?

39. Explain the difference between $\frac{1}{4}$ of a quantity and $\frac{1}{4}\%$ of a quantity.

40. Match the percents in column A with their equivalent fractions in column B.

Column A	Column B
(a) $37\frac{1}{2}\%$	(1) $\frac{3}{5}$
(b) 5%	(2) $\frac{5}{8}$
(c) $33\frac{1}{3}\%$	(3) $\frac{1}{20}$
(d) $83\frac{1}{3}\%$	(4) $\frac{3}{8}$
(e) 60%	(5) $\frac{5}{6}$
(f) $62\frac{1}{2}\%$	(6) $\frac{1}{3}$

23. _____

24. _____

25. _____

26. _____

27. _____

28. _____

29. _____

30. _____

31. _____

32. _____

33. _____

34. _____

35. _____

36. _____

37. _____

38. _____

39. _____

40. _____

41. Statistics Use the percentages given in the bar graph to complete the table below.

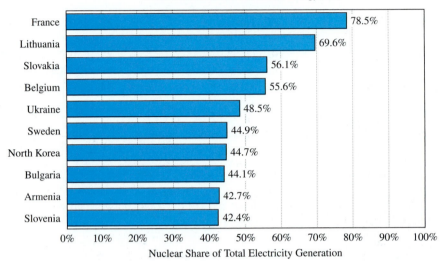

Nations Most Reliant on Nuclear Energy, 2005

Nuclear Share of Total Electricity Generation

Source: International Atomic Energy Agency, July 2006

Nation	Decimal Equivalent	Fraction Equivalent
France		
Lithuania		
Slovakia		
Belgium		
Ukraine		
Sweden		
North Korea		
Bulgaria		
Armenia		
Slovenia		

42. The minimum daily values (MDV) for certain foods are given. They are based on a 2,000 calorie per day diet. Find decimal and fractional notation for the percent notation in each sentence.

(a) 1 ounce of Tostitos provides 9% of the MDV of fat.

(b) $\frac{1}{2}$ cup of B & M baked beans contains 15% of the MDV of sodium.

(c) $\frac{1}{2}$ cup of Campbells' New England clam chowder provides 6% of the MDV of iron.

(d) 2 ounces of Star Kist tuna provide 27% of the MDV of protein.

(e) Four 4-in. Aunt Jemima pancakes provide 33% of the MDV of sodium.

(f) 36 grams of Pop-Secret butter popcorn provide 2% of the MDV of sodium.

43.

44.

45.

46.

47.

48.

49.

50.

51.

52.

53.

54.

55.

56.

57.

58.

59.

60.

61.

62.

63.

64.

65.

66.

Write each decimal as a percent.

43. 0.05 **ALEKS**

44. 0.13 **ALEKS**

45. 0.18 **ALEKS**

46. 0.63 **ALEKS**

47. 0.7 **ALEKS**

48. 0.6 **ALEKS**

49. 1.10 **ALEKS**

50. 2.50 **ALEKS**

51. 4.40   **ALEKS**

52. 5 **ALEKS**

53. 0.004 **VIDEO** **ALEKS**

54. 0.001 **ALEKS**

Write each fraction as a percent.

55. $\dfrac{1}{4}$ **ALEKS**

56. $\dfrac{4}{5}$ **ALEKS**

57. $\dfrac{2}{5}$ **ALEKS**

58. $\dfrac{1}{2}$ **ALEKS**

59. $3\dfrac{1}{2}$ **ALEKS**

60. $\dfrac{2}{3}$ **ALEKS**

61. $\dfrac{1}{6}$ **ALEKS**

62. $\dfrac{3}{16}$ **ALEKS**

63. $\dfrac{7}{9}$ (to nearest tenth of a percent)

64. $\dfrac{5}{11}$ (to nearest tenth of a percent)

VIDEO

In exercises 65 to 68, partially shaded decimal squares are given. Express the partially shaded region as (a) a decimal (b) a fraction, and (c) a percent.

65.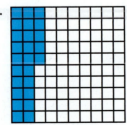

66.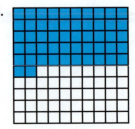

67.

68.

69. Complete the table of equivalents. Round decimals to the nearest ten thousandth. Round percents to the nearest hundredth of a percent.

Fraction	Decimal	Percent
$\dfrac{7}{12}$		
	0.08	
		35%
	0.265	
		$4\dfrac{3}{8}\%$
$\dfrac{11}{18}$		

Business and Finance Business travelers were asked how much they spent on different items during a business trip. The circle shows the results for every $1,000 spent. Use this information to answer exercises 70 to 73.

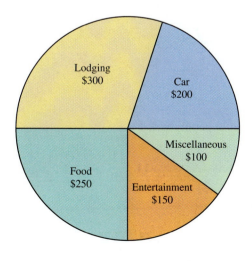

70. What percent was spent on car expenses?

71. _____

72. _____

73. _____

74. _____

75. _____

76. _____

77. _____

78. _____

79. _____

80. _____

71. What percent was spent on food?

72. Where was the least amount of money spent? What percent was this?

73. What percent was spent on food and lodging together?

In exercises 74 and 75, fill in the blank with always, sometimes, or never.

74. A decimal that is greater than 1 will _____ convert to a percent that is greater than 100%.

75. A mixed number is _____ equivalent to a decimal between 0 and 1.

76. Describe a real-world situation in which you would be likely to express a number in percent form.

77. Describe a real-world situation in which you would be likely to express a number in decimal form.

78. Describe a real-world situation in which you would be likely to express a number in fraction form.

79. When the national census occurs, people are given forms to fill out voluntarily and return. In 1970, the nonresponse rate was 15%, that is, 15% of the households did not initially comply. Express this percent as a simplified fraction.

80. In 1990, $\frac{1}{3}$ of the households did not initially respond with completed census forms. Express this fraction as a percent.

Answers

1. 35%   **3.** 75%   **5.** 53% of the eligible people voted.

7. 74% register for English composition.

9. $\frac{5}{20} = \frac{25}{100} = 25\%$; 25% of the students receive As.   **11.** $\frac{13}{20}$ **13.** $\frac{1}{2}$

15. $\frac{23}{50}$ **17.** $\frac{33}{50}$ **19.** $1\frac{1}{2}$ **21.** $1\frac{2}{3}$ **23.** 0.2   **25.** 0.05

27. 1.35   **29.** 0.236   **31.** 0.064   **33.** 0.002   **35.** 0.075   **37.** $\frac{17}{20}$

39.

41.

Nation	Decimal Equivalent	Fraction Equivalent
France	0.785	$\dfrac{157}{200}$
Lithuania	0.696	$\dfrac{87}{125}$
Slovakia	0.561	$\dfrac{561}{1,000}$
Belgium	0.556	$\dfrac{139}{250}$
Ukraine	0.485	$\dfrac{97}{200}$
Sweden	0.449	$\dfrac{449}{1,000}$
North Korea	0.447	$\dfrac{447}{1,000}$
Bulgaria	0.441	$\dfrac{441}{1,000}$
Armenia	0.427	$\dfrac{427}{1,000}$
Slovenia	0.424	$\dfrac{53}{125}$

43. 5% **45.** 18% **47.** 70% **49.** 110% **51.** 440%

53. 0.4% or $\dfrac{2}{5}$% **55.** $\dfrac{1}{4} = 0.25 = 25\%$ **57.** 40% **59.** 350%

61. $16\dfrac{2}{3}$% **63.** 77.8% **65.** 0.25, $\dfrac{1}{4}$, 25% **67.** 0.47, $\dfrac{47}{100}$, 47%

69. Fractions: $\dfrac{2}{25}$, $\dfrac{7}{20}$, $\dfrac{53}{200}$, $\dfrac{7}{160}$; decimals: 0.5833, 0.35, 0.0438, 0.6111; percents: 58.33%, 8%, 26.5%, 61.11% **71.** 25% **73.** 55%

75. Never **77.** ✍ **79.** $\dfrac{3}{20}$

7.2 Solving Percent Problems Using Proportions

There are many practical applications of our work with percents. All of these problems have three basic parts that need to be identified. We will look at some definitions that will help with that process.

> **Definition:** Base, Amount, and Rate
>
> The **base** is the whole in a problem. It is the standard used for comparison.
>
> The **amount** is the part of the whole being compared to the base.
>
> The **rate** is the ratio of the amount to the base. It is written as a percent.

We now look at an example of identifying the three parts in a percent problem.

OBJECTIVE 1

Example 1 **Identifying the Rate, Base, and Amount**

Determine the rate, base, and amount in this problem:

12% of 800 is what number?

Finding the *rate* is not difficult. Just look for the percent symbol or the word *percent.* In this exercise, 12% is the rate.

The *base* is the whole. Here, it follows the word *of.* 800 is the whole or the base.

The *amount* remains after the rate and the base have been found. Here, the amount is the unknown. It follows the word *is.* "What number" asks for the unknown amount.

CHECK YOURSELF 1

Find the rate, base, and amount in the statements or questions.

(a) 75 is 25% of 300.   **(b)** 20% of what number is 50?

We use percents to solve a variety of applied problems. In all of these situations, you have to identify the three parts of the problem. We work through some examples intended to help you build that skill.

Example 2 **Identifying the Rate, Base, and Amount**

Determine the rate, base, and amount in the application: In an algebra class of 35 students, 7 received a grade of A. What percent of the class received an A?

The *base* is the whole in the problem, or the number of students in the class. 35 is the base.

The *amount* is the portion of the base, here the number of students that receive the A grade. 7 is the amount.

The *rate* is the unknown in this example. "What percent" asks for the unknown rate.

CHECK YOURSELF 2

Determine the rate, base, and amount in the application: In a shipment of 150 parts, 9 of the parts were defective. What percent were defective?

We have now seen how to determine the amount, the rate, and the base. As you will see in the remainder of this section, our work in Chapter 6 with proportions allows us to solve many types of percent problems in an identical fashion.

First, we write what is called the **percent proportion.**

> **Property: The Percent Proportion**
>
> $$\frac{\text{Amount}}{\text{Base}} = \frac{r}{100}$$
>
> In symbols,
>
> $$\frac{A}{B} = \frac{r}{100}$$

NOTE $R = \dfrac{r}{100}$ is the rate.

In any percent problem we know two of the three quantities (A, B, R), so we can always solve for the unknown term. Consider the use of the percent proportion in Example 3.

OBJECTIVE 2

Example 3   Solving a Problem Involving an Unknown Amount

NOTE This is an *unknown-amount problem.*

_____ is 30% of 150.

$$\uparrow \qquad \uparrow \qquad \uparrow$$
$$A \qquad R \quad B$$

Substitute the values into the percent proportion.

RECALL If

$$\frac{a}{b} = \frac{c}{d}$$

then

$$ad = bc$$

$$\frac{A}{150} = \frac{30}{100} \;\; r$$

The amount A is the unknown quantity of the proportion.

We solve the proportion with the methods of Section 6.3.

$$100A = 150 \cdot 30$$

$$100A = 4{,}500$$

Divide by the coefficient, 100.

$$\frac{\overset{1}{\cancel{100}A}}{\cancel{100}} = \frac{4,500}{100} = 45$$

$$A = 45$$

The amount is 45. This means that 45 is 30% of 150.

 CHECK YOURSELF 3 _____

Use the percent proportion to answer this question: What is 24% of 300?

The same percent proportion works if you want to find the rate.

Example 4 Solving a Problem Involving an Unknown Rate

NOTE This is an *unknown-rate problem.*

_____% of 400 is 72.

   ↑   ↑   ↑

  R   *B*   *A*

Substitute the known values into the percent proportion.

A

$$\frac{72}{400} = \frac{r}{100} \qquad \text{\textit{r}, the rate, is the unknown quantity in this case.}$$

B

Solving, we get

$$400r = 7,200$$

$$\frac{\overset{1}{\cancel{400}r}}{\cancel{400}} = \frac{7,200}{400} = 18$$

$$r = 18$$

The rate is 18%. So 18% of 400 is 72.

 CHECK YOURSELF 4 _____

Use the percent proportion to answer this question: What percent of 50 is 12.5?

Finally, we use the same proportion to find an unknown base.

Example 5 Solving a Problem Involving an Unknown Base

NOTE This is an *unknown-base problem.*

40% of _____ is 200.

↑   ↑   ↑

R   *B*   *A*

Substitute the known values into the percent proportion.

$$\underset{A}{\searrow} \quad \underset{r}{\swarrow}$$

$$\frac{200}{B} = \frac{40}{100}$$ In this case B, the base, is the unknown quantity in the proportion.

Solving gives

$$40B = 200 \cdot 100$$

$$\frac{\overset{1}{\cancel{40}}B}{\underset{1}{\cancel{40}}} = \frac{20,000}{40} = 500$$

$$B = 500$$

The base is 500, and 40% of 500 is 200.

 CHECK YOURSELF 5 _____

288 is 60% of what number?

Remember that a percent (the rate) can be greater than 100.

Example 6   Solving a Percent Problem

NOTE The rate is 125%; the base is 300. We want to find the amount.

NOTE When the rate is greater than 100%, the amount will be *greater than* the base.

What is 125% of 300?

In the percent proportion, we have

$$\frac{A}{300} = \frac{125}{100}$$

So $100A = 300 \cdot 125$.

Dividing by 100 yields

$$A = \frac{37,500}{100} = 375$$

So 375 is 125% of 300.

 CHECK YOURSELF 6 _____

Find 150% of 500.

We next look at two examples of solving percent problems involving fractions of a percent.

Example 7   Solving a Percent Problem

34 is 8.5% of what number?

Using the percent proportion yields

NOTE The amount is 34; the rate is 8.5%. We want to find the base.

$$\frac{34}{B} = \frac{8.5}{100}$$

Solving, we have

$$8.5B = 34 \cdot 100$$

or

$$B = \frac{3,400}{8.5} = 400 \qquad \text{Divide by 8.5.}$$

So 34 is 8.5% of 400.

CHECK YOURSELF 7 _____

12.5% of what number is 75?

     Estimating is an important and useful skill. When using a calculator, it is always a good idea to check to see whether the answer you get is reasonable. That is done by estimation. Sometimes operations with fractions can be used to estimate the result to a problem involving percents. Example 8 illustrates such a case.

OBJECTIVE 3 | **Example 8**   **Estimating Percentages**

Estimate 19.3% of 500.

Round the rate to 20% $\left(\text{as a fraction, } \frac{1}{5}\right)$. An estimate of the amount is then

$$\frac{1}{5} \cdot 500 = 100$$

Rounded rate   Base   Estimate of amount

CHECK YOURSELF 8 _____

Estimate the amount.

20.2% of 800

READING YOUR TEXT _____

The following fill-in-the-blank exercises are designed to assure that you understand the key vocabulary used in this section. Each sentence usually comes directly from the section. You will find the correct answers in Appendix C.

Section 7.2

(a) In a percent problem, the _____ is the whole.

(b) In a percent problem, the _____ is the part of the whole being compared to the base.

(c) In a percent problem, the _____ is the ratio of the amount to the base.

(d) $\dfrac{A}{B} = \dfrac{r}{100}$ is called the percent _____.

CHECK YOURSELF ANSWERS

1. **(a)** $R = 25\%$, $B = 300$, $A = 75$; **(b)** $R = 20\%$, $B =$ "what number," $A = 50$

2. $B = 150$, $A = 9$, $R =$ "what percent" (the unknown)

3.

$$\underset{B}{\overset{r}{\frac{A}{300} = \frac{24}{100}}}$$

$100A = 7{,}200$; $A = 72$

4.

$$\underset{B}{\overset{A}{\frac{12.5}{50} = \frac{r}{100}}}$$

$50r = 1{,}250$; $r = 25$ so $R = 25\%$

5.

$$\overset{A}{\underset{}{\frac{288}{B}}} = \overset{r}{\frac{60}{100}}$$

$60B = 28{,}800$; $B = 480$

6. 750   7. 600   8. 160

Name _____

Section _____ Date _____

ANSWERS

1. _____	2. _____
3. _____	4. _____
5. _____	6. _____
7. _____	8. _____
9. _____	10. _____
11. _____	12. _____
13. _____	14. _____
15. _____	16. _____
17. _____	18. _____
19. _____	20. _____
21. _____	22. _____
23. _____	24. _____
25. _____	
26. _____	
27. _____	
28. _____	
29. _____	
30. _____	
31. _____	
32. _____	
33. _____	
34. _____	

7.2 Exercises

Solve each of the problems involving percent.

1. What is 35% of 600?

2. 20% of 400 is what number?

3. 45% of 200 is what number?

4. What is 40% of 1,200? **ALEKS**

5. Find 40% of 2,500.

6. What is 75% of 120? **ALEKS**

7. What percent of 50 is 4?

8. 51 is what percent of 850?

9. What percent of 500 is 45?

10. 14 is what percent of 200?

11. What percent of 200 is 340?

12. 392 is what percent of 2,800?

13. 46 is 8% of what number?

14. 7% of what number is 42?

15. Find the base if 11% of the base is 55.

16. 16% of what number is 192?

17. 58.5 is 13% of what number?

18. 21% of what number is 73.5?

19. Find 110% of 800.

20. What is 115% of 600?

21. What is 108% of 4,000?

22. Find 160% of 2,000.

23. 210 is what percent of 120?

24. What percent of 40 is 52?

25. 360 is what percent of 90?

26. What percent of 15,000 is 18,000?

27. 625 is 125% of what number?

28. 140% of what number is 350?

29. Find the base if 110% of the base is 935.

30. 130% of what number is 1,170?

31. Find 8.5% of 300.

32. $8\frac{1}{4}$% of 800 is what number?

33. Find $11\frac{3}{4}$% of 6,000.

34. What is 3.5% of 500?

35. What is 5.25% of 3,000?

36. What is 7.25% of 7,600?

37. 60 is what percent of 800?

38. 500 is what percent of 1,500?

39. What percent of 180 is 120?

40. What percent of 800 is 78?

41. What percent of 1,200 is 750?

42. 68 is what percent of 800?

43. 10.5% of what number is 420?

44. Find the base if $11\frac{1}{2}$% of the base is 46.

45. 58.5 is 13% of what number?

46. 6.5% of what number is 325?

47. 195 is 7.5% of what number?

48. 21% of what number is 73.5?

Estimate the amount in each of the exercises.

49. Find 25.8% of 4,000.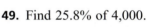

50. What is 48.3% of 1,500?

51. 74.7% of 600 is what number?

52. 9.8% of 1,200 is what number?

53. Find 152% of 400.

54. What is 118% of 5,000?

55. It is customary when eating in a restaurant to leave a 15% tip.

    (a) Outline a method to do a quick approximation for the amount of tip to leave.

    (b) Use this method to figure a 15% tip on a bill of $47.76.

56. Business and Finance If the restaurant tax in a certain city is 9%, how might you use this to find a reasonable amount to leave as a tip?

35. _____

36. _____

37. _____

38. _____

39. _____

40. _____

41. _____

42. _____

43. _____

44. _____

45. _____

46. _____

47. _____

48. _____

49. _____

50. _____

51. _____

52. _____

53. _____

54. _____

55. _____

56. _____

57. _____

58. _____

59. _____

60. _____

61. _____

62. _____

In exercises 57 to 60, fill in the blank with less than, equal to, or greater than.

57. 95% of a number is _____ the number.

58. 150% of a number is _____ the number.

59. 100% of a number is _____ the number.

60. 105% of a number is _____ the number.

In exercises 61 and 62, fill in the blank with always, sometimes, or never.

61. In a percent problem, the base is _____ larger than the amount.

62. In a percent problem, the number attached to the percent symbol is _____ the rate.

Answers

1. 210   **3.** 90   **5.** 1,000   **7.** 8%   **9.** 9%   **11.** 170%   **13.** 575
15. 500   **17.** 450   **19.** 880   **21.** 4,320   **23.** 175%   **25.** 400%
27. 500   **29.** 850   **31.** 25.5   **33.** 705   **35.** 157.5   **37.** 7.5%

39. $66\frac{2}{3}$%   **41.** 62.5%   **43.** 4,000   **45.** 450   **47.** 2,600

49. 1,000   **51.** 450   **53.** 600   **55.**   **57.** Less than

59. Equal to   **61.** Sometimes

7.3 Solving Percent Applications Using Equations

7.3 OBJECTIVES

1. Solve percent applications using equations
2. Solve percent applications involving commission
3. Solve percent applications involving sales tax
4. Solve percent applications involving discount or markup

The concept of percent is perhaps the most frequently encountered mathematical idea that we will consider in this text. In this section and Sections 7.4 and 7.5, we discuss a few of the most common applications. We also discuss the special terms that are used in these applications. First, we look at the equations generated from the percent proportion.

In Section 7.2, we found the answer to the question, "What is 30% of 150?" by setting up the proportion

$$\frac{A}{150} = \frac{30}{100}$$

Using the multiplication rule for equations from Chapter 4, we can rewrite this statement as

$$A = \frac{30}{100} \cdot 150$$

and solve for A. We could also obtain the same equation directly from the question: What is 30% of 150?

$$A = \frac{30}{100} \cdot 150$$

When we write an equation directly from a statement or question, we are using the **method of translation.**

OBJECTIVE 1

> ### Example 1 Translating an Application into an Equation
>
> A salesman sells a used car for $9,500. His commission rate is 4%. What will be his commission for the sale?
>
> First we must rephrase the problem as a single question. Here we have, "What is 4% of $9,500?" Translate the question into an equation.

NOTE "Of" indicates multiplication in this case.

What is 4% of $9,500?

$$A = \frac{4}{100} \cdot 9,500 \qquad \text{Divide the numerator and denominator by 100.}$$

$$A = 4 \cdot 95$$

$$A = 380$$

His commission is $380.

 CHECK YOURSELF 1 _____

Jenny sells a $36,000 building lot. If her real estate commission rate is 5%, what commission will she receive for the sale?

If the question that is to be translated is not obvious, the equation can always be derived from the percent proportion.

OBJECTIVE 2

Example 2   Solving a Percent Problem

A clerk sold $3,500 in merchandise during one week. If he received a commission of $140, what was the commission rate?

The base is $3,500, and the amount is the commission of $140. Using the percent proportion we have

$$\frac{140}{3,500} = \frac{r}{100} \quad \text{or} \quad \frac{r}{100} = \frac{140}{3,500}$$

Multiplying both sides by 100, we get

$$r = \frac{14,000}{3,500}$$

$$r = 4$$

The commission rate is 4%.

 CHECK YOURSELF 2 _____

On a purchase of $500 you pay a sales tax of $21. What is the tax rate?

Example 3, involving a commission, shows how to find the total sold.

Example 3   Solving a Percent Problem

A saleswoman has a commission rate of 3.5%. To earn $280, how much must she sell?

The rate is 3.5%. The amount is the commission, $280. We want to find the base. In this case, this is the quantity that the saleswoman needs to sell.

By the percent proportion

$$\frac{280}{B} = \frac{3.5}{100} \quad \text{or} \quad 3.5B = 280 \cdot 100$$

$$B = \frac{28,000}{3.5} = 8,000$$

The saleswoman must sell $8,000 to earn $280 in commissions.

 CHECK YOURSELF 3 _____

Kerri works with a commission rate of 5.5%. If she wants to earn $825 in commissions, find the total sales that she must make.

OBJECTIVE 3

Example 4   Solving a Percent Problem

A state taxes sales at 5.5%. How much sales tax will you pay on a purchase of $48?

The tax you pay is the amount (the part of the whole). Here the base is the purchase price, $48, and the rate is the tax rate, 5.5%. The question is,

NOTE In an application involving taxes, the tax paid is usually the amount.

What is 5.5% of $48?

$$A = \frac{5.5}{100} \cdot 48$$

$$A = \frac{264}{100} = 2.64 \qquad 48 \cdot 5.5 = 264$$

The sales tax paid is $2.64.

CHECK YOURSELF 4

Suppose that a state has a sales tax rate of $6\frac{1}{2}$%. If you buy a used car for $1,200, how much sales tax must you pay?

Percents are also used to deal with store markups or discounts. Consider Example 5.

OBJECTIVE 4

Example 5   Solving a Percent Problem

A store marks up items to make a 30% profit. If an item costs $7.50 from the supplier, what will the selling price be?

The base is the cost of the item, $7.50, and the rate is 30%. In the percent proportion, the markup is the amount in this application.

The markup is 30% of $7.50.

$$A = \frac{30}{100} \cdot 7.5$$

$$A = 2.25$$

NOTE

Selling price = original cost + markup

The markup is $2.25. Finally, we have

Selling price = $7.50 + $2.25 = $9.75   Add the cost and the markup to find the selling price.

CHECK YOURSELF 5

A store wants to discount (or mark down) an item by 25% for a sale. If the original price of the item was $45, find the sale price. [Hint: Find the discount (the amount the item will be marked down) and subtract that from the original price.]

One of the most common (and desirable!) applications of percents is the discount.

Example 6   Solving an Application Involving Discount

One of the benefits Vlade gets for working at worlddom.com is a 30% discount on all on-line purchases. Last month he ordered $230 worth of software on-line. What was his total discount?

We rephrase the problem as a single question. Here, we have, "What is 30% of $230?" Translate the question into an equation.

What is 30% of $230?

$$A = \frac{30}{100} \cdot 230$$

$$A = \frac{3\cancel{0}}{10\cancel{0}} \cdot 23\cancel{0}$$

$$A = 3 \cdot 23$$

$$A = 69$$

His discount was $69.

CHECK YOURSELF 6

Indira receives a 26% discount for all of her tickets on AirGeorgia flights. What would be her discount on a $580 flight?

In many everyday applications of percent, the computations required become quite lengthy, and so your calculator can be a great help. We will look at some examples.

Example 7	Solving a Problem Involving an Unknown Rate

In a test, 41 of 720 lightbulbs burn out before their advertised life of 700 h. What percent of the bulbs failed to last the advertised life?

We know the amount and base but want to find the percent (a rate). We will use the percent proportion for the solution.

$$\frac{\overset{A}{41}}{\underset{B}{720}} = \frac{r}{100}$$

$$720r = 4,100$$

$$r = \frac{4,100}{720}$$

Now use your calculator to divide

4,100 ÷ 720 = 5.6944444

5.7% of the lightbulbs failed.   We round the result to the nearest tenth of a percent.

CHECK YOURSELF 7

Last month, 35 of the 475 emergency calls received by the local police department were false alarms. What percent of the calls were false alarms?

Example 8   Solving a Problem Involving an Unknown Base

The price of a particular model of sofa has increased $48.20. If this represents an increase of 9.65%, what was the price before the increase?

We want to find the base (the original price). Again, we will use the percent proportion for the solution.

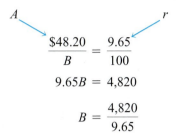

$$\frac{\$48.20}{B} = \frac{9.65}{100}$$

$$9.65B = 4,820$$

$$B = \frac{4,820}{9.65}$$

Using the calculator gives

4,820 ÷ 9.65 = 499.48187

The original price was $499.48.   Round to the nearest cent.

 CHECK YOURSELF 8 _____

The cost for medical insurance increased $818.40 last year. If this represents a 12.35% increase, what was the cost before the increase?

Example 9 demonstrates an alternative method for solving percent problems.

Example 9   Solving a Problem Involving Markup

A store marks up items 22.5% to allow for profit. If an item costs a store $36.40, what will be the selling price?

We can diagram the problem:

Cost	Markup
100%	22.5%
$36.40	$?

We could have done this example in two steps, finding the markup and then adding that amount to the cost.

This method allows you to do the problem in one step.

Selling price

122.5%

NOTE This approach may lead to time-consuming hand calculations, but using a calculator reduces the amount of work involved.

Now the base is $36.40 and the rate is 122.5%, and we want to find the amount (the selling price).

$$\frac{A}{36.40} = \frac{122.5}{100}$$

so

$$A = \frac{122.5 \cdot 36.40}{100} = \$44.59$$

The selling price should be \$44.59.

CHECK YOURSELF 9

An item costs \$75.40. If the markup is 36.2%, what is the selling price?

A similar approach will allow us to solve problems that involve a decrease in one step.

> **Example 10   Solving a Problem Involving an Unknown Amount**

NOTE We could have done a problem like this by finding the decrease and then subtracting from the original value. Again, using this method requires just one step.

Paul invests \$5,250 in a piece of property. At the end of a 6-month period, the value has decreased 7.5%. What is the property worth at the end of the period?

Again, we will diagram the problem.

Original value

100%
or \$5,250

Decrease	Ending value
7.5%	100% − 7.5% = 92.5%

So the amount (ending value) is found as

$$\frac{A}{5,250} = \frac{92.5}{100}$$

$$A = \frac{92.5 \cdot \$5,250}{100} = \$4,856.25 \qquad \text{The ending value is \$4,856.25}$$

CHECK YOURSELF 10

Tom buys a baseball card collection for \$750. After 1 year, the value has decreased 8.2%. What is the value of the collection after 1 year?

READING YOUR TEXT

The following fill-in-the-blank exercises are designed to assure that you understand the key vocabulary used in this section. Each sentence usually comes directly from the section. You will find the correct answers in Appendix C.

Section 7.3

(a) When we write an equation directly from a statement or question, we are using the method of _____ .

(b) In a _____ problem, the base is the quantity that a saleswoman needs to sell.

(c) In an application involving taxes, the tax paid is always the _____ .

(d) Selling price = original cost + _____ .

CHECK YOURSELF ANSWERS

1. $1,800   **2.** 4.2%   **3.** $15,000   **4.** $78   **5.** $33.75   **6.** $150.80
7. 7.4%   **8.** $6,626.70   **9.** $102.69   **10.** $688.50

Name _____

Section _____ Date _____

ANSWERS

1. _____

2. _____

3. _____

4. _____

5. _____

6. _____

7. _____

8. _____

7.3   Exercises

Solve each of the applications.

1. **Business and Finance**   If a salesman is paid a $560 commission on the sale of a $11,200 sailboat, what is his commission rate?

2. **Business and Finance**   A real estate agent's commission rate is 6%. What will be the amount of the commission on the sale of an $215,000 home?

3. **Business and Finance**   A state sales tax is levied at a rate of 6.4%. How much tax would one pay on a purchase of $260?

4. **Business and Finance**   A state sales tax rate is 3.5%. If the tax on a purchase is $7, what was the price of the purchase?

5. **Business and Finance**   If a house sells for $325,000 and the commission rate is $6\frac{1}{2}$%, how much will the salesperson make for the sale?

6. **Business and Finance**   A saleswoman is working on a 5% commission basis. If she wants to make $1,800 in 1 month, how much must she sell?

7. **Business and Finance**   An appliance dealer marks up refrigerators 22% (based on cost). If the cost of one model was $600, what will be its selling price?

8. **Business and Finance**   A television set is marked down $75, to be placed on sale. If this is a 12.5% decrease from the original price, what was the selling price before the sale?

9. **Business and Finance**   A stereo system is marked down from $450 to $382.50. What is the discount rate?

10. **Business and Finance**   At True Grip hardware, you pay $10 in tax for a barbecue grill, which is 6% of the purchase price. At Loose Fit hardware, you pay $10 in tax for the same grill, but it is 8% of the purchase price. At which store do you get the better buy? Why?

11. **Business and Finance**   A store marks up merchandise 25% to allow for profit. If an item costs the store $11, what will be its selling price?

12. **Business and Finance**   What were Jamal's total sales in a given month if he earned a commission of $2,458 at a commission rate of 1.6%?

![calculator] **Calculator Exercises**

Solve each of the percent problems.

13. What percent is 648 of 8,640?

14. 53.1875 is 9.25% of what number?

15. Find 7.65% of 375.

16. 17.4 is what percent (to the nearest tenth) of 81.5?

17. Find the base if 18.2% of the base is 101.01.

18. What is 3.52% of 2,450?

19. What percent (to the nearest tenth) of 1,625 is 182?

20. 22.5% of what number is 3,762?

ANSWERS

9. _____

10. _____

11. _____

12. _____

13. _____

14. _____

15. _____

16. _____

17. _____

18. _____

19. _____

20. _____

21. _____

22. _____

23. _____

24. _____

25. _____

26. _____

27. _____

28. _____

21. **Business and Finance** A dealer marks down the last year's model appliances 22.5% for a sale. If the regular price of an air conditioner was $279.95, how much will it be discounted (to the nearest cent)?

22. **Social Science** The population of a town increases 4.2% in 1 year. If the original population was 19,500, what is the population after the increase?

23. **Information Technology** An Ethernet network transmits a maximum packet size of 1,500 bytes. 1.7% of each packet is lost as "overhead." How much information (in bytes) in a maximum-sized packet is overhead?

24. **Environmental Technology** In some communities, "green" laws require that 40% of a lot remain green (covered by grass or other vegetation). How much green space would be required if a lot measured 12,680 ft^2?

In exercises 25 and 26, fill in the blank with always, sometimes, or never.

25. The discount rate for an item on sale will _____ be less than 100%.

26. The markup rate for an item will _____ be more than 100%.

27. In the 2000 census, 120 million households were given forms to fill out. Four months after they were to be returned, 8.6 million households had still not responded. What percent of the households did not respond? Round to the nearest tenth of a percent. (7)

28. Census data showed that in 2002 there were 15.5 million postsecondary students in the U.S., compared to 15.1 million high school students. The number of high school students was what percent of the number of postsecondary students? Round to the nearest tenth of a percent. (7)

Answers

1. 5%   **3.** $16.64   **5.** $21,125   **7.** $732   **9.** 15%   **11.** $13.75
13. 7.5%   **15.** 28.6875   **17.** 555   **19.** 11.2%   **21.** $62.99
23. 25.5 bytes   **25.** Always   **27.** 7.2%

7.4 Applications: Simple and Compound Interest

7.4 OBJECTIVES

1. Solve percent applications involving simple interest
2. Solve percent applications involving compound interest

NOTE The money borrowed, invested, or saved is called the **principal.**

As we said earlier, there are many applications of percents. One that almost all of us encounter involves *interest*. When you borrow money, you pay interest. When you place money in a savings account, you earn interest. **Interest** is a percent of the whole (in this case, the *principal*), and the percent is called the **interest rate.**

OBJECTIVE 1

Example 1 Solving a Problem Involving an Unknown Amount

Find the interest you must pay if you borrow $2,000 for 1 year with an interest rate of $9\frac{1}{2}\%$.

The base (the principal) is $2,000, the rate is $9\frac{1}{2}\%$, and we want to find the interest (the amount). The question becomes,

RECALL

$9\frac{1}{2}\% = 9.5\%$

What is 9.5% of $2,000?

$$A = \frac{9.5}{100} \cdot 2,000 = 0.095 \cdot 2,000$$

$$A = 190 \qquad \text{The interest (amount) is } \$190.$$

CHECK YOURSELF 1

You invest $5,000 for 1 year at $8\frac{1}{2}\%$. How much interest will you earn?

Example 2 Solving a Problem Involving an Unknown Base

Ms. Hobson agrees to pay 11% interest on a loan for her new automobile. She is charged $2,200 interest on a loan for 1 year. How much did she borrow?

The rate is 11%. The amount, or interest, is $2,200. We want to find the base, which is the principal, or the size of the loan. To solve the problem, we have

$$\frac{2,200}{B} = \frac{11}{100}$$

$$11B = 2,200 \cdot 100$$

$$B = \frac{220,000}{11} = 20,000$$

She borrowed $20,000.

CHECK YOURSELF 2

Sue paid $210 interest for a 1-year loan at 10.5%. What was the size of her loan?

When looking at the total amount paid for a loan, it is helpful to be able to actually calculate the interest rate.

Example 3 Finding the Rate of Interest

Jaime borrowed $1,200 from his brother to make a down payment on his car. At the end of a year, he paid back $1,450. What was the rate of interest?

When finding a rate, it is usually easiest to use the percent proportion. First, notice that the amount of interest (*A*) was $250. He paid back $250 more than he borrowed. With that in mind, we have the proportion

$$\frac{250}{1,200} = \frac{r}{100} \qquad \text{so} \qquad r = \frac{250}{1,200} \cdot 100$$

$$r \approx 20.83 \qquad \text{The rate was 20.83\%.}$$

CHECK YOURSELF 3

Dean borrowed $500 from his sister to pay his fall tuition. He paid her back $580 a year later. What was the rate of interest?

NOTE

$1,000 $\xrightarrow{\text{At 5\%}}$ $1,050
Start   Year 1

$1,050 $\xrightarrow{\text{At 5\%}}$ $1,102.50
Year 1   Year 2

The three examples so far in this section involve **simple interest.** Many percent problems involve calculating what is known as **compound interest.**

Suppose that you invest $1,000 at 5% in a savings account for 1 year. For year one, the interest is 5% of $1,000, or 0.05 · $1,000 = $50. At the end of year one, you will have $1,050 in the account.

Now if you leave that amount in the account for a second year, the interest will be calculated on the original principal, $1,000, plus the first year's interest, $50, that is, *compound interest.*

For year two, the interest is 5% of $1,050, or 0.05 · $1,050 = $52.50. At the end of the second year, you will have $1,102.50 in the account.

OBJECTIVE 2

Example 4 Calculating Compound Interest

Brittney invested $500 in a certificate that will pay her 7.5% interest per year. What will be the value of her investment in 2 years? To find the total, calculate the interest for each year and then add it to the previous year's total.

$ 500.00	Start
+ 37.50	First-year interest = 500.00 · 0.075 = 37.50
537.50	Year 1
+ 40.31	Second-year interest = 537.50 · 0.075 = 40.31
577.81	Year 2

Her investment will have a value of $577.81.

CHECK YOURSELF 4

Dominic invested $1,000 at 8.5% per year. What will be the value of his investment in 3 years?

READING YOUR TEXT

The following fill-in-the-blank exercises are designed to assure that you understand the key vocabulary used in this section. Each sentence usually comes directly from the section. You will find the correct answers in Appendix C.

Section 7.4

(a) When you borrow money, you pay _____ .

(b) In an application involving interest, the money borrowed or saved is called the _____ .

(c) Interest is a percent of the whole, and the percent is called the _____ .

(d) When the interest is calculated on the original principal plus the previous year's interest, this is called _____ interest.

CHECK YOURSELF ANSWERS

1. $425   **2.** $2,000   **3.** 16%   **4.** $1,277.29

Name _____

Section _____ Date _____

7.4   Exercises

Solve each of the applications.

1. **Business and Finance**   What interest will you pay on a $3,400 loan for 1 year if the interest rate is 12%?

2. **Business and Finance**   Ms. Jordan has been given a loan of $2,500 for 1 year. If the interest charged is $275, what is the interest rate on the loan?

3. **Business and Finance**   Joan was charged $18 interest for 1 month on a $1,200 credit card balance. What was the monthly interest rate?

4. **Business and Finance**   Patty pays $525 interest for a 1-year loan at 10.5%. How much was her loan?

5. **Business and Finance**   A time-deposit savings plan gives an interest rate of 6.42% on deposits. If the interest on an account for 1 year was $545.70, how much was deposited?

Complete the table for exercises 6 to 14.

	Principal	Interest Rate	Interest
6.	$1,000	8.75%	
7.	$3,000	9.25%	
8.	$12,000	6.5%	
9.	$3,450		$258.75
10.	$36,500		$3,467.50
11.		6.75%	$238.95
12.		5.75%	$727.09
13.	$14,640		$1,207.80
14.	$9,850		$418.63

In exercises 15 to 18, assume the interest is compounded annually (at the end of each year) and find the amount in an account with the given interest rate and principal.

15. $4,000, 6%, 2 years

16. $3,000, 7%, 2 years

17. $4,000, 5%, 3 years

18. $5,000, 6%, 3 years

A more general formula for simple interest is $I = P \cdot R \cdot T$, where T is the time in years. To find interest for a portion of a year, use an appropriate fraction for T. For example, if we want to find the interest for six months, we use $\dfrac{6}{12}\left(\text{or } \dfrac{1}{2}\right)$ for T.

In exercises 19 to 22, use this information to find the interest for the given principal, rate, and time.

19. $1,000, 8%, 3 months

20. $4,000, 6%, 8 months

21. $5,000, 4%, 10 months

22. $2,000, 5%, 2 months

Answers

1. $408   **3.** 1.5%   **5.** $8,500   **7.** $277.50   **9.** 7.5%   **11.** $3,540
13. 8.25%   **15.** $4,494.40   **17.** $4,630.50   **19.** $20   **21.** $166.67

7.5 More Applications of Percent

7.5 OBJECTIVE

1. Solve a variety of percent applications

Percents are used in too many ways for us to list. Look at the variety in the examples presented, which illustrate some additional situations in which you will find percents.

OBJECTIVE 1

Example 1   Solving a Problem Involving an Unknown Amount

NOTE A *rate, base,* and *amount* will appear in *all* problems involving percents.

A student needs 70% to pass an examination containing 50 questions. How many questions must she answer correctly?

The *rate* is 70%. The *base* is the number of questions on the test, here 50. The *amount* is the number of questions that must be correct.

To find the amount, we use the percent proportion from Section 7.2.

NOTE Substitute 50 for *B* and 70 for *r*.

$$B \longrightarrow \frac{A}{50} = \frac{70}{100} \longleftarrow r$$

so

$$100A = 50 \cdot 70$$

Dividing by 100 gives

$$A = \frac{3,500}{100} = 35$$

She must answer 35 questions correctly to pass.

CHECK YOURSELF 1

Generally, 72% of the students in a chemistry class pass the course. If there are 150 students in the class, how many can be expected to pass?

Next, we look at an application that requires finding the rate.

Example 2   Solving a Problem Involving an Unknown Rate

Simon works at a restaurant called La Catalana. The $45 tip he received from a family on Friday was the largest tip Simon has ever received. If the bill totaled $250 before the tip was added, what percent of the total was the tip?

The base is the total of the bill, $250. The amount is the $45 tip. To find the desired percent, we again use the percent proportion.

$$\frac{45}{250} = \frac{r}{100} \qquad \text{We are finding the rate.}$$

$$250 \cdot r = 45 \cdot 100$$

$$250r = 4,500$$

$$r = \frac{4,500}{250} = 18$$

The tip was 18% of the bill.

CHECK YOURSELF 2 _____

Last year, Xian reported an income of $27,500 on her tax return. Of that, she paid $6,600 in taxes. What percent of her income went to taxes?

Increases and decreases are often stated in terms of percents, as Examples 3 to 5 illustrate.

> **Example 3   Solving a Percent Problem**

The population of a town increased 15% during a 3-year period. If the original population was 12,000, what was the population at the end of the 3 years?

First we find the increase in the population. That increase is the amount in the problem.

$$\frac{A}{12,000} = \frac{15}{100} \quad \text{so} \quad 100A = 15 \cdot 12,000$$

$$A = \frac{180,000}{100}$$

$$= 1,800$$

To find the population at the end of the period, we add

$$12,000 + 1,800 = 13,800$$

Original population   Increase   New population

CHECK YOURSELF 3 _____

A school's enrollment decreased by 8% from a given year to the next. If the enrollment was 550 students the first year, how many students were enrolled the second year?

> **Example 4   Solving a Percent Problem**

Enrollment at a school increased from 800 to 888 students from a given year to the next. What was the rate of increase?

First we must subtract to find the amount of the increase.

Increase: $888 - 800 = 88$ students

Now to find the rate, we have

NOTE We use the *original* enrollment, 800, as our base.

$$\frac{88}{800} = \frac{r}{100} \quad \text{so} \quad 800r = 88 \cdot 100$$

$$r = \frac{8,800}{800} = 11$$

The enrollment increased at a rate of 11%.

CHECK YOURSELF 4 _____

Car sales at a dealership decreased from 350 units one year to 322 units the next. What was the rate of decrease?

Example 5 Solving a Percent Problem

A company hired 18 new employees in 1 year. If this was a 15% increase, how many employees did the company have before the increase?

The rate is 15%. The amount is 18, the number of new employees. The base in this problem is the number of employees *before the increase*. So

$$\frac{18}{B} = \frac{15}{100}$$

$$15B = 18 \cdot 100 \quad \text{or} \quad B = \frac{1,800}{15} = 120$$

The company had 120 employees before the increase.

CHECK YOURSELF 5 _____

A school had 54 new students in one term. If this was a 12% increase over the previous term, how many students were there before the increase?

There are many computer-related applications that include percentages as either part of the problem or as part of the solution.

Example 6 Solving a Computer Application

A computer loaded 60% of a new program in 120 seconds. How long should it take to load the entire program?

The rate is 60%; the amount is 120 seconds. The base is the total time taken to load the program.

$$\frac{120}{B} = \frac{60}{100}$$

$$60B = 12,000$$

$$B = 200$$

It will take 200 seconds to load the program.

CHECK YOURSELF 6

A virus-scanning program is checking every computer file for viruses. It checked 30% of the files in 240 seconds. How long should it take to check all of the files?

READING YOUR TEXT

The following fill-in-the-blank exercises are designed to assure that you understand the key vocabulary used in this section. Each sentence usually comes directly from the section. You will find the correct answers in Appendix C.

Section 7.5

(a) A rate, base, and amount will appear in all problems involving _____ .

(b) In an application involving a restaurant bill, the _____ is the tip.

(c) In an application involving population increase, the original population is the _____ .

(d) In an application involving population increase, the actual increase is the _____ in the problem.

CHECK YOURSELF ANSWERS

1. 108   **2.** 24%   **3.** 506   **4.** 8%   **5.** 450   **6.** 800 s (13 min 20 s)

Name _____

Section _____ Date _____

ANSWERS

1. _____

2. _____

3. _____

4. _____

5. _____

7.5 Exercises

Solve each of the applications.

1. **Business and Finance** Roberto has 26% of his pay withheld for deductions. If he earns $550 per week, what amount is withheld?

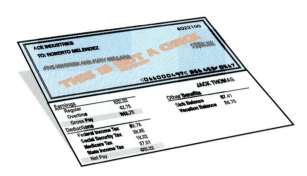

2. **Science and Medicine** A chemist has 300 mL of solution that is 18% acid. How many milliliters of acid are in the solution?

3. **Statistics** On a test, Alice had 80% of the problems right. If she had 20 problems correct, how many questions were on the test?

4. **Business and Finance** Betty must make a $9\frac{1}{2}$% down payment on the purchase of a $2,000 motorcycle. How much must she pay down?

5. **Science and Medicine** There are 117 mL (milliliters) of acid in 900 mL of a solution of acid and water. What percent of the solution is acid?

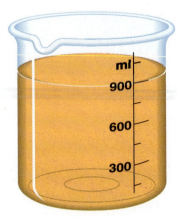

6. **Statistics**   Marla needs 70% on a final exam to pass the course. If the exam has 120 questions, how many questions must she answer correctly?

7. **Social Science**   A study has shown that 102 of the 1,200 people in the workforce of a small town are unemployed. What is the town's unemployment rate?

8. **Statistics**   A survey of 400 people found that 66 were left-handed. What percent of those surveyed were left-handed?

9. **Social Science**   Of 60 people who started a training program, 45 completed the course. What was the dropout rate?   **ALEKS**

10. **Business and Finance**   In a shipment of 250 parts, 40 are found to be defective. What percent of the parts are faulty?

11. **Statistics**   In a recent survey, 65% of those responding were in favor of a freeway improvement project. If 780 people were in favor of the project, how many people responded to the survey?

12. **Social Science**   A college finds that 42% of the students taking a foreign language are enrolled in Spanish. If 1,512 students are taking Spanish, how many foreign language students are there?

13. **Business and Finance**   22% of Samuel's monthly salary is deducted for withholding. If those deductions total $627, what is his salary?   **VIDEO**

14. **Business and Finance**   The Townsend's budget 36% of their monthly income for food. If they spend $864 on food, what is their monthly income?

15. **Business and Finance**   A home lot purchased for $26,000 increased in value by 25% over 3 years. What was the lot's value at the end of the period?   **VIDEO**

16. **Business and Finance**   New cars depreciate an average of 28% in their first year of use. What will an $18,000 car be worth after 1 year?

17. **Social Science**   A school's enrollment was up from 950 students in 1 year to 1,064 students in the next. What was the rate of increase?

18. **Business and Finance**   Under a new contract, the salary for a position increases from $44,000 to $47,560. What rate of increase does this represent?

ANSWERS

6. _____
7. _____
8. _____
9. _____
10. _____
11. _____
12. _____
13. _____
14. _____
15. _____
16. _____
17. _____
18. _____

19. _____

20. _____

21. _____

22. _____

23. _____

24. _____

25. _____

26. _____

27. _____

28. _____

19. Business and Finance The price of a new van has increased $4,060, which amounts to a 14% increase. What was the price of the van before the increase?

20. Business and Finance The electricity costs of a business decreased from $12,000 one year to $10,920 the next. What was the rate of decrease?

21. Business and Finance A company had 66 fewer employees in July 2001 than in July 2000. If this represents a 5.5% decrease, how many employees did the company have in July 2000?

22. Business and Finance Carlotta received a monthly raise of $162.50. If this represented a 6.5% increase, what was her monthly salary before the raise?

23. Business and Finance Mr. Hernandez buys stock for $15,000. At the end of 6 months, the stock's value has decreased 7.5%. What is the stock worth at the end of the period?

24. Social Science The population of a town increases 14% in 2 years. If the population was 6,000 originally, what is the population after the increase?

25. Statistics A computer loaded 80% of a new program in 80 seconds. How long should it take to load the entire program?

26. Business and Finance Tranh's pay is $450 per week. If deductions from his paycheck average 25%, what is the amount of his weekly paycheck (after deductions)?

27. Business and Finance The two ads pictured appeared last week and this week in the local paper. Is this week's ad accurate?

LAST WEEK

CHICKEN $2 75 lb.
QUANTITIES LIMITED

CHICKEN $1 97 lb.
SAVE 40%

THIS WEEK

28. Statistics A virus-scanning program is checking every file for viruses. It has completed checking 40% of the files in 300 seconds. How long should it take to check all the files?

29. Business and Finance A pair of shorts is advertised for $48.75 and as being 25% off the original price. What was the original price?

30. Business and Finance If the total bill at a restaurant, including a 15% tip, is $65.32, what was the cost of the meal alone?

The chart shows U.S. trade with Mexico from 1992 to 1997. Use this information for exercises 31 to 34.

U.S. Trade with Mexico, 1992–97 (Millions of Dollars)			
Year	**Exports**	**Imports**	**Trade Balance[1]**
1992	$40,592	$35,211	$5,381
1993	41,581	39,917	1,664
1994[2]	50,844	49,494	1,350
1995	46,292	61,685	−15,393
1996	56,792	74,297	−17,506
1997	71,388	85,938	−14,549

(1) Totals may not tally because of rounding.
(2) NAFTA provisions began to take effect Jan. 1, 1994.
Source: Office of Trade and Economic Analysis. U.S. Dept. of Commerce.

31. What is the rate of increase (to the nearest whole percent) of exports from 1992 to 1997?

32. What is the rate of increase (to the nearest whole percent) of imports from 1992 to 1997?

33. By what percent did exports exceed imports in 1992?

34. By what percent did imports exceed exports in 1997?

© 2010 McGraw-Hill Companies

Automotive Technology   Between 1980 and 2001, the average fuel efficiency of new U.S. cars increased from 24.3 to 24.6 $\dfrac{mi}{gal}$. During this time, the average fuel efficiency for the entire fleet of U.S. cars rose from 16.0 to 22.1 $\dfrac{mi}{gal}$. Use this information for the next two problems.

35. The increase in fuel efficiency for new cars is given by

$$\frac{28.6 - 24.3}{24.3}$$

Report the increase, as a percent (to the nearest tenth of a percent).

36. The increase in fuel efficiency for entire fleet is given by

$$\frac{22.1 - 16.0}{16.0}$$

Report the increase, as a percent (to the nearest tenth of a percent).

37. Social Science   In 1989, transportation accounted for 63% of U.S. petroleum consumption. If 10.85 million bbl of petroleum are used each day for transportation in the United States, what is the total daily petroleum consumption by all sources in the United States?   **VIDEO**

38. Agricultural Technology   There were 52 in. of rainfall during one growing season. If 6.5 in. fell in August, what percentage of the season's rainfall fell in August?

39. The progress of the local Lions club is shown. What percent of the goal has been achieved so far?

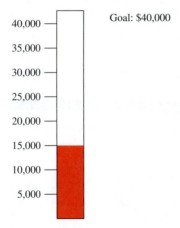

Goal: $40,000

40. According to the 1990 and 2000 census, the Hispanic and Latino populations nationwide rose 58% during the intervening decade, increasing by 13 million. What were the Hispanic and Latino populations in 1990? Express your answer to the nearest tenth of a million.

41. The number of students enrolled in postsecondary education rose from 13.8 million in 1997 to 15.5 million in 2002. What was the rate of increase? Express your answer to the nearest tenth of a percent.

Answers

1. $143 **3.** 25 questions **5.** 13% **7.** 8.5% **9.** 25%
11. 1,200 people **13.** $2,850 **15.** $32,500 **17.** 12% **19.** $29,000
21. 1,200 employees **23.** $13,875 **25.** 100 s **27.** **29.** $65

31. 76% **33.** +15% **35.** 17.7% **37.** 17.22 million bbl **39.** 37.5%
41. 12.3%

ACTIVITY 7: POPULATION CHANGES

Each chapter in this text includes an activity. The activity is related to the vignette you encountered in the chapter opening. The activity provides you with the opportunity to apply the math you studied in the chapter to a relevant topic.

Your instructor will determine how best to use this activity in your instructional setting. You may find yourself working in class or outside of class; you may find yourself working alone or in small groups; or you may even be asked to perform research in a library or on the Internet.

The table below gives the population for the United States and each of the six largest states from both the 1990 and 2000 census. Use this table to answer the questions that follow. Round all computations of percents to the nearest tenth of a percent.

	1990 Population	2000 Population
United States	248,709,873	281,421,906
California	29,760,021	33,871,648
Texas	16,986,510	20,851,820
New York	17,990,455	18,976,457
Florida	12,937,926	15,982,378
Illinois	11,430,602	12,419,293
Pennsylvania	11,881,643	12,281,054

1. Find the percent increase in the U.S. population from 1990 to 2000.

2. By examining the table (no actual calculations yet!), predict which state had the greatest percent increase.

3. Predict which state had the smallest percent increase.

4. Now find the percent increase in population during this period for each state.

   California: Texas:
   New York: Florida:
   Illinois: Pennsylvania:

5. Which state had the greatest percent increase during this period? Which had the smallest?

6. The population of the six largest states combined represented what percent of the U.S. population in 1990?

7. The population of the six largest states combined represented what percent of the U.S. population in 2000?

8. Determine the percent increase in population from 1990 to 2000 for the combined six largest states.

If you are interested in exploring some of the results of the last census, visit the following website: www.census.gov/main/www/cen2000.html

7 **Summary**

DEFINITION/PROCEDURE	EXAMPLE	REFERENCE
Percents, Decimals, and Fractions		**Section 7.1**
Percent Another way of naming parts of a whole. Percent means per hundred.	Fractions and decimals are other ways of naming parts of a whole. $$21\% = \frac{21}{100} = 0.21$$	*p. 573*
1. *To convert a percent to a fraction,* divide by 100.	$$37\% = \frac{37}{100}$$	*p. 574*
2. *To convert a percent to a decimal,* remove the percent symbol and move the decimal point two places to the left.	$37\% = 0.37$	*p. 575*
3. *To convert a decimal to a percent,* move the decimal point two places to the right and attach the percent symbol.	$0.58 = 58\%$	*p. 575*
4. *To convert a fraction to a percent,* write the decimal equivalent of the fraction and then change that decimal to a percent.	$$\frac{3}{5} = 0.60 = 60\%$$	*p. 576*
Solving Percent Problems Using Proportions		**Section 7.2**
Every percent problem has the following three parts: **1.** *The base.* This is the whole in the problem. It is the standard used for comparison. Label the base B. **2.** *The amount.* This is the part of the whole being compared to the base. Label the amount A. **3.** *The rate.* This is the ratio of the amount to the base. The rate is written as a percent. Label the rate $\frac{r}{100}$ or R.	45 is 30% of 150. $\uparrow$ $\uparrow$ $\uparrow$ A R B	*p. 588*
Using the Percent Proportion The percent proportion is $$\frac{A}{B} = \frac{r}{100}$$ To solve a percent problem using this proportion: **1.** Substitute the two known values into the proportion. **2.** Solve the proportion to find the unknown value.	What is 24% of 300? $$\frac{A}{300} = \frac{24}{100}$$ $$100A = 7,200$$ $$A = 72$$	*p. 589*

Continued

DEFINITION/PROCEDURE	EXAMPLE	REFERENCE
Solving Percent Applications Using Equations		**Section 7.3**
Translating a question into an equation: What is $r\%$ of B? $$A = \frac{r}{100} \cdot B$$	What is 23% of 300? $$A = \frac{23}{100} \cdot 300$$ $$A = 23 \cdot 3$$ $$= 69$$	*p. 597*
Sales tax, commissions, and **discounts** are all rates. Applications can be solved by the percent proportion or by translating the question into an equation.	If a 12% commission is \$3,000, what is the total? $$\frac{3,000}{B} = \frac{12}{100}$$ $$12B = 300,000$$ $$B = \$25,000$$	*p. 598*
Applications: Simple and Compound Interest		**Section 7.4**
Simple interest is a rate. **Compound interest** is interest on an amount that includes previous interest.		*p. 607*

Summary Exercises

This summary exercise set is provided to give you practice with each of the objectives of this chapter. Each exercise is keyed to the appropriate chapter section. When you are finished, you can check your answers to the odd-numbered exercises against those presented in the back of the text. If you have difficulty with any of these questions, go back and reread the examples from that section. The answers to the even-numbered exercises appear in the *Instructor's Manual*. Your instructor will give you guidelines on how to best use these exercises in your class.

[7.1]

1. Use a percent to name the shaded portion of the diagram.

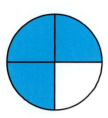

In exercises 2 to 7, write the percent as a common fraction or a mixed number.

2. 2%

3. 20%

4. 37.5%

5. 150%

6. $233\frac{1}{3}\%$

7. 300%

In exercises 8 to 13, write the percents as decimals.

8. 75%

9. 4%

10. 6.25%

11. 13.5%

12. 0.6%

13. 225%

In exercises 14 to 25, write as percents.

14. 0.06

15. 0.375

16. 2.4

17. 7

18. 0.035

19. 0.005

20. $\dfrac{43}{100}$

21. $\dfrac{7}{10}$

22. $\dfrac{2}{5}$

23. $1\dfrac{1}{4}$

24. $2\dfrac{2}{3}$

25. $\dfrac{3}{11}$ (to nearest tenth of a percent)

[7.2]   In exercises 26 to 31, find the unknown using the percent proportion.

26. 80 is 4% of what number?

27. 70 is what percent of 50?

28. 11% of 3,000 is what number?

29. 24 is what percent of 192?

30. Find the base if 12.5% of the base is 625.

31. 90 is 120% of what number?

[7.3]   In exercises 32 to 37, find the unknown using an equation.

32. What is 9.5% of 700?

33. Find 150% of 50.

34. Find the base if 130% of the base is 780.

35. 350 is what percent of 200?

36. What is 225% of 48?

37. 28.8 is what percent of 960?

Estimate the amount in the exercises.

38. 24.3% of 810

39. 109% of 592

[7.4]   In exercises 40 to 43, solve the applications.

40. Business and Finance   Joan works on a 4% commission basis. She sold $45,000 in merchandise during 1 month. What was the amount of her commission?

41. **Business and Finance**   David buys a dishwasher that is marked down $77 from its original price of $350. What is the discount rate?

42. **Business and Finance**   A store advertises, "Buy the red-tagged items at 25% off their listed price." If you buy a coat marked $136, what will you pay for the coat during the sale?

43. **Business and Finance**   A state sales tax rate is 7.5%. If the tax on a purchase is $9.75, what was the price of the purchase?

[7.5]   In exercises 44 to 46, solve the applications.

44. **Business and Finance**   A savings bank offers 5.25% on 1-year time deposits. If you place $3,000 in an account, how much will you have at the end of the year?

45. **Business and Finance**   If you received $285 in interest on a CD that paid 9.5%, what was the original investment?

46. **Business and Finance**   If $500 is compounded at 8% for 3 years, what will the final total be?

[7.6]   In exercises 47 to 55, solve the applications.

47. **Science and Medicine**   A chemist prepares a 400-mL acid-water solution. If the solution contains 30 mL of acid, what percent of the solution is acid?

48. **Business and Finance**   The price of a new compact car has increased $819 over the previous year. If this amounts to a 4.5% increase, what was the price of the car before the increase?

49. **Business and Finance**   Tom has 6% of his salary deducted for a retirement plan. If that deduction is $168, what is his monthly salary?

50. **Social Science**   A college finds that 35% of its science students take biology. If there are 252 biology students, how many science students are there altogether?

51. **Business and Finance**   A company finds that its advertising costs increased from $72,000 to $76,680 in 1 year. What was the rate of increase?

52. **Business and Finance**   Maria's company offers her a 4% pay raise. This will amount to a $126 per month increase in her salary. What is her monthly salary before and after the raise?

53. Statistics   A virus-scanning program is checking every file for viruses. It has completely scanned 30% of the files in 150 seconds. How long should it take to check all the files?

54. Business and Finance   If the total bill at a restaurant for 10 people is $572.89, including an 18% tip, what was the cost of the food itself?

55. Business and Finance   A pair of running shoes, which was advertised as selling at 30% off the original price, sold for $80.15. What was the original price?

Self-Test for Chapter 7

Name _____

Section _____ Date _____

The purpose of this self-test is to help you check your progress so that you can find sections and concepts that you need to review before the next in-class exam. Allow yourself about an hour to take this test. At the end of that hour, check your answers against those given in the back of this text. If you missed a question, notice the section reference that accompanies the question. Go back to that section and reread the examples until you have mastered that particular concept.

ANSWERS

1. Use a percent to name the shaded portion of the diagram.

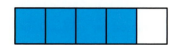

In exercises 2 and 3, write as fractions.

2. 7%   **3.** 72%

In exercises 4 to 6, write as decimals.

4. 42%   **5.** 6%   **6.** 160%

In exercises 7 to 10, write as percents.

7. 0.03   **8.** 0.042

9. $\dfrac{2}{5}$ **10.** $\dfrac{5}{8}$

In exercises 11 to 13, identify the rate, base, and amount. *Do not solve* at this point.

11. 50 is 25% of 200.   **12.** What is 8% of 500?

13. Business and Finance   A state sales tax rate is 6%. If the tax on a purchase is $30, what is the amount of the purchase?

In exercises 14 to 21, solve the percent problems.

14. What is 4.5% of 250?   **15.** $33\dfrac{1}{3}$% of 1,500 is what number?

16. Find 125% of 600.   **17.** What percent of 300 is 60?

18. 4.5 is what percent of 60?   **19.** 875 is what percent of 500?

20. 96 is 12% of what number?   **21.** 8.5% of what number is 25.5?

1. _____
2. _____
3. _____
4. _____
5. _____
6. _____
7. _____
8. _____
9. _____
10. _____
11. _____
12. _____
13. _____
14. _____
15. _____
16. _____
17. _____
18. _____
19. _____
20. _____
21. _____

22. _____

23. _____

24. _____

25. _____

26. _____

27. _____

28. _____

29. _____

30. _____

31. _____

32. _____

In exercises 22 to 32, solve the applications.

22. Business and Finance A state taxes sales at 6.2%. What tax will you pay on an item that costs $80?

23. Statistics You receive a grade of 75% on a test of 80 questions. How many questions did you answer correctly?

24. Business and Finance An item that costs a store $54 is marked up 30% (based on cost). Find its selling price.

25. Business and Finance Mrs. Sanford pays $300 in interest on a $2,500 loan for 1 year. What is the interest rate for the loan?

26. Business and Finance Jovita's monthly salary is $2,200. If the deductions for taxes from her monthly paycheck are $528, what percent of her salary goes for these deductions?

27. Business and Finance A car is marked down $1,552 from its original selling price of $19,400. What is the discount rate?

28. Business and Finance Sarah earned $540 in commissions in 1 month. If her commission rate is 3%, what were her total sales?

29. Social Science A community college has 480 more students in fall 2006 than in fall 2005. If this is a 7.5% increase, what was the fall 2005 enrollment?

30. Business and Finance Shawn arranges financing for his new car. The interest rate for the financing plan is 12%, and he will pay $2,220 interest for 1 year. How much money did he borrow to finance the car?

31. Business and Finance Jalene earns a commission of 7% on property that she lists and sells. What is her commission on a $250,000 piece of property that she listed and sold?

32. Business and Finance If $2,650 is compounded at 7% for 3 years, what will be the final total?

Name _____

Section _____ Date _____

ANSWERS

The following exercises are presented to help you review concepts from earlier chapters that you may have forgotten. This section is meant as review material and not as a comprehensive exam. The answers are presented in the back of the text. If you have difficulty with any of these exercises, be certain to at least read through the summary related to that section.

1. What is the place value of 4 in the number 234,768?

Perform the indicated operations.

2. $56 \cdot 203$

3. $3,026 \div 34$

Evaluate the expressions.

4. $-4 + 10$

5. $-5 - (-1)$

6. $\dfrac{-8 + 6}{-8 - (-10)}$

7. $5 + 3 \cdot (4 - 6)^2$

8. Write the prime factorization of 260.

9. Find the greatest common factor (GCF) of 84 and 140.

10. Find the least common multiple (LCM) of 18, 20, and 30.

Perform the indicated operations.

11. $3\dfrac{2}{5} \cdot 2\dfrac{1}{2}$

12. $5\dfrac{1}{3} \div 4$

13. $4\dfrac{3}{4} + 3\dfrac{5}{6}$

14. $7\dfrac{1}{6} - 2\dfrac{3}{8}$

15. A kitchen measures $5\dfrac{1}{2}$ yd by $3\dfrac{1}{4}$ yd. If vinyl flooring costs $16 per square yard, what will it cost to cover the floor?

16. If you drive 180 mi in $3\dfrac{1}{3}$ h, what is your average speed?

17. A bookshelf that is $54\dfrac{5}{8}$ in. long is cut from a board that is 8 ft long. If $\dfrac{1}{8}$ in. is wasted in the cut, what length board remains?

Find the indicated place values.

18. 8 in 4.2835

19. 4 in 6.09743

1. _____
2. _____
3. _____
4. _____
5. _____
6. _____
7. _____
8. _____
9. _____
10. _____
11. _____
12. _____
13. _____
14. _____
15. _____
16. _____
17. _____
18. _____
19. _____

20.
_____

21.
_____

22.
_____

23.
_____

24.
_____

25.
_____

26.
_____

27.
_____

28.
_____

29.
_____

30.
_____

31.
_____

32.
_____

33.
_____

34.
_____

35.
_____

36.
_____

37.
_____

38.
_____

39.
_____

40.
_____

Complete each statement using the symbol $<$, $=$, or $>$.

20. 6.28 _____ 6.3

21. 3.75 _____ 3.750

Write as a common fraction or a mixed number. Simplify.

22. 0.36

23. 5.125

Perform the indicated operations.

24. $2.8 \cdot 4.03$

25. $54.528 \div 3.2$

26. A television set has an advertised price of $599.95. You buy the set and agree to make payments of $29.50 per month for 2 years. How much extra are you paying on this installment plan?

Solve each equation and check your result.

27. $9x + (-5) = 8x$

28. $\dfrac{2}{3}x = -24$

29. $2x + 3 = 7x + 5$

30. $\dfrac{4}{3}x + (-6) = 4 + \left(-\dfrac{2}{3}x\right)$

Solve for the unknown.

31. $\dfrac{3}{7} = \dfrac{8}{x}$

32. $\dfrac{1.9}{y} = \dfrac{5.7}{1.2}$

33. On a map the scale is $\dfrac{1}{4}$ in. $= 25$ mi. How many miles apart are two towns that are $3\dfrac{1}{2}$ in. apart on the map?

34. Diane worked 23.5 h on a part-time job and was paid $131.60. She is asked to work 25 h the next week at the same pay rate. What salary will she receive?

35. Write 34% as a decimal and as a fraction.

36. Write $\dfrac{11}{20}$ as a decimal and as a percent.

37. Find 18% of 250.

38. 11% of what number is 55?

39. A company reduced the number of employees by 8% this year. There are now 115 employees. How many were there last year?

40. The sales tax on an item priced at $72 is $6.12. What percent is the tax rate?

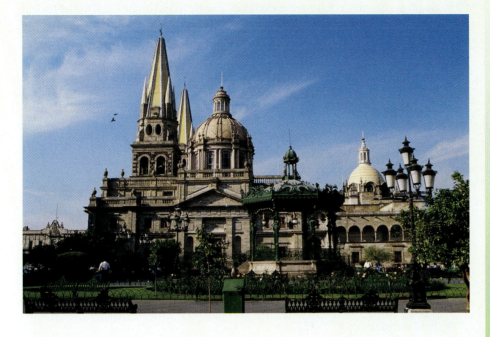

GEOMETRY

CHAPTER 8   OUTLINE

Section 8.1   Lines and Angles   page 635

Section 8.2   Perimeter and Circumference   page 652

Section 8.3   Area and Volume   page 662

INTRODUCTION

Since the Norman conquests, Norman architecture has been extremely popular throughout Europe. In the eleventh century (ca. 1066), institutions such as churches, castles, and universities began including Norman windows in their design, as shown in the photograph. Later designs, such as Gothic architecture, grew out of the Norman designs.

Norman windows are one example of composite geometric figures. These are figures that incorporate several simpler geometric figures. Another composite figure, race tracks, are formed by attaching a semicircle to either end of a rectangle.

Whether determining the amount of fertilizer needed for the grass inside a race track, the amount of glass needed for a custom window, the amount of paint needed for the trim of a home design, or the amount of fluid that a bottle can hold, you will find that composite geometric figures are all around us.

The properties of geometric figures, such as perimeter and area, are critical for success in many trades. Architects, craftspeople, and contractors frequently work with two- and three-dimensional figures.

We examine the geometric properties of figures throughout this chapter. You can find exercises relating to Norman windows in Sections 8.2 and 8.3. At the end of the chapter, before the chapter summary, you will find a more in-depth activity on Norman windows and architecture.

Pretest Chapter 8

This pretest will provide a preview of the types of exercises you will encounter in each section of this chapter. The answers for these exercises can be found in the back of the text. If you are working on your own or ahead of the class, this pretest can help you identify the sections in which you should focus more of your time.

[8.1]   **1.** Draw the line $\overleftrightarrow{AB}$.

A   B

2. Find the measure of $\angle BCA$.

3. Find $m\angle A$.

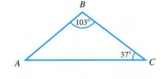

4. Find the measure of $\angle ABC$.

[8.2]   **5.** Find the perimeter of the triangle.

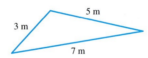

6. Find the circumference of the circle. Round to the nearest tenth.

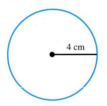

[8.3]   **7.** Determine the area of the triangle.

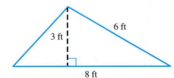

8.1 Lines and Angles

8.1 OBJECTIVES

1. Identify lines, line segments, and angles
2. Determine when lines are parallel or perpendicular
3. Determine whether an angle is right, acute, obtuse, or straight
4. Use a protractor to measure an angle
5. Determine when pairs of angles are complementary or supplementary
6. Identify vertical angles and adjacent angles
7. Find the measure of the third angle of a triangle

NOTE *Geo* means earth, just as it does in the words *geography* and *geology*.

Once early communities had mastered the counting of their animals, they became interested in measuring their land. This is the foundation of geometry. Literally translated, *geometry* means earth measurement. Many of the topics we consider in geometry, such as angles, perimeter, and area, were first studied as part of surveying.

As is usually the case, we start the study of a new topic by learning some vocabulary. Most of the terms we discuss will be familiar to you. It is important that you understand what we mean when we use these words in the context of geometry.

We begin with a **point,** which is a location; it has no size and covers no area.

If we string points together forever, we create a **line.** In our studies, we consider only straight lines. We use arrowheads to indicate that a line goes on forever.

A piece of a line that has two endpoints is called a **line segment.**

OBJECTIVE 1

Example 1   Recognizing Lines and Line Segments

NOTE Capital letters are used to name points.

Label each of the parts as a line or a line segment.

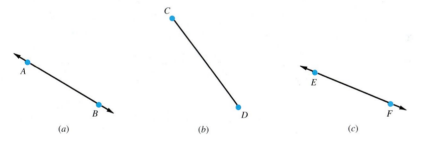

The lines in both (*a*) and (*c*) continue forever in both directions. They are lines. We write $\overleftrightarrow{AB}$ and $\overleftrightarrow{EF}$, respectively, to indicate the lines through those points. Part (*b*) has two endpoints. It is a line segment. We designate that it is a segment by writing $\overline{CD}$.

 CHECK YOURSELF 1

Label each as a line or a line segment.

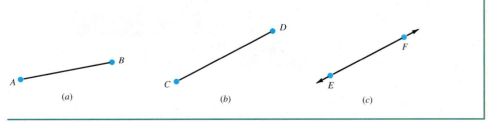

> **Definition: Angle**
>
> An **angle** is a geometric figure consisting of two line segments that share a common endpoint.

This surveyor is using a transit to locate a property line.

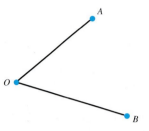

$\overline{OA}$ and $\overline{OB}$ are line segments. O is the **vertex** of the angle.

Surveyors use an instrument called a **transit.** A transit allows surveyors to measure angles so that, from a mathematical description, they can determine exactly where a property line is.

> **Definition: Perpendicular Lines**
>
> When two lines cross (or intersect) they form four angles. If the lines intersect such that four equal angles are formed, we say that the two lines are **perpendicular.**

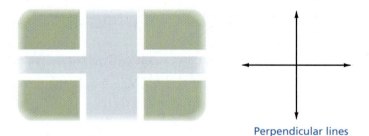

Perpendicular lines

At most street intersections, the two roads are perpendicular.

> **Definition: Parallel Lines**
>
> If two lines are drawn so that they never intersect (even if we extend the lines forever), we say that the two lines are **parallel.**

Parallel lines

Parallel parking gets its name from the fact that the parking spot is parallel to the traffic lane.

OBJECTIVE 2

Example 2   Recognizing Parallel and Perpendicular Lines

Label each pair of lines as parallel, perpendicular, or neither.

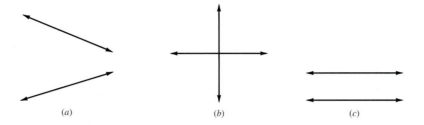

(a)   (b)   (c)

Although we don't see the lines in (a) intersecting, if they were extended as the arrowheads indicate, they would. They are neither parallel nor perpendicular. The lines in (b) are perpendicular because the four angles formed are equal. Only the lines in (c) are parallel.

 CHECK YOURSELF 2

Label each pair of lines as parallel, perpendicular, or neither.

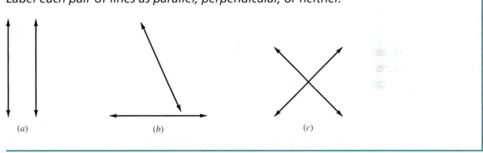

(a)   (b)   (c)

RECALL We used this small square in Chapter 5.

We call the angle formed by two perpendicular lines or line segments a **right angle.** We designate a right angle by drawing a small square.

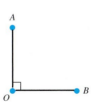

We can refer to a specific angle by naming three points. The middle point is the vertex of the angle. We can call this ∠*AOB* or, if there is no confusion, we can call it ∠*O*.

OBJECTIVE 1

Example 3   Naming an Angle

Name the highlighted angle.

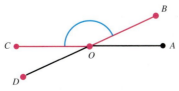

NOTE We could also call this angle ∠*BOC*.

The vertex of the angle is *O*, and the angle begins at *C* and ends at *B*, so we would name the angle ∠*COB*. We cannot refer to this as ∠*O* because it would not be clear to which angle we were referring.

© 2010 McGraw-Hill Companies

CHECK YOURSELF 3 _____

Name the highlighted angle.

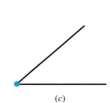

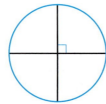

One way to measure an angle is to use a unit that is called a **degree.** There are 360 degrees (we write this as 360°) in a complete circle. Note in the picture on the left that there are four right angles in a circle. If we divide 360° by 4, we find that each **right angle** must measure 90°. Here are some other angles with their measurements.

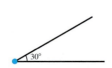

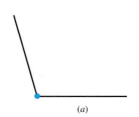 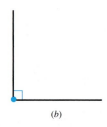  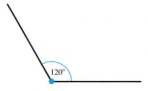

An **acute angle** measures between 0° and 90°. An **obtuse angle** measures between 90° and 180°. A **straight angle** measures 180°.

OBJECTIVE 3   **Example 4   Labeling Types of Angles**

Label each of the angles as an acute, an obtuse, a right, or a straight angle.

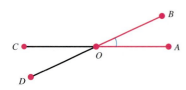

Part (*a*) is obtuse (the angle is more than 90°). Part (*b*) is a right angle (designated by the small square). Part (*c*) is an acute angle (it is less than 90°), and part (*d*) is a straight angle.

CHECK YOURSELF 4 _____

Label each angle as an acute, an obtuse, a right, or a straight angle.

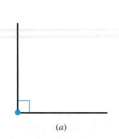 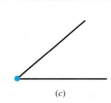 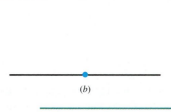 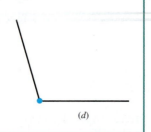

NOTE Your protractor may show the degree measures in both directions.

When assigning a measurement to an angle, we usually use a tool called a **protractor**.

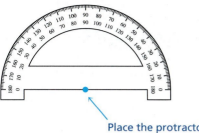

Place the protractor so that the vertex of the angle is here.

We read the protractor by placing one line segment of the angle at 0°. We then read the number that the other line segment passes through. This number represents the degree measurement of the angle. The point at the center of the protractor, the endpoint of the two line segments, is the vertex of the angle.

OBJECTIVE 4   **Example 5   Measuring an Angle**

Use the protractor to estimate the measurement for each angle.

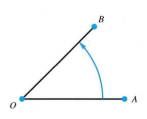
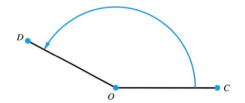
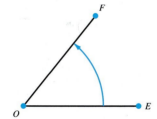

The measure of ∠AOB is 45°. The measure of ∠COD is between 150° and 155°. We could estimate it as 152°. The measure of ∠EOF is between 50° and 55°. We could estimate that it is a 52° angle.

CHECK YOURSELF 5

Use a protractor to estimate the measurement for each angle.

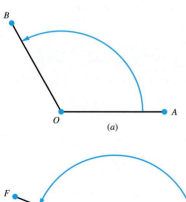

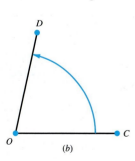

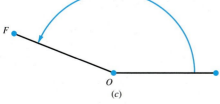

If we wish to refer to the *degree measure* of $\angle ABC$, we use $m\angle ABC$.

Example 6   **Measuring an Angle**

Find $m\angle AOB$.

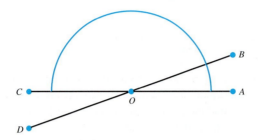

NOTE $m\angle AOB = 20°$ is read as, "the measure of angle *AOB* is 20 degrees."

Using a protractor, we find $m\angle AOB = 20°$.

CHECK YOURSELF 6

Find $m\angle AOC$.

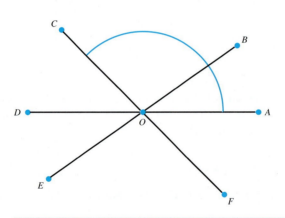

Definition: **Complementary Angles**

Two angles are **complementary** if the sum of their angles is 90°.

OBJECTIVE 5   **Example 7**   **Using Complementary Angles**

A and *B* are complementary angles. The measure of $\angle A = 42°$; find the measure of $\angle B$.

Because they are complementary, the sum of the angles is 90°.

$$m\angle A + m\angle B = 90°$$

$$42° + m\angle B = 90°$$

$$m\angle B = 90° - 42°$$

$$m\angle B = 48°$$

 CHECK YOURSELF 7

A and B are complementary angles. The measure of ∠A = 71°; find the measure of ∠B.

Definition: Supplementary Angles

Two angles are **supplementary** if the sum of their measures is 180°.

Example 8   Using Supplementary Angles

A and *B* are supplementary angles. The measure of ∠*B* is 67°. Find the measure of ∠*A*.

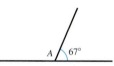

Because they are supplementary, the sum of the angles is 180°.

$$m\angle A + m\angle B = 180°$$

$$m\angle A + 67° = 180°$$

$$m\angle A = 180° - 67°$$

$$m\angle A = 113°$$

 CHECK YOURSELF 8

A and B are supplementary angles. m∠A = 75°. Find the measure of ∠B.

Property: Intersecting Lines

When two lines intersect, the adjacent angles are supplementary.

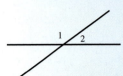

∠1 and ∠2 are supplementary.

OBJECTIVE 6 | **Example 9** **Finding the Measure of an Adjacent Angle**

Find the measure of ∠A.

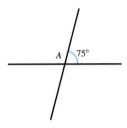

$m\angle A + 75° = 180°$

$m\angle A = 180° - 75°$

$m\angle A = 105°$

 CHECK YOURSELF 9

Find the measure of ∠A.

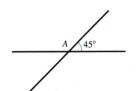

Definition: Vertical Angles

When two lines intersect at a single point, the angles that are not adjacent are called **vertical angles.**

Property: Vertical Angles

The measures of any two vertical angles are equal.

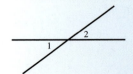

∠1 and ∠2 are vertical angles.

Example 10   Finding the Measure of Vertical and Adjacent Angles

Find the measures of $\angle A$, $\angle B$, and $\angle C$.

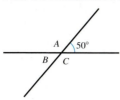

$\angle A$ is adjacent to a 50° angle, so the measure of $\angle A$ is 130°. (180 − 50 = 130)
$\angle B$ and the 50° angle are vertical angles, so $m\angle B = 50°$.
$\angle C$ is adjacent to a 50° angle, so $m\angle C = 130°$.

  CHECK YOURSELF 10 _____

Find the measures of $\angle A$, $\angle B$, and $\angle C$.

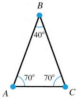

Now that you know something about angles, it is interesting to again look at triangles. Why is this shape called a triangle?

Literally, *triangle* means "three angles." Each of the triangles shown has three angles. For each triangle, the measure of each of its three angles is shown.

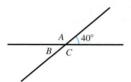 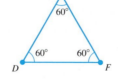 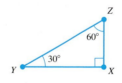

Now we will determine the sum of the angles inside each of the triangles. You will note that they always add up to 180°. No matter how we draw a triangle, the sum of the three angles inside the triangle is *always* 180°.

Here is an experiment that might convince you that this is always the case.

1. Using a straight edge, draw any triangle you wish on a sheet of paper.

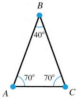

2. Use scissors to cut out the triangle.

3. Rip the three vertices off of the triangle.

4. Lay the three vertices (with the points of the triangle touching) together. They always form a straight angle, which we saw earlier in this section has a measure of 180°.

Property: Angles of a Triangle

For any triangle *ABC,*

$m\angle A + m\angle B + m\angle C = 180°$

OBJECTIVE 7   **Example 11   Finding an Angle Measure**

Find the measure of the third angle in this triangle.

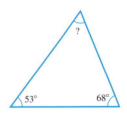

    We need the three measurements to add to 180°, so we add the two given measurements ($53° + 68° = 121°$). Then we subtract that from 180° ($180° - 121° = 59°$). This gives us the measure of the third angle, 59°.

CHECK YOURSELF 11

Find the measure of ∠ABC.

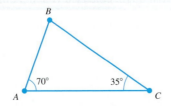

READING YOUR TEXT

The following fill-in-the-blank exercises are designed to assure that you understand the key vocabulary used in this section. Each sentence comes directly from the section. You will find the correct answers in Appendix C.

Section 8.1

(a) A line segment has _____ endpoints.

(b) Two _____ lines never intersect.

(c) Two angles are _____ if the sum of their measures is 90°.

(d) The measures of any two vertical angles are _____.

CHECK YOURSELF ANSWERS

1. **(a)** Line segment; **(b)** line segment; **(c)** line
2. **(a)** Parallel; **(b)** neither; **(c)** perpendicular   3. $\angle BOA$ or $\angle AOB$
4. **(a)** Right; **(b)** straight; **(c)** acute; **(d)** obtuse   5. **(a)** 120°; **(b)** 80°; **(c)** 160°
6. 135°   7. 19°   8. 105°   9. 135°   10. $m\angle A = 140°$ 11. 75°
$m\angle B = 40°$
$m\angle C = 140°$

Name _____

Section _____ Date _____

ANSWERS

1. _____

2. _____

3. _____

4. _____

5. _____

6. _____

7. _____

8. _____

9. _____

10. _____

11. _____

12. _____

13. _____

14. _____

15. _____

16. _____

17. _____

18. _____

8.1   Exercises

1. Draw line segment $\overline{AB}$.

$\dot{A}$ $\dot{B}$

2. Draw line $\overleftrightarrow{EF}$.

$\dot{E}$ $\dot{F}$

3. Draw line $\overleftrightarrow{AC}$.

$\dot{A}$ $\dot{C}$

4. Draw line segment $\overline{BC}$.

$\dot{B}$ $\dot{C}$

Identify each object as a line or line segment.

5.

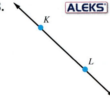

6.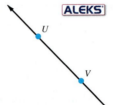

7.

8.

9.

10.

11.

12.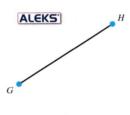

Label exercises 13 to 22 as true or false.

13. There are exactly two distinct line segments that can be drawn using two points.

14. There are exactly two distinct lines that can be drawn through two points.

15. Two opposite sides of a square are parallel line segments.

16. Two adjacent sides of a square are perpendicular line segments.

17. $\angle ABC$ will always have the same measure as $\angle CAB$.

18. Two acute angles have the same measure.

19. The sum of two vertical angles is 180°.

20. The sum of two complementary angles is 90°.

21. Two intersecting lines create two pairs of vertical angles.

22. Two intersecting lines create pairs of complementary angles.

23. Are the two lines parallel, perpendicular, or neither?

24. Are the two lines parallel, perpendicular, or neither?

In exercises 25 to 32 label each indicated angle.

25.

26.

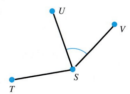

27.

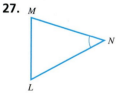

28.

29.

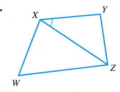

30.

ANSWERS

19. _____

20. _____

21. _____

22. _____

23. _____

24. _____

25. _____

26. _____

27. _____

28. _____

29. _____

30. _____

31.

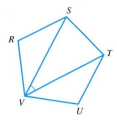

32.

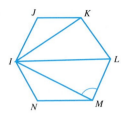

For each angle described, give its measure in degrees. One revolution is a full circle. Sketch the angle.

33. $\angle A$ represents $\dfrac{1}{6}$ of a revolution

34. $\angle B$ represents $\dfrac{1}{3}$ of a revolution

35. $\angle C$ represents $\dfrac{7}{12}$ of a revolution

36. $\angle D$ represents $\dfrac{11}{12}$ of a revolution

Measure each angle with a protractor. Identify the angle as acute, right, obtuse, or straight.

37.

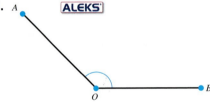

38.

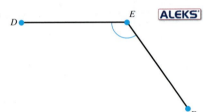

39.

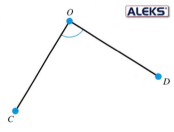

40.

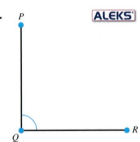

41.

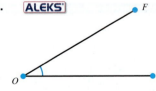

42.

In the figure, two parallel lines are intersected by a third line, forming eight angles. Draw lines like these on your paper.

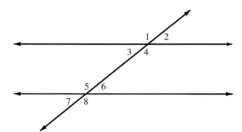

43. Use your protractor to measure ∠2 and ∠6. What do you notice?

44. Use your protractor to measure ∠3 and ∠6. What do you notice?

In exercises 45 to 50, find the missing angle.

45. **ALEKS**

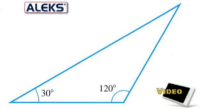

30°   120°

46. **ALEKS**

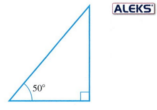

50°

47. **ALEKS**

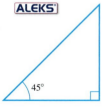

45°

48. **ALEKS**

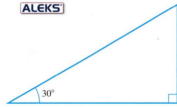

30°

49. **ALEKS**

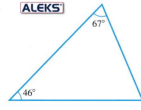

67°

46°

50. **ALEKS**

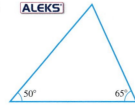

50°   65°

© 2010 McGraw-Hill Companies

For each triangle shown, find the indicated angle.

51. Find $m\angle C$.

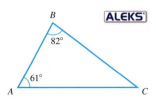

52. Find $m\angle B$.

53. Find $m\angle A$.

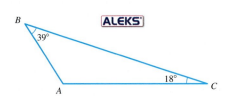

54. Find $m\angle B$.

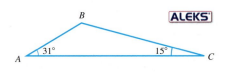

55. Find $m\angle B$.

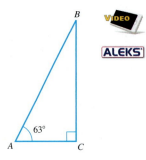

56. Find $m\angle A$.

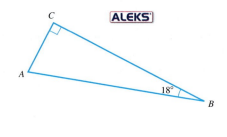

In exercises 57 to 59, one side of the triangle has been extended, forming what is called an **exterior angle.** In each case, find the measure of the indicated exterior angle.

57.

58.

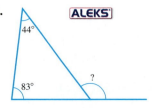

59.

60. What do you observe from exercises 57 to 59? Write a general conjecture about the exterior angle of a triangle.

61. Draw any triangle using a ruler. With your protractor, carefully measure the three interior angles and find their sum. Do this again with two more triangles of different shapes. What do you notice about the sums of the angles? Does it fit the rule of triangles?

62. A **quadrilateral** is a four-sided figure. Draw any quadrilateral and measure the four interior angles with a protractor. Record these and find their sum. Make a conjecture concerning the sum of the interior angles of *any* quadrilateral. Test your conjecture on another quadrilateral.

63. A **pentagon** is a five-sided figure. Draw any pentagon and measure the five interior angles with a protractor. Record these and find their sum. Make a conjecture concerning the sum of the interior angles of *any* pentagon. Test your conjecture on another pentagon.

64. A **hexagon** is a six-sided figure. Draw any hexagon and measure the six interior angles with a protractor. Record these and find their sum. Make a conjecture concerning the sum of the interior angles of *any* hexagon. Test your conjecture on another hexagon.

65. Argue that a triangle cannot have more than one obtuse angle.

66. Create an argument to support the following statement:

If $\triangle ABC$ is a right triangle, with $m\angle C = 90°$, then $\angle A$ and $\angle B$ must be acute and complementary.

Answers

1. A •———————• B   **3.** ←——•————•——→   **5.** Line
                                          A   C
7. Line segment   **9.** Line segment   **11.** Line   **13.** False   **15.** True
17. False   **19.** False   **21.** True   **23.** Parallel   **25.** $\angle POQ$
27. $\angle MNL$ **29.** $\angle FEG$ **31.** $\angle SVT$ **33.** 60°   **35.** 210°
37. 135°; obtuse   **39.** 90°; right   **41.** 30°; acute   **43.** 40°; 40°
45. 30°   **47.** 45°   **49.** 67°   **51.** 37°   **53.** 123°   **55.** 27°
57. 143°   **59.** 159°   **61.**   **63.**   **65.**

8.2 Perimeter and Circumference

© 2010 McGraw-Hill Companies

8.2 OBJECTIVES

1. Determine perimeters of rectangles, squares, triangles, parallelograms, and other polygons
2. Determine the approximate circumference of a circle

We discussed squares and rectangles in Chapter 1 of this text. Each of these shapes is an example of a polygon.

> **Definition: Polygon**
>
> A **polygon** is a simple closed geometric figure with three or more sides in which each side is a line segment.

In Section 3.2, we introduced the idea of perimeter. In this section, we expand our definition to include all polygons.

> **Definition: Perimeter**
>
> The **perimeter** of a polygon is the sum of the lengths of its sides. We usually designate perimeter with the letter P.

The formulas given here can be used to find the perimeters of a triangle, a square, and a rectangle, respectively. Note that each is simply a way of finding the sum of the lengths of the sides.

> **Property: Formula for Perimeter of a Triangle**
>
> $P = a + b + c$
>
> in which a, b, and c are the lengths of the three sides.

> **Property: Formula for Perimeter of a Square**
>
> $P = 4s$
>
> in which s is the length of a side.

> **Property: Formula for Perimeter of a Rectangle**
>
> $P = 2L + 2W$
>
> in which L is the length and W is the width.

OBJECTIVE 1   **Example 1   Finding the Perimeter of a Polygon**

Find the perimeter of each polygon.

(a)

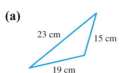

To find the perimeter of a triangle, we add the three lengths:

$P = 23 \text{ cm} + 15 \text{ cm} + 19 \text{ cm} = 57 \text{ cm}$

(b)

9 in.

The perimeter of a square is four times the length of one side.

$P = 4(9 \text{ in.}) = 36 \text{ in.}$

(which can be written as 3 ft or 1 yd).

(c)

1.7 ft

3.6 ft

The perimeter of a rectangle is twice the length plus twice the width.

$P = 2(3.6 \text{ ft}) + 2(1.7 \text{ ft}) = 7.2 \text{ ft} + 3.4 \text{ ft} = 10.6 \text{ ft}$

CHECK YOURSELF 1

Find the perimeter of each polygon.

(a)

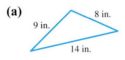

9 in.   8 in.   14 in.

(b)

5.3 m

(c)

5 cm   14 cm

To find the perimeter of other polygons, we add the length of each side of the polygon. In some cases, we can multiply the length of one side by the number of sides, as with a square. A polygon in which all of the sides are the same length and all interior angles have the same measure is said to be *regular.*

In general, polygons are named by the number of sides. A five-sided polygon is a **pentagon.** A six-sided one is called a **hexagon,** and a ten-sided polygon is a **decagon.**

Four-sided polygons are very common in the real world. We have already mentioned the square and the rectangle. Other four-sided polygons include the parallelogram and the trapezoid.

Definition: Parallelogram

A **parallelogram** is a four-sided polygon in which opposite sides are parallel. Opposite sides of a parallelogram have the same length.

Definition: Trapezoid

A **trapezoid** is a four-sided polygon in which exactly one pair of opposite sides is parallel.

Find the perimeter of each polygon.

(a) Parallelogram

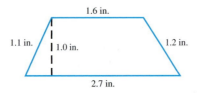

Opposite sides of a parallelogram have the same length:

$P = 2(6 \text{ m}) + 2(3 \text{ m})$

$ = 12 \text{ m} + 6 \text{ m}$

$ = 18 \text{ m}$

(b) Trapezoid

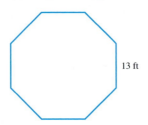

We can add the length of each of the four sides:

$P = 2.7 \text{ in.} + 1.1 \text{ in.} + 1.6 \text{ in.} + 1.2 \text{ in.}$

$ = 6.6 \text{ in.}$

(c) Regular octagon

13 ft

We are told that this octagon is regular. Therefore, each of the 8 sides is 13 feet in length.

$P = 8 \cdot 13 \text{ ft} = 104 \text{ ft}$

CHECK YOURSELF 2

Find the perimeter of each polygon

(a) Trapezoid   **(b)** Trapezoid   **(c)** Regular hexagon

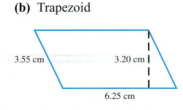

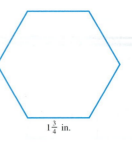

The distance around the outside of a circle is closely related to the concept of perimeter. We call the perimeter of a circle its **circumference.**

Definition: Circumference of a Circle

The **circumference** of a circle is the distance around that circle.

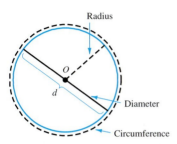

We begin by defining some terms. In the given circle, d represents the **diameter.** This is the distance across the circle through its center (labeled with the letter O, for **origin**). The **radius** r is the distance from the center to a point on the circle. Note that, for a given circle, every radius is the same length. The diameter is always twice the radius.

It was discovered long ago that the ratio of the circumference of a circle to its diameter is always the same. That is, no matter how big or small a circle is, if we divide its circumference by its diameter, we get a number slightly larger than three.

In 1706, William Jones named this ratio with the Greek letter π, **pi.** Leonhard Euler adopted the symbol 40 years later, and its popularity was assured.

Pi is an **irrational number,** which means that we cannot give an exact value of pi. It is approximately 3.14, rounded to two decimal places.

If we "wrap" a circle's diameter around the circumference pi times, we cover the entire circle exactly once, as illustrated in the diagrams below.

NOTE Formulations of π appear as early as 3000 B.C. when the Egyptians constructed the pyramids.

Archimedes computed π to two decimal places in 250 B.C.

Zu Chongzhi of China computed seven decimal places in the fifth century A.D.

Al-Kashi of Samarkand computed 14 decimal places in the fifteenth century.

Today, we have computed pi to over six billion decimal places.

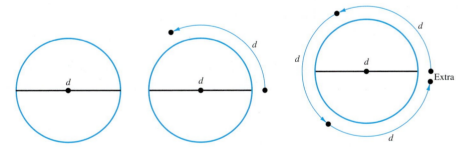

Using the formulation pictured, and the ratio $\pi = \dfrac{C}{d}$, we get formulas to compute the circumference of a circle.

Property: Formula for the Circumference of a Circle

The circumference C of a circle with diameter d is

$$C = \pi d$$

Because the diameter of a circle is twice the length of its radius ($d = 2r$), we can also write the formula as

$$C = 2\pi r$$

NOTE One U.S. state attempted to legislate an incorrect value for π in 1897.

OBJECTIVE 2

Example 3 **Finding the Circumference of a Circle**

(a) A circle has a diameter of 4.5 ft. Find its circumference, using 3.14 for π. If your calculator has a $\boxed{\pi}$ key, use that key instead of 3.14.

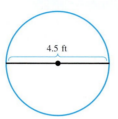

4.5 ft

NOTE Because 3.14 is an approximation for pi, we can only say that the circumference is approximately 14.1 ft. The symbol $\approx$ means approximately.

$C = \pi d$

$\approx 3.14 \cdot 4.5 \text{ ft}$

$\approx 14.1 \text{ ft}$ (rounded to one decimal place)

(b) A circle has a radius of 8 cm. Find its circumference using 3.14 for π.

8 cm

NOTE If we use the π button on a calculator and round to one decimal place, we get 50.3 cm.

$C = 2\pi r$

$\approx 2 \cdot 3.14 \cdot 8 \text{ cm}$

$\approx 50.2 \text{ cm}$ (rounded to one decimal place)

 CHECK YOURSELF 3

(a) A circle has a diameter of $3\frac{1}{2}$ in. Find its circumference.

(b) Find the circumference of a circle with a radius of 2.5 cm.

Sometimes we will want to combine the ideas of perimeter and circumference to solve a problem.

Example 4 **Finding a Perimeter**

We wish to build a wrought-iron frame gate according to the diagram. How many feet of material will be needed? (Round to the nearest tenth of a foot.)

NOTE The distance around the semicircle is $\frac{1}{2}\pi d$.

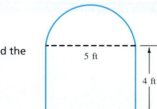

5 ft

4 ft

The problem can be broken into two parts. The upper part of the frame is a semicircle (half a circle). The remaining part of the frame is just three sides of a rectangle.

NOTE Using a calculator with a $\boxed{\pi}$ key,

$1 \boxed{\div} 2 \boxed{\times} \boxed{\pi} \boxed{\times} 5$

Circumference (upper part) $\approx \dfrac{1}{2} \cdot 3.14 \cdot 5$ ft ≈ 7.9 ft

Perimeter (lower part) $= 4 + 5 + 4 = 13$ ft

Adding, we have

$7.9 + 13 = 20.9$ ft

We need approximately 20.9 ft of material.

 CHECK YOURSELF 4 _____

Find the perimeter of the figure. Round to the nearest tenth of a meter.

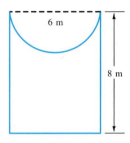

READING YOUR TEXT _____

The following fill-in-the-blank exercises are designed to assure that you understand the key vocabulary used in this section. Each sentence comes directly from the section. You will find the correct answers in Appendix C.

Section 8.2

(a) The _____ of a polygon is the sum of the lengths of its sides.

(b) The distance around a _____ is called its circumference.

(c) The length of the _____ of a circle is twice the length of its radius.

(d) Pi is the ratio of the circumference of a circle to its _____.

CHECK YOURSELF ANSWERS _____

1. (a) 31 in.; **(b)** 21.2 m; **(c)** 38 cm   **2. (a)** 41 yd; **(b)** 19.6 cm; **(c)** $10\dfrac{1}{2}$ in.

3. (a) $C \approx 11.0$ in.; **(b)** $C \approx 15.7$ cm   **4.** $P \approx 31.4$ m

Name _____

Section _____ Date _____

ANSWERS

1. _____
2. _____
3. _____
4. _____
5. _____
6. _____
7. _____
8. _____
9. _____
10. _____
11. _____
12. _____
13. _____
14. _____

8.2 Exercises

Find the perimeter for each triangle.

1.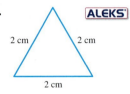
2 cm, 2 cm, 2 cm

2.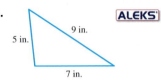
9 in., 5 in., 7 in.

3.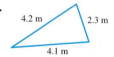
4.2 m, 2.3 m, 4.1 m

4.
$8\frac{1}{2}$ ft, 3 ft, $7\frac{1}{4}$ ft

Find the perimeter for each polygon.

5.
2.7 mm

6.
$3\frac{1}{2}$ ft

7.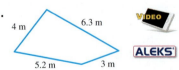
8.3 cm, 15.6 cm

8.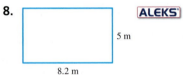
5 m, 8.2 m

9.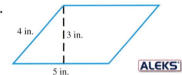
4 in., 3 in., 5 in.

10.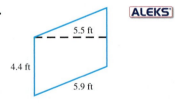
5.5 ft, 4.4 ft, 5.9 ft

11.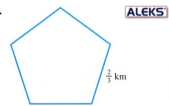
4 m, 6.3 m, 5.2 m, 3 m

12.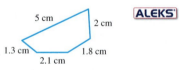
5 cm, 2 cm, 1.3 cm, 2.1 cm, 1.8 cm

13.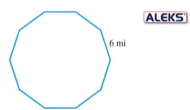
$\frac{2}{3}$ km

14.
6 mi

15. _____

16. _____

17. _____

18. _____

19. _____

20. _____

21. _____

22. _____

23. _____

24. _____

25. _____

Find the circumference of each figure. Use 3.14 for π, and round your answer to one decimal place.

15.

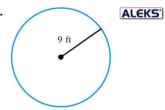

9 ft · ALEKS

16.

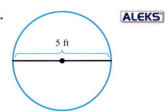

5 ft · ALEKS

17.

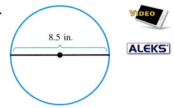

8.5 in. · VIDEO · ALEKS

18.

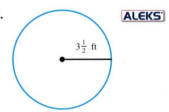

3.75 ft · ALEKS

19.

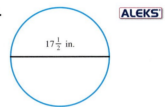

$17\frac{1}{2}$ in. · ALEKS

20.

$3\frac{1}{2}$ ft · ALEKS

Find the perimeter of each figure. Use 3.14 to approximate π and round answers to one decimal place.

21.

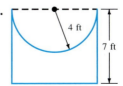

9 ft · 7 ft · VIDEO · ALEKS

22.

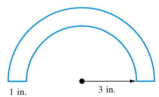

1 in. · 3 in.

23.

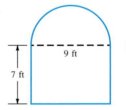

4 ft · 7 ft

24.

10 in.

Solve the applications.

25. Science and Medicine   A path runs around a circular lake with a diameter of 1,000 yd. Robert jogs around the lake three times for his morning run. How far has he run?

26. **Crafts** A circular rug is 6 ft in diameter. Binding for the edge costs $1.50 per yard. What will it cost to bind around the rug?

27. The shape of a Norman window is shown in the figure below. Determine the length of trim needed to frame the window. **(8)** **ALEKS**

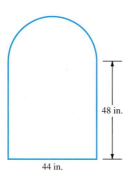

48 in.

44 in.

28. Many Norman windows include a trim piece separating the semicircle from the rectangle. Find the length of trim needed for the window shown below. **(8)**

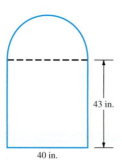

43 in.

40 in.

29. What happens to the circumference of a circle if you double the radius? If you double the diameter? If you triple the radius? Create some examples to demonstrate your answers.

30. The distance from Philadelphia to Sea Isle City is 100 mi. A car was driven this distance using tires with a radius of 14 in. How many revolutions of each tire occurred on the trip? **ALEKS**

Answers

1. 6 cm   **3.** 10.6 m   **5.** 10.8 mm   **7.** 47.8 cm   **9.** 18 in.

11. 18.5 m   **13.** $10\frac{1}{3}$ km   **15.** 56.5 ft   **17.** 26.7 in.   **19.** 55 in.

21. 37.1 ft   **23.** 34.6 ft   **25.** 9,420 yd   **27.** $\approx$209.1 in.

29. Doubled; doubled; tripled

8.3 OBJECTIVES

1. Determine areas of rectangles, squares, triangles, and parallelograms
2. Convert between units of area
3. Determine the area of a circle
4. Determine volumes of rectangular solids, cubes, spheres, and cylinders

In Section 1.5, we examined the area of a rectangle. In this section, we build on that idea. We begin with a definition.

> **Definition: Area**
>
> The **area** of an object is the number of unit squares that cover its surface. We usually designate area with the letter A.

For a rectangle with length L and width W, we can cover the rectangle with $L \cdot W$ unit squares, as illustrated below.

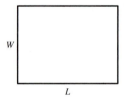
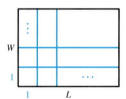

This leads to the following formulas for the areas of squares and rectangles.

> **Property: Formula for Area of a Square**
>
> $A = s^2$
>
> in which s is the length of a side.

NOTE A square is a rectangle in which a side $s = L = W$.

> **Property: Formula for Area of a Rectangle**
>
> $A = L \cdot W$
>
> in which L is the length and W is the width.

OBJECTIVE 1

Example 1 Finding the Area of a Square

Find the area of the square.

3.2 cm

The area of a square is the square of the length of one side. Here we have

$$A = s^2 = (3.2 \text{ cm})^2 = 10.24 \text{ cm}^2$$

It is important to note that the units for area are *always* square units.

 CHECK YOURSELF 1

Find the area of the square.

5.3 m

In Example 2, we find the area of a rectangle.

Example 2   Finding the Area of a Rectangle

Find the area of the rectangle.

$2\frac{2}{3}$ ft

$3\frac{1}{4}$ ft

Using the formula $A = L \cdot W$, we get

$$A = 3\frac{1}{4} \text{ ft} \cdot 2\frac{2}{3} \text{ ft} = \frac{13}{4} \cdot \frac{8}{3} \text{ ft}^2 = \frac{104}{12} \text{ ft}^2 = \frac{26}{3} \text{ ft}^2 = 8\frac{2}{3} \text{ ft}^2$$

Note that the area is just under 9 ft².

 CHECK YOURSELF 2

Find the area of the rectangle.

1.5 cm

3.2 cm

Two other figures that are frequently encountered are parallelograms and triangles.

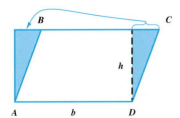

© 2010 McGraw-Hill Companies

In the figure, *ABCD* is a *parallelogram* (its opposite sides are parallel and of equal length). We can draw a line from *D* that forms a right angle with side *BC*. This cuts off one corner of the parallelogram. Now imagine that we move that corner over to the left side of the figure, as shown. This gives us a rectangle instead of a parallelogram. Because the area of the figure is not changed by moving the corner, the parallelogram has the same area as the rectangle, the product of the base and the height.

> **Property:** Formula for the Area of a Parallelogram
>
> $A = b \cdot h$
>
> in which *b* is the base and *h* is the height.

Example 3   Finding the Area of a Parallelogram

A parallelogram has the dimensions shown in the figure. What is its area?

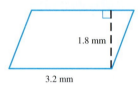

Use the formula for the area of a parallelogram, with $b = 3.2$ mm and $h = 1.8$ mm.

$A = b \cdot h$

$\quad = 3.2 \text{ mm} \cdot 1.8 \text{ mm} = 5.76 \text{ mm}^2$

 CHECK YOURSELF 3 —————————————————————

If the base of a parallelogram is $3\frac{1}{2}$ in. and its height is $1\frac{1}{2}$ in., what is its area?

As we saw in Section 8.1, another common geometric figure is the *triangle.* It has three sides. Triangle *ABC* is shown in the figure.

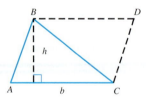

b is the base of the triangle.
h is the height, or the
altitude, of the triangle.

Once we have a formula for the area of a parallelogram, it is not hard to find the area of a triangle. If we draw the dotted lines from *B* to *D* and from *C* to *D* parallel to two sides of the triangle, we form a parallelogram. The area of the triangle is then one-half the area of the parallelogram (which is $b \cdot h$).

Property: Formula for the Area of a Triangle

$$A = \frac{1}{2} \cdot b \cdot h$$

Example 4   Finding the Area of a Triangle

A triangle has an altitude of 2.3 in., and its base is 3.4 in. What is its area?

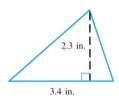

2.3 in.

3.4 in.

Using the formula for the area of a triangle, with $b = 3.4$ in. and $h = 2.3$ in.

$$A = \frac{1}{2} \cdot b \cdot h$$

$$= \frac{1}{2} \cdot 3.4 \text{ in.} \cdot 2.3 \text{ in.} = 3.91 \text{ in.}^2$$

CHECK YOURSELF 4

A triangle has a base of 10 kilometers (km) and an altitude of 6 km. Find its area.

Sometimes we want to convert from one square unit to another. For instance, look at 1 yd^2 in the figure.

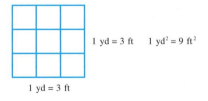

1 yd = 3 ft   1 yd^2 = 9 ft^2

1 yd = 3 ft

The table gives some useful relationships.

Square Units and Equivalents	
1 square foot (ft²)	= 144 square inches (in.²)
1 square yard (yd²)	= 9 ft²
1 acre	= 4,840 yd² = 43,560 ft²
1 square mile (mi²)	= 640 acres

NOTE Originally, an **acre** was the area that could be plowed by a team of oxen in a day!

OBJECTIVE 2

Example 5 **Converting between Square Feet and Square Yards in Finding Area**

A room has the dimensions 12 ft by 15 ft. How many square yards of linoleum will be needed to cover the floor?

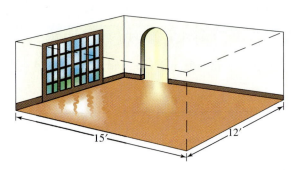

NOTE We first find the area in square feet and then convert to square yards.

$A = 12 \text{ ft} \cdot 15 \text{ ft} = 180 \text{ ft}^2$

$$= \overset{20}{\cancel{180}} \text{ ft}^2 \cdot \frac{1 \text{ yd}^2}{\underset{1}{\cancel{9}} \text{ ft}^2}$$ Recall that $\dfrac{1 \text{ yd}^2}{9 \text{ ft}^2}$ is a units fraction.

$$= 20 \text{ yd}^2$$

 CHECK YOURSELF 5

A hallway is 27 ft long and 4 ft wide. How many square yards of carpeting will be needed to carpet the hallway?

Example 6 illustrates the use of a common unit of area, the acre.

Example 6 **Converting between Square Yards and Acres in Finding Area**

A rectangular field is 220 yd long and 110 yd wide. Find its area in acres.

$A = 220 \text{ yd} \cdot 110 \text{ yd} = 24,200 \text{ yd}^2$

$$= \overset{5}{\cancel{24,200}} \text{ yd}^2 \cdot \frac{1}{\underset{1}{\cancel{4840}}} \frac{\text{acre}}{\text{yd}^2}$$

$$= 5 \text{ acres}$$

 CHECK YOURSELF 6

A proposed site for an elementary school is 220 yd long and 198 yd wide. Find its area in acres.

The number pi (π), which we used to find circumference in Section 8.2, is also used in finding the area of a circle. If r is the radius of a circle, we have this formula.

NOTE This is read as, "Area equals pi *r* squared." You can multiply the radius by itself and then by pi.

> **Property:** Formula for the Area of a Circle
>
> $A = \pi r^2$

OBJECTIVE 3

> **Example 7**   **Find the Area of a Circle**

A circle has a radius of 7 in. What is its area?

Use the formula for the area of a circle, 3.14 for π, and $r = 7$ in.

$A \approx 3.14 \cdot (7 \text{ in.})^2$

$\approx 3.14 \cdot (49 \text{ in.}^2)$ Again the area is an approximation because we use 3.14, an approximation for π.

$\approx 153.9 \text{ in.}^2$

 CHECK YOURSELF 7

Find the area of a circle whose diameter is 4.8 cm. Remember that the formula refers to the radius. Use 3.14 for π and round your result to the nearest tenth of a square centimeter.

Our next measurement deals with finding the **volume** of a solid, which is the measure of the space contained in the solid.

> **Definition:** Solids
>
> A **solid** is a three-dimensional figure. It has length, width, and height.

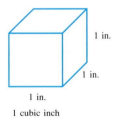

1 in.

1 in.

1 in.

1 cubic inch

Volume is measured in **cubic units.** Examples include cubic inches, cubic feet, and cubic centimeters. A cubic inch, for instance, is the measure of the space contained in a cube that is 1 in. on each edge.

In finding the volume of a figure, we want to know how many cubic units are contained in that figure. We will start with a simple example, a **rectangular solid.** A rectangular solid is a very familiar figure. A box, a crate, and most rooms are rectangular solids. Say that the dimensions of the solid are 5 in. by 3 in. by 2 in. as pictured. If we divide the solid into units of 1 in.^{3}, we have two layers, each containing 3 units by 5 units, or 15 in.^{3} Because there are two layers, the volume is 30 in.^{3}

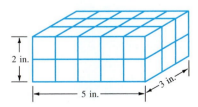

2 in.

5 in.

3 in.

In general, we can see that the volume of a rectangular solid is the product of its length, width, and height.

> **Property:** Formula for the Volume of a Rectangular Solid
>
> $V = L \cdot W \cdot H$

OBJECTIVE 4

Example 8 **Finding a Volume**

A crate has dimensions 4 ft by 2 ft by 3 ft. Find its volume.

Use the formula for the volume of a rectangular solid, with $L = 4$ ft, $W = 2$ ft, and $H = 3$ ft.

NOTE We are not particularly worried about which is the length, which is the width, and which is the height, because the order in which we multiply does not change the result.

$$V = L \cdot W \cdot H$$
$$= 4 \text{ ft} \cdot 2 \text{ ft} \cdot 3 \text{ ft}$$
$$= 24 \text{ ft}^3$$

 CHECK YOURSELF 8

A room is 15 ft long, 10 ft wide, and 8 ft high. What is its volume?

The formulas for the volume of a sphere (the shape of a ball) and a right circular cylinder (the shape of a can) both involve the number π.

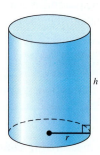

The process of finding the volume of a cylinder is similar to a rectangular solid.

A one-unit high "slice" gives a piece with a volume equal to the area of the base. Because the base is a circle, we get an area of πr^2 for this slice. There are h of these slices, as illustrated below.

NOTE Archimedes of Syracuse studied both cylinders and spheres in the third century B.C.

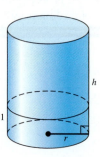 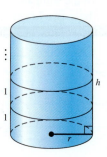

The formula for the area of a sphere is less intuitive. You may learn its derivation in a more advanced math class.

Property: Volume of a Sphere

$$V = \frac{4}{3}\pi r^3$$

in which r is the radius.

Property: Volume of a Right Circular Cylinder

$$V = \pi r^2 h$$

in which r is the radius and h is the height.

Example 9   Finding a Volume

Find the volume of each solid shape.

(a)

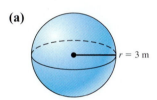

$r = 3$ m

This is a sphere with radius 3 m. Using the formula for the volume of a sphere,

$$V = \frac{4}{3}\pi r^3$$

$$= \frac{4}{3}\pi (3 \text{ m})^3$$

$$= \frac{4}{3}\pi \cdot 27 \text{ m}^3$$

$$= \frac{4}{3} \cdot \pi \cdot \frac{27}{1} \text{ m}^3$$

$$= 36\pi \text{ m}^3$$

$$\approx 36(3.14) \text{ m}^3$$

$$\approx 113.0 \text{ m}^3$$

(b)

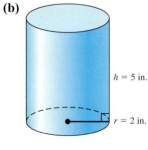

$h = 5$ in.

$r = 2$ in.

This is a right circular cylinder with radius 2 in. and height 5 in. Using the formula for the volume of a right circular cylinder,

$$V = \pi r^2 h$$

$$= \pi (2 \text{ in.})^2 (5 \text{ in.})$$

$$= \pi 4 \text{ in.}^2 (5 \text{ in.})$$

$$= 20\pi \text{ in.}^3$$

$$\approx 20(3.14) \text{ in.}^3$$

$$\approx 62.8 \text{ in.}^3$$

CHECK YOURSELF 9

Find the volume of each solid shape.

(a)

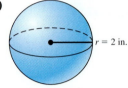

$r = 2$ in.

(b)

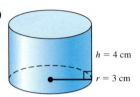

$h = 4$ cm

$r = 3$ cm

READING YOUR TEXT

The following fill-in-the-blank exercises are designed to assure that you understand the key vocabulary used in this section. Each sentence comes directly from the section. You will find the correct answers in Appendix C.

Section 8.3

(a)   Area is always measured in _____ units.

(b)   A _____ has the same area as a rectangle with the same base and height.

(c)   A diagonal in a parallelogram divides the parallelogram into two _____.

(d)   _____ is always measured in cubic units.

CHECK YOURSELF ANSWERS

1. 28.09 m^2 **2.** 4.8 cm^2 **3.** $A = 5\dfrac{1}{4}$ in.^{2}

4. $A = 30$ km^2 **5.** 12 yd^2 **6.** 9 acres   **7.** ≈ 18.1 cm^2

8. 1,200 ft^3 **9. (a)** ≈ 33.5 in.^{3}; **(b)** ≈ 113.0 cm^3

8.3 Exercises

Name _____

Section _____ Date _____

ANSWERS

Find the area of each figure.

1.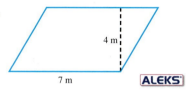

2.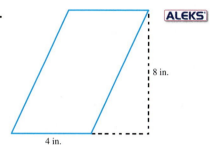

3.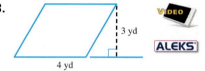

4.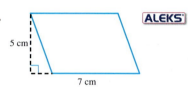

5.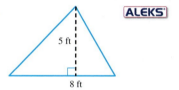

6.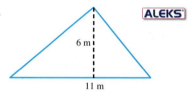

7.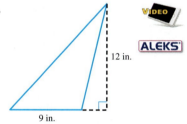

8.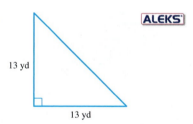

9.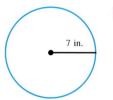

10.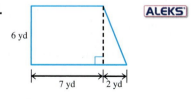

1. _____

2. _____

3. _____

4. _____

5. _____

6. _____

7. _____

8. _____

9. _____

10. _____

11. _____

12. _____

Find the area of each figure. Use 3.14 for π and round your answer to one decimal place.

11.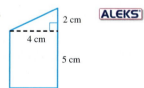

12.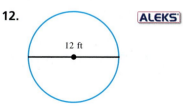

ANSWERS

13. _____

14. _____

15. _____

16. _____

17. _____

18. _____

19. _____

20. _____

21. _____

13. 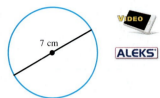 7 cm **ALEKS**

14. **ALEKS** 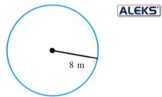 8 m

15. **ALEKS** 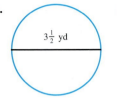 $3\frac{1}{2}$ yd

16. **ALEKS** 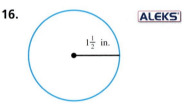 $1\frac{1}{2}$ in.

Solve the applications.

17. Crafts A circular piece of lawn has a radius of 28 ft. You have a bag of fertilizer that will cover 2,500 ft² of lawn. Do you have enough?

18. Crafts A circular coffee table has a diameter of 5 ft. What will it cost to have the top refinished if the company charges $3 per square foot for the refinishing?

19. Construction A circular terrace has a radius of 6 ft. If it costs $1.50 per square foot to pave the terrace with brick, what will the total cost be? **VIDEO**

20. Construction A house addition is in the shape of a semicircle (a half circle) with a radius of 9 ft. What is its area?

21. Crafts A Tetra-Kite uses 12 triangular pieces of plastic for its surface. Each triangle has a base of 12 in. and a height of 12 in. How much material is needed for the kite?

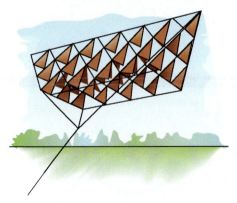

ANSWERS

22. _____

23. _____

24. _____

25. _____

26. _____

27. _____

28. _____

29. _____

30. _____

31. _____

22. **Construction**   You buy a square lot that is 110 yd on each side. What is its size in acres?   **ALEKS**

23. **Crafts**   You are making rectangular posters 12 in. by 15 in. How many square feet of material will you need for four posters?   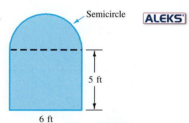   **ALEKS**

24. **Crafts**   Andy is carpeting a recreation room 18 ft long and 12 ft wide. If the carpeting costs $15 per square yard, what will be the total cost of the carpet?

25. **Construction**   A shopping center is rectangular, with dimensions of 550 yd by 440 yd. What is its size in acres?

26. **Construction**   An A-frame cabin has a triangular front with a base of 30 ft and a height of 20 ft. If the front is to be glass that costs $3 per square foot, what will the glass cost?

Find the area of the shaded part in each figure. Round your answers to one decimal place.

27. 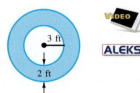   **VIDEO**   **ALEKS**

3 ft

2 ft

28. **ALEKS**   Semicircle

5 ft

6 ft

29. 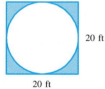   **VIDEO**   **ALEKS**

20 ft

20 ft

30. **ALEKS**

10 in.

10 in.

31. Papa Doc's delivers pizza. The price of an 8-in.-diameter pizza is $8.99, and the price of a 16-in.-diameter pizza is $17.98. Write a plan to determine which is the better buy.

32. An indoor track layout is shown.

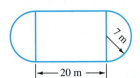

7 m

20 m

How much would it cost to lay down hardwood floor to cover the entire track if the hardwood floor costs $10.50 per square meter?

33. Find the area and the circumference (or perimeter) of each item:

(a) a penny   **(b)** a nickel   **(c)** a dime   **(d)** a quarter   **(e)** a half-dollar
(f) a silver dollar   **(g)** a Sacagawea dollar   **(h)** a dollar bill
(i) one face of the pyramid on the back of a $1 bill.

34. How would you determine the cross-sectional area of a Douglas fir tree (at, say, 3 ft above the ground) without cutting it down? Use your method to solve this problem:

If the circumference of a Douglas fir is 6 ft 3 in., measured at a height of 3 ft above the ground, compute the cross-sectional area of the tree at that height.

35. What is the effect on the area of a triangle if the base is doubled and the altitude is cut in half? Create some examples to demonstrate your ideas.

36. What happens to the area of a circle if you double the radius? If you double the diameter? If you triple the radius? Create some examples to demonstrate your answers.

Find the volume of each solid shown. Where appropriate, use 3.14 as an approximation for π and round to the nearest tenth.

ANSWERS

37. _____

38. _____

39. _____

40. _____

41. _____

42. _____

43. _____

44. _____

45. _____

46. _____

47. _____

48. _____

37.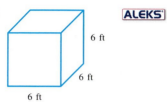

6 ft
6 ft
6 ft

38.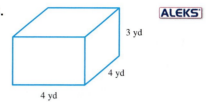

3 yd
4 yd
4 yd

39.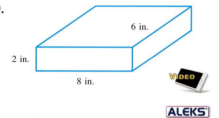

6 in.
2 in.
8 in.

40.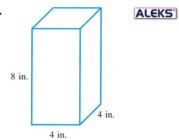

8 in.
4 in.
4 in.

41.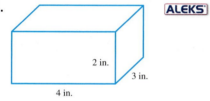

2 in.
3 in.
4 in.

42.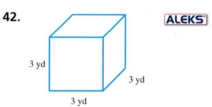

3 yd
3 yd
3 yd

43.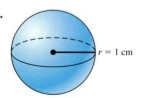

$r = 1$ cm

44.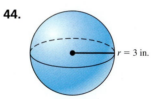

$r = 3$ in.

45.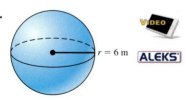

$r = 6$ m

46.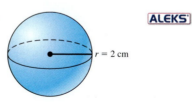

$r = 2$ cm

47.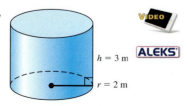

$h = 3$ m
$r = 2$ m

48.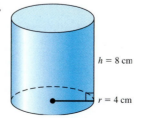

$h = 8$ cm
$r = 4$ cm

Solve the applications.

49. **Business and Finance**   A shipping container is 5 ft by 3 ft by 2 ft. What is its volume?

50. **Crafts**   A cord of wood is 4 ft by 4 ft by 8 ft. What is its volume?

51. **Crafts**   The inside dimensions of a meat market's cooler are 3 m by 3 m by 2 m. What is the capacity of the cooler in cubic meters?

52. **Crafts**   A storage bin is 6 m long, 2 m wide, and 2 m high. What is its volume in cubic meters?

53. Determine the area of the two pieces of the Norman window shown below, separately. Then add them together to get the total amount of glass needed for the window.   (8)

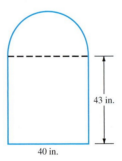

43 in.

40 in.

54. Determine the area of the Norman window shown below.   (8)

48 in.

44 in.

Answers

1. 28 m^2 **3.** 12 yd^2 **5.** 20 ft^2 **7.** 54 in.^2 **9.** 24 cm^2
11. 153.9 in.^2 **13.** 38.5 cm^2 **15.** 9.6 yd^2 **17.** yes; $\approx 2{,}461.8 \text{ ft}^2$
19. $\approx \$169.56$ **21.** 864 in.^2 **23.** 5 ft^2 **25.** 50 acres   **27.** 50.2 ft^2
29. 86 ft^2 **31.** 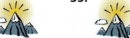   **33.**   **35.**   **37.** 216 ft^3

39. 96 in.^3 **41.** 24 in.^3 **43.** $\approx 4.2 \text{ cm}^3$ **45.** 904.3 m^3 **47.** $\approx 37.7 \text{ m}^3$
49. 30 ft^3 **51.** 18 m^3 **53.** 1,720 sq in.; ≈ 628.3 sq in.; $\approx 2{,}348.3$ sq in.

ACTIVITY 8: NORMAN WINDOWS

Each chapter in this text includes an activity. The activity is related to the vignette you encountered in the chapter opening. The activity provides you with the opportunity to apply the math you studied in the chapter to a relevant topic.

Your instructor will determine how best to use this activity with your class. You may find yourself working in class or outside of class; you may find yourself working alone or in small groups; or you may even be asked to perform research in a library or on the Internet.

When first introduced to geometry in your math class, you worked with fairly straightforward figures (squares, circles, triangles, etc.). Most real-world objects do not fit so neatly into such simple geometric figures. Consider the chair that you are using right now. The chair is probably made up of several shapes put together. None of these individual shapes is quite the perfect figure in textbooks (is there a curve to the back of the chair?).

Composite geometric figures are figures formed by combining several simple geometric figures. One example of a composite geometric figure is the Norman window. Norman windows are windows constructed by combining a rectangle with a half-circle (see the figure). Given such a figure, there are several questions we could ask. We might ask an area question (the amount of glass used), a perimeter question (the amount of frame), or a combination question to determine the amount of wall space necessary to accommodate such a window.

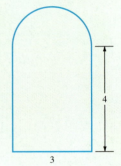

Assume all measurements are in feet and round answers to two decimal places.

1. Find the area of the rectangular piece of glass in the Norman window pictured.

2. Find the area of the half-circle piece of glass.

3. Find the total area of the glass.

4. If the glass costs $3 per square foot for a rectangular piece and $4.75 per square foot for the circular piece, find the total cost of the glass.

5. Find the outer perimeter of the figure.

6. Find the length of framework needed for the Norman window (do not forget to include the strip of frame separating the rectangle and half-circle).

7. Find the cost of the framework if it costs $1.25 per foot for straight pieces and $3.25 per foot for curved pieces.

8. Use your answers to exercises 4 and 7 to determine the total cost of the Norman window pictured.

9. Find the dimensions of the smallest rectangle that would completely contain the Norman window.

10. Find the area of the "leftover" rectangle created by cutting the Norman window from the rectangle found in exercise 9.

8 Summary

DEFINITION/PROCEDURE	EXAMPLE	REFERENCE
Lines and Angles		**Section 8.1**
Line A series of points that goes on forever.		*p. 635*
Line segment A piece of a line that has two endpoints.		*p. 635*
Angle A geometric figure consisting of two line segments that share a common endpoint.		*p. 636*
Perpendicular lines Lines are *perpendicular* if they intersect to form four equal angles.		*p. 636*
Parallel lines Lines are *parallel* if they never intersect.		*p. 636*
Right angles have a measure of 90°.		*p. 637*
Acute angles have a measure less than 90°.		*p. 638*
Obtuse angles have a measure between 90° and 180°.		*p. 638*

Continued

DEFINITION/PROCEDURE	EXAMPLE	REFERENCE
Perimeter and Circumference		**Section 8.2**

A **polygon** is a closed figure with three or more sides. Each side is a line segment.
    The **perimeter** is the sum of the lengths of its sides:
 	
Triangle $P = a + b + c$
Square $P = 4s$
Rectangle $P = 2L + 2W$
Circumference
Circle $C = 2\pi r$
	Find the perimeter of the figure:
	
	1.6 cm
	5 cm
	
	$P = 2L + 2W$
	   $= 2(1.6 \text{ cm}) + 2(5 \text{ cm})$
	   $= 3.2 \text{ cm} + 10 \text{ cm}$
	   $= 13.2 \text{ cm}$
		p. 652–653

Area and Volume		**Section 8.3**

The area of an object is the number of unit squares needed to cover its surface:
 	
Square $A = s^2$
Rectangle $A = L \cdot W$
Triangle $A = \dfrac{1}{2} \cdot b \cdot h$
Parallelogram $A = b \cdot h$
Circle $A = \pi r^2$
	Find the area of the figure:
	
	2.4 in.
	6 in.
	
	$A = \dfrac{1}{2}(6 \text{ in.})(2.4 \text{ in.})$
	   $= (3 \text{ in.})(2.4 \text{ in.})$
	   $= 7.2 \text{ in.}^2$
		p. 662

Some common volume formulas:
 	
Cube $V = s^3$
Rectanglar solid $V = L \cdot W \cdot H$
Sphere $V = \dfrac{4}{3}\pi r^3$
Cylinder $V = \pi r^2 h$
	Find the volume:
	
	$h = 4$ cm
	$r = 3$ cm
	
	$V = \pi r^2 h$
	   $= \pi(3 \text{ cm})^2(4 \text{ cm})$
	   $= 36\pi \text{ cm}^3$
	   $\approx 113.1 \text{ cm}^3$
		p. 667–669

Summary Exercises

This summary exercise set is provided to give you practice with each of the objectives of this chapter. Each exercise is keyed to the appropriate chapter section. When you are finished, you can check your answers to the odd-numbered exercises against those presented in the back of the text. If you have difficulty with any of these questions, go back and reread the examples from that section. The answers to the even-numbered exercises appear in the *Instructor's Manual*. Your instructor will give you guidelines on how to best use these exercises in your class.

[8.1]   For exercises 1 to 6, name the angle; label it as acute, obtuse, right, or straight; and then estimate its measure with a protractor to the nearest 10 degrees.

1.

2.

3.

4.

5.

6.

Give the measure of each angle in degrees. **Note:** A revolution is 360°.

7. $\angle A$ represents $\dfrac{3}{8}$ of a revolution

8. $\angle B$ represents $\dfrac{7}{10}$ of a revolution

[8.2]   Find the perimeter or circumference of each figure.

9.

3.4 in.

10.

5.1 cm

11.

4 m

12.

3 ft

8.2 ft

[8.3]   Find the area of each figure.

13.

3.4 in.

14.

5.1 cm

15.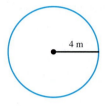

4 m

16.

3 ft

8.2 ft

[8.2]   Find the perimeter or circumference of each figure. Round the answers to the nearest tenth.

17.

12 ft

18.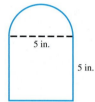

5 in.

5 in.

[8.3]

19. Find the area of a circle with radius 10 ft.

In exercises 20 and 21, find the area of each figure.

20.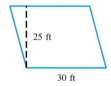

25 ft

30 ft

21.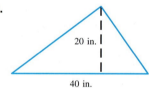

20 in.

40 in.

In exercises 22 and 23, solve the applications.

22. Crafts   How many square feet of vinyl floor covering will be needed to cover the floor of a room that is 10 ft by 18 ft? How many square yards will be needed? (*Hint:* How many square feet are in a square yard?)

23. Construction   A rectangular roof for a house addition measures 15 ft by 30 ft. A roofer will charge $175 per "square" (100 ft^2). Find the cost of the roofing for the addition.

In exercises 24 to 26, find the volume of each figure.

24.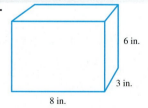

6 in.

3 in.

8 in.

25.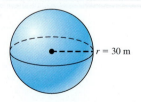

r = 30 m

26.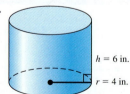

h = 6 in.

r = 4 in.

The purpose of this self-test is to help you check your progress so that you can find sections and concepts that you need to review before the next in-class exam. Allow yourself about an hour to take this test. At the end of that hour, check your answers against those given in the back of this text. If you missed an answer, notice the section reference that accompanies the question. Go back to that section and reread the examples until you have mastered that particular concept.

Name _____

Section _____ Date _____

ANSWERS

1. Label each pair of lines as parallel, perpendicular, or neither.

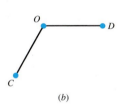   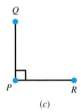

(a)   (b)   (c)   (d)

1. _____

2. Label each of the angles as acute, obtuse, right, or straight.

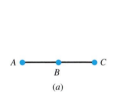   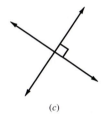   

(a)   (b)   (c)

2. _____

3. _____

4. _____

5. _____

6. _____

7. _____

In exercises 3 and 4, use a protractor to estimate the measure of each angle.

3.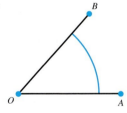

4.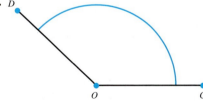

5. If $\angle A$ represents $\dfrac{5}{6}$ of a revolution, give the measure of $\angle A$ in degrees. (Note that one revolution is equal to $360°$.)

Find the perimeter or circumference of each figure.

6.  $s = 4.2$ cm

7.  $s = 5\frac{1}{4}$ in.

8. _____

9. _____

10. _____

11. _____

12. _____

13. _____

14. _____

15. _____

16. _____

17. _____

18. _____

19. _____

20. _____

21. _____

8.

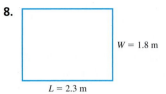

$W = 1.8$ m

$L = 2.3$ m

9.

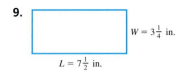

$W = 3\frac{1}{4}$ in.

$L = 7\frac{1}{2}$ in.

10.

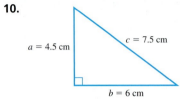

$a = 4.5$ cm

$c = 7.5$ cm

$b = 6$ cm

11.

$r = 5$ ft

Find the area of each figure.

12.

$s = 4.2$ cm

13.

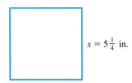

$s = 5\frac{1}{4}$ in.

14.

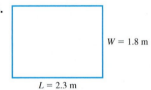

$W = 1.8$ m

$L = 2.3$ m

15.

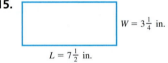

$W = 3\frac{1}{4}$ in.

$L = 7\frac{1}{2}$ in.

16.

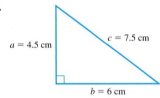

$a = 4.5$ cm

$c = 7.5$ cm

$b = 6$ cm

17.

$r = 5$ ft

Find the volume of each figure.

18.

$s = 2.5$ m

19.

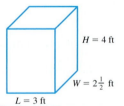

$H = 4$ ft

$W = 2\frac{1}{2}$ ft

$L = 3$ ft

20.

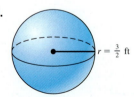

$r = \frac{3}{2}$ ft

21.

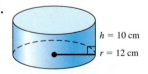

$h = 10$ cm

$r = 12$ cm

The following exercises are presented to help you review concepts from earlier chapters that you may have forgotten. This is meant as review material and not as a comprehensive exam. The answers are presented in the back of the text. Beside the questions is the section from which the topic relevant to the questions was first presented. If you have difficulty with any of these exercises, be certain to at least read through the summary related to that section.

Name _____

Section _____ Date _____

ANSWERS

1. A classroom is 7 yd wide by 8 yd long. If the room is to be recarpeted with material costing $16 per square yard, find the cost of the carpeting.

2. Michael bought a washer-dryer combination that, with interest, cost $959. He paid $215 down and agreed to pay the balance in 12 monthly payments. Find the amount of each payment.

Evaluate the expressions.

3. $(-3) \cdot (-8)$

4. $-6 - 3 \cdot 5$

5. -4^2

6. $\dfrac{4 - (-3)^2}{-7 \cdot 4 + 3}$

7. Write the prime factorization for 168.

8. Find the greatest common factor of 12 and 20.

9. Arrange in order from smallest to largest: $\dfrac{5}{8}, \dfrac{3}{5}, \dfrac{2}{3}$

10. Multiply: $\dfrac{2}{3} \cdot 1\dfrac{4}{5} \cdot \dfrac{5}{8}$

11. Divide: $4\dfrac{1}{6} \div 10$

12. Find the least common multiple of 6, 15, and 20.

13. Add: $\dfrac{3}{5} + \dfrac{1}{6} + \dfrac{4}{15}$

14. Subtract: $7\dfrac{3}{8} - 3\dfrac{5}{6}$

ANSWERS

1. _____
2. _____
3. _____
4. _____
5. _____
6. _____
7. _____
8. _____
9. _____
10. _____
11. _____
12. _____
13. _____
14. _____

15. _____

16. _____

17. _____

18. _____

19. _____

20. _____

21. _____

22. _____

23. _____

24. _____

25. _____

26. _____

Solve each equation and check your result.

15. $5 - 4x = 25$

16. $6x - 5 = 3x - 29$

17. $\dfrac{5}{7}x + 4 = 14$

18. $7x + 3(2x + 5) = 10x + 17$

19. You pay for purchases of $14.95, $18.50, $11.25, and $7 with a $70 check. How much cash will you have left?

20. Find the area of a rectangle with length 6.4 cm and width 4.35 cm.

21. Find the perimeter of a rectangle with length 6.4 cm and width 4.35 cm.

22. Find the circumference of a circle whose radius is 3.2 ft. Use 3.14 for π and round the result to the nearest tenth of a foot.

23. Find the decimal equivalent of $\dfrac{9}{16}$.

24. Write the decimal form of $\dfrac{7}{13}$. Round to the nearest thousandth.

25. If two legs of a right triangle have lengths 8 ft and 15 ft, find the length of the hypotenuse.

26. Solve for the unknown in the following proportion: $\dfrac{15}{x} = \dfrac{10}{16}$

27. _____

28. _____

29. _____

30. _____

31. _____

32. _____

33. _____

34. _____

35. _____

36. _____

37. _____

38. _____

27. If the scale on a map is $\frac{1}{4}$ in. equals 20 mi, how far apart are two towns that are 5 in. apart on the map?

28. Felipe traveled 342 mi using 19 gal of gas. At this rate, how far can he travel on 25 gal?

29. Write as a simplified fraction: 12.5%

30. What is 43% of 8,200?

31. 315 is what percent of 140?

32. 120% of what number is 180?

33. A home that was purchased for $125,000 increased in value by 14% over a 3-year period. What was its value at the end of that period?

34. Find the sum of 8 lb 14 oz and 12 lb 13 oz.

35. Find the difference between 7 ft 2 in. and 4 ft 5 in.

Complete each statement.

36. 62 kg = _____ g

37. 740 mm = _____ cm

38. Use a protractor to find the measure of the given angle.

39. Find the missing angle.

40. The given triangles are similar. Find x.

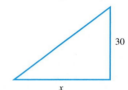

chapter

9

GRAPHING AND INTRODUCTION TO STATISTICS

CHAPTER 9   OUTLINE

Section 9.1   Tables and Graphs of Data   page 693

Section 9.2   The Rectangular Coordinate System   page 715

Section 9.3   Linear Equations in Two Variables   page 729

Section 9.4   Mean, Median, and Mode   page 753

INTRODUCTION

The twenty-first century has been dubbed the information age, in part, because data is everywhere. The worlds of business, sports, politics, marketing, science, and others all thrive on the tremendous amount of data gathered.

The 2000 census estimated the U.S. population to be 281,421,906. According to the Census Bureau, 25,458,208 Americans were 70 or older. While the overall population figure represented a 13.2% increase over the 1990 census, the increase in elderly Americans was greater than 20%.

Population projections predict that this trend will become even more lopsided as we move toward the future. Political and social decisions need to take into account data, trends, and projections. How else can we determine when to build more schools, hospitals, and sewer lines? Waiting until changes have already taken place guarantees that these institutions will not be ready when they are needed.

An educated person needs to understand how to make use of data, how to read graphs and tables, and how to make decisions on the basis of the information presented. Many people will try to convince you to share their beliefs. Less scrupulous individuals might present a graph in which features have been altered to better convince you of their position. As an informed citizen, you need to be on your guard for these abuses.

In this chapter, we cover some of the important uses of statistics and data. When you have completed it, you should be better able to make decisions based on graphs and information, and you should have a better idea of how to distinguish good data from bad.

You can find exercises related to graphs in Sections 9.1 to 9.3. At the end of the chapter, before the chapter summary, you will find a more in-depth activity on the uses and abuses of data, statistics, and graphs.

Pretest Chapter 9

This pretest will provide a preview of the types of exercises you will encounter in each section of this chapter. The answers for these exercises can be found in the back of the text. If you are working on your own or ahead of the class, this pretest can help you identify the sections in which you should focus more of your time.

[9.1]   **Social Science**   Use the bar graph to answer exercises 1 and 2.

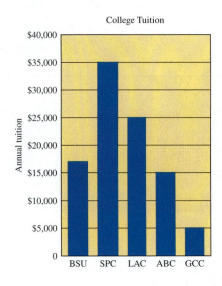

College Tuition

1. What is the difference in tuition between GCC and BSU?

2. What is the average (mean) tuition for these five schools?

Social Science   Use the line graph to answer exercises 3 to 5.

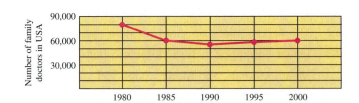

3. How many fewer family doctors were there in the United States in 1990 than in 1980?

4. What was the total change in family doctors between 1980 and 2000?

5. In what 5-year period was the decrease in family doctors the greatest? What was the decrease?

Statistics   Use the pictograph to answer Exercises 6 and 7.

Visitors to Yellowstone National Park

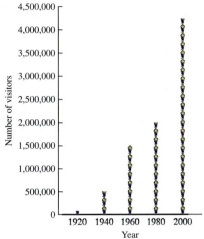

6. How many visitors did Yellowstone have in 1960?

7. What was the percent increase in visitors between 1940 and 2000?

[9.3]   Determine which of the ordered pairs are solutions for the given equations.

8. $x - y = 12$ (15, 3), (9, 6), (18, 6)

9. $3x + 2y = 6$ (1, 2), (0, 3), (2, 0)

10. Complete the ordered pairs so that each is a solution for the given equation.
$2x + y = 5$ (1,), (0,), (, 11)

11. Find three solutions for each of the equations.
$2x - 3y = 8$ $6x + y = 11$

[9.2]   **12.** Give the coordinates of the graphed points.

13. Plot the points with the given coordinates. $S(-1, 2)$, $T(3, 0)$

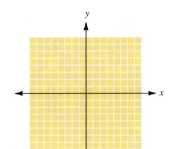

14. _____

15. _____

16. _____

17. _____

[9.3] Graph each of the equations.

14. $x + y = 5$

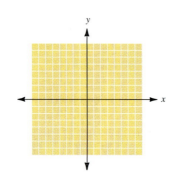

15. $y = \dfrac{1}{2}x - 1$

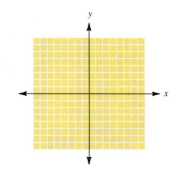

16. $y = -2$

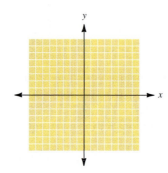

[9.4] **17.** Find the mean, median, and mode for the set of numbers.

12, 16, 17, 18, 24, 24, 29, 33, 42, 42, 42

9.1 Tables and Graphs of Data

9.1 OBJECTIVES

1. Interpret information from a table of data
2. Interpret and construct a bar graph
3. Interpret a pictograph
4. Interpret a line graph
5. Make predictions from a line graph
6. Construct a line graph

A **table** is a display of information in parallel rows or columns. Tables can be used anywhere that information is to be summarized.

The given table describes land area and world population. Each entry in the table is called a **cell.**

Continent or Region	Land Area (1,000 mi^2)	% of Earth	Population 1900	Population 1950	Population 2000
North America	9,400	16.2	106,000,000	221,000,000	477,000,000
South America	6,900	11.9	38,000,000	111,000,000	339,000,000
Europe	3,800	6.6	400,000,000	392,000,000	729,000,000
Asia (including Russia)	17,400	30.1	932,000,000	1,591,000,000	3,788,000,000
Africa	11,700	20.2	118,000,000	229,000,000	771,000,000
Oceania (including Australia)	3,300	5.7	6,000,000	12,000,000	31,000,000
Antarctica	5,400	9.3	Uninhabited		
World total	57,900		1,600,000,000	2,556,000,000	6,135,000,000

(*Source:* Bureau of the Census, U.S. Dept. of Commerce.)

OBJECTIVE 1   **Example 1   Reading a Table**

Using the land area and world population table, answer each question.

(a) What was the population of Africa in 1950?

Looking at the cell that is in the row labeled "Africa" and the column labeled "Population 1950," we find a population of 229,000,000.

(b) What is the land area of Asia in square miles?

The cell in the row Asia and column Land Area says 17,400. But note that the column is labeled "1,000 mi^2." The land area is 17,400 thousand square miles, or 17,400,000 mi^2.

CHECK YOURSELF 1

Use the previous table to answer each question.

(a) What was the population of South America in 1900?

(b) What is the land area of Africa as a percent of Earth's land area?

Often data is presented using graphs. Graphs make it easy to see relevant information. They can also help us to recognize trends or important aspects that are difficult to see in a table.

Some kinds of graphs show the relationship between two sets of data. Perhaps the most common is the **bar graph.**

A bar graph is read in two ways. You can look at the heights (or lengths) of the bars, and compare them to each other. You can also read the values represented by individual bars as marked off on the vertical axis.

NOTE The vertical line on the left and the horizontal line at the bottom are called **axes.**

OBJECTIVE 2

Example 2 Reading a Bar Graph

The given bar graph represents the response to a 2001 Gallup poll that asked people what their favorite spectator sport was. In the graph, the information at the bottom describes the sport, and the information along the side describes the percent of people surveyed. The height of the bar indicates the percent of people who favor that particular sport.

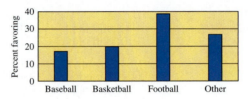

(a) Find the percent of people for whom football is their favorite spectator sport.

We frequently have to estimate our answer when reading a bar graph. In this case, 38% would be a good estimate.

(b) Find the percent of people for whom baseball is their favorite spectator sport.

Again, we can only estimate our answer. It appears to be approximately 17% of the people responding who favored baseball.

CHECK YOURSELF 2

This bar graph represents the number of students who majored in each of five areas at Berndt Community College (BCC).

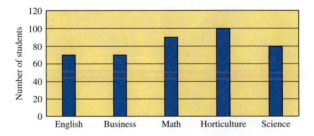

(a) How many mathematics majors were there?

(b) How many English majors were there?

Some bar graphs display additional information by using different colors or shading for different bars. With such graphs, it is important to read the legend. The **legend** is the key that describes what each color or shade of the bar represents.

Example 3   Reading the Legend of a Graph

The bar graph represents the average student age at BCC.

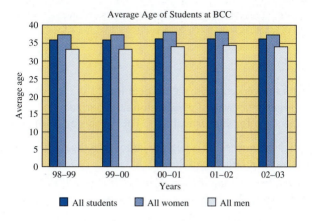

Average Age of Students at BCC

(a) What was the average age of female students in 2002–2003?

The legend tells us that the ages of all women are represented as the medium blue color. Looking at the height of the medium blue column for the year 02–03, we see the average age was about 38.

(b) Who tends to be older, male students or female students?

The medium blue bar is higher than the light blue bar in every year. Female students tend to be older than male students at BCC.

CHECK YOURSELF 3 _____

Use the graph in Example 3 to answer the questions.

(a) Did the average age of female students increase or decrease between 01–02 and 02–03?

(b) What was the average age of male students in 00–01?

We just learned to read a bar graph. In Example 4, we create one.

Example 4   Creating a Bar Graph

The table represents the 2000 population of the six most populated urban areas in the world. Each population is the population of the city plus the population of all of its suburbs. Create a bar graph from the information in the table.

NOTE Mumbai was formerly called Bombay.

Population of the World's Largest Urban Areas	
City	**2000 Population**
Tokyo, Japan	26,400,000
Mexico City, Mexico	18,100,000
Mumbai, India	18,100,000
São Paulo, Brazil	17,800,000
New York City, USA	16,600,000
Lagos, Nigeria	13,400,000

Source: United Nations Human Settlements Programme

We let the vertical axis, the vertical line to the left of the graph, represent population. The six urban areas will be placed along the horizontal axis. To create a graph, we must decide on the scale for the vertical axis. The given steps will accomplish that.

1. Pick a number that is slightly larger than the biggest number we are to graph. 30,000,000 is slightly larger than 26,400,000.

2. Decide how long the axis will be. It is best if this length easily divides into the number of step 1. To accomplish this division, we will choose 3 in.

NOTE Therefore, each $\frac{1}{2}$ in. represents 5,000,000 people.

3. Scale the axis by dividing it with tick marks. Label each tick mark with the appropriate number. In this graph, each inch will represent 10,000,000 people (the 30,000,000 divided by the 3 in. results in 10,000,000 people per inch).

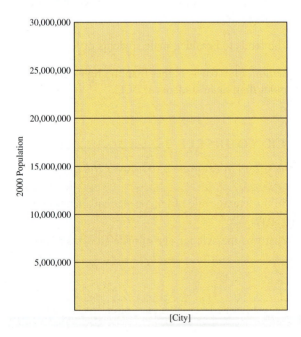

Now, the height of each bar is determined by using the scale created for the axis. Remembering that we are using 10,000,000 people per inch, we divide each population by 10,000,000. The result is the height of each bar. The height for Mexico City is 1.81 in. That would be approximately $1\frac{3}{4}$ in. Remember, all we can get from a bar graph is a rough approximation of the actual number.

NOTE Often, you will use technology such as spreadsheets to get accurately-sized bars.

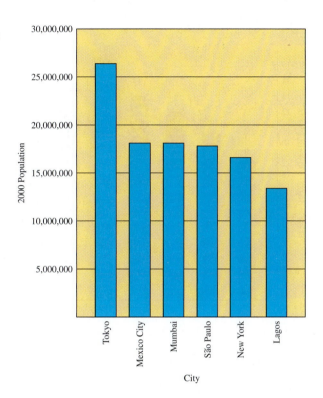

CHECK YOURSELF 4

The table represents the 2000 population of the six most populated cities in the United States. Each population is the population within the city limits, which is why the New York population is so different from that in the table in Example 4. Create a bar graph from the information in the table.

Population of the Largest Cities in the United States	
City	**2000 Population**
New York City, NY	8,008,000
Los Angeles, CA	3,695,000
Chicago, IL	2,896,000
Houston, TX	1,954,000
Philadelphia, PA	1,518,000
Phoenix, AZ	1,321,000
Source: Bureau of the Census; U.S. Dept. of Commerce	

A **pictograph** is very closely related to a bar graph. Pictographs are frequently used in newspapers.

OBJECTIVE 3 | **Example 5** Reading a Pictograph

The pictograph shown gives the number of cars registered in the U.S. by year. Each car pictured represents 10,000,000 registrations. To read from the pictograph, count the number of cars in each column.

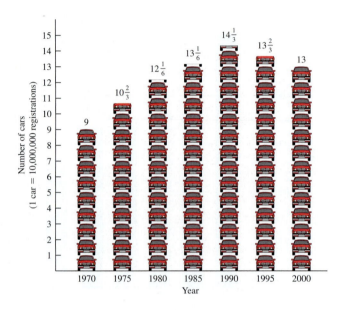

(a) How many cars were registered in 2000? 1995?

There are 13 cars pictured in the 2000 column.

$13 \cdot 10,000,000 = 130,000,000$

There were 130,000,000 cars registered in the U.S. in 2000.

There were $13\frac{2}{3} \cdot 10,000,000 \approx 137,000,000$ cars registered (rounded) in 1995.

NOTE Always use the earlier quantity for the base in percentage increase or percentage decrease problems.

(b) What was the percentage decrease in the number of cars registered between 1995 and 2000?

The decrease is $137,000,000 - 130,000,000 = 7,000,000$ cars.

$7,000,000 \div 137,000,000 \approx 0.051$

Compared to 1995, there were 5% fewer cars registered in 2000.

 CHECK YOURSELF 5

Use the pictograph above to answer the following questions.

(a) How many cars were registered in 1970? 1990?
(b) What was the percentage decrease in the number of cars registered during the 1990s?
(c) What was the average yearly decrease in the number of cars registered in the 1990s?

Another useful type of graph is called a **line graph.** In a line graph, one of the pieces of information is almost always related to time (clock time, days, months, or years).

OBJECTIVE 4   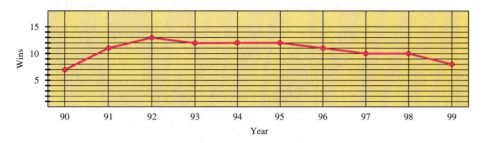 **Example 6   Reading a Line Graph**

This graph represents the number of regular season games won by the Dallas Cowboys each year of the 1990s. Note that the information across the bottom indicates the year and the information along the side indicates the number of victories.

RECALL You first saw line graphs in the Chapter 2 Activity.

(a) How many games did they win in 1994?

We look across the bottom until we find 1994. We then look straight up until we see the line of the graph. Following across to the left, we see that they won 12 games in 1994.

NOTE We further develop the idea of the mean in Section 9.4.

(b) Find the average (mean) number of games won by the Cowboys in the 1990s by adding the games won in each of the 10 years and then dividing by 10.

For each x (or dot) on the line, we look to the left to find how many victories it represents. We then add, so

$$\bar{x} = \frac{7 + 11 + 13 + 12 + 12 + 12 + 11 + 10 + 10 + 8}{10}$$

$\bar{x}$ is a symbol that represents the "mean."

NOTE As a decimal, we could write this as 10.6.

$$= \frac{106}{10} = 10\frac{6}{10} = 10\frac{3}{5}$$

The mean number of games won is $10\frac{3}{5}$.

CHECK YOURSELF 6

The graph indicates the high temperatures in Baltimore, Maryland, for a week in September.

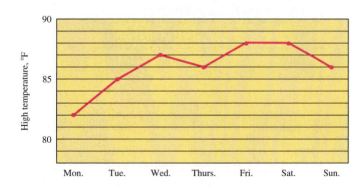

(a) What was the high temperature on Friday?

(b) Find the average (mean) high temperature for that week.

It is often tempting, and sometimes useful, to use a line graph to predict a future value. Using an earlier trend to predict a future value is called **extrapolation.** This is something that statisticians warn us to be cautious about.

OBJECTIVE 5

Example 7   Making a Prediction Using a Line Graph

Use the given line graph and table to "predict" the number of Social Security Beneficiaries in the year 2005.

Social Security Admin.

Year	Beneficiaries
1955	6,000,000
1965	18,000,000
1975	28,000,000
1985	36,000,000
1995	41,000,000
2005	?

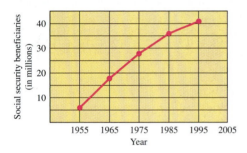

From the shape of the line graph, it would be reasonable to guess that the next point on the graph would continue on the same "curve."

Social Security Admin.

Year	Beneficiaries
2005	44,000,000

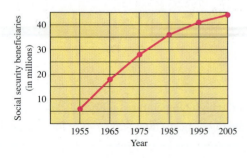

This point indicates that, by 2005, we should expect about 44,000,000 Social Security beneficiaries. This number closely matches more sophisticated predictions for the number of beneficiaries in the year 2005.

CHECK YOURSELF 7

The graph and table show the amount spent on health care (in billions of dollars) in the United States every 5 years from 1965 to 2000. Use that information to predict the amount spent in the year 2005.

National Center for Health Stats

Year	Health Care Expenditures (in Billions of $)
1965	41
1970	73
1975	130
1980	247
1985	428
1990	700
1995	991
2000	1,285
2005	?

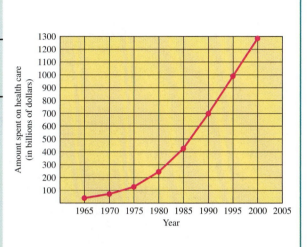

Now, we will look at an example that yields an interesting result.

Example 8   Making a Prediction Using a Line Graph

Use the line graph and table to predict the cost of a first-class stamp on January 1, 2005.

Year	Cost, Cents
1960	4
1965	5
1970	6
1975	10
1980	15
1985	22
1990	25
1995	32
2000	33
2005	?

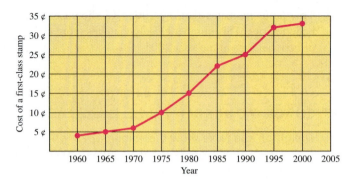

Place a straightedge so that it comes close to going through all of the points. The result looks like:

Year	Cost, Cents
1960	4
1965	5
1970	6
1975	10
1980	15
1985	22
1990	25
1995	32
2000	33
2005	34

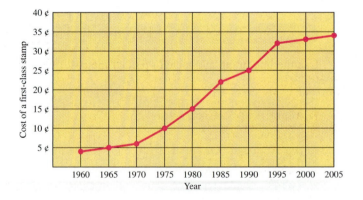

The cost of postage will increase again by 2006.

From the line graph, it would be reasonable to guess that the cost would be 34 cents. In fact, on June 30, 2005, the cost was 37 cents. This is evidence of the danger of extrapolation, a danger we mentioned before Example 7.

CHECK YOURSELF 8 _____

The graph represents the number of larceny-theft cases in the United States every 5 years from 1980 to 1995. Use the graph to predict the number of cases in 2000.

FBI Uniform Crime Report	
Year	**Larceny-Theft Cases (in Hundred Thousands)**
1980	66
1985	73
1990	79
1995	80
2000	?

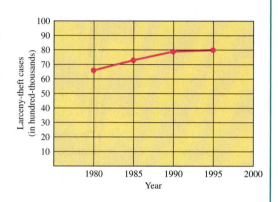

We have seen how to read a line graph and use it to make predictions. Now we create one from a table.

OBJECTIVE 6   **Example 9   Creating a Line Graph**

Create a line graph from the table, which gives the average low temperature in Fairbanks, Alaska, for each calendar month.

Month	Jan	Feb	Mar	Apr	May	Jun	Jul	Aug	Sep	Oct	Nov	Dec
Temp., in °F	−8	−6	19	28	37	42	45	48	34	27	15	−1

The first step in creating a line graph is to set up and label the two axes. We must be certain that all of our values will fit on the graph. The vertical axis must accommodate all of the numbers from −8 to 48. We will sketch axes so that all 12 months will be labeled, and the vertical axis marks every 10° from −10 to 50.

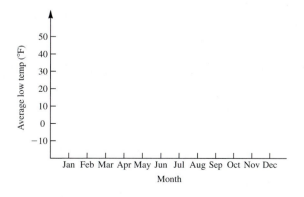

Now we mark each of the 12 points indicated on the table.

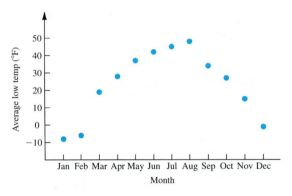

Finally, we connect each adjacent pair of points.

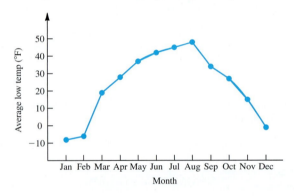

 CHECK YOURSELF 9

Create a line graph from the table, which gives the average high temperature in Johannesburg, South Africa, for each calendar month.

Month	Jan	Feb	Mar	Apr	May	Jun	Jul	Aug	Sep	Oct	Nov	Dec
Temp., in °F	83	87	78	65	55	48	45	48	55	62	71	78

READING YOUR TEXT

The following fill-in-the-blank exercises are designed to assure that you understand the key vocabulary used in this section. Each sentence comes directly from the section. You will find the correct answers in Appendix C.

Section 9.1

(a) The entries in a _____ are sometimes called cells.

(b) Bar graphs show the relationship between _____ sets of data.

(c) In bar graphs, the _____ is the key that describes what each bar represents.

(d) In a _____, one of the pieces of information is almost always related to time.

CHECK YOURSELF ANSWERS

1. (a) 38,000,000; (b) 20.2%   2. (a) 90; (b) 70   3. (a) It decreased; (b) 34

4.

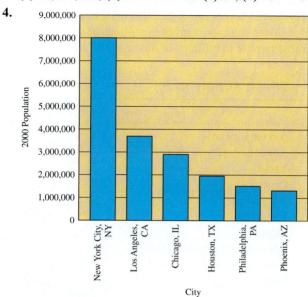

5. (a) 90,000,000; 143,000,000; (b) 9.3%; (c) 1,300,000 per year

6. (a) 88°F; (b) 86°F   7. $1,600 billion   8. 8,500,000

9.

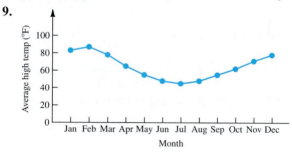

Name _____

Section _____ Date _____

ANSWERS

1. _____

2. _____

3. _____

4. _____

9.1   Exercises

Business and Finance   The table below describes characteristics of the U.S. wheat supply over a recent period. Use this table to answer exercises 1 to 4.

Note: Acreage is given in millions of acres. Yield is given in bushels per acre. All other measures are in millions of bushels. The 2001 and 2002 data are estimates.

Wheat Supply (U.S.)

	1998	1999	2000	2001	2002
Acreage planted	65.8	62.7	62.6	59.6	60.4
Acreage harvested	59.0	53.8	53.1	48.6	45.8
Yield	43.2	42.7	42.0	40.2	35.3
Production	2,547.3	2,299.0	2,232.5	1,957.0	1,616.4
Beginning stocks	722.5	945.9	949.7	876.2	777.1
Imports	103.0	94.5	89.8	107.5	70.0
Supply	3,372.8	3,339.4	3,272.0	2,940.0	2,463.6
Food and industrial use	909.1	921.0	949.6	926.3	935.0
Feed, seed, and residual use	471.8	380.1	384.2	276.1	209.0
Exports	1,046.0	1,088.6	1,062.0	961.3	875.0
Total use	2,426.9	2,389.7	2,395.9	2,163.7	2,019.0
Ending stocks	945.9	949.7	876.2	777.1	444.6

Source: Farm Service Agency; U.S. Department of Agriculture (8/03)

1. How much wheat did the United States export in 2000?

2. What was the difference in the quantity of wheat exported between 2000 and 2002?

3. Find the percentage change in the quantity of wheat exported from 2000 to 2002.

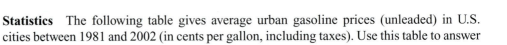

4. What percent of the acreage planted was actually harvested in 2000?

Statistics   The following table gives average urban gasoline prices (unleaded) in U.S. cities between 1981 and 2002 (in cents per gallon, including taxes). Use this table to answer exercises 5 to 10.

Average Urban Gasoline Prices (Retail)

Year	Regular	Premium	Year	Regular	Premium
1981	137.8	147.0	**1992**	112.7	131.6
1982	129.6	141.5	**1993**	110.8	130.2
1983	124.1	138.3	**1994**	111.2	130.5
1984	121.2	136.6	**1995**	114.7	133.6
1985	120.2	134.0	**1996**	123.1	141.3
1986	92.7	108.5	**1997**	123.4	141.6
1987	94.8	109.3	**1998**	105.9	125.0
1988	94.6	110.7	**1999**	116.5	135.7
1989	102.1	119.7	**2000**	151.0	169.3
1990	116.4	134.9	**2001**	146.1	165.7
1991	114.0	132.1	**2002**	135.5	157.8

Source: Energy Information Administration, U.S. Department of Energy Monthly Energy Review (10/03)

5. How much did regular unleaded gasoline cost a motorist in 1981?

6. How much did regular gasoline cost a motorist in 1991?

7. What was the percent decrease in the price of regular gasoline from 1981 to 1991?

8. What was the percent increase in the price of regular gasoline from 1991 to 2001?

9. How much extra did a motorist pay for premium gasoline than for regular gasoline in 1981? 2001?

10. Find the percentage change in the difference in price between regular and premium gasoline from 1981 to 2001.

Business and Finance   Use the bar graph below, showing the total U.S. motor vehicle production for the years 1996 to 2002, to answer exercises 11 to 14.

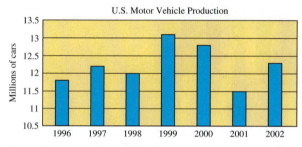

Source: Automotive News Market Data Book

11. What was the production in 2001?   **ALEKS**

12. In what year did the greatest production occur?   **ALEKS**

13. How much did production increase from 2001 to 2002?   **VIDEO**   **ALEKS**

14. In what year did production decline the most from the previous year?   **ALEKS**

Business and Finance   Use the bar graph, showing the attendance at a circus for seven days in August, to solve exercises 15 to 18.

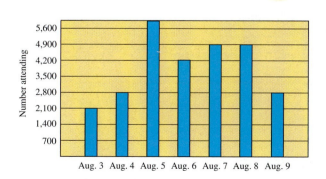

© 2010 McGraw-Hill Companies

15. _____

16. _____

17. _____

18. _____

19. _____

20. _____

21. _____

22. _____

23. _____

15. Find the attendance on August 4. ALEKS

16. Which day had the greatest attendance? ALEKS

17. Which day had the lowest attendance? ALEKS

18. How much did attendance drop from August 8 to August 9? ALEKS

Business and Finance   In 1993, 1,327,507 sport utility vehicles (SUVs) were sold in the United States, accounting for 26.3% of all sales of light vehicles (SUVs, minivans, vans, pickup trucks, and trucks under 14,000 lb). By 2002, sales of SUVs in the U.S. increased to 4,186,698, accounting for 48.3% of total light vehicle sales.

Source: Office of Transportation Technologies

19. How many SUVs were sold in 2000? ALEKS

20. In what year did the greatest sales occur? ALEKS

21. What was the percent increase in sales from 1993 to 2002? ALEKS

22. In what year did the greatest increase in sales occur? ALEKS

23. Statistics   The following table gives the number of Senate members with military service in the 106th U.S. Congress by branch.

ALEKS

Branch	Count
Army	17
Navy	10
Air Force	4
Marines	6
Coast Guard	1
National Guard	2

Construct a bar graph from this information.

24. Social Science The table below gives the number of foreign-born residents of the U.S. by region of birth in the year 2000.

Region	Number of Residents
Europe	4,400,000
North America	700,000
Latin America	14,500,000
Asia	7,200,000
Other Areas	1,600,000

Source: U.S. Census Bureau

Construct a bar graph from this table.

24. _____

25. _____

26. _____

Social Science The pictograph represents the world population by continent in 2000.

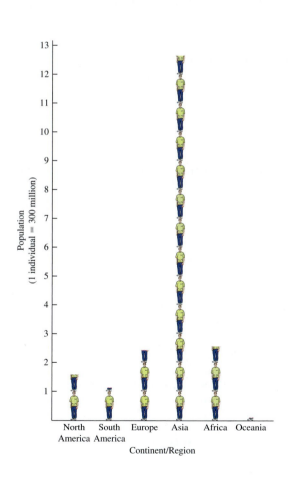

25. What was the population of Africa?

26. What was the total population of North and South America?

Use the line graph, showing the annual utility costs for a given family, to answer exercises 27 to 30.

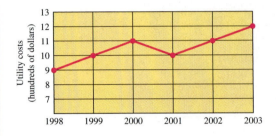

27. What were utility costs in 2001? **VIDEO** **ALEKS**

28. Identify the years in which utility costs were the same (two answers). **ALEKS**

29. What was the decrease in the cost of utilities from 2000 to 2001? **ALEKS**

30. In what year was the cost of utilities the smallest? **ALEKS**

Social Science Use the line graph, showing the number of robberies in a town during the last 6 months of a year, to answer exercises 31 to 34.

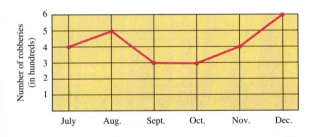

31. In which month did the greatest number of robberies occur? **ALEKS**

32. How many robberies occurred in November? **ALEKS**

33. Find the decrease in the number of robberies between August and September. **ALEKS**

34. What was the total number of robberies for the six months? **ALEKS**

35. **Business and Finance** The line graph and table show the income to the Hospital Insurance Trust Fund. Use this information to predict the income in the year 2005.

Year	Total Income (in billions)
1975	$12.568
1980	$25.415
1990	$79.563
1995	$114.847
2000	$130.559
2005	?

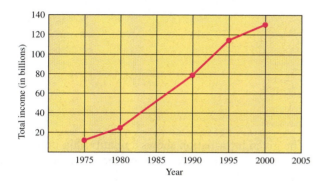

36. **Business and Finance** The line graph and table give the monthly principal and interest payments for a mortgage from 1999 to 2000. Use this information to predict the payment for 2005.

Year	Payment
1999	$578
2000	$613
2001	$654
2002	$675
2003	$706
2004	$730
2005	?

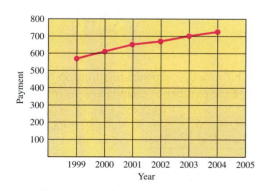

37. Social Science The given information shows a relationship between years of formal education and typical income (in thousands of dollars) at age 30:

Years	8	10	12	14	16
Income ($\times 1,000$)	16	21	23	28	31

Use this information to make a line graph and then predict the yearly income associated with 18 years of formal education.

38. Business and Finance The given information shows a relationship between the amount spent per week on advertising by a small fast-food shop and the total sales per week:

Amount spent (in $)	10	20	30	40
Sales (in $)	200	380	625	790

Use this information to make a line graph and then predict the weekly sales associated with the expenditure of $50 (per week) on advertising.

39. Science and Medicine The given information shows a relationship between the number of weeks on a special diet and the number of pounds lost during that time:

Number of weeks	2	4	6	8
Number of pounds	2	5	9	11

Use this information to make a line graph and then predict the number of pounds lost associated with 10 weeks on the special diet.

40. **Technology**   The given information shows a relationship between the speed of a certain car over a 100-mi test trip and the gas mileage (in miles per gallon) obtained:

Speed	40	45	50	55	60
Gas mileage	28	25	21	18	16

Use this information to make a line graph and then predict the gas mileage associated with a speed of 65 $\frac{\text{mi}}{\text{h}}$.

41. **Activity**   A company claims that in independent tests the flavor in their Flav'rBurst brand chewing gum lasts longer than that of the competition. They provide the results of tests in the bar graph below to support their claim. How is the graph shown misleading?

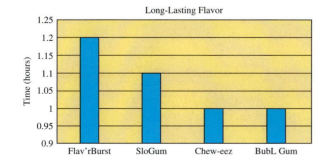

42.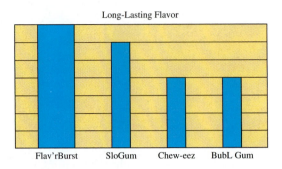

42. Activity   Describe how the bar graph, showing the same situation (with time measured in minutes) as exercise 41, might be misleading.

Long-Lasting Flavor

| Flav'rBurst | SloGum | Chew-eez | BubL Gum |

Answers

1. 1,062,000,000 bu   **3.** 17.6%   **5.** $1.378 per gal   **7.** 17.3%
9. 9.2¢ per gal; 19.6¢ per gal   **11.** 11,500,000   **13.** 800,000
15. 2,800   **17.** August 3   **19.** 3,600,000   **21.** 215.4%
23.   **25.** ≈750,000,000

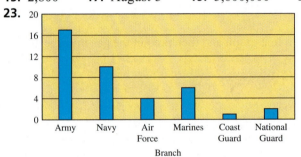

27. $1,000   **29.** $100   **31.** December   **33.** 200
35. $158,000,000,000   **37.** $34,000   **39.** 14 lb   **41.**

9.2 The Rectangular Coordinate System

9.2 OBJECTIVES

1. Determine the coordinates of a plotted point
2. Plot a point corresponding to an ordered pair

In Chapter 3, we looked at equations such as

$$2x + 3 = 7$$

For such an equation, we found a solution (here $x = 2$), which is a value that makes the equation a true statement.

In algebra, we also encounter equations with two variables, such as

$$y = 2x + 1$$

What would the solution to such an equation be? If $x = 0$ and $y = 1$, the equation would be a true statement. However, if $x = 1$ and $y = 3$, the equation would also be a true statement. Because solutions with two variables require two numbers, we need a way to indicate that two numbers together make a solution. To do this, we write the numbers as an ordered pair.

Instead of $x = 0$ and $y = 1$, we write (0, 1).

Instead of $x = 1$ and $y = 3$, we write (1, 3).

Each is called an **ordered pair** because the order is important (the x always comes before the y). The pair (1, 0) is different from the pair (0, 1). For the pair (1, 0), $x = 1$ and $y = 0$, and we can check that (1, 0) is *not* a solution for the equation $y = 2x + 1$. In this section, we will learn to graph information that comes in ordered pairs.

Because there are two numbers (one for x and one for y), we will need two number lines. One line is drawn horizontally, and the other is drawn vertically; their point of intersection (at their respective zero points) is called the **origin.** The horizontal line is called the **x-axis,** and the vertical line is called the **y-axis.** Together the lines form the **rectangular coordinate system.**

The axes divide the plane into four regions called **quadrants,** which are numbered (usually by Roman numerals) counterclockwise from the upper right.

NOTE We call the flat surface on which we are working the **plane.**

NOTE This system is also called the **Cartesian coordinate system,** named in honor of its inventor, René Descartes (1596–1650), a French mathematician and philosopher.

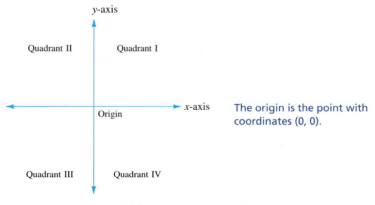

The origin is the point with coordinates (0, 0).

We now want to establish correspondences between ordered pairs of numbers (x, y) and points in the plane.

For any ordered pair

$$(x, y)$$

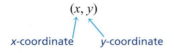

x-coordinate y-coordinate

the following are true:

1. If the x-coordinate is

Positive, the point corresponding to that pair is located x units to the *right* of the y-axis.
Negative, the point is x units to the *left* of the y-axis.
Zero, the point is on the y-axis.

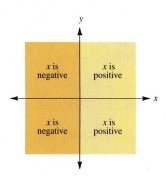

2. If the y-coordinate is

Positive, the point is y units *above* the x-axis.
Negative, the point is y units *below* the x-axis.
Zero, the point is on the x-axis.

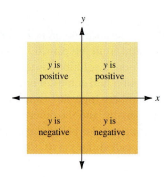

Example 1 illustrates how to use these guidelines to give coordinates to points in the plane.

OBJECTIVE 1

Example 1   Identifying the Coordinates for a Given Point

NOTE The x-coordinate gives the *horizontal* distance from the y-axis. The y-coordinate gives the *vertical* distance from the x-axis.

Give the coordinates for the given point.

(a)

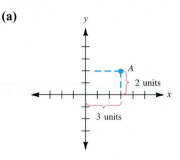

Point A is 3 units to the *right* of the y-axis and 2 units *above* the x-axis. Point A has coordinates (3, 2).

(b)

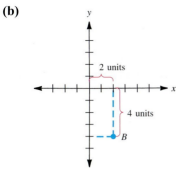

Point *B* is 2 units to the *right* of the *y*-axis and 4 units *below* the *x*-axis. Point *B* has coordinates (2, −4).

(c)

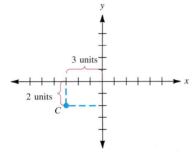

Point *C* is 3 units to the *left* of the *y*-axis and 2 units *below* the *x*-axis. Point *C* has coordinates (−3, −2).

(d)

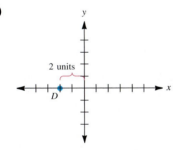

Point *D* is 2 units to the *left* of the *y*-axis and *on* the *x*-axis. Point *D* has coordinates (−2, 0).

CHECK YOURSELF 1

Give the coordinates of points P, Q, R, and S.

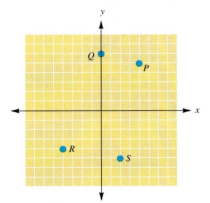

P _____

Q _____

R _____

S _____

Reversing this process allows us to graph (or plot) a point in the plane given the coordinates of the point. You can use these steps.

Step by Step: To Graph a Point in the Plane

Step 1   Start at the origin.
Step 2   Move right or left according to the value of the *x*-coordinate.
Step 3   Move up or down according to the value of the *y*-coordinate.

OBJECTIVE 2

Example 2   Graphing Points

(a) Graph the point corresponding to the ordered pair (4, 3).

Move 4 units to the right of the origin on the *x*-axis. Then move 3 units up from the point you stopped at on the *x*-axis. This locates the point corresponding to (4, 3).

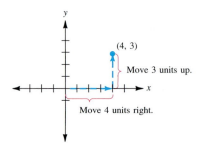

(b) Graph the point corresponding to the ordered pair (−5, 2).

In this case, move 5 units *left* of the origin (because the *x*-coordinate is negative) and then 2 units *up*.

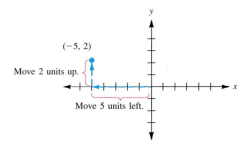

(c) Graph the point corresponding to (−4, −2).

Here move 4 units *left* of the origin and then 2 units *down* (the *y*-coordinate is negative).

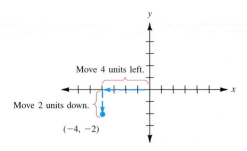

(d) Graph the point corresponding to $(0, -3)$.

There is *no* horizontal movement because the *x*-coordinate is 0. Move 3 units *down* from the origin.

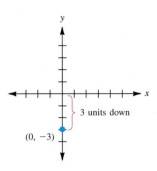

(e) Graph the point corresponding to $(5, 0)$.

Move 5 units *right* of the origin. The desired point is on the *x*-axis because the *y*-coordinate is 0.

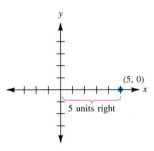

 CHECK YOURSELF 2

Graph the points corresponding to M(4, 3), N(−2, 4), P(−5, −3), and Q(0, −3).

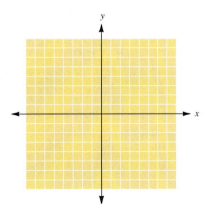

READING YOUR TEXT

The following fill-in-the-blank exercises are designed to assure that you understand the key vocabulary used in this section. Each sentence comes directly from the section. You will find the correct answers in Appendix C.

Section 9.2

(a) The point at which the *x*-axis and *y*-axis intersect is called the _____.

(b) The *x*-coordinate is _____ than zero in the fourth quadrant.

(c) The *y*-coordinate measures _____ distance from the origin.

(d) Points on the axes have _____ for at least one of their coordinates.

CHECK YOURSELF ANSWERS

1. $P(4, 5)$, $Q(0, 6)$, $R(-4, -4)$, and $S(2, -5)$

2.

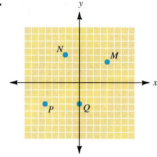

Name _____

Section _____ Date _____

ANSWERS

Give the coordinates of the graphed points.

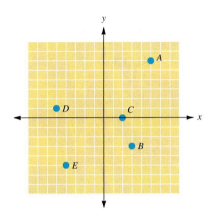

1. *A*   **ALEKS**

2. *B*   **ALEKS**

3. *C*   **ALEKS**

4. *D*   **ALEKS**

5. *E*   **ALEKS**

Give the coordinates of the graphed points.

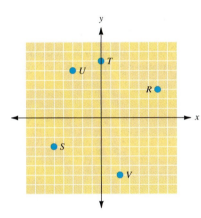

6. *R*   **ALEKS**

7. *S*   **ALEKS**

8. *T*   **ALEKS**

9. *U*   **ALEKS**

10. *V*   **ALEKS**

Plot points with the given coordinates on the graph provided.

11. $M(5, 3)$ **ALEKS**

12. $N(0, -3)$ **ALEKS**

13. $P(-2, 6)$ **ALEKS**

14. $Q(5, 0)$ **ALEKS**

15. $R(-4, -6)$ **ALEKS**

16. $S(-3, -4)$ **ALEKS**

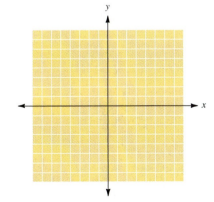

1. _____

2. _____

3. _____

4. _____

5. _____

6. _____

7. _____

8. _____

9. _____

10. _____

11. _____

12. _____

13. _____

14. _____

15. _____

16. _____

Plot points with the given coordinates on the graph provided.

17. $F(-3, -1)$ **ALEKS**   **18.** $G(4, 3)$ **ALEKS**

19. $H(5, -2)$ **ALEKS**   **20.** $I(-3, 0)$ **ALEKS**

21. $J(-5, 3)$ **ALEKS**   **22.** $K(0, 6)$ **ALEKS**

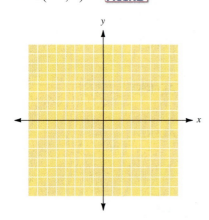

23. Graph points with coordinates $(2, 3)$, $(3, 4)$, and $(4, 5)$ on the graph provided. What do you observe? Can you give the coordinates of another point with the same property?

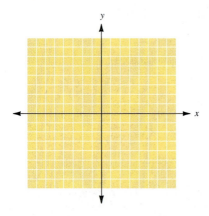

24. Graph points with coordinates $(-1, 4)$, $(0, 3)$, and $(1, 2)$ on the graph provided. What do you observe? Can you give the coordinates of another point with the same property?

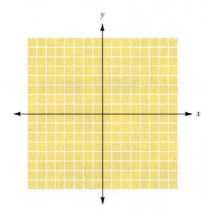

25. Graph points with coordinates $(-1, 3)$, $(0, 0)$, and $(1, -3)$ on the graph provided. What do you observe? Can you give the coordinates of another point with the same property?

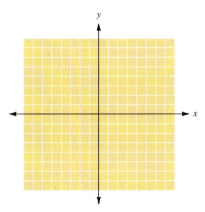

26. Graph points with coordinates $(1, 5)$, $(-1, 3)$, and $(-3, 1)$ on the graph provided. What do you observe? Can you give the coordinates of another point with the same property?

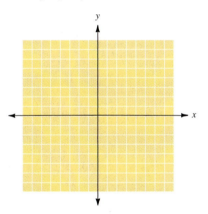

27. Technology   A local plastics company is sponsoring a plastics recycling contest for the local community. The focus of the contest is collecting plastic milk, juice, and water jugs. The company will award $200 plus the current market price of the jugs collected to the group that collects the most jugs in a single month. The number of jugs collected and the amount of money won can be represented as an ordered pair.

    (a) In April, group A collected 1,500 lb of jugs to win first place. The prize for the month was $350. On the graph, x represents the pounds of jugs and y represents the amount of money that the group won. Graph the point that represents the winner for April.

(b) In May, group *B* collected 2,300 lb of jugs to win first place. The prize for the month was $430. Graph the point that represents the May winner on the same coordinate system you used in part (a).

(c) In June, group *C* collected 1,200 lb of jugs to win the contest. The prize for the month was $320. Graph the point that represents the June winner on the same coordinate system as used before.

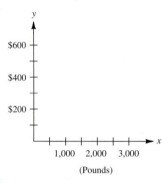

(Pounds)

28. Social Science The table gives the hours, *x*, that Damien studied for five different math exams and the resulting grades, *y*. Plot the data given in the table.

x	4	5	5	2	6
y	83	89	93	75	95

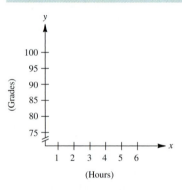

(Hours)

29. Science and Medicine The table gives the average temperature *y* (in degrees Fahrenheit) for the first 6 months of the year, *x*. The months are numbered 1 through 6, with 1 corresponding to January. Plot the data given in the table.

x	1	2	3	4	5	6
y	4	14	26	33	42	51

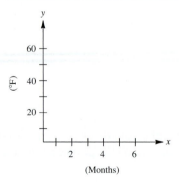

(Months)

30. **Business and Finance**   The table gives the total salary of a salesperson, y, for each of the four quarters of the year, x. Plot the data given in the table.

x	1	2	3	4
y	$6,000	$5,000	$8,000	$9,000

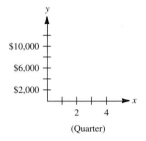

31. **Statistics**   The table shows the number of runs scored by the Anaheim Angels in each game of the 2002 World Series.

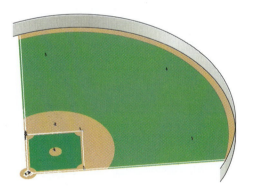

Game	1	2	3	4	5	6	7
Runs	3	11	10	3	4	6	4

Source: Major League Baseball

Plot the data given in the table.

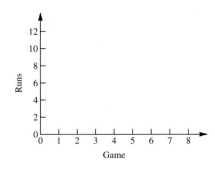

32. Statistics The table shows the number of wins and total points for the five teams in the Atlantic Division of the National Hockey League in the early part of the 1999–2000 season.

Team	Wins	Points
New Jersey Devils	5	12
Philadelphia Flyers	4	10
New York Rangers	4	9
Pittsburgh Penguins	2	6
New York Islanders	2	5

Plot the data given in the table.

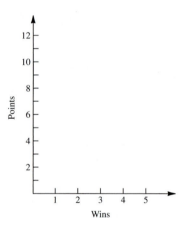

33. How would you describe a rectangular coordinate system? Explain what information is needed to locate a point in a coordinate system.

34. Some newspapers have a special day that they devote to automobile ads. Use this special section or the Sunday classified ads from your local newspaper to find all the ads for a particular automobile model. Make a list of the model year and asking price for 10 ads, being sure to get a variety of ages for this model. After collecting the information, make a graph of the age and the asking price for the car.

Describe your graph, including an explanation of how you decided which variable to put on the vertical axis and which on the horizontal axis. What trends or other information are given by the graph?

35. Technology The map shown below uses letters and numbers to label a grid that helps to locate a city. For instance, Salem is located at E-4.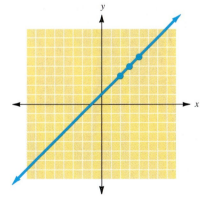

(a) Find the coordinates for the following: White Swan, Newport, and Wheeler.

(b) What cities correspond to the following coordinates: A2, F4, and A5?

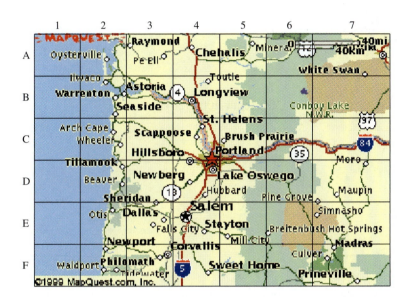

Answers

1. (5, 6)   **3.** (2, 0)   **5.** (−4, −5)   **7.** (−5, −3)   **9.** (−3, 5)

11–21.

23. The points lie on a line; sample answer is (1, 2)

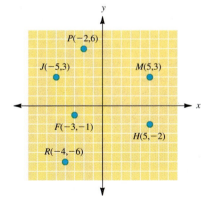

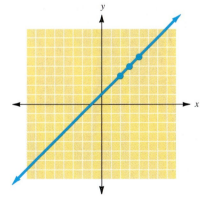

25. The points lie on a line; sample answer is $(2, -6)$

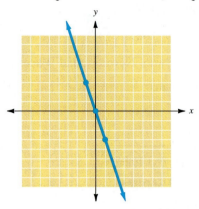

27. (a) $(1,500, 350)$; **(b)** $(2,300, 430)$; **(c)** $(1,200, 320)$

29.   **31.**

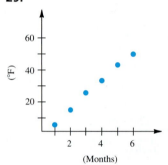
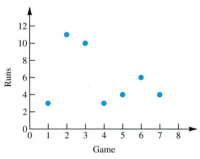

33.   **35. (a)** A7, F2, C2; **(b)** Oysterville, Sweet Home, Mineral

9.3 Linear Equations in Two Variables

9.3 OBJECTIVES

1. Find solutions for an equation in two variables
2. Use ordered pair notation to write solutions for an equation in two variables
3. Graph a linear equation by plotting points
4. Graph vertical and horizontal lines

We discussed finding solutions for equations in Section 2.9. Recall that a solution is a value for the variable that "satisfies" the equation, or makes the equation a true statement. For instance, we know that 4 is a solution of the equation

$$2x + 5 = 13$$

We know this is true because, when we replace x with 4, we have

$$2(4) + 5 \overset{?}{=} 13$$
$$8 + 5 \overset{?}{=} 13$$
$$13 = 13 \qquad \text{A true statement}$$

We now want to study **equations in two variables.** An example is

$$x + y = 5$$

As we saw in Section 9.2, a solution is not going to be a single number because there are two variables. Here a solution will be a pair of numbers—one value for each of the variables, x and y. Suppose that x has the value 3. In the equation $x + y = 5$, you can substitute 3 for x.

$$(3) + y = 5$$

Solving for y gives

$$y = 2$$

NOTE An equation in two variables "pairs" two numbers, one for x and one for y.

So the pair of values $x = 3$ and $y = 2$ satisfies the equation because

$$(3) + (2) = 5$$

That pair of numbers is then a *solution* for the equation in two variables.

How many such pairs are there? Choose any value for x (or for y). You can always find the other *paired* or *corresponding* value in an equation of this form. We say that there are an *infinite* number of pairs that will satisfy the equation. Each of these pairs is a solution. In Example 1, we find some other solutions for the equation $x + y = 5$.

OBJECTIVE 1

Example 1 Solving for Corresponding Values

For the equation $x + y = 5$, find (a) y if $x = 5$ and (b) x if $y = 4$.

(a) If $x = 5$

$$(5) + y = 5 \qquad \text{or} \qquad y = 0$$

(b) If $y = 4$,

$$x + (4) = 5 \qquad \text{or} \qquad x = 1$$

So the pairs $x = 5$, $y = 0$ and $x = 1$, $y = 4$ are both solutions.

CHECK YOURSELF 1

For the equation $2x + 3y = 26$,

(a) If $x = 4$, $y = ?$ **(b)** If $y = 0$, $x = ?$

RECALL We introduced ordered pairs in Section 9.2.

To simplify writing the pairs that satisfy an equation, we use the **ordered-pair notation.** Recall that the numbers are written in parentheses and are separated by a comma. For example, we know that the values $x = 3$ and $y = 2$ satisfy the equation $x + y = 5$. So we write the pair as

$$(3, 2)$$

The x-coordinate   The y-coordinate

CAUTION

$(3, 2)$ means $x = 3$ and $y = 2$. $(2, 3)$ means $x = 2$ and $y = 3$. $(3, 2)$ and $(2, 3)$ are entirely different. This is why we call them **ordered pairs.**

The first number of the pair is *always* the value for x and is called the **x-coordinate.** The second number of the pair is *always* the value for y and is the **y-coordinate.**

Using this ordered-pair notation, we can say that $(3, 2)$, $(5, 0)$, and $(1, 4)$ are all *solutions* for the equation $x + y = 5$. Each pair gives values for x and y that will satisfy the equation.

OBJECTIVE 2   **Example 2   Identifying Solutions of Two-Variable Equations**

Which of the ordered pairs (a) $(2, 5)$, (b) $(5, -1)$, and (c) $(3, 4)$ are solutions for the equation $2x + y = 9$?

(a) To check whether $(2, 5)$ is a solution, let $x = 2$ and $y = 5$ and see if the equation is satisfied.

$$2x + y = 9 \qquad \text{The original equation.}$$

$$2(2) + (5) \stackrel{?}{=} 9 \qquad \text{Substitute 2 for } x \text{ and 5 for } y.$$
$$4 + 5 \stackrel{?}{=} 9$$
$$9 = 9 \qquad \text{A true statement}$$

NOTE $(2, 5)$ is a solution because a *true* statement results when the values are inserted in the equation.

$(2, 5)$ is a solution for the equation.

(b) For $(5, -1)$, let $x = 5$ and $y = -1$.

$$2(5) + (-1) \stackrel{?}{=} 9$$
$$10 - 1 \stackrel{?}{=} 9$$
$$9 = 9 \qquad \text{A true statement}$$

So $(5, -1)$ is a solution.

(c) For (3, 4), let $x = 3$ and $y = 4$. Then

$$2(3) + (4) \overset{?}{=} 9$$
$$6 + 4 \overset{?}{=} 9$$
$$10 = 9 \qquad \textit{Not a true statement}$$

So (3, 4) is *not* a solution for the equation.

 ### CHECK YOURSELF 2 _____

Which of the ordered pairs (3, 4), (4, 3), (1, −2), and (0, −5) are solutions for the equation?

$3x - y = 5$

If the equation contains only one variable, then the missing variable can take on any value.

Example 3   Identifying Solutions of One-Variable Equations

Which of the ordered pairs (2, 0), (0, 2), (5, 2), (2, 5), and (2, −1) are solutions for the equation $x = 2$?

A solution is any ordered pair in which the *x*-coordinate is 2. That makes (2, 0), (2, 5), and (2, −1) solutions for the given equation.

 ### CHECK YOURSELF 3 _____

Which of the ordered pairs (3, 0), (0, 3), (3, 3), (−1, 3), and (3, −1) are solutions for the equation y = 3?

Remember that, when an ordered pair is presented, the first number is always the *x*-coordinate and the second number is always the *y*-coordinate.

Example 4   Completing Ordered Pair Solutions

Complete the ordered pairs (a) (9,), (b) (, −1), (c) (0,), and (d) (, 0) for the equation $x - 3y = 6$.

NOTE The *x*-coordinate is sometimes called the **abscissa** and the *y*-coordinate the **ordinate**.

(a) The first number, 9, appearing in (9,) represents the *x*-value. To complete the pair (9,), substitute 9 for *x* and then solve for *y*.

$$(9) - 3y = 6$$
$$-3y = -3$$
$$y = 1$$

(9, 1) is a solution.

(b) To complete the pair (, −1), let y be −1 and solve for x.

$$x - 3(-1) = 6$$
$$x + 3 = 6$$
$$x = 3$$

$(3, -1)$ is a solution.

(c) To complete the pair $(0,)$, let x be 0.

$$(0) - 3y = 6$$
$$-3y = 6$$
$$y = -2$$

$(0, -2)$ is a solution.

(d) To complete the pair (, 0), let y be 0.

$$x - 3(0) = 6$$
$$x - 0 = 6$$
$$x = 6$$

$(6, 0)$ is a solution.

CHECK YOURSELF 4

Complete the ordered pairs so that each is a solution for the equation 2x + 5y = 10.

(10,), (, 4), (0,), and (, 0)

| Example 5 | Finding Some Solutions of a Two-Variable Equation |

Find four solutions for the equation

$$2x + y = 8$$

NOTE Generally, you want to pick values for *x* (or for *y*) so that the resulting equation in one variable is easy to solve.

In this case, the values used to form the solutions are *up to you*. You can assign any value for x (or for y). We will demonstrate with some possible choices.

(a) Solution with $x = 2$:

$$2(2) + y = 8$$
$$4 + y = 8$$
$$y = 4$$

$(2, 4)$ is a solution.

(b) Solution with $y = 6$:

$$2x + (6) = 8$$
$$2x = 2$$
$$x = 1$$

$(1, 6)$ is a solution.

(c) Solution with $x = 0$:

$$2(0) + y = 8$$
$$y = 8$$

NOTE The solutions (0, 8) and (4, 0) have special significance later in graphing. They are also easy to find!

(0, 8) is a solution.

(d) Solution with $y = 0$:

$$2x + (0) = 8$$
$$2x = 8$$
$$x = 4$$

(4, 0) is a solution.

CHECK YOURSELF 5

Find four solutions for $x - 3y = 12$.

We are now ready to combine our work of Section 9.2 with our knowledge of two-variable equations. You just learned to write the solutions of equations in two variables as ordered pairs. In Section 9.2, these ordered pairs were graphed in the plane. Putting these ideas together will let us graph certain equations. Example 6 illustrates this approach.

OBJECTIVE 3

Example 6   Graphing a Linear Equation

Graph $x + 2y = 4$.

NOTE We are going to find *three* solutions for the equation. We will point out why shortly.

Step 1   Find some solutions for $x + 2y = 4$. To find solutions, we choose any convenient values for x, say $x = 0$, $x = 2$, and $x = 4$. Given these values for x, we can substitute and then solve for the corresponding value for y. So

If $x = 0$, then $y = 2$, so (0, 2) is a solution.
If $x = 2$, then $y = 1$, so (2, 1) is a solution.
If $x = 4$, then $y = 0$, so (4, 0) is a solution.

A handy way to show this information is in a table such as this:

NOTE A table is just a convenient way to display the information. It is the same as writing (0, 2), (2, 1), and (4, 0).

x	y
0	2
2	1
4	0

Step 2   We now graph the solutions found in step 1.

$x + 2y = 4$

x	y
0	2
2	1
4	0

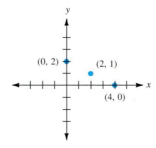

What pattern do you see? It appears that the three points lie on a straight line, and that is in fact the case.

Step 3   Draw a straight line through the three points graphed in step 2.

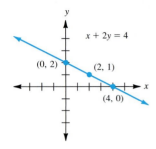

NOTE The arrows on the end of the line mean that the line extends indefinitely in either direction.

NOTE The graph is a "picture" of the solutions for the given equation.

The line shown is the **graph** of the equation $x + 2y = 4$. It represents *all* of the ordered pairs that are solutions (an infinite number) for that equation.

Every ordered pair that is a solution to the equation is plotted on this line. Any point on the line will have coordinates that are a solution for the equation.

Note: Why did we suggest finding *three* solutions in step 1? Two points determine a line, so technically you need only two. The third point that we find is a check to catch any possible errors.

CHECK YOURSELF 6

Graph $2x - y = 6$ using the steps shown in Example 6.

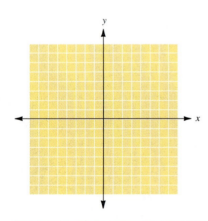

An equation that can be written in the form

$$Ax + By = C$$

in which A, B, and C are real numbers and A and B cannot both be 0 is called a **linear equation in two variables.** The graph of this equation is a *straight line*.

> **Step by Step:** To Graph a Linear Equation
>
> **Step 1** Find at least three solutions for the equation and put your results in tabular form.
> **Step 2** Graph the solutions found in step 1.
> **Step 3** Draw a straight line through the points determined in step 2 to form the graph of the equation.

Example 7   Graphing a Linear Equation

Graph $y = 3x$.

Step 1 Some solutions are

NOTE Let $x = 0$, 1, and 2, and substitute to determine the corresponding y-values. Again the choices for x are simply convenient. Other values for x would serve the same purpose.

x	y
0	0
1	3
2	6

Step 2 Graph the points.

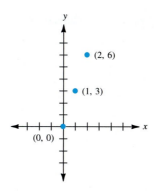

NOTE Connecting any two of these points produces the same line.

Step 3 Draw a line through the points.

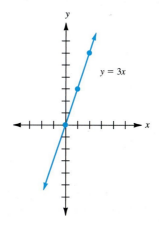

CHECK YOURSELF 7

Graph the equation y = −2x after completing the table of values.

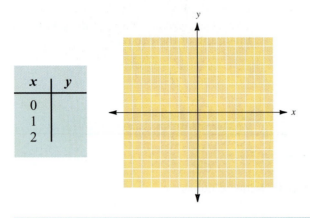

x	y
0	
1	
2	

Here is another example of graphing a line from its equation.

Example 8   Graphing a Linear Equation

Graph $y = 2x + 3$.

Step 1   Some solutions are

x	y
0	3
1	5
2	7

Step 2   Graph the points corresponding to these values.

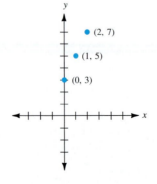

Step 3   Draw a line through the points.

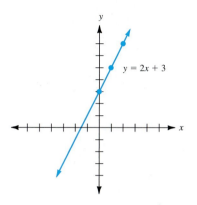

$y = 2x + 3$

CHECK YOURSELF 8

Graph the equation y = 3x − 2 after completing the table of values.

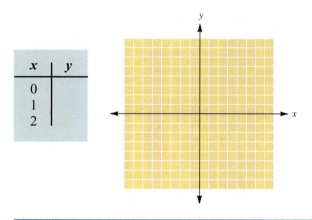

x	y
0	
1	
2	

In graphing equations, particularly when fractions are involved, a careful choice of values for *x* can simplify the process. Consider Example 9.

Example 9   Graphing a Linear Equation

Graph

$$y = \frac{3}{2}x - 2$$

As before, we want to find solutions for the given equation by picking convenient values for *x*. Note that in this case, choosing *multiples of 2* avoids fractional values for *y* and make the plotting of those solutions much easier. For instance, here we might choose values of −2, 0, and 2 for *x*.

Step 1   If $x = -2$:

$$y = \frac{3}{2}(-2) - 2$$

$$= -3 - 2 = -5$$

If $x = 0$:

NOTE Suppose we do *not* choose a multiple of 2, say, $x = 3$. Then

$$y = \frac{3}{2}(3) - 2$$

$$= \frac{9}{2} - 2$$

$$= \frac{5}{2}$$

$\left(3, \dfrac{5}{2}\right)$ is still a valid solution, but we must graph a point with fractional coordinates.

$$y = \frac{3}{2}(0) - 2$$

$$= 0 - 2 = -2$$

If $x = 2$:

$$y = \frac{3}{2}(2) - 2$$

$$= 3 - 2 = 1$$

In tabular form, the solutions are

x	y
-2	-5
0	-2
2	1

Step 2   Graph the points determined in step 1.

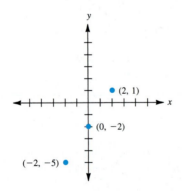

Step 3   Draw a line through the points.

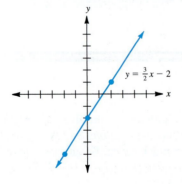

CHECK YOURSELF 9

Graph the equation $y = -\dfrac{1}{3}x + 3$ *after completing the table of values.*

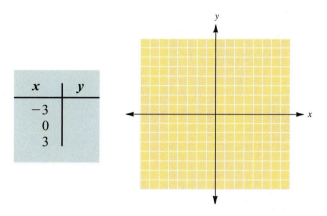

x	y
-3	
0	
3	

Some special cases of linear equations are illustrated in Examples 10 and 11.

OBJECTIVE 4

Example 10   Graphing an Equation That Results in a Vertical Line

Graph $x = 3$.

The equation $x = 3$ is equivalent to $x + 0 \cdot y = 3$. We will look at some solutions.

If $y = 1$:	If $y = 4$:	If $y = -2$:
$x + 0(1) = 3$	$x + 0(4) = 3$	$x + 0(-2) = 3$
$x = 3$	$x = 3$	$x = 3$

In tabular form,

x	y
3	1
3	4
3	-2

What do you observe? The variable x has the value 3, regardless of the value of y. Look at the given graph.

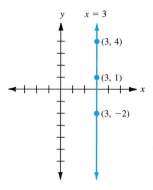

The graph of $x = 3$ is a vertical line crossing the x-axis at $(3, 0)$.

Note that graphing (or plotting) points in this case is not really necessary. Simply recognize that the graph of $x = 3$ *must* be a vertical line (parallel to the y-axis) that intercepts the x-axis at $(3, 0)$.

 CHECK YOURSELF 10

Graph the equation $x = -2$.

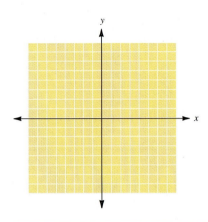

Example 11 is a related example involving a horizontal line.

Example 11 Graphing an Equation That Results in a Horizontal Line

Graph $y = 4$.

Because $y = 4$ is equivalent to $0x + y = 4$, any value for x paired with 4 for y will form a solution. A table of values might be

x	y
-2	4
0	4
2	4

Here is the graph.

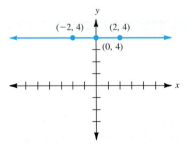

This time the graph is a horizontal line that crosses the *y*-axis at (0, 4). Again the graphing of points is not required. The graph of *y* = 4 *must* be horizontal (parallel to the *x*-axis) and intercepts the *y*-axis at (0, 4).

CHECK YOURSELF 11

Graph the equation y = −3.

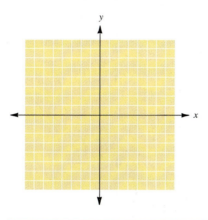

The Property box summarizes our work in Examples 10 and 11.

Property: Vertical and Horizontal Lines

1. The graph of *x* = *a* is a *vertical line* crossing the *x*-axis at (*a*, 0).
2. The graph of *y* = *b* is a *horizontal line* crossing the *y*-axis at (0, *b*).

READING YOUR TEXT

The following fill-in-the-blank exercises are designed to assure that you understand the key vocabulary used in this section. Each sentence comes directly from the section. You will find the correct answers in Appendix C.

Section 9.3

(a) The _____ number in an ordered pair is the *y*-coordinate.

(b) An _____ is a solution to an equation in two variables if substituting the coordinates for the variables results in a true statement.

(c) The graph of a linear equation in two variables is always a _____ line.

(d) The graph of *y* = *b* is a _____ line.

CHECK YOURSELF ANSWERS

1. **(a)** $y = 6$; **(b)** $x = 13$ **2.** $(3, 4), (1, -2)$, and $(0, -5)$ are solutions

3. $(0, 3), (3, 3)$, and $(-1, 3)$ are solutions   **4.** $(10, -2), (-5, 4), (0, 2)$, and $(5, 0)$

5. $(6, -2), (3, -3), (0, -4)$, and $(12, 0)$ are four possibilities

6.

x	y
1	−4
2	−2
3	0

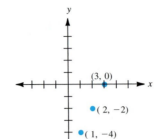

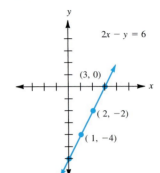

7.

x	y
0	0
1	−2
2	−4

8.

x	y
0	−2
1	1
2	4

9.

x	y
−3	4
0	3
3	2

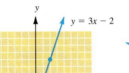

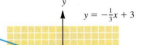

10.

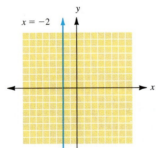

11.

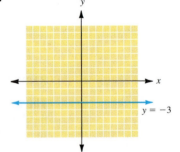

Determine which of the ordered pairs are solutions for the given equation.

1. $x + y = 6$ $(4, 2), (-2, 4), (0, 6), (-3, 9)$ **ALEKS**

2. $x - y = 12$ $(13, 1), (13, -1), (12, 0), (6, 6)$ **ALEKS**

3. $2x - y = 8$ $(5, 2), (4, 0), (0, 8), (6, 4)$ **ALEKS**

4. $x + 5y = 20$ $(10, -2), (10, 2), (20, 0), (25, -1)$ **ALEKS**

5. $3x - 2y = 12$ $(4, 0), \left(\dfrac{2}{3}, -5\right), (0, 6), \left(5, \dfrac{3}{2}\right)$ **ALEKS**

6. $3x + 4y = 12$ $(-4, 0), \left(\dfrac{2}{3}, \dfrac{5}{2}\right), (0, 3), \left(\dfrac{2}{3}, 2\right)$ **ALEKS**

7. $y = 4x$ $(0, 0), (1, 3), (2, 8), (8, 2)$

8. $y = 2x - 1$ $(0, -2), (0, -1), \left(\dfrac{1}{2}, 0\right), (3, -5)$

9. $x = 3$ $(3, 5), (0, 3), (3, 0), (3, 7)$

10. $y = 5$ $(0, 5), (3, 5), (-2, -5), (5, 5)$

Complete the ordered pairs so that each is a solution for the given equation.

11. $x + y = 12$ $(4,\ ), (\ , 5), (0,\ ), (\ , 0)$

12. $x - y = 7$ $(\ , 4), (15,\ ), (0,\ ), (\ , 0)$

13. $3x + y = 9$ $(3,\ ), (\ , 9), (\ , -3), (0,\ )$

14. $x + 5y = 20$ $(0,\ ), (\ , 2), (10,\ ), (\ , 0)$

15. $3x - 2y = 12$ $(\ , 0), (\ , -6), (2,\ ), (\ , 3)$

16. $2x + 5y = 20$ $(0,\ ), (5,\ ), (\ , 0), (\ , 6)$

17. $y = 3x - 4$ $(0,\ ), (\ , 5), (\ , 0), \left(\dfrac{5}{3},\ \right)$

18. $y = -2x + 5$ $(0,\ ), (\ , 5), \left(\dfrac{3}{2},\ \right), (\ , 1)$

Find four solutions for each of the equations. **Note:** Your answers may vary from those shown in the answer section.

19. $x - y = 7$ **ALEKS**

20. $x + y = 18$ **ALEKS**

21. $x + 4y = 8$ **ALEKS**

22. $x + 3y = 12$ **ALEKS**

23. $2x - 5y = 10$

24. $2x + 7y = 14$

25. $y = 2x + 3$

26. $y = 8x - 5$

27. $x = -5$

28. $y = 8$

An equation in three variables has an ordered triple as a solution. For example, $(1, 2, 2)$ is a solution to the equation $x + 2y - z = 3$. Complete the ordered-triple solutions for each equation.

29. $x + y + z = 0$ $(2, -3, \ )$

30. $2x + y + z = 2$ $(, -1, 3)$

31. $x + y + z = 0$ $(1, \ , 5)$

32. $x + y - z = 1$ $(-2, 1, \ )$

33. Science and Medicine The number of programs for the disabled in the United States from 1993 to 1997 is approximated by the equation $y = 162x + 4{,}365$ in which x is the number of years after 1993. Complete the table.   **VIDEO**

x	1	2	3	4	6
y					

34. Business and Finance Your monthly pay as a car salesman is determined using the equation $S = 200x + 1{,}500$ in which x is the number of cars you can sell each month.

(a) Complete the table.

x	12	15	17	18
S				

(b) You are offered a job at a salary of $56,400 per year. How many cars would you have to sell per month to equal this salary?

35. You now have had practice solving equations with one variable and equations with two variables. Compare equations with one variable to equations with two variables. How are they alike? How are they different?

36. Each of the sentences describes pairs of numbers that are related. After completing the sentences in parts (a) to (g), write two of your own sentences in (h) and (i).

(a) The *number of hours you work* determines the *amount you are* _____.

(b) The *number of gallons of gasoline* you put in your car determines *the amount you* _____.

(c) The *amount of the* _____ in a restaurant is related to *the amount of the tip.*

(d) The *sales amount of a purchase in a store* determines _____.

(e) The *age of an automobile* is related to _____.

(f) The *amount of electricity you use in a month* determines _____.

(g) The *cost of food for a family of four* is affected by _____.

Think of two more:

(h) _____.

(i) _____.

Graph each of the equations by plotting points and finding the line.

37. $x + y = 6$ **ALEKS**

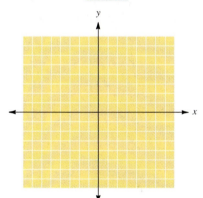

38. $x - y = 5$ **ALEKS**

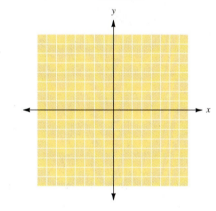

39. $x - y = -3$ **ALEKS**

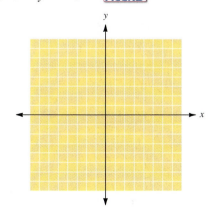

40. $x + y = -3$ **ALEKS**

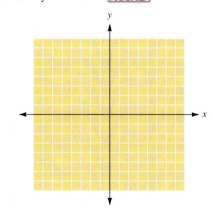

35. _____

36. _____

37. _____

38. _____

39. _____

40. _____

41. $2x + y = 2$ ALEKS

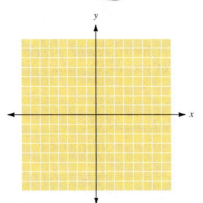

42. $x - 2y = 6$ ALEKS

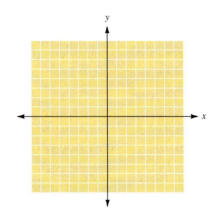

43. $3x + y = 0$ ALEKS

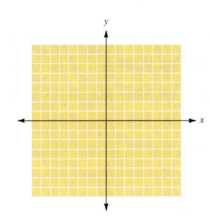

44. $2x - 3y = 6$ ALEKS

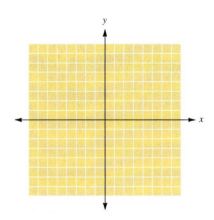

45. $y = 5x$

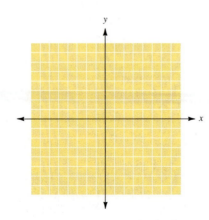

46. $y = -4x$

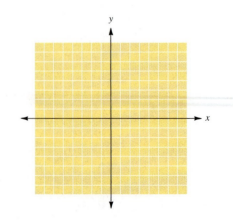

47. $y = 2x - 1$ ALEKS

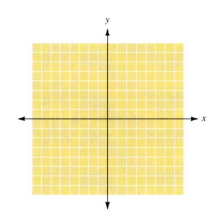

48. $y = -3x - 3$ ALEKS

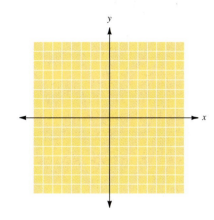

49. $y = \dfrac{1}{3}x$

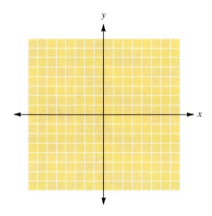

50. $y = -\dfrac{1}{4}x$

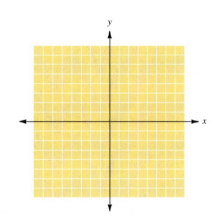

51. $y = \dfrac{2}{3}x - 3$ ALEKS

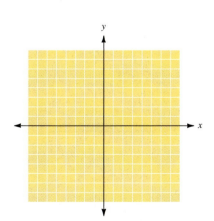

52. $y = \dfrac{3}{4}x + 2$ ALEKS

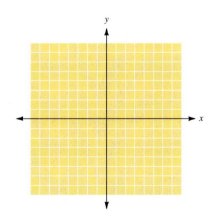

ANSWERS

47. _____

48. _____

49. _____

50. _____

51. _____

52. _____

53. $x = 5$ ALEKS

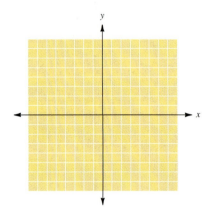

54. $y = -3$ ALEKS

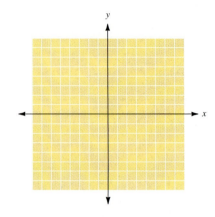

55. $y = 1$ ALEKS

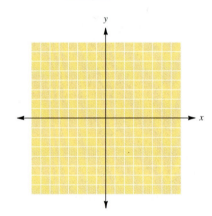

56. $x = -2$ ALEKS

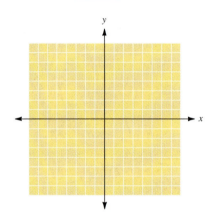

Write an equation that describes the given relationships between x and y. Then graph each relationship.

57. y is twice x.

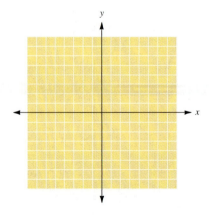

58. y is 3 times x.

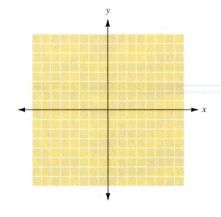

59. y is 3 more than x.

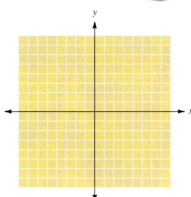

60. y is 2 less than x.

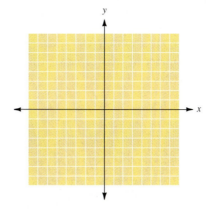

61. y is 3 less than 3 times x.

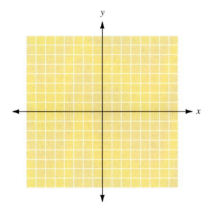

62. y is 4 more than twice x.

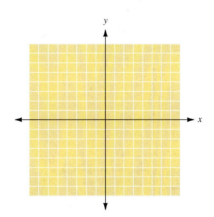

Graph each pair of equations on the same axes. Estimate the coordinates of the point where the lines intersect.

63. $x + y = 4$
$x - y = 2$

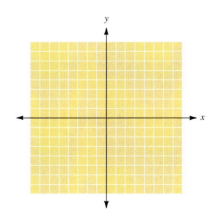

64. $x - y = 3$
$x + y = 5$

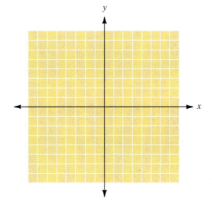

ANSWERS

59. _____

60. _____

61. _____

62. _____

63. _____

64. _____

65. Technology   A high school class wants to raise some money by recycling newspapers. They decide to rent a truck for a weekend and to collect the newspapers from homes in the neighborhood. The market price for recycled newsprint is currently $11 per ton. The equation $y = 11x - 100$ describes the amount of money the class will make, in which y is the amount of money made in dollars, x is the number of tons of newsprint collected, and 100 is the cost in dollars to rent the truck.

(a) Using the axes below, draw a graph that represents the relationship between newsprint collected and money earned.

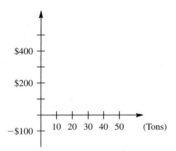

(b) The truck is costing the class $100. How many tons of newspapers must the class collect to break even on this project?

(c) If the class members collect 16 tons of newsprint, how much money will they earn?

(d) Six months later the price of newsprint is $17 a ton, and the cost to rent the truck has risen to $125. Write the equation that describes the amount of money the class might make at that time.

66. Business and Finance   The cost of producing a number of items x is given by $C = mx + b$, in which b is the fixed cost and m is the variable cost (the cost of producing one more item).

(a) If the fixed cost is $40 and the variable cost is $10, write the cost equation.

(b) Graph the cost equation.

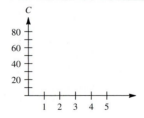

(c) The revenue generated from the sale of x items is given by $R = 50x$. Graph the revenue equation on the same set of axes as the cost equation.

(d) How many items must be produced for the revenue to equal the cost (the breakeven point)?

67. Activity The two graphs shown depict the same linear equation. Briefly discuss the differences in their appearance. Explain the reasons the graphs look different.

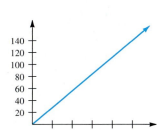
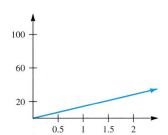

Answers

1. $(4, 2), (0, 6), (-3, 9)$ **3.** $(5, 2), (4, 0), (6, 4)$ **5.** $(4, 0), \left(\frac{2}{3}, -5\right), \left(5, \frac{3}{2}\right)$

7. $(0, 0), (2, 8)$ **9.** $(3, 5), (3, 0), (3, 7)$ **11.** $8, 7, 12, 12$ **13.** $0, 0, 4, 9$

15. $4, 0, -3, 6$ **17.** $-4, 3, \dfrac{4}{3}, 1$ **19.** $(0, -7), (2, -5), (4, -3), (6, -1)$

21. $(8, 0), (-4, 3), (0, 2), (4, 1)$ **23.** $(-5, -4), (0, -2), (5, 0), (10, 2)$

25. $(0, 3), (1, 5), (2, 7), (3, 9)$ **27.** $(-5, 0), (-5, 1), (-5, 2), (-5, 3)$

29. $(2, -3, 1)$ **31.** $(1, -6, 5)$ **33.** $4,527; 4,689; 4,851; 5,013; 5,337$

35.   **37.** $x + y = 6$ **39.** $x - y = -3$

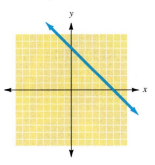
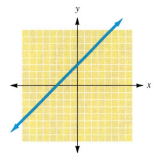

41. $2x + y = 2$ **43.** $3x + y = 0$ **45.** $y = 5x$

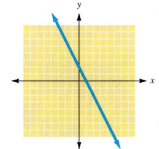
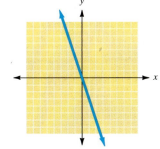
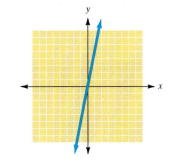

47. $y = 2x - 1$

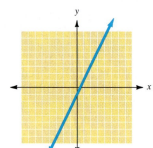

49. $y = \dfrac{1}{3}x$

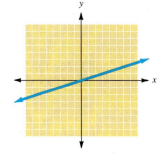

51. $y = \dfrac{2}{3}x - 3$

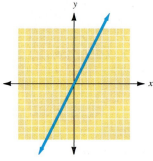

53. $x = 5$

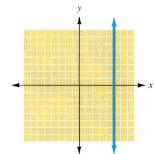

55. $y = 1$

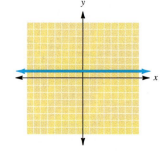

57. $y = 2x$

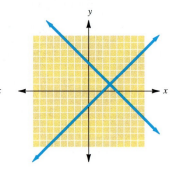

59. $y = x + 3$

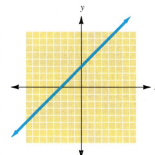

61. $y = 3x - 3$

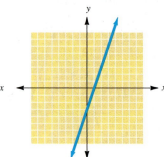

63. $(3, 1)$

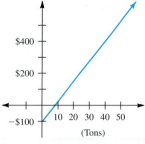

65. (a) Graph; **(b)** $\dfrac{100}{11}$ or ≈ 9 tons; **(c)** \$76; **(d)** $y = 17x - 125$

67.

9.4 Mean, Median, and Mode

© 2010 McGraw-Hill Companies

9.4 OBJECTIVES

1. Calculate and interpret a mean
2. Find and interpret a median
3. Find and interpret a mode
4. Determine the most appropriate measure of center for a data set

A very useful concept is the **average** of a group of numbers. An average is a number that is typical of a larger group of numbers. In mathematics, we have several different kinds of averages that we can use to represent a larger group of numbers. The first of these is the **mean.**

RECALL We first computed a mean in Section 9.1.

Step by Step: Finding the Mean

To find the mean for a group of numbers, follow these two steps:

Step 1   Add all of the numbers in the group.
Step 2   Divide that sum by the number of items in the group.

OBJECTIVE 1

Example 1   Finding the Mean

Find the mean of the group of numbers 12, 19, 15, and 14.

Step 1   Add all of the numbers.

$12 + 19 + 15 + 14 = 60$

Step 2   Divide that sum by the number of items.

$60 \div 4 = 15$ There are four items in this group.

The mean of this group of numbers is 15.

 CHECK YOURSELF 1 _____

Find the mean of the group of numbers 17, 24, 19, *and* 20.

Next, we apply the mean to a word problem.

Example 2   Finding the Mean

The ticket prices (in dollars) for the nine concerts held at the Civic Arena this school year were

33, 31, 30, 59, 30, 35, 32, 36, 56

What was the mean price for these tickets?

Step 1   Add all the numbers.

$33 + 31 + 30 + 59 + 30 + 35 + 32 + 36 + 56 = 342$

NOTE Divide by 9 because there are 9 ticket prices.

Step 2 Divide by 9.

$342 \div 9 = 38$

The mean ticket price was $38.

CHECK YOURSELF 2

The costs (in dollars) of the six textbooks that Aaron needs for the fall quarter are

75, 69, 57, 87, 76, 80

Find the mean cost of these books.

Although the mean is probably the most common way to find an average for a group of numbers, it is not always the most representative. Another kind of average is called the **median.**

Step by Step: Finding the Median

The median is the number for which there are as many instances that are to the right of that number as there are instances to the left of it. To find the median, follow these steps:

Step 1 Rewrite the numbers in order from smallest to largest.
Step 2 Count from both ends to find the number in the middle.
Step 3 If there are two numbers in the middle, add them together and find their mean.

OBJECTIVE 2

Example 3 Finding the Median

Find the median for the following groups of numbers.

(a) 35, 18, 27, 38, 19, 63, 22

Step 1 Rewrite the numbers in order from smallest to largest.

18, 19, 22, 27, 35, 38, 63

Step 2 Count from both ends to find the number in the middle.

Counting from both ends, we find that 27 is the median. There are three numbers to the right of 27 and three numbers to the left of it.

(b) 29, 88, 74, 81, 62, 37

Step 1 Rewrite the numbers in order from smallest to largest.

29, 37, 62, 74, 81, 88

Step 2 Count from both ends to find the number in the middle.

Counting from both ends, we find that there are two numbers in the middle, 62 and 74. We go on to step 3.

Step 3   If there are two numbers in the middle, find their mean.

$(62 + 74) \div 2 = 136 \div 2 = 68$

CHECK YOURSELF 3

Find the median for each group of numbers.

(a) 8, 6, 19, 4, 21, 5, 27   **(b)** 43, 29, 13, 37, 29, 53

There are times in which the median is a better representative of a group of numbers than the mean is. Example 4 illustrates such a case.

Example 4   Comparing the Mean and the Median

The given numbers represent the hourly wage of seven employees of a local chip manufacturing plant.

12, 11, 14, 16, 32, 13, 14

(a) Find the mean hourly wage.

Step 1   Add all of the numbers in the group.

$12 + 11 + 14 + 16 + 32 + 13 + 14 = 112$

Step 2   Divide that sum by the number of items in the group.

$112 \div 7 = 16$

The mean wage is $16 an hour.

(b) Find the median wage for the seven workers.

Step 1   Rewrite the numbers in order from smallest to largest.

11, 12, 13, 14, 14, 16, 32

Step 2   Count from both ends to find the number in the middle.

The middle number is 14. There are three numbers to the right of it and three numbers to the left of it. The median salary is $14 per hour. Which salary do you think is more typical of the workers? Why?

CHECK YOURSELF 4

The following are Jessica's phone bills for each month of 2000.

26, 67, 31, 24, 15, 17, 41, 27, 17, 22, 26, 47

(a) Find the mean amount of her phone bills.
(b) Find the median amount of her phone bills.

Another measure used as an average is called the **mode.**

> **Definition:** Mode
>
> The **mode** of a set of data is the item or number that appears most frequently.

OBJECTIVE 3

Example 5   Finding a Mode

Find the mode for the set of numbers given.

22, 24, 24, 24, 24, 27, 28, 32, 32

The mode, 24, is the number that appears most frequently.

CHECK YOURSELF 5

Find the mode for the set of numbers given.

7, 7, 7, 9, 11, 13, 13, 15, 15, 15, 15, 21

One advantage of the mode is that it can be used with data that are not a set of numbers.

Example 6   Finding a Mode

NOTE If each eye color appeared three times, there would be no mode! Not every data set has a mode.

The following are the eye colors of a class of 12 students. Which color is the mode?

blue, brown, hazel, blue, brown, brown, brown, brown, blue, brown, hazel, green

Because brown occurs most frequently, it is the mode.

CHECK YOURSELF 6

The given types of computers were available in the lab. Which type was the mode?

Apple, IBM, Compaq, Dell, Apple, IBM, Apple, Compaq, Dell, Apple, IBM, Apple, Dell, Apple, Compaq

NOTE *Qualitative data* can be separated into categories. Usually, qualitative data does not consist of numbers.

We opened this section by saying that an "average is a number that is typical of a larger group of numbers." We extended this idea to sets that did not consist of numbers, such as eye color.

In situations where a set consists of qualitative data, the mode is the best measure of the average.

But when our data is quantitative (numerical), which should we use, the mean or the median?

Mathematicians and statisticians prefer to use the mean when possible. If you take a course in statistics, you will see that many analytical techniques use the mean. There are times, though, when the median gives a result that is more typical of the larger set.

OBJECTIVE 4

NOTE We call an average a "measure of the center" because it represents a typical (or *center*) member of a data set.

| Example 7 | Choosing an Appropriate Measure of Center |

Consider the annual salaries earned 10 years after graduation by the 6 graduates of a university geology department.

$35,000 (teaches middle school)

$18,000 (in graduate school)

$12,000 (in graduate school)

$75,000 (works in industry)

$31,000 (works for a city parks department)

$35,000,000 (professional athlete with endorsement income)

The mean annual salary 10 years after graduation from the university geology department is

$$\frac{35,000 + 18,000 + 12,000 + 75,000 + 31,000 + 35,000,000}{6} = \$5,861,833.33$$

The median is

$$\frac{31,000 + 35,000}{2} = \$33,000$$

A graduate from the university geology department is much more likely to earn (in the neighborhood of) $33,000 than nearly $6 million.

In this case, the median is more typical of the salaries than the mean.

 CHECK YOURSELF 7

Home prices for 10 houses in a lakeside community are given. Find the mean and median home price.

Which is more typical of the data set? Explain your answer.

In thousands:

$485	$510
$320	$430
$615	$2,500
$390	$360
$400	$485

In many cases, the mean and median of a set are fairly close to each other. In Example 7, there was a large difference between the two. This occurred because there was one salary that was vastly different from the others.

The median usually represents the more typical member of a group when one number is vastly different from the other numbers. This number is said to *skew* the data set.

Using Your Calculator to Find an Average

Most electronic calculators have statistical functions that allow you to calculate means, medians, and other statistical values. Because these features vary so much from one calculator to another, you will need to consult your owner's manual to learn how to access these features.

Here, we focus on using a calculator to compute a mean.

Example 8 Calculating a Mean

Find the mean for the set of numbers.

45, 48, 53, 59, 67, 76

When entering these numbers in your calculator, keep in mind the order of operations. There are two different techniques you may use.

Method 1

Add the numbers

$45 + 48 + 53 + 59 + 67 + 76 = 348$

then divide the sum by 6 (there are six numbers)

$348 \div 6 = 58$

Method 2

Use parentheses to find the mean.

$(45 + 48 + 53 + 59 + 67 + 76) \div 6 = 58$

 CHECK YOURSELF 8 _____

Find the mean for the set of numbers.

132, 144, 156, 158, 279, 337

Unfortunately, not all calculations result in answers that are whole numbers.

Example 9 Calculating a Mean

Find the mean of the set of numbers below.

13, 15, 16, 19, 26, 28, 38

Entering this as a single expression, we have

$(13 + 15 + 16 + 19 + 26 + 28 + 38) \div 7 \approx 22$

Your calculator is probably displaying something like 22.14285714. Remember that this is just another (although closer) approximation.

We can approximate the answer as 22, as we indicated. We could also subtract 22 from the calculator display, leaving only the calculator's internal decimal approximation, which is approximately

0.142857142

If we multiply this by the divisor, 7, we get the remainder, which is 1. The exact answer is

$$22\frac{1}{7}$$

CHECK YOURSELF 9

Find an approximate and exact mean for the set of numbers.

17, 23, 33

READING YOUR TEXT

The following fill-in-the-blank exercises are designed to assure that you understand the key vocabulary used in this section. Each sentence comes directly from the section. You will find the correct answers in Appendix C.

Section 9.4

(a) An _____ is a number that is typical of a larger group of numbers.

(b) The _____ is the number for which there are as many instances that are to the right of the number as there are instances to the left of it.

(c) A set with two different modes is called _____.

(d) One advantage of the _____ is that it can be used with nonnumerical data.

CHECK YOURSELF ANSWERS

1. 20   **2.** $74   **3. (a)** 8; **(b)** 33   **4. (a)** $30; **(b)** $26   **5.** 15   **6.** Apple
7. Mean: $649,500; median: $457,500; median   **8.** 201

9. Approximately 24, exactly $24\frac{1}{3}$

Name _____

Section _____ Date _____

ANSWERS

1. _____ 2. _____

3. _____ 4. _____

5. _____ 6. _____

7. _____

8. _____

9. _____

10. _____

11. _____

12. _____

13. _____

14. _____

15. _____

16. _____

17. _____

18. _____

19. _____

20. _____

21. _____

22. _____

23. _____

24. _____

25. _____

9.4   Exercises

Find the mean for each set of numbers.

1. 6, 9, 10, 8, 12   ALEKS

2. 13, 15, 17, 17, 18   ALEKS

3. 13, 15, 17, 19, 24, 25   ALEKS

4. 41, 43, 56, 67, 69, 72   ALEKS

5. 12, 14, 15, 16, 16, 16, 17, 22, 25, 27   ALEKS

6. 21, 25, 27, 32, 36, 37, 43, 43, 44, 51   ALEKS

7. 5, 8, 9, 11, 12   VIDEO   ALEKS

8. 7, 18, 11, 7, 12   ALEKS

9. 9, 8, 11, 14, 9   ALEKS

10. 21, 23, 25, 27, 22, 20   ALEKS

Find the median for each set of numbers.

11. 2, 3, 5, 6, 10   ALEKS

12. 12, 13, 15, 17, 18   ALEKS

13. 23, 24, 27, 31, 36, 38, 41   ALEKS

14. 1, 4, 9, 16, 25, 36, 49   ALEKS

15. 46, 13, 47, 25, 68, 51   ALEKS

16. 26, 71, 33, 69, 71, 25, 75   ALEKS

Find the mode for each set of numbers.

17. 17, 13, 16, 18, 17   ALEKS

18. 41, 43, 56, 67, 69, 72   ALEKS

19. 21, 44, 25, 27, 32, 36, 37, 44   ALEKS

20. 9, 8, 10, 9, 9, 10, 8   ALEKS

21. 12, 13, 7, 14, 4, 11, 9   VIDEO   ALEKS

22. 8, 2, 3, 3, 4, 9, 9, 3   ALEKS

Solve the applications.

23. Science and Medicine   High temperatures (in °F) of 86, 91, 92, 103, and 98 were recorded for the first 5 days of July. What was the mean high temperature?

24. Statistics   A salesperson drove 238, 159, 87, 163, and 198 mi on a 5-day trip. What was the mean number of miles driven per day?

25. Technology   Highway mileage ratings for seven new diesel cars were 43, 29, 51, 36, 33, 42, and 32 $\frac{\text{mi}}{\text{gal}}$. What was the mean rating?

26. Social Science   The enrollments in the four elementary schools of a district are 278, 153, 215, and 198 students. What is the mean enrollment?

27. Statistics   To get an A in history, you must have a mean of 90 on five tests. Your scores thus far are 83, 93, 88, and 91. How many points must you have on the final test to receive an A? (*Hint:* First find the total number of points you need to get an A.)   **VIDEO**   **ALEKS**

28. Statistics   To pass biology, you must have a mean of 70 on six quizzes. So far your scores have been 65, 78, 72, 66, and 71. How many points must you have on the final quiz to pass biology?   **ALEKS**

29. Statistics   Louis had scores of 87, 82, 93, 89, and 84 on five tests. Tamika had scores of 92, 83, 89, 94, and 87 on the same five tests. Who had the higher mean score? By how much?

30. Business and Finance   The Wong family had heating bills of $105, $110, $90, and $67 in the first 4 months of 1999. The bills for the same months of 2000 were $110, $95, $75, and $76. In which year was the mean monthly bill higher? By how much?

Technology   Monthly energy use, in kilowatt-hours (kWh), by appliance type for four typical U.S. families is shown in the table.

	Wong Family	McCarthy Family	Abramowitz Family	Gregg Family
Electric range	97	115	80	96
Electric heat	1,200	1,086	1,103	975
Water heater	407	386	368	423
Refrigerator	127	154	98	121
Lights	75	99	108	94
Air conditioner	123	117	96	120
Color TV	39	45	21	47

© 2010 McGraw-Hill Companies

31. _____

32. _____

33. _____

34. _____

35. _____

36. _____

37. _____

38. _____

39. _____

40. _____

41. _____

42. _____

43. _____

31. What is the mean number of kilowatt-hours used each month by the four families for heating their homes?

32. What is the mean number of kilowatt-hours used each month by the four families for hot water?

33. What is the mean number of kilowatt-hours used per appliance by the McCarthy family?

34. What is the mean number of kilowatt-hours used per appliance by the Gregg family?

35. Business and Finance Fred kept the following records of his utility bills for 12 months: $53, $51, $43, $37, $32, $29, $34, $41, $58, $55, $49, and $58. What was the mean monthly bill?

36. Statistics The following scores were recorded on a 200-point final examination: 193, 185, 163, 186, 192, 135, 158, 174, 188, 172, 168, 183, 195, 165, 183. What was the mean of the scores?

37. Science and Medicine The following are eye colors from a class of eight students. Which color is the mode?

Hazel, green, brown, brown, blue, green, hazel, green

38. Science and Medicine The weather in Philadelphia over the last 7 days was as follows:

Rain, sunny, cloudy, rain, sunny, rain, rain. What type of weather was the mode?

Find the mean, median, and mode for the following sets. Which best describes the typical member of the set? Why?

39. 42, 56, 42, 48, 101, 52, 50, 47   **40.** 12, 8, 10, 18, 435, 15, 16, 14

41. List the advantages and disadvantages of the mean, median, and mode.

42. In a certain math class, you take four tests and the final, which counts as two tests. Your grade is the average of the six test scores. At the end of the course, you compute both the mean and the median.

(a) You want to convince the teacher to use the mean to compute your average. Write a note to your teacher explaining why this is a better choice. Choose numbers that make a convincing argument.

(b) You want to convince the teacher to use the median to compute your average. Write a note to your teacher explaining why this is a better choice. Choose numbers that make a convincing argument.

43. Create a set of five numbers such that the mean is equal to the median.

44. Create a set of five numbers such that the mean is greater than the median.

45. Create a set of five numbers such that the mean is less than the median.

 ## Calculator Exercises

In exercises 46 to 52, find the mean for each set of numbers.

46. 20, 18, 17, 24, 22, 19

47. 108, 113, 109, 113, 110, 101, 112, 114

48. 211, 213, 215, 208, 209, 220, 215, 221

49. 2,357, 2,361, 2,372, 2,371, 2,357, 2,375, 2,364, 2,371

50. 16,430, 16,211, 16,149, 16,232, 16,317, 16,113

51. 24,637, 24,251, 24,454, 24,580, 24,324, 24,478

52. 311,431, 286,356, 356,090, 292,007, 301,857, 299,005

In exercises 53 to 58, find the approximate and exact mean for each set of numbers.

53. 18, 21, 20, 22

54. 36, 41, 43, 39, 40, 37, 39

55. 125, 121, 129, 126, 128, 123

56. 356, 371, 366, 373, 359, 363

57. 1,898, 1,913, 1,875, 1,937

58. 15,865, 16,270, 16,090, 15,904

Solve the applications.

59. Business and Finance The revenue for the leading apparel companies in the United States in 1997 is given in the table.

Company	Revenue (in Millions)
Nike	$9,187
Vanity Fair	$5,222
Liz Claiborne	$2,413
Reebok	$3,637
Fruit of the Loom	$2,140
Nine West	$1,865
Kellwood	$1,521
Warmaio	$1,437
Jones Apparel	$1,387

What is the mean revenue taken in by these companies?

ANSWERS

44. _____

45. _____

46. _____

47. _____

48. _____

49. _____

50. _____

51. _____

52. _____

53. _____

54. _____

55. _____

56. _____

57. _____

58. _____

59. _____

60. Social Science

Unemployment in the United States (in Thousands)		
Year	**Employed**	**Unemployed**
1993	120,259	8,940
1994	123,060	7,996
1995	124,900	7,404
1996	126,708	7,236
1997	129,558	6,739
1998	131,463	6,210
1999	133,488	5,880
2000	136,891	5,692
2001	136,933	6,801
2002	136,485	6,378

Source: U.S. Department of Labor, Bureau of Labor Statistics

Find the mean number of employed and unemployed people per year from 1993 to 2002.

Business and Finance   The work stoppages (strikes and lockouts) in the United States from 1995 to 2002 are given in the table.

Year	**No. of Stoppages**	**Work Days Idle**
1995	192	5,771
1996	273	4,889
1997	339	4,497
1998	387	5,116
1999	73	1,996
2000	394	20,419
2001	99	1,151
2002	46	6,596

Source: U.S. Department of Labor, Bureau of Labor Statistics

61. Find the mean number of work stoppages per year from 1995 to 2002.

62. Find the mean number of work days idle from 1995 to 2002.

63. Activity   A distributor sells power saws for $499. They sell replacement blades for $19 and belts for $4.

(a) Compute the average selling price for the three items.

(b) Can the distributor determine revenues by multiplying the number of items sold by the average selling price given in **(a)**? Why or why not?

64. Activity   Grades on a class exam are shown below.   `ALEKS`

80	84	51	0
75	92	97	87
75	97	72	91
81	65	0	76
81	88	78	82

(a) Find the mean grade on the exam.

(b) Find the median grade on the exam.

(c) Which is a better measure of the typical exam grade? Explain.

Answers

1. 9   **3.** $18\frac{5}{6}$ **5.** 18   **7.** 9   **9.** 10.2   **11.** 5   **13.** 31

15. $46\frac{1}{2}$ **17.** 17   **19.** 44   **21.** No mode   **23.** 94°F   **25.** $38\,\frac{\text{mi}}{\text{gal}}$

27. 95 points   **29.** Louis's mean score was 87; Tamika's was 89. Tamika's average score was 2 points higher than Louis's   **31.** 1,091 kWh   **33.** 286 kWh

35. $45   **37.** Green   **39.** Mean: 53.75; median: 50; mode: 42; mean

41.   **43.**   **45.**   **47.** 110   **49.** 2,366

51. 24,454   **53.** $20; 20\frac{1}{4}$ **55.** $125; 125\frac{1}{3}$ **57.** $1,906; 1,905\frac{3}{4}$

59. $3,201,000,000   **61.** 225   **63. (a)** $174; **(b)**

ACTIVITY 9: GRAPHS IN THE MEDIA

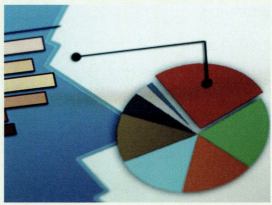

Each chapter in this text concludes with a chapter activity such as this one. These activities provide you with the opportunity to apply the math you studied in that chapter to a relevant topic.

Your instructor will determine how best to use these activities with your class. You may find yourself working in class or outside of class; you may find yourself working alone or in small groups; or you may even be asked to perform research in a library or on the Internet.

At times, it seems that we are being bombarded with images and graphs conveying information.

Some of these pictures display information in an easy-to-read format. Others seek to convince the reader of some position. Occasionally, someone will even construct a graph to mislead the reader in order to get them to believe the designer's position.

As a math student, you should begin to critically assess graphs and images that you see in the media. This activity will introduce you to some of the questions you should ask when looking at such a graph or image.

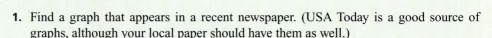

1. Find a graph that appears in a recent newspaper. (USA Today is a good source of graphs, although your local paper should have them as well.)
2. What information does the graph convey? How well does the graph present its data? That is, how easy is it to determine the graph's content?
3. Is the graph honest? Does it present its information in a straightforward manner, without trying to "fool" the reader, or is it "sensationalistic"?
4. Find another graph in the media. This time, find one that "abuses" the data. That is, find a graph that leads the reader to some conclusion by altering aspects of the image.
5. How would you change the graph found in (4) to make it more "honest"?
6. Do you reach the same conclusion with your graph from (5) as the author did with theirs?
7. Many graphs in the media provide information that allows you to find the source data. For one of the graphs you found, locate the source data. Construct your own graph that shows this information more clearly than the one in the newspaper.
8. Find some data that interests you. For example, you could look at the daily high temperature for a week, the prices of a stock of a company, or some statistics about political parties or sports. Now construct your own graph to clearly represent this data.
9. Create a new graph using your data from (8). This time, alter your graph so that readers reach some conclusion that the data do not support. Do not "lie." Rather, change the scales of the axes, the thickness of the bars, or some other properties so that the readers come to this false conclusion on their own.
10. Write a paragraph describing some of the things one should look for to determine when a graph is describing information in a less than honest manner.

9 Summary

DEFINITION/PROCEDURE	EXAMPLE	REFERENCE
Tables and Graphs of Data		**Section 9.1**
A **table** is a display of information in parallel columns or rows.		*p. 693*
A **graph** is a diagram that relates two different pieces of information. One of the most common graphs is the **bar graph.**		*p. 694*
In **line graphs,** one of the axes is usually related to time.		*p. 698*
The Rectangular Coordinate System		**Section 9.2**
The Rectangular Coordinate System The rectangular coordinate system is a system formed by two perpendicular axes that intersect at a point called the **origin.** The horizontal line is called the *x*-**axis.** The vertical line is called the *y*-**axis.**		*p. 715*
Graphing Points from Ordered Pairs The coordinates of an ordered pair allow you to associate a point in the plane with every ordered pair. To graph a point in the plane, **Step 1** Start at the origin. **Step 2** Move right or left according to the value of the *x*-coordinate: to the right if *x* is positive or to the left if *x* is negative. **Step 3** Then move up or down according to the value of the *y*-coordinate: up if *y* is positive and down if *y* is negative.	To graph the point corresponding to (2, 3): 	*p. 718*
Linear Equations in Two Variables		**Section 9.3**
Solutions of Linear Equations A pair of values that satisfies the equation. Solutions for linear equations in two variables are written as *ordered pairs.* An ordered pair has the form (x, y) *x*-coordinate *y*-coordinate	If $2x - y = 10$, (6, 2) is a solution for the equation, because substituting 6 for *x* and 2 for *y* gives a true statement.	*p. 730*

Continued

© 2010 McGraw-Hill Companies

DEFINITION/PROCEDURE	EXAMPLE	REFERENCE
Linear Equations in Two Variables		**Section 9.3**
Linear Equation An equation that can be written in the form $$Ax + By = C$$ in which A and B are not both 0.	$2x - 3y = 4$ is a linear equation.	*p. 734*
Graphing Linear Equations **Step 1** Find at least three solutions for the equation and put your results in tabular form. **Step 2** Graph the solutions found in step 1. **Step 3** Draw a straight line through the points determined in step 2 to form the graph of the equation.		*p. 735*
Mean, Median, and Mode		**Section 9.4**
Finding the Mean To find the **mean** for a group of numbers follow these two steps: **Step 1** Add all the numbers in the group. **Step 2** Divide that sum by the number of items in the group.	Given the numbers 4, 8, 17, 23 $$4 + 8 + 17 + 23 = 52$$ $$\text{Mean} = \frac{52}{4} = 13$$	*p. 753*
Finding the Median The **median** is the number for which there are as many instances that are to the right of that number as there are instances to the left of it. To find the median follow these steps: **Step 1** Rewrite the numbers in order from smallest to largest. **Step 2** Count from both ends to find the number in the middle. **Step 3** If there are two numbers in the middle, add them together and find their mean.	Given the numbers 9, 2, 5, 13, 7, 3 Rewrite them $$2, 3, 5, 7, 9, 13$$ The middle numbers are 5 and 7 $$\frac{5 + 7}{2} = \frac{12}{2} = 6$$ $$\text{Median} = 6$$	*p. 754*
Finding the Mode The **mode** is the number that occurs most frequently in a set of numbers.	Given the numbers 2, 3, 3, 3, 5, 5, 7, 7, 9, 11. 3 is the mode.	*p. 756*

Summary Exercises

This summary exercise set is provided to give you practice with each of the objectives of this chapter. Each exercise is keyed to the appropriate chapter section. When you are finished, you can check your answers to the odd-numbered exercises against those presented in the back of the text. If you have difficulty with any of these questions, go back and reread the examples from that section. The answers to the even-numbered exercises appear in the *Instructor's Solutions Manual*. Your instructor will give you guidelines on how to best use these exercises in your class.

[9.1] The table below shows the U.S. motor vehicle production in thousands, by source, in 2002.

Source	Number (in Thousands)	Percent
General Motors	4,093	33%
Ford	3,413	28%
Chrysler	1,837	15%
Foreign-based domestics	2,985	24%

Source: Automotive News Market Data Books

1. Which company produces the most motor vehicles?

2. How many more motor vehicles did Ford produce than Chrysler?

3. Chrysler's production is equal to what percent of Ford's production?

4. What percent of the domestic motor vehicle production does General Motors account for?

The table represents the 10 most expensive markets for home prices in the United States in 2003.

City	Average Price ($)
La Jolla, CA	1,362,375
Palo Alto, CA	1,179,000
Greenwich, CT	1,170,600
Beverly Hills, CA	1,097,250
San Francisco, CA	971,750
New Canaan, CT	963,750
Wellesley, MA	959,048
Kailua Kona, HI	906,250
Manhattan Beach, CA	904,500

Source: Home Price Comparison Index

5. Which market is the most expensive?

6. How much more expensive are home prices in Greenwich, CT, than New Canaan, CT?

7. What is the percent difference in home prices between Greenwich, CT, and New Canaan, CT?

8. What is the percent difference in home prices between the two most expensive markets?

Use the bar graph for exercises 9 and 10.

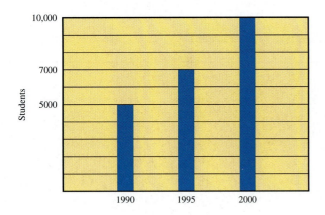

9. How many more students were enrolled in 2000 than in 1990?

10. What was the percent increase from 1990 to 2000?

11. Create a bar graph from the table used for Exercises 1 to 4 on page 769.

Use the line graph for exercises 12 to 14.

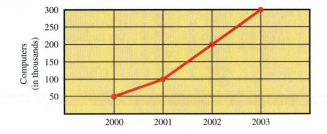

12. How many more personal computers were sold in 2003 than in 2000?

13. What was the percent increase in sales from 2000 to 2003?

14. Predict the sales of personal computers in the year 2004.

15. Create a line graph from the given data showing attendance at a county fair.

Day	Attendance
Monday	4,600
Tuesday	5,100
Wednesday	4,800
Thursday	4,900
Friday	5,700
Saturday	6,500
Sunday	6,200

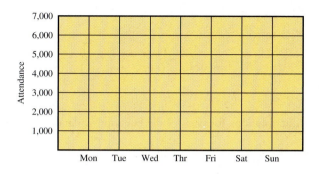

[9.2] Give the coordinates of the graphed points.

16. A

17. B

18. E

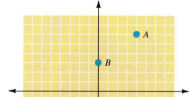
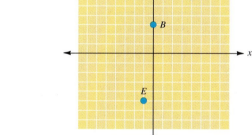

Plot points with the coordinates shown.

19. $P(6, 0)$

20. $Q(5, 4)$

21. $T(-2, 4)$

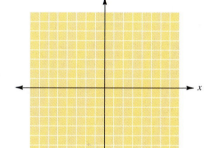

[9.3] Tell whether the number shown in parentheses is a solution for the given equation.

22. $7x + 2 = 16$ (2)

23. $5x - 8 = 3x + 2$ (4)

24. $7x - 2 = 2x + 8$ (2)

25. $\dfrac{2}{3}x - 2 = 10$ (21)

Determine which of the ordered pairs are solutions for the given equations.

26. $2x + y = 8$ (4, 0), (2, 2), (2, 4), (4, 2)

27. $2x + 3y = 6$ (3, 0), (6, 2), (-3, 4), (0, 2)

28. $2x - 5y = 10$ (5, 0), $\left(\dfrac{5}{2}, -1\right)$, $\left(2, \dfrac{2}{5}\right)$, (0, -2)

Complete the ordered pairs so that each is a solution for the given equation.

29. $x - 2y = 10$ $(0, \ )$, $(12, \ )$, $(\ , -2)$, $(8, \ )$

30. $2x + 3y = 6$ $(3, \ )$, $(6, \ )$, $(\ , -4)$, $(-3, \ )$

31. $y = 3x + 4$ $(2, \ )$, $(\ , 7)$, $\left(\dfrac{1}{3}, \ \right)$, $\left(\dfrac{4}{3}, \ \right)$

Find four solutions for each of the equations. (Answers may vary.)

32. $2x + y = 8$ **33.** $2x - 3y = 6$

34. $y = -\dfrac{3}{2}x + 2$

Graph each of the equations.

35. $x + y = 5$ **36.** $x - y = 6$ **37.** $y = 2x$

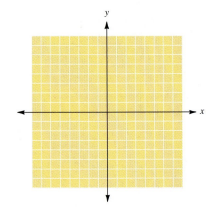   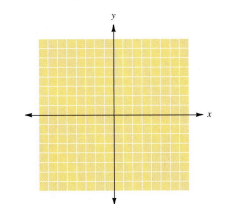   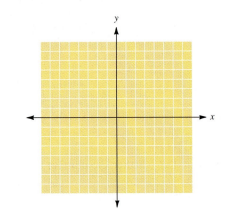

38. $y = -3x$ **39.** $y = \dfrac{3}{2}x$ **40.** $y = 3x + 2$

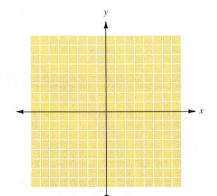   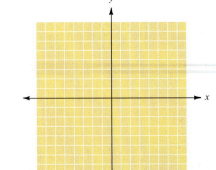   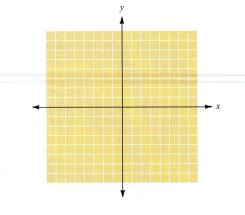

41. $y = -3x + 4$

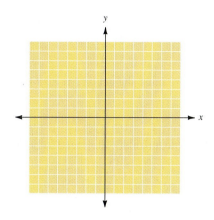

42. $y = \dfrac{2}{3}x + 2$

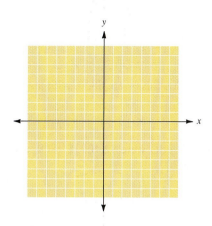

43. $3x - y = 3$

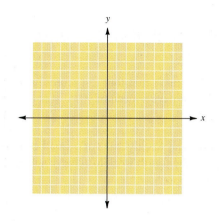

44. $3x + 2y = 12$

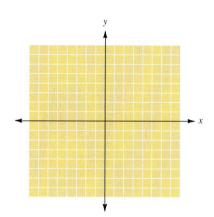

45. $x = 3$

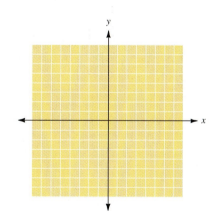

46. $y = -2$

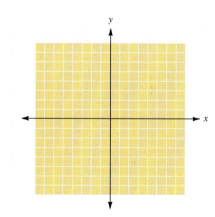

[9.4] In exercises 47 to 50, find the mean for each set of numbers.

47. 8, 6, 7, 4, 5

48. 12, 14, 17, 19, 13

49. 117, 121, 122, 118, 115, 125, 123, 119

50. 134, 126, 128, 129, 133, 125, 122, 127

51. Elmer had test scores of 89, 71, 93, and 87 on his four math tests. What was his mean score?

52. The costs (in dollars) of the seven textbooks that Jacob needs for the spring semester are 77, 66, 55, 49, 85, 80, and 78. Find the mean cost of these books.

In exercises 53 to 56, find the median and the mode for each set of data.

53. 16, 20, 20, 19, 18

54. 8, 9, 9, 11, 11, 8, 7, 11, 12, 14, 10

55. 26, 31, 28, 35, 27, 28, 31, 30, 28, 30

56. 15, 18, 21, 23, 17, 19, 30, 35, 15, 32

57. Anita's first four test scores in her mathematics class were 88, 91, 86, and 93. What score must she get on her next test to have a mean of 90?

58. The weekly sales of a small company for 3 weeks were $2,400, $2,800, and $3,300. How much do sales need to be in the fourth week to achieve a mean of $3,000?

Self-Test for Chapter 9

The purpose of this self-test is to help you check your progress so that you can find sections and concepts that you need to review before the next in-class exam. Allow yourself about an hour to take this test. At the end of that hour, check your answers against those given in the back of this text. If you missed an answer, notice the section reference that accompanies the question. Go back to that section and reread the examples until you have mastered that particular concept.

The table below gives home prices in 10 of the most affordable markets in the United States in 2003.

City	Average Price ($)
Binghamton, NY	121,400
Killeen, TX	127,175
Minot, ND	129,075
Oklahoma City, OK	132,670
Topeka, KS	136,266
Tulsa, OK	136,625
Aberdeen, SD	138,000
Billings, MT	138,725
Sioux City, IA	139,500
Parkersburg, WV	141,250

Source: Home Price Comparison Index

1. How much more affordable is Oklahoma City than Tulsa?

2. What is the percent difference in home prices in Oklahoma City compared to Tulsa?

3. What is the difference in home prices between the two most affordable markets?

4. What is the percent difference between the two most affordable markets?

The bar graph represents the number of bankruptcy filings during a recent 5-year period.

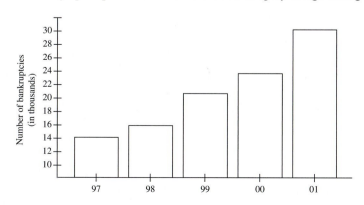

Name _____

Section _____ Date _____

ANSWERS

1. _____

2. _____

3. _____

4. _____

5. How many people filed for bankruptcy in 1998?

6. What was the increase in filings from 1997 to 2001?

7. Between which 2 years did the greatest increase in filings occur?

8. The given information represents a relationship between age and college education. Create a bar graph from this information.

Age Group	Percent with 4 Years of College
25–34	24
35–44	27
45–54	21
55–64	15
65+	11

8. _____

9. _____

10. _____

11. _____

9. The pictograph shown represents the U.S. motor vehicle production, by source, in 2002.

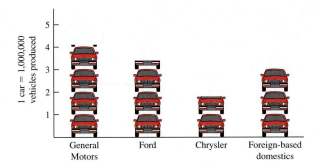

(a) How many more vehicles were produced by General Motors than Ford?
(b) What is the percent difference in the number of vehicles produced by General Motors compared to Ford?

The line graph shows ticket sales for the last 6 months of the year.

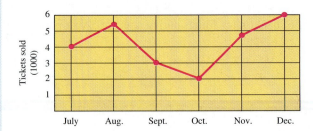

10. What month had the greatest number of ticket sales?

11. Between what 2 months did the greatest decrease in ticket sales occur?

12. The information in the table shows a relationship between the number of workers absent from the assembly line and the number of defects coming off the line. Use this information to create a line graph and then predict the number of defects coming off the line if five workers are absent.

Number of Workers Absent	Number of Defects
0	9
1	10
2	12
3	16
4	18

Give the coordinates of the graphed points.

13. *A*

14. *B*

15. *C*

Plot points with the coordinates shown.

16. $S(1, -2)$

17. $T(0, 3)$

18. $U(-2, -3)$

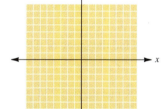

Determine which of the ordered pairs are solutions for the given equation.

19. $4x - y = 16$ $(4, 0), (3, -1), (5, 4)$

Complete the ordered pairs so that each is a solution for the given equation.

20. $x + 3y = 12$ $(3, \ ), (\ , 2), (9, \ )$

Find four solutions for the given equation.

21. $x - y = 7$

Graph each of the given equations.

22. $x + y = 4$ **23.** $y = 3x$

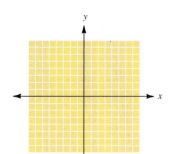

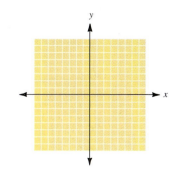

12. _____

13. _____

14. _____

15. _____

16. _____

17. _____

18. _____

19. _____

20. _____

21. _____

22. _____

23. _____

24. $2x + 5y = 10$

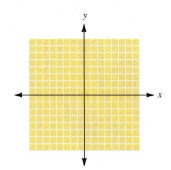

25. $y = -4$

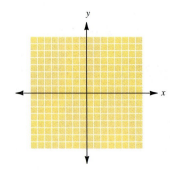

26. Find the mean of the numbers 12, 19, 15, 20, 11, and 13.

27. Find the median of the numbers 12, 18, 9, 10, 16, and 6.

28. Find the mode of the numbers 6, 2, 3, 6, 2, 9, 2, 6, and 6.

29. A bus carried 234 passengers on the first day of a newly scheduled route. The next 4 days there were 197, 172, 203, and 214 passengers. What was the mean number of riders per day?

30. The following hair colors are from a class of eight students. What color is the mode?

Brown, black, red, blonde, brown, brown, blue, gray

Cumulative Review
for Chapters 1 to 9

The following exercises are presented to help you review concepts from earlier chapters that you may have forgotten. This is meant as review material and not as a comprehensive exam. The answers are presented in the back of the text. Beside the questions is the section from which the topic relevant to the questions was first presented. If you have difficulty with any of these exercises, be certain to at least read through the summary related to that section.

Name _____

Section _____ Date _____

ANSWERS

1. What is the place value of 6 in the numeral 126,489?

In exercises 2 to 7, perform the indicated operation.

2. $20 \div (-4)$

3. $15 \div 0$

4. $2.45 \cdot 30.7$

5. $\dfrac{4}{7} \cdot \dfrac{28}{24}$

6. $\dfrac{11}{15} \div \dfrac{121}{90}$

7. $3\dfrac{2}{3} + 5\dfrac{5}{6} - 2\dfrac{5}{12}$

Evaluate the following expressions if $x = 5$, $y = 2$, and $w = -4$.

8. $2xy$

9. $4(x + 3w)$

Combine like terms.

10. $7x - x$

11. $6x + 5y - 2x - 8y$

Solve each equation and check your result.

12. $-\dfrac{3}{4}x = 18$

13. $6x - 8 = 2x - 3$

In exercises 14 and 15, solve for the unknown.

14. $\dfrac{4}{7} = \dfrac{8}{x}$

15. $\dfrac{3}{5} = \dfrac{x}{15}$

16. Write 18% as a decimal and fraction.

17. Write $\dfrac{17}{40}$ as a decimal and percent.

In exercise 18, do the indicated operation.

18.   7 lb   9 oz
   + 3 lb 12 oz

In exercises 19 to 22, complete each statement.

19. 8 km = _____ m

20. 3,000 mg = _____ g

21. 500 cm = _____ m

22. 25 cL = _____ mL

1. _____
2. _____
3. _____
4. _____
5. _____
6. _____
7. _____
8. _____
9. _____
10. _____
11. _____
12. _____
13. _____
14. _____
15. _____
16. _____
17. _____
18. _____
19. _____ 20. _____
21. _____ 22. _____

ANSWERS

23.
_____

24.
_____

25.
_____

26.
_____

27.
_____

28.
_____

29.
_____

30.
_____

31.
_____

32.
_____

33.
_____

34.
_____

35.
_____

23. According to the line graph, between what 2 years was the increase in benefits the greatest?

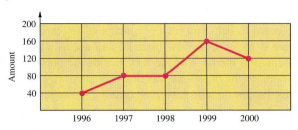

24. Construct a bar graph to represent the following data.

Type of Stock	Number of Stocks
Industrial capital gains	110
Industrial consumer's goods	184
Public utilities	60
Railroads	15
Banks	25
Property liability insurance	16

25. Calculate the mean, median, and mode for the given data.

11, 9, 3, 6, 7, 9, 8, 11, 12, 13, 11, 11, 4, 8, 12

26. If a boat uses 14 gal of gas to go 102 mi, how many gallons would be needed to go 510 mi?

27. The floor of a room that is 12 ft by 18 ft is to be carpeted. If the price of the carpet is $17 per square yd, what will the carpet cost?

28. If you drive 152 mi in $3\frac{1}{6}$ h, what is your average speed?

29. A rectangle has length $8\frac{3}{5}$ cm and width $5\frac{7}{10}$ cm. Find the perimeter.

30. The sides of a square each measure $13\frac{5}{6}$ ft. Find the perimeter of the square.

31. What is $9\frac{1}{2}\%$ of 1,400?

32. 15 is what percent of 7,500?

33. 111 is 60% of what number?

34. Find $\frac{2}{3}$ of $6\frac{1}{2}$.

35. The number of students attending a small college increased 6% since last year. This year there are 2,968 students. How many students attended last year?

POLYNOMIALS

CHAPTER 10 OUTLINE

Section 10.1   Properties of Exponents   page 783

Section 10.2   Introduction to Polynomials   page 790

Section 10.3   Addition and Subtraction of Polynomials   page 797

Section 10.4   Multiplying Polynomials   page 807

Section 10.5   Introduction to Factoring Polynomials   page 817

INTRODUCTION

In the seventeenth century, Sir Isaac Newton came up with a nice way of modeling the way an object moves in a free fall due to gravity.

Throughout history, people have used mathematics to model situations that arise in the world. Ancient civilizations used geometry to model massive construction projects. In modern times, architects use mathematical models in their constructions, business managers use models to make predictions, and the list goes on.

Today, some of the more fascinating uses of models involve attempts to predict the weather, trends in the stock market, and the impact of political actions.

Polynomials are a common tool used in many modeling applications. Polynomials are relatively easy to work with because their behavior is very predictable. In this chapter, you will be introduced to some of the properties and uses of polynomials.

You can find exercises related to Newton's model in Sections 10.2, 10.3, and 10.5. At the end of the chapter, before the chapter summary, you will find a more in-depth activity relating to falling objects.

© 2010 McGraw-Hill Companies

Pretest Chapter 10

This pretest will provide a preview of the types of exercises you will encounter in each section of this chapter. The answers for these exercises can be found in the back of the text. If you are working on your own or ahead of the class, this pretest can help you identify the sections in which you should focus more of your time.

ANSWERS

(a)
1. (b) _____

(a)
2. (b) _____

3. _____

4. _____

5. _____

6. _____

7. _____

8. _____

9. _____

10. _____

11. _____

12. _____

13. _____

14. _____

15. _____

[10.1] **1.** Simplify each expression.

    (a) $x^2 \cdot x^5$ **(b)** $(m^5 \cdot m^2 \cdot n^4)^3$

[10.2] **2.** Classify each polynomial as a monomial, binomial, or trinomial.

    (a) $6x^2 - 7x$ **(b)** $-4x^3 + 5x - 9$

    3. Evaluate $3x^2 + 2x - 7$ when $x = -1$.

[10.3] Add.

    4. $4x^2 - 7x + 5$ and $-2x^2 + 5x - 7$

Subtract.

    5. $-2x^2 + 3x - 1$ from $7x^2 - 8x + 5$

[10.1] Simplify.

    6. $(2x^3y^2)(4x^2y^5)$

[10.4] Multiply.

    7. $3xy(4x^2y^2 - 2xy + 7xy^3)$ **8.** $(3x + 2)(2x - 5)$

    9. $(x + 2y)(x - 2y)$ **10.** $(4m + 5)(4m + 5)$

    11. $(3x - 2y)(x^2 - 4xy + 3y^2)$

[10.5] Completely factor each of the polynomials.

    12. $15c + 35$ **13.** $8q^4 - 20q^3$

    14. $6x^2 - 12x + 24$ **15.** $7c^3d^2 - 21cd + 14cd^3$

10.1 Properties of Exponents

10.1 OBJECTIVES

1. Use the product property of exponents
2. Use the power property of exponents
3. Use the power of a product property of exponents
4. Use the quotient properties of exponents

RECALL We first introduced exponential notation in Section 1.7.

Before we learn about polynomials, we need to learn more about exponents and exponential expressions. There are several properties associated with exponents that make it easy to use these expressions.

Exponent notation indicates repeated multiplication; the exponent tells us how many times a base is to be used as a factor.

Exponent
$$3^5 = \underbrace{3 \cdot 3 \cdot 3 \cdot 3 \cdot 3}_{5 \text{ factors}} = 243$$
Base

Now, we look at the properties of exponents.

Consider the expression $4^2 \cdot 4^3$. Since this represents $\underbrace{(4 \cdot 4)}_{2 \text{ factors}} \cdot \underbrace{(4 \cdot 4 \cdot 4)}_{3 \text{ factors}}$, we see that

$$4^2 \cdot 4^3 = \underbrace{4 \cdot 4 \cdot 4 \cdot 4 \cdot 4}_{5 \text{ factors}} = 4^5$$

This suggests the first property of exponents. It is used when multiplying two values with the same base.

Property: Product Property of Exponents

For any real number a and positive integers m and n,

$$a^m \cdot a^n = a^{m+n}$$

For example,

$$2^5 \cdot 2^7 = 2^{12}$$

OBJECTIVE 1

Example 1 Using the Product Property of Exponents

Simplify each expression.

(a) $8^5 \cdot 8^9 = 8^{5+9} = 8^{14}$

(b) $x^6 \cdot x^2 = x^{6+2} = x^8$

(c) $5y^3 \cdot 2y^4 = 5 \cdot 2 \cdot y^3 \cdot y^4 = 10 \cdot y^{3+4} = 10y^7$

 CHECK YOURSELF 1

Simplify each expression.

(a) $7^3 \cdot 7^6$ **(b)** $z^5 \cdot z^3$ **(c)** $4x^6 \cdot 7x^5$

© 2010 McGraw-Hill Companies

Now consider the following:

$$(x^2)^4 = x^2 \cdot x^2 \cdot x^2 \cdot x^2 = x^8$$

This leads us to our second property for exponents.

Property: Power Property of Exponents

For any real number a and positive integers m and n,

$$(a^m)^n = a^{m \cdot n}$$

For example,

$$(2^3)^2 = 2^{3 \cdot 2} = 2^6$$

The use of this new property is illustrated in Example 2.

OBJECTIVE 2

Example 2   **Using the Power Property of Exponents**

Simplify each expression.

C A U T I O N

Be sure to distinguish between the correct use of the product property and the power property.

$(x^4)^5 = x^{4 \cdot 5} = x^{20}$

but

$x^4 \cdot x^5 = x^{4+5} = x^9$

(a) $(x^4)^5 = x^{4 \cdot 5} = x^{20}$ Multiply the exponents.

(b) $(2^3)^4 = 2^{3 \cdot 4} = 2^{12}$

CHECK YOURSELF 2

Simplify each expression.

(a) $(m^5)^6$ **(b)** $(m^5)(m^6)$

(c) $(3^2)^4$ **(d)** $(3^2)(3^4)$

NOTE Here the base is $3x$.

Suppose we now have a product raised to a power. Consider an expression such as $(3x)^4$. We know that

NOTE We apply the commutative and associative properties of multiplication.

$$(3x)^4 = (3x)(3x)(3x)(3x)$$
$$= (3 \cdot 3 \cdot 3 \cdot 3)(x \cdot x \cdot x \cdot x)$$
$$= 3^4 \cdot x^4 = 81x^4$$

Note that the power, here 4, has been applied to each factor, 3 and x. In general, we have

---

Property: Power of a Product Property of Exponents

For any real numbers a and b and positive integer m,

$$(ab)^m = a^m b^m$$

For example,

$$(3x)^3 = 3^3 \cdot x^3 = 27x^3$$

The use of this property is shown in Example 3.

Example 3   Using the Power of a Product Property of Exponents

Simplify each expression.

NOTE $(2x)^5$ and $2x^5$ are entirely different expressions. For $(2x)^5$, the base is $2x$, so we raise each factor to the fifth power. For $2x^5$, the base is x, and so the exponent applies only to x.

(a) $(2x)^5 = 2^5 \cdot x^5 = 32x^5$

(b) $(3ab)^4 = 3^4 \cdot a^4 \cdot b^4 = 81a^4b^4$

(c) $5(2r)^3 = 5 \cdot 2^3 \cdot r^3 = 40r^3$

 CHECK YOURSELF 3

Simplify each expression.

(a) $(3y)^4$ **(b)** $(2mn)^6$
(c) $3(4x)^2$ **(d)** $5x^3$

We may have to use more than one of our properties in simplifying an expression involving exponents. Consider Example 4.

Example 4   Using the Properties of Exponents

NOTE To help you understand each step of the simplification, we refer to the property being applied.

Simplify each expression.

(a) $(r^4s^3)^3 = (r^4)^3 \cdot (s^3)^3$ Power of a product property of exponents

$= r^{12}s^9$ Power property of exponents

(b) $(3x^2)^2 \cdot (2x^3)^3$

$= 3^2(x^2)^2 \cdot 2^3 \cdot (x^3)^3$ Power of a product property of exponents

$= 9x^4 \cdot 8x^9$ Power property of exponents

$= 72x^{13}$ Multiply the coefficients and apply the product property of exponents.

 CHECK YOURSELF 4

Simplify each expression.

(a) $(m^5n^2)^3$ **(b)** $(2p)^4(4p^2)^2$

The final two properties of exponents are used when dividing. We mention them and provide an example, but you will not use them until you take an algebra class.

First, when dividing two exponential expressions that have the same base, we can rewrite the expression by subtracting the exponents. This is similar to the product property of exponents.

$$\frac{3^5}{3^3} = \frac{3 \cdot 3 \cdot 3 \cdot 3 \cdot 3}{3 \cdot 3 \cdot 3} = \frac{\overset{1}{\cancel{3}} \cdot \overset{1}{\cancel{3}} \cdot \overset{1}{\cancel{3}} \cdot 3 \cdot 3}{\underset{1}{\cancel{3}} \cdot \underset{1}{\cancel{3}} \cdot \underset{1}{\cancel{3}}} = 3 \cdot 3 = 3^2$$

Notice that $3^{5-3} = 3^2$. Do you see that it will always work out like this?

Second, when taking a quotient to a power, we raise both the numerator and the denominator to that power. This is similar to the power property of exponents.

$$\left(\frac{2}{3}\right)^3 = \frac{2}{3} \cdot \frac{2}{3} \cdot \frac{2}{3} = \frac{2^3}{3^3}$$

Property: The Quotient Properties of Exponents

For real numbers a and b, b not zero, and positive integers m and n with $m > n$, we have

$$\frac{b^m}{b^n} = b^{m-n} \qquad \text{and} \qquad \left(\frac{a}{b}\right)^m = \frac{a^m}{b^m}$$

For example,

$$\frac{2^5}{2^2} = 2^3 \qquad \text{and} \qquad \left(\frac{2}{3}\right)^4 = \frac{2^4}{3^4}$$

OBJECTIVE 4

Example 5   Using the Quotient Properties of Exponents

Simplify each expression.

(a) $\dfrac{x^7}{x^3} = x^{7-3} = x^4$

(b) $\left(\dfrac{3}{4}\right)^3 = \dfrac{3^3}{4^3} = \dfrac{27}{64}$

✔ CHECK YOURSELF 5

Simplify each expression.

(a) $\dfrac{7^{12}}{7^9}$ **(b)** $\left(\dfrac{x^2}{y}\right)^5$

The table summarizes the properties of exponents for your convenience.

General Form	Example
1. $a^m a^n = a^{m+n}$	$x^2 \cdot x^3 = x^5$
2. $(a^m)^n = a^{mn}$	$(x^2)^3 = x^6$
3. $(ab)^m = a^m b^m$	$(4x)^3 = 4^3 x^3 = 64x^3$
4. $\dfrac{a^m}{a^n} = a^{m-n}$ $(a \neq 0)$	$\dfrac{x^7}{x^2} = x^5$
5. $\left(\dfrac{a}{b}\right)^m = \dfrac{a^m}{b^m}$ $(b \neq 0)$	$\left(\dfrac{x}{y}\right)^3 = \dfrac{x^3}{y^3}$

READING YOUR TEXT

The following fill-in-the-blank exercises are designed to assure that you understand the key vocabulary used in this section. Each sentence comes directly from the section. You will find the correct answers in Appendix C.

Section 10.1

(a) Three is the _____ in the expression 3^5.

(b) The _____ tells us how many times the base is to be used as a factor.

(c) To simplify $(2^3)^2$, _____ the exponents.

(d) To simplify $\dfrac{x^4}{x^2}$, _____ the exponents.

CHECK YOURSELF ANSWERS

1. (a) 7^9; **(b)** z^8; **(c)** $28x^{11}$ **2. (a)** m^{30}; **(b)** m^{11}; **(c)** 3^8; **(d)** 3^6

3. (a) $81y^4$; **(b)** $64m^6n^6$; **(c)** $48x^2$; **(d)** $5x^3$ **4. (a)** $m^{15}n^6$; **(b)** $256p^8$

5. (a) 7^3 or 343; **(b)** $\dfrac{x^{10}}{y^5}$

10.1 Exercises

Simplify each expression.

1. $x^2 \cdot x^3$

2. $a^5 \cdot a^3$

3. $5^7 \cdot 5^8$

4. $8^4 \cdot 8^6$

5. $2y^4 \cdot 6y^3$

6. $5m^9 \cdot 3m^2$

7. $(x^2)^3$

8. $(a^5)^3$

9. $(2^4)^2$

10. $(3^3)^2$

11. $(5^3)^5$

12. $(7^2)^4$

13. $(3x)^3$

14. $(4m)^2$

15. $(2xy)^4$

16. $(5pq)^3$

17. $5(3ab)^3$

18. $4(2rs)^4$

19. $(a^8b^6)^2$

20. $(p^3q^4)^2$

21. $(4x^2y)^3$

22. $(4m^4n^4)^2$

23. $\dfrac{6^8}{6^5}$

24. $\dfrac{x^7}{x^3}$

25. $\left(\dfrac{m}{n^2}\right)^4$

26. $\left(\dfrac{2}{3}\right)^2$

Solve each problem.

27. Write x^{12} as a power of x^2.

28. Write y^{15} as a power of y^3.

29. Write a^{16} as a power of a^2.

30. Write m^{20} as a power of m^5.

31. Write each as a power of 8. (Remember that $8 = 2^3$.)

$2^{12}, 2^{18}, (2^5)^3, (2^7)^6$

32. Write each as a power of 9.

$3^8, 3^{14}, (3^5)^8, (3^4)^7$

33. Write an explanation of why $(x^3)(x^4)$ is *not* x^{12}.

34. Your algebra study partners are confused. "Why isn't $x^2 \cdot x^3 = 2x^5$?" they ask you. Write an explanation that will convince them.

Answers

1. x^5　**3.** 5^{15}　**5.** $12y^7$　**7.** x^6　**9.** 2^8　**11.** 5^{15}
13. $27x^3$　**15.** $16x^4y^4$　**17.** $135a^3b^3$　**19.** $a^{16}b^{12}$　**21.** $64x^6y^3$
23. 6^3 or 216　**25.** $\dfrac{m^4}{n^8}$　**27.** $(x^2)^6$　**29.** $(a^2)^8$　**31.** $8^4, 8^6, 8^5, 8^{14}$

33.

ANSWERS

27. _____

28. _____

29. _____

30. _____

31. _____

32. _____

33. _____

34. _____

10.2 Introduction to Polynomials

10.2 OBJECTIVES

1. Identify polynomials, terms, and coefficients
2. Identify types of polynomials
3. Evaluate a polynomial

The rest of this chapter deals with the most common kind of algebraic expression, a *polynomial*. To define a polynomial, we should first define the word *term*.

Definition: Term

A **term** is a number or the product of a number and one or more variables.

For example, x^5, $3x$, $-4xy^2$, and 8 are terms.

Definition: Polynomial

RECALL The whole numbers are 0, 1, 2, 3, . . . , and so on.

A **polynomial** is a single term, or the sum of two or more terms, in which the only allowable exponents of any variables are whole numbers.

NOTE Now consider a nonpolynomial term such as $\dfrac{4}{x}$. As you will see in a later algebra class, when x appears in the denominator, we cannot write this as a product and still have whole-number exponents. Thus, an expression containing a term such as $\dfrac{4}{x}$ is not a polynomial.

Consider the expression $x^3 - 2x^2 - 7$. Recall that subtraction may always be rewritten as addition of an opposite. So we can rewrite the expression as $x^3 + (-2x^2) + (-7)$. This expression is then a sum of three terms. The exponents are whole numbers, so the expression is a polynomial. In general, we can say that the terms of a polynomial are separated by addition signs.

Definition: Numerical Coefficient

In each term of a polynomial, the number in front of the variable is called the **numerical coefficient,** or more simply the **coefficient,** of that term.

OBJECTIVE 1

Example 1 Identifying Polynomials, Terms, and Coefficients

NOTE We can rewrite x as $1 \cdot x$. Thus, the coefficient of x is 1.

(a) $x + 3$ is a polynomial. The terms are x and 3. The coefficients are 1 and 3.

(b) $3x^2 - 2x + 5$, or $3x^2 + (-2x) + 5$, is also a polynomial. Its terms are $3x^2$, $-2x$, and 5. The coefficients are 3, -2, and 5.

(c) $5x^3 + 2 - \dfrac{3}{x}$ is *not* a polynomial because of the presence of x in the denominator.

CHECK YOURSELF 1

Which of the given expressions are polynomials?

(a) $5x^2$ **(b)** $3y^3 - 2y + \dfrac{5}{y}$ **(c)** $4x^2 - 2x + 3$

NOTE The prefix *mono-* means one. The prefix *bi-* means two. The prefix *tri-* means three. There are no special names for polynomials with four or more terms.

Definition: Monomial, Binomial, and Trinomial

A polynomial with exactly one term is called a **monomial.**
A polynomial with exactly two terms is called a **binomial.**
A polynomial with exactly three terms is called a **trinomial.**

OBJECTIVE 2   **Example 2   Identifying Types of Polynomials**

(a) $3x^2y$ is a monomial. It has one term.
(b) $2x^3 + 5x$ is a binomial. It has two terms, $2x^3$ and $5x$.
(c) $5x^2 - 4x + 3$, or $5x^2 + (-4x) + 3$, is a trinomial. Its three terms are $5x^2$, $-4x$, and 3.

 CHECK YOURSELF 2

Classify each of these as monomial, binomial, or trinomial.

(a) $5x^4 - 2x^3$ **(b)** $4x^7$ **(c)** $2x^2 + 5x - 3$

The value of a polynomial depends on the value given to the variable.

OBJECTIVE 3   **Example 3   Evaluating Polynomials**

Given the polynomial

$3x^3 - 2x^2 - 4x + 1$

Find the value of the polynomial when $x = 2$.

Substituting 2 for x, we have

RECALL You must apply the order of operations properly, as we learned in Section 1.8.

$3(2)^3 - 2(2)^2 - 4(2) + 1$
$= 3(8) - 2(4) - 4(2) + 1$
$= 24 - 8 - 8 + 1$
$= 9$

CHECK YOURSELF 3

Find the value of the polynomial

$4x^3 - 3x^2 + 2x - 1 \quad when \quad x = 3$

READING YOUR TEXT

The following fill-in-the-blank exercises are designed to assure that you understand the key vocabulary used in this section. Each sentence comes directly from the section. You will find the correct answers in Appendix C.

Section 10.2

(a)   A term is a number or the product of a number and one or more _____.

(b)   The _____ in front of the variable(s) of a term is called the coefficient.

(c)   The _____ of a polynomial are separated by addition signs.

(d)   A polynomial with exactly three terms is called a _____.

CHECK YOURSELF ANSWERS

1. (a) and **(c)** are polynomials.   **2. (a)** binomial; **(b)** monomial; **(c)** trinomial
3. 86

Name _____

Section _____ Date _____

ANSWERS

Which of the expressions are polynomials?

1. $7x^3$

2. $5x^3 - \dfrac{3}{x}$

3. $4x^4y^2 - 3x^3y$

4. 7

5. -7

6. $4x^3 + x$

7. $\dfrac{3 + x}{x^2}$

8. $5a^2 - 2a + 7$

For each of the polynomials, list the terms and the coefficients.

9. $2x^2 - 3x$

10. $5x^3 + x$

11. $4x^3 - 3x + 2$

12. $7x^2$

Classify each as monomial, binomial, trinomial, or polynomial where possible.

13. $7x^3 - 3x^2$

14. $4x^7$

15. $7y^2 + 4y + 5$

16. $2x^2 + 3xy + y^2$

17. $2x^4 - 3x^2 + 5x - 2$

18. $x^4 + \dfrac{5}{x} + 7$

19. $6y^8$

20. $4x^4 - 2x^2 + 5x - 7$

21. $x^5 - \dfrac{3}{x^2}$

22. $4x^2 - 9$

Find the values of each of the polynomials for the given values of the variable.

23. $6x + 1$, $x = 1$ and $x = -1$ **ALEKS**

24. $5x - 5$, $x = 2$ and $x = -2$ **ALEKS**

25. $x^3 - 2x$, $x = 2$ and $x = -2$ **ALEKS**

26. $3x^2 + 7$, $x = 3$ and $x = -3$ **ALEKS**

1. _____

2. _____

3. _____

4. _____

5. _____

6. _____

7. _____

8. _____

9. _____

10. _____

11. _____

12. _____

13. _____

14. _____

15. _____

16. _____

17. _____

18. _____

19. _____

20. _____

21. _____

22. _____

23. _____   24. _____

25. _____   26. _____

27. $3x^2 + 4x - 2$, $x = 4$ and $x = -4$
ALEKS

28. $2x^2 - 5x + 1$, $x = 2$ and $x = -2$
ALEKS

29. $-x^2 - 2x + 3$, $x = 1$ and $x = -3$
ALEKS

30. $-x^2 - 5x - 6$, $x = -3$ and $x = -2$
ALEKS

Indicate whether each of the statements is always true, sometimes true, or never true.

31. A monomial is a polynomial.

32. A binomial is a trinomial.

33. A polynomial has four or more terms.

34. A binomial must have two coefficients.

Capital italic letters such as P or Q are often used to name polynomials. For example, we might write $P(x) = 3x^3 - 5x^2 + 2$ in which $P(x)$ is read "P of x." The notation permits a convenient shorthand. We write $P(2)$, read "P of 2," to indicate the value of the polynomial when $x = 2$. Here

$$P(2) = 3(2)^3 - 5(2)^2 + 2$$
$$= 3 \cdot 8 - 5 \cdot 4 + 2$$
$$= 6$$

Use this to complete exercises 35–48. If $P(x) = x^3 - 2x^2 + 5$ and $Q(x) = 2x^2 + 3$, find:

35. $P(1)$

36. $P(-1)$

37. $Q(2)$

38. $Q(-2)$

39. $P(3)$

40. $Q(-3)$

41. $P(0)$

42. $Q(0)$

43. $P(2) + Q(-1)$

44. $P(-2) + Q(3)$

45. $P(3) - Q(-3) \div Q(0)$

46. $Q(-2) \div Q(2) \cdot P(0)$

47. $\left| Q(4) \right| - \left| P(4) \right|$

48. $\dfrac{P(-1) + Q(0)}{P(0)}$

Solve the applications.

49. **Business and Finance** The cost, in dollars, of typing a term paper is given as 3 times the number of pages plus 20. Use y as the number of pages to be typed and write a polynomial to describe this cost. Find the cost of typing a 50-page paper.

50. **Business and Finance** The cost, in dollars, of making suits is described as 20 times the number of suits plus 150. Use s as the number of suits and write a polynomial to describe this cost. Find the cost of making seven suits.

51. **Business and Finance** The revenue, in dollars, when x pairs of shoes are sold is given by $3x^2 - 95$. Find the revenue when 12 pairs of shoes are sold. What is the average (mean) revenue per pair of shoes?

52. **Business and Finance** The cost in dollars of manufacturing w wing nuts is given by the expression $0.07w + 13.3$. Find the cost when 375 wing nuts are made. What is the average (mean) cost to manufacture one wing nut?

53. The velocity, v, based on gravity, of an object t seconds after it begins to fall, is given by $v = 32t$. How fast is a ball falling (in feet per second) 3 seconds after it is dropped? $\left(\begin{smallmatrix}10\end{smallmatrix}\right)$

54. Use the equation in exercise 53 to determine how fast a ball is falling after 5 seconds. $\left(\begin{smallmatrix}10\end{smallmatrix}\right)$

55. The height of an object dropped from an initial height of 1,000 ft is given by $h = -16t^2 + 1,000$. Find the height of a ball after 3 seconds. $\left(\begin{smallmatrix}10\end{smallmatrix}\right)$

56. What is the height of the ball in exercise 55 after 5 seconds?

ANSWERS

49. _____

50. _____

51. _____

52. _____

53. _____

54. _____

55. _____

56. _____

Answers

1. Polynomial   **3.** Polynomial   **5.** Polynomial   **7.** Not a polynomial

9. $2x^2, -3x; 2, -3$ **11.** $4x^3, -3x, 2; 4, -3, 2$ **13.** Binomial

15. Trinomial   **17.** Polynomial   **19.** Monomial   **21.** Not a polynomial

23. $7, -5$ **25.** $4, -4$ **27.** $62, 30$ **29.** $0, 0$ **31.** Always

33. Sometimes   **35.** 4   **37.** 11   **39.** 14   **41.** 5

43. 10   **45.** 7   **47.** -2 **49.** $3y + 20$, \$170   **51.** \$337, \$28.08

53. 96 ft/s   **55.** 856 ft

10.3 Addition and Subtraction of Polynomials

10.3 OBJECTIVES

1. Add two polynomials
2. Subtract two polynomials

Addition is always a matter of combining like quantities (two apples plus three apples, four books plus five books, and so on). If you keep that basic idea in mind, adding polynomials is easy. It is just a matter of combining like terms. Suppose that you want to add

NOTE To review combining like terms, see Section 2.8.

$$5x^2 + 3x + 4 \qquad \text{and} \qquad 4x^2 + 5x - 6$$

Parentheses are sometimes used in adding, so for the sum of these polynomials, we can write

NOTE The plus sign between the parentheses indicates the addition.

$$(5x^2 + 3x + 4) + (4x^2 + 5x - 6)$$

Now what about the parentheses? You can use this rule.

Property: Removing Signs of Grouping

If a plus sign (+) or nothing at all appears in front of parentheses, just remove the parentheses. No other changes are necessary.

Now we will return to the addition.

NOTE Just remove the parentheses. No other changes are necessary.

$$(5x^2 + 3x + 4) + (4x^2 + 5x - 6)$$
$$= 5x^2 + 3x + 4 + 4x^2 + 5x - 6$$

Note that the addition sign in front of the parentheses remains.

Like terms   Like terms   Like terms

NOTE Use the associative and commutative properties to reorder and regroup.

Collect like terms. (*Remember:* Like terms have the same variables raised to the same power.)

$$= (5x^2 + 4x^2) + (3x + 5x) + (4 - 6)$$

Combine like terms for the result:

NOTE Here we use the distributive property. For example,

$$5x^2 + 4x^2 = (5 + 4)x^2$$
$$= 9x^2$$

$$= 9x^2 + 8x - 2$$

As should be clear, much of this work can be done mentally. You can then write the sum directly by locating like terms and combining. Example 1 illustrates this approach.

OBJECTIVE 1

NOTE We call this the **horizontal method** because the entire problem is written on one line.
$3 + 4 = 7$ is the horizontal method.

$$\begin{array}{r} 3 \\ +\,4 \\ \hline 7 \end{array}$$

is the vertical method.

Example 1   Combining Like Terms

Add $3x - 5$ and $2x + 3$.
    Write the sum.

$$(3x - 5) + (2x + 3)$$
$$= 3x - 5 + 2x + 3 = 5x - 2$$

Like terms   Like terms

CHECK YOURSELF 1 ⎯⎯⎯⎯⎯⎯⎯⎯⎯⎯⎯⎯⎯⎯⎯⎯⎯⎯⎯⎯⎯

Add $6x^2 + 2x$ and $4x^2 - 7x$.

The same technique is used to find the sum of two trinomials.

> **Example 2   Adding Polynomials Using the Horizontal Method**

Add $4a^2 - 7a + 5$ and $3a^2 + 3a - 4$.
  Write the sum.

RECALL Only the like terms are combined in the sum.

$(4a^2 - 7a + 5) + (3a^2 + 3a - 4)$

$= 4a^2 - 7a + 5 + 3a^2 + 3a - 4 = 7a^2 - 4a + 1$

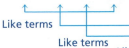

Like terms
Like terms
Like terms

 CHECK YOURSELF 2 ⎯⎯⎯⎯⎯⎯⎯⎯⎯⎯⎯⎯⎯⎯⎯⎯⎯⎯⎯⎯⎯

Add $5y^2 - 3y + 7$ and $3y^2 - 5y - 7$.

> **Example 3   Adding Polynomials Using the Horizontal Method**

Add $2x^2 + 7x$ and $4x - 6$.
  Write the sum.

$(2x^2 + 7x) + (4x - 6)$

$= 2x^2 + \underbrace{7x + 4x} - 6$

These are the only like terms;
$2x^2$ and -6 cannot be combined.

$= 2x^2 + 11x - 6$

 CHECK YOURSELF 3 ⎯⎯⎯⎯⎯⎯⎯⎯⎯⎯⎯⎯⎯⎯⎯⎯⎯⎯⎯⎯⎯

Add $5m^2 + 8$ and $8m^2 - 3m$.

It is sometimes helpful to rewrite polynomials in descending-exponent form before adding. Descending-exponent form requires that the degree of each term is greater than the degree of any term that follows.

> **Example 4   Adding Polynomials Using the Horizontal Method**

Add $3x - 2x^2 + 7$ and $5 + 4x^2 - 3x$.

Write the polynomials in descending-exponent form and then add.

$(-2x^2 + 3x + 7) + (4x^2 - 3x + 5)$

$= 2x^2 + 12$

CHECK YOURSELF 4 _____

Add $8 - 5x^2 + 4x$ and $7x - 8 + 8x^2$.

Subtracting polynomials requires another rule for removing signs of grouping.

> **Property:** Removing Signs of Grouping
>
> If a minus sign (−) appears in front of a set of parentheses, the parentheses can be removed by changing the sign of each term inside the parentheses.

The use of this rule is illustrated in Example 5.

OBJECTIVE 2   **Example 5   Removing Parentheses**

Remove the parentheses.

NOTE We use the distributive property, because

$-(2x + 3y) = (-1)(2x + 3y)$
$= -2x - 3y$

(a) $-(2x + 3y) = -2x - 3y$ Change each sign to remove the parentheses.

(b) $m - (5n - 3p) = m \underbrace{- 5n + 3p}_{\text{Signs change.}}$ Note that the subtraction sign in front of the parentheses is dropped.

(c) $2x - (-3y + z) = 2x \underbrace{+ 3y - z}_{\text{Signs change.}}$

CHECK YOURSELF 5 _____

Remove the parentheses.

(a) $-(3m + 5n)$ **(b)** $-(5w - 7z)$
(c) $3r - (2s - 5t)$ **(d)** $5a - (-3b - 2c)$

Subtracting polynomials is now a matter of using the property above to remove the parentheses and then combining the like terms. Consider Example 6.

Example 6   Subtracting Polynomials Using the Horizontal Method

(a) Subtract $5x - 3$ from $8x + 2$.

Write

NOTE The expression following "from" is written first in the problem.

$(8x + 2) - (5x - 3)$

$= 8x + 2 \underbrace{- 5x + 3}_{\text{Signs change.}}$

$= 3x + 5$

(b) Subtract $4x^2 - 8x + 3$ from $8x^2 + 5x - 3$.

Write

$(8x^2 + 5x - 3) - (4x^2 - 8x + 3)$

$= 8x^2 + 5x - 3 \underbrace{- 4x^2 + 8x - 3}_{\text{Signs change.}}$

$= 4x^2 + 13x - 6$

CHECK YOURSELF 6

(a) Subtract $7x + 3$ from $10x - 7$.

(b) Subtract $5x^2 - 3x + 2$ from $8x^2 - 3x - 6$.

Again, writing all polynomials in descending-exponent form will make locating and combining like terms much easier. Look at Example 7.

Example 7　Subtracting Polynomials Using the Horizontal Method

(a) Subtract $4x^2 - 3x^3 + 5x$ from $8x^3 - 7x + 2x^2$.

Write

$(8x^3 + 2x^2 - 7x) - (-3x^3 + 4x^2 + 5x)$

$= 8x^3 + 2x^2 - 7x \underbrace{+ 3x^3 - 4x^2 - 5x}_{\text{Signs change.}}$

$= 11x^3 - 2x^2 - 12x$

(b) Subtract $8x - 5$ from $-5x + 3x^2$.

Write

$(3x^2 - 5x) - (8x - 5)$

$= 3x^2 \underbrace{- 5x - 8x}_{} + 5$

↖ Only the like terms can be combined.

$= 3x^2 - 13x + 5$

CHECK YOURSELF 7

(a) Subtract $7x - 3x^2 + 5$ from $5 - 3x + 4x^2$.

(b) Subtract $3a - 2$ from $5a + 4a^2$.

If you think back to addition and subtraction in arithmetic, you will remember that the work was arranged vertically. That is, the numbers being added or subtracted were placed under one another so that each column represented the same place value. This meant that in adding or subtracting columns, you were always dealing with "like quantities."

It is also possible to use a vertical method for adding or subtracting polynomials. First, rewrite the polynomials in descending-exponent form and then arrange them one under another so that each column contains like terms. Then add or subtract in each column.

Example 8   Adding Using the Vertical Method

Add $2x^2 - 5x$, $3x^2 + 2$, and $6x - 3$.

Like terms

$$
\begin{array}{r}
2x^2 - 5x \\
3x^2 \quad\quad + 2 \\
6x - 3 \\
\hline
5x^2 + \;\; x - 1
\end{array}
$$

CHECK YOURSELF 8

Add $3x^2 + 5$, $x^2 - 4x$, and $6x + 7$.

Example 9 illustrates subtraction by the vertical method.

Example 9   Subtracting Using the Vertical Method

(a) Subtract $5x - 3$ from $8x - 7$.

Write

$$
\begin{array}{r}
8x - 7 \\
(-)\;\; 5x - 3 \\
\hline
3x - 4
\end{array}
$$

To subtract, change each
sign of $5x - 3$ to get
$-5x + 3$ and then add.

$$
\begin{array}{r}
8x - 7 \\
(+)\; -5x + 3 \\
\hline
3x - 4
\end{array}
$$

(b) Subtract $5x^2 - 3x + 4$ from $8x^2 + 5x - 3$.

Write

$$
\begin{array}{r}
8x^2 + 5x - 3 \\
(-)\;\; 5x^2 - 3x + 4 \\
\hline
3x^2 + 8x - 7
\end{array}
$$

To subtract, change each
sign of $5x^2 - 3x + 4$ to get
$-5x^2 + 3x - 4$ and then add.

$$
\begin{array}{r}
8x^2 + 5x - 3 \\
(+)\; -5x^2 + 3x - 4 \\
\hline
3x^2 + 8x - 7
\end{array}
$$

Subtracting using the vertical method takes some practice. Take time to study the method carefully.

CHECK YOURSELF 9

Subtract using the vertical method.

(a) $4x^2 - 3x$ from $8x^2 + 2x$ **(b)** $8x^2 + 4x - 3$ from $9x^2 - 5x + 7$

In practice, we often encounter the addition or subtraction of polynomials without directions to add or subtract. Although such expressions may appear to be complicated, we just keep in mind the following: (1) remove parentheses according to our earlier properties and (2) combine like terms.

> **Example 10** **Simplifying Expressions Involving Addition or Subtraction**
>
> Perform the indicated operations.
>
> **(a)** $(x^2 - 6x + 5) + (3x^2 - 2x - 4)$
> $= x^2 - 6x + 5 + 3x^2 - 2x - 4$ Use the addition property to remove parentheses.
> $= 4x^2 - 8x + 1$ Collect and combine like terms.
>
> **(b)** $(2x^3 - x^2 + 7) - (3x^3 + 5x - 2)$
> $= 2x^3 - x^2 + 7 - 3x^3 - 5x + 2$ The second set of parentheses are removed using the subtraction property.
> $= -x^3 - x^2 - 5x + 9$ Collect and combine like terms.

 CHECK YOURSELF 10 _____

Perform the indicated operations.

(a) $(5x^2 + 3x - 7) + (x^2 - 6x + 7)$
(b) $(x^3 - 2x^2 + 4x - 5) - (-2x^3 + 3x^2 - 2x + 3)$

READING YOUR TEXT

The following fill-in-the-blank exercises are designed to assure that you understand the key vocabulary used in this section. Each sentence comes directly from the section. You will find the correct answers in Appendix C.

Section 10.3

(a) If a _____ sign appears in front of the parentheses, simply remove the parentheses and add.

(b) It is sometimes helpful to rewrite polynomials in _____ form before adding.

(c) Removing parentheses when subtracting polynomials requires the _____ property.

(d) When adding and subtracting polynomials, you can only combine _____ terms.

CHECK YOURSELF ANSWERS _____

1. $10x^2 - 5x$ **2.** $8y^2 - 8y$ **3.** $13m^2 - 3m + 8$ **4.** $3x^2 + 11x$
5. **(a)** $-3m - 5n$; **(b)** $-5w + 7z$; **(c)** $3r - 2s + 5t$; **(d)** $5a + 3b + 2c$
6. **(a)** $3x - 10$; **(b)** $3x^2 - 8$ **7.** **(a)** $7x^2 - 10x$; **(b)** $4a^2 + 2a + 2$
8. $4x^2 + 2x + 12$ **9.** **(a)** $4x^2 + 5x$; **(b)** $x^2 - 9x + 10$
10. **(a)** $6x^2 - 3x$; **(b)** $3x^3 - 5x^2 + 6x - 8$

10.3 Exercises

Add.

1. $6a - 5$ and $3a + 9$

2. $9x + 3$ and $3x - 4$

3. $8b^2 - 11b$ and $5b^2 - 7b$

4. $2m^2 + 3m$ and $6m^2 - 8m$

5. $3x^2 - 2x$ and $-5x^2 + 2x$

6. $3p^2 + 5p$ and $-7p^2 - 5p$

7. $2x^2 + 5x - 3$ and $3x^2 - 7x + 4$

8. $4d^2 - 8d + 7$ and $5d^2 - 6d - 9$

9. $2b^2 + 8$ and $5b + 8$

10. $4x - 3$ and $3x^2 - 9x$

11. $8y^3 - 5y^2$ and $5y^2 - 2y$

12. $9x^4 - 2x^2$ and $2x^2 + 3$

13. $2a^2 - 4a^3$ and $3a^3 + 2a^2$

14. $9m^3 - 2m$ and $-6m - 4m^3$

15. $4x^2 - 2 + 7x$ and $5 - 8x - 6x^2$

16. $5b^3 - 8b + 2b^2$ and $3b^2 - 7b^3 + 5b$

Remove the parentheses in each of the expressions and simplify when possible.

17. $-(2a + 3b)$

18. $-(7x - 4y)$

19. $5a - (2b - 3c)$

20. $7x - (4y + 3z)$

21. $9r - (3r + 5s)$

22. $10m - (3m - 2n)$

23. $5p - (-3p + 2q)$

24. $8d - (-7c - 2d)$

25.

26.

27.

28.

29.

30.

31.

32.

33.

34.

35.

36.

37.

38.

39.

40.

41.

42.

43.

44.

45.

46.

47.

48.

49.

50.

Subtract.

25. $x + 4$ from $2x - 3$

26. $x - 2$ from $3x + 5$

27. $3m^2 - 2m$ from $4m^2 - 5m$

28. $9a^2 - 5a$ from $11a^2 - 10a$

29. $6y^2 + 5y$ from $4y^2 + 5y$

30. $9n^2 - 4n$ from $7n^2 - 4n$

31. $x^2 - 4x - 3$ from $3x^2 - 5x - 2$

32. $3x^2 - 2x + 4$ from $5x^2 - 8x - 3$

33. $3a + 7$ from $8a^2 - 9a$

34. $3x^3 + x^2$ from $4x^3 - 5x$

35. $4b^2 - 3b$ from $5b - 2b^2$

36. $7y - 3y^2$ from $3y^2 - 2y$

37. $x^2 - 5 - 8x$ from $3x^2 - 8x + 7$

38. $4x - 2x^2 + 4x^3$ from $4x^3 + x - 3x^2$

Perform the indicated operations.

39. Subtract $3b + 2$ from the sum of $4b - 2$ and $5b + 3$.

40. Subtract $5m - 7$ from the sum of $2m - 8$ and $9m - 2$.

41. Subtract $3x^2 + 2x - 1$ from the sum of $x^2 + 5x - 2$ and $2x^2 + 7x - 8$.

42. Subtract $4x^2 - 5x - 3$ from the sum of $x^2 - 3x - 7$ and $2x^2 - 2x + 9$.

43. Subtract $2x^2 - 3x$ from the sum of $4x^2 - 5$ and $2x - 7$.

44. Subtract $5a^2 - 3a$ from the sum of $3a - 3$ and $5a^2 + 5$.

45. Subtract the sum of $3y^2 - 3y$ and $5y^2 + 3y$ from $2y^2 - 8y$.

46. Subtract the sum of $7r^3 - 4r^2$ and $-3r^3 + 4r^2$ from $2r^3 + 3r^2$.

Add, using the vertical method.

47. $2w^2 + 7$, $3w - 5$, and $4w^2 - 5w$

48. $3x^2 - 4x - 2$, $6x - 3$, and $2x^2 + 8$

49. $3x^2 + 3x - 4$, $4x^2 - 3x - 3$, and $2x^2 - x + 7$

50. $5x^2 + 2x - 4$, $x^2 - 2x - 3$, and $2x^2 - 4x - 3$

Subtract, using the vertical method.

51. $3a^2 - 2a$ from $5a^2 + 3a$

52. $6r^3 + 4r^2$ from $4r^3 - 2r^2$

53. $5x^2 - 6x + 7$ from $8x^2 - 5x + 7$

54. $8x^2 - 4x + 2$ from $9x^2 - 8x + 6$

55. $5x^2 - 3x$ from $8x^2 - 9$

56. $7x^2 + 6x$ from $9x^2 - 3$

Perform the indicated operations.

57. $(5x - 7) + (3x + 2)$

58. $(6x + 1) - (3x - 5)$

59. $(x^2 - 3x - 2) - (3x^2 - x + 5)$

60. $(4x^2 - x + 6) + (2x^2 - x - 9)$

61. $(3x^3 - 7x + 5) + (x^3 + x^2 - x)$

62. $(2x^3 + x^2 - 4) - (4x^3 - 2x^2 + x)$

63. $[(9x^2 - 3x + 5) - (3x^2 + 2x - 1)] - (x^2 - 2x - 3)$ **ALEKS**

64. $[(5x^2 + 2x - 3) - (-2x^2 + x - 2)] - (2x^2 + 3x - 5)$ **ALEKS**

Find values for a, b, c, and d so that the equations are true.

65. $3ax^4 - 5x^3 + x^2 - cx + 2 = 9x^4 - bx^3 + x^2 - 2d$

66. $(4ax^3 - 3bx^2 - 10) - 3(x^3 + 4x^2 - cx - d) = x^2 - 6x + 8$

67. Geometry A rectangle has sides of $8x + 9$ and $6x - 7$. Find the polynomial that represents its perimeter.

$6x - 7$

$8x + 9$

68. Geometry A triangle has sides $3x + 7$, $4x - 9$, and $5x + 6$. Find the polynomial that represents its perimeter.

$5x + 6$ $3x + 7$

$4x - 9$

ANSWERS

51. _____

52. _____

53. _____

54. _____

55. _____

56. _____

57. _____

58. _____

59. _____

60. _____

61. _____

62. _____

63. _____

64. _____

65. _____

66. _____

67. _____

68. _____

69. Business and Finance The cost of producing x units of an item is $C = 150 + 25x$. The revenue for selling x units is $R = 90x - x^2$. The profit is given by the revenue minus the cost. Find the polynomial that represents the profit.

70. Business and Finance The revenue for selling y units is $R = 3y^2 - 2y + 5$ and the cost of producing y units is $C = y^2 + y - 3$. Find the polynomial that represents the profit.

In the exercises of Section 10.2, you learned that the height of an object, h, with an initial height of 1,000 ft at time t is given by

$$h = -16t^2 + 1,000$$

If the object has an initial velocity of 30 ft/s upward, then you need to add the polynomial $30t$ to the height polynomial to determine the height after t seconds.

71. Find the height of an object 3 seconds after it is thrown upward from an initial height of 1,000 ft with an initial velocity of 30 ft/s.

72. Find the height of the ball in exercise 71 after 8 seconds.

You need to subtract the polynomial $30t$ from the height polynomial if the initial velocity is 30 ft/s downward.

73. Find the height of an object 3 seconds after it is thrown downward from an initial height of 1,000 ft with an initial velocity of 30 ft/s.

74. Find the height of the ball in exercise 73 after 8 seconds.

Answers

1. $9a + 4$ **3.** $13b^2 - 18b$ **5.** $-2x^2$ **7.** $5x^2 - 2x + 1$
9. $2b^2 + 5b + 16$ **11.** $8y^3 - 2y$ **13.** $-a^3 + 4a^2$ **15.** $-2x^2 - x + 3$
17. $-2a - 3b$ **19.** $5a - 2b + 3c$ **21.** $6r - 5s$ **23.** $8p - 2q$
25. $x - 7$ **27.** $m^2 - 3m$ **29.** $-2y^2$ **31.** $2x^2 - x + 1$
33. $8a^2 - 12a - 7$ **35.** $-6b^2 + 8b$ **37.** $2x^2 + 12$ **39.** $6b - 1$
41. $10x - 9$ **43.** $2x^2 + 5x - 12$ **45.** $-6y^2 - 8y$ **47.** $6w^2 - 2w + 2$
49. $9x^2 - x$ **51.** $2a^2 + 5a$ **53.** $3x^2 + x$ **55.** $3x^2 + 3x - 9$
57. $8x - 5$ **59.** $-2x^2 - 2x - 7$ **61.** $4x^3 + x^2 - 8x + 5$
63. $5x^2 - 3x + 9$ **65.** $a = 3, b = 5, c = 0, d = -1$ **67.** $28x + 4$
69. $-x^2 + 65x - 150$ **71.** 946 ft **73.** 766 ft

10.4 Multiplying Polynomials

10.4 OBJECTIVES

1. Find the product of a monomial and a polynomial
2. Find the product of two binomials
3. Find the product of two polynomials

We are ready to combine what we learned about the properties of exponents in Section 10.1 with our work on polynomials in Sections 10.2 and 10.3. We begin by finding the product of two monomials.

Step by Step: To Find the Product of Two Monomials

RECALL The product property of exponents:

$a^m \cdot a^n = a^{m+n}$

Step 1 Multiply the coefficients.
Step 2 Use the product property of exponents to combine the variables.

OBJECTIVE 1

Example 1 Multiplying Monomials

Multiply $3x^2y$ and $2x^3y^5$.
   Write

NOTE We use the commutative and associative properties to rewrite the problem.

$(3x^2y)(2x^3y^5)$

$= \underbrace{(3 \cdot 2)}\underbrace{(x^2 \cdot x^3)}\underbrace{(y \cdot y^5)}$

Multiply   Add the exponents.
the coefficients.

$= 6x^5y^6$

✔ CHECK YOURSELF 1

Multiply.

(a) $(5a^2b)(3a^2b^4)$ **(b)** $(-3xy)(4x^3y^5)$

NOTE You might want to review Section 1.5 before going on.

Our next task is to find the product of a monomial and a polynomial. Here we use the distributive property, which we introduced in Section 1.5. That property leads us to this rule for multiplication.

Property: To Multiply a Polynomial by a Monomial

NOTE Distributive property:

$a(b + c) = ab + ac$

Use the distributive property to multiply each term of the polynomial by the monomial.

Example 2 **Multiplying a Monomial and a Binomial**

(a) Multiply $2x + 3$ by x.

Write

$x(2x + 3)$

$= x \cdot 2x + x \cdot 3$

$= 2x^2 + 3x$

Multiply x by $2x$ and then by 3, the terms of the polynomial. That is, "distribute" the multiplication over the sum.

(b) Multiply $2a^3 + 4a$ by $3a^2$.

Write

$3a^2(2a^3 + 4a)$

$= 3a^2 \cdot 2a^3 + 3a^2 \cdot 4a = 6a^5 + 12a^3$

CHECK YOURSELF 2

Multiply.

(a) $2y(y^2 + 3y)$ **(b)** $3w^2(2w^3 + 5w)$

The patterns of Example 2 extend to *any* number of terms.

Example 3 **Multiplying a Monomial and a Polynomial**

Multiply the following.

(a) $3x(4x^3 + 5x^2 + 2)$

$= 3x \cdot 4x^3 + 3x \cdot 5x^2 + 3x \cdot 2 = 12x^4 + 15x^3 + 6x$

(b) $5y^2(2y^3 - 4)$

$= 5y^2 \cdot 2y^3 - 5y^2 \cdot 4 = 10y^5 - 20y^2$

(c) $-5c(4c^2 - 8c)$

$= (-5c)(4c^2) - (-5c)(8c) = -20c^3 + 40c^2$

(d) $3c^2d^2(7cd^2 - 5c^2d^3)$

$= 3c^2d^2 \cdot 7cd^2 - 3c^2d^2 \cdot 5c^2d^3 = 21c^3d^4 - 15c^4d^5$

CHECK YOURSELF 3

Multiply.

(a) $3(5a^2 + 2a + 7)$ **(b)** $4x^2(8x^3 - 6)$

(c) $-5m(8m^2 - 5m)$ **(d)** $9a^2b(3a^3b - 6a^2b^4)$

In order to multiply two binomials, we need to apply the distributive property repeatedly. This is illustrated in the next example.

OBJECTIVE 2

Example 4   Multiplying Binomials

(a) Multiply $x + 2$ by $x + 3$.

NOTE This ensures that each term, x and 2, of the first binomial is multiplied by each term, x and 3, of the second binomial.

We can think of $x + 2$ as a single quantity and apply the distributive property.

$\overset{\frown}{(x + 2)}(x + 3)$ Multiply $x + 2$ by x and then by 3.

$= (x + 2)x + (x + 2)3$ Apply the distributive property two more times.

$= x \cdot x + 2 \cdot x + x \cdot 3 + 2 \cdot 3$

$= x^2 + 2x + 3x + 6$ Combine like terms.

$= x^2 + 5x + 6$

(b) Multiply $a - 3$ by $a - 4$. (Think of $a - 3$ as a single quantity and distribute.)

$(a - 3)(a - 4)$

$= (a - 3)a - (a - 3)(4)$

$= a \cdot a - 3 \cdot a - [(a \cdot 4) - (3 \cdot 4)]$ Use the distributive property two more times.

$= a^2 - 3a - (4a - 12)$ Note that the parentheses are needed here because a *minus* sign precedes the binomial.

$= a^2 - 3a - 4a + 12$

$= a^2 - 7a + 12$

CHECK YOURSELF 4 _____

Multiply.

(a) $(x + 4)(x + 5)$ **(b)** $(y + 5)(y - 6)$

Fortunately, there is a pattern to this kind of multiplication that allows you to write the product of the two binomials directly without going through all these steps. We call it the **FOIL method** of multiplying. The reason for this name will be clear as we look at the process in more detail.

To multiply $(x + 2)(x + 3)$:

NOTE Remember this by F.

1. $(x + 2)(x + 3)$ Find the product of the *first* terms of the factors.

$x \cdot x$

NOTE Remember this by O.

2. $(x + 2)(x + 3)$ Find the product of the *outer* terms.

$x \cdot 3$

NOTE Remember this by I.

3. $(x + 2)(x + 3)$ Find the product of the *inner* terms.

$2 \cdot x$

NOTE Remember this by L.

4. $(x + 2)(x + 3)$ Find the product of the *last* terms.

$2 \cdot 3$

Combining the four steps, we have

NOTE These are the same four
terms found in Example 4a.

$(x + 2)(x + 3)$

$= x^2 + 3x + 2x + 6$

$= x^2 + 5x + 6$

NOTE FOIL gives you an easy
way of remembering the steps:
First, Outer, Inner, and Last.

With practice, the FOIL method will let you write the products quickly and easily. Consider Example 5, which illustrates this approach.

Example 5   Using the FOIL Method

Find the following products, using the FOIL method.

(a) $(x + 4)(x + 5)$

NOTE When possible, you
should combine the outer and
inner products mentally and
write just the final product.

$= x^2 + 5x + 4x + 20$

$= x^2 + 9x + 20$

(b) $(x - 7)(x + 3)$

Combine the outer and inner products as $-4x$.

$= x^2 - 4x - 21$

✔ CHECK YOURSELF 5

Multiply.

(a) $(x + 6)(x + 7)$ **(b)** $(x + 3)(x - 5)$ **(c)** $(x - 2)(x - 8)$

Using the FOIL method, you can also find the product of binomials with coefficients other than 1 or with more than one variable.

Example 6   Using the FOIL Method

Find the following products, using the FOIL method.

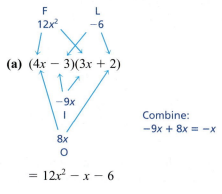

(a) $(4x - 3)(3x + 2)$

Combine:
$-9x + 8x = -x$

$= 12x^2 - x - 6$

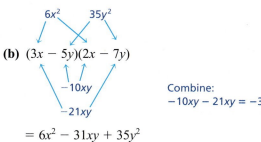

(b) $(3x - 5y)(2x - 7y)$

Combine:
$-10xy - 21xy = -31xy$

$= 6x^2 - 31xy + 35y^2$

CHECK YOURSELF 6

Multiply.

(a) $(5x + 2)(3x - 7)$ **(b)** $(4a - 3b)(5a - 4b)$

(c) $(3m + 5n)(2m + 3n)$

This rule summarizes our work in multiplying binomials.

Step by Step: To Multiply Two Binomials

Step 1   Find the first term of the product of the binomials by multiplying the first terms of the binomials (F).

Step 2   Find the outer and inner terms of the product. If they are like terms, combine them to create the middle term (O + I).

Step 3   Find the last term of the product by multiplying the last terms of the binomials (L).

Sometimes, especially with larger polynomials, it is easier to use the vertical method to find their product. This is the same method you originally learned when multiplying two large integers.

Example 7 **Multiplying Using the Vertical Method**

Use the vertical method to find the product of $(3x + 2)(4x - 1)$.
First, we rewrite the multiplication in vertical form.

NOTE We chose to write $4x - 1$ as $4x + (-1)$.

$$\begin{array}{r} 3x + 2 \\ 4x + (-1) \end{array}$$

Multiplying the quantity $3x + 2$ by -1 yields

NOTE We are using the distributive property here.

$$\begin{array}{r} 3x + 2 \\ 4x + (-1) \\ \hline -3x + (-2) \end{array}$$

Note that we maintained the columns of the original binomial when we found the product. We will continue with those columns as we multiply by the $4x$ term.

NOTE We use the distributive property again.

$$\begin{array}{r} 3x + 2 \\ 4x + (-1) \\ \hline -\,3x + (-2) \\ 12x^2 + 8x \\ \hline 12x^2 + 5x + (-2) \end{array}$$

We could write the product as $(3x + 2)(4x - 1) = 12x^2 + 5x - 2$.

CHECK YOURSELF 7

Use the vertical method to find the product of $(5x - 3)(2x + 1)$.

We use the vertical method again in Example 8. This time, we multiply a binomial and a trinomial. Note that the FOIL method only works with the product of two binomials.

OBJECTIVE 3 | **Example 8** **Using the Vertical Method**

Multiply $x^2 - 5x + 8$ by $x + 3$.

NOTE You should recognize the repeated uses of the distributive property.

Step 1
$$\begin{array}{r} x^2 - 5x + 8 \\ x + 3 \\ \hline 3x^2 - 15x + 24 \end{array}$$
Multiply each term of $x^2 - 5x + 8$ by 3.

Step 2
$$\begin{array}{r} x^2 - 5x + 8 \\ x + 3 \\ \hline 3x^2 - 15x + 24 \\ x^3 - 5x^2 + 8x \end{array}$$
Now multiply each term by x.

We shift this line over so that like terms are in the same columns.

NOTE Using this vertical method ensures that each term of one factor multiplies each term of the other.

Step 3
$$\begin{array}{r} x^2 - 5x + 8 \\ x + 3 \\ \hline 3x^2 - 15x + 24 \\ x^3 - 5x^2 + 8x \\ \hline x^3 - 2x^2 - 7x + 24 \end{array}$$
Now add like terms to write the product.

CHECK YOURSELF 8

Multiply $2x^2 - 5x + 3$ by $3x + 4$.

READING YOUR TEXT

The following fill-in-the-blank exercises are designed to assure that you understand the key vocabulary used in this section. Each sentence comes directly from the section. You will find the correct answers in Appendix C.

Section 10.4

(a) When finding the product of monomials, _____ the coefficients.

(b) Use the _____ property to multiply each term of the polynomial by a monomial.

(c) Use the FOIL method to multiply _____ .

(d) The _____ method is often easier to use when multiplying polynomials with several terms.

CHECK YOURSELF ANSWERS

1. **(a)** $15a^4b^5$; **(b)** $-12x^4y^6$ 2. **(a)** $2y^3 + 6y^2$; **(b)** $6w^5 + 15w^3$
3. **(a)** $15a^2 + 6a + 21$; **(b)** $32x^5 - 24x^2$; **(c)** $-40m^3 + 25m^2$; **(d)** $27a^5b^2 - 54a^4b^5$
4. **(a)** $x^2 + 9x + 20$; **(b)** $y^2 - y - 30$
5. **(a)** $x^2 + 13x + 42$; **(b)** $x^2 - 2x - 15$; **(c)** $x^2 - 10x + 16$
6. **(a)** $15x^2 - 29x - 14$; **(b)** $20a^2 - 31ab + 12b^2$; **(c)** $6m^2 + 19mn + 15n^2$
7. $10x^2 - x - 3$ 8. $6x^3 - 7x^2 - 11x + 12$

Name _____

Section _____ Date _____

ANSWERS

1. _____   2. _____

3. _____   4. _____

5. _____

6. _____

7. _____

8. _____

9. _____

10. _____

11. _____

12. _____

13. _____

14. _____

15. _____

16. _____

17. _____

18. _____

19. _____

20. _____

21. _____

22. _____

23. _____

24. _____

814 SECTION 10.4

10.4 **Exercises**

Multiply.

1. $(-2b^2)(14b^8)$

2. $(14y^4)(-4y^6)$

3. $(-10p^6)(-4p^7)$

4. $(-6m^8)(9m^7)$

5. $(-3m^5n^2)(2m^4n)$ **ALEKS**

6. $(7a^3b^5)(-6a^4b)$ **ALEKS**

7. $5(2x + 6)$

8. $4(7b - 5)$

9. $3a(4a + 5)$

10. $5x(2x - 7)$

11. $3s^2(4s^2 - 7s)$

12. $9a^2(3a^3 + 5a)$

13. $2x(4x^2 - 2x + 1)$

14. $5m(4m^3 - 3m^2 + 2)$

15. $3xy(2x^2y + xy^2 + 5xy)$

16. $5ab^2(ab - 3a + 5b)$

17. $(x + 3)(x + 2)$ **ALEKS**

18. $(a - 3)(a - 7)$ **ALEKS**

19. $(m - 5)(m - 9)$ **ALEKS**

20. $(b + 7)(b + 5)$ **ALEKS**

21. $(p - 8)(p + 7)$ **ALEKS**

22. $(x - 10)(x + 9)$ **ALEKS**

23. $(3x - 5)(x - 8)$

24. $(w + 5)(4w - 7)$

25. $(2x - 3)(3x + 4)$

26. $(5a + 1)(3a + 7)$

27. $(3p - 4q)(7p + 5q)$

28. $(5x - 4y)(2x - y)$

29. $(x + 5)(x + 5)$

30. $(y + 8)(y + 8)$

31. $(y - 9)(y - 9)$

32. $(2a + 3)(2a + 3)$

33. $(6m + n)(6m + n)$

34. $(7b - c)(7b - c)$

35. $(a - 5)(a + 5)$

36. $(x - 7)(x + 7)$

37. $(x - 2y)(x + 2y)$

38. $(7x + y)(7x - y)$

Multiply, using the vertical method.

39. $(x + 2)(3x + 5)$

40. $(a - 3)(2a + 7)$

41. $(2m - 5)(3m + 7)$

42. $(5p + 3)(4p + 1)$

43. $(a^2 + 3ab - b^2)(a^2 - 5ab + b^2)$ ALEKS

44. $(m^2 - 5mn + 3n^2)(m^2 + 4mn - 2n^2)$ ALEKS

45. $(x - 2y)(x^2 + 2xy + 4y^2)$

46. $(m + 3n)(m^2 - 3mn + 9n^2)$

Label each statement as true or false.

47. $(x + y)^2 = x^2 + y^2$

48. $(x - y)^2 = x^2 - y^2$

49. $(x + y)^2 = x^2 + 2xy + y^2$

50. $(x - y)^2 = x^2 - 2xy + y^2$

25. _____

26. _____

27. _____

28. _____

29. _____

30. _____

31. _____

32. _____

33. _____

34. _____

35. _____

36. _____

37. _____

38. _____

39. _____

40. _____

41. _____

42. _____

43. _____

44. _____

45. _____

46. _____

47. _____

48. _____

49. _____

50. _____

51. Geometry The length of a rectangle is given by $3x + 5$ cm and the width is given by $2x - 7$ cm. Express the area of the rectangle in terms of x.

52. Geometry The base of a triangle measures $3y + 7$ in. and the height is $2y - 3$ in. Express the area of the triangle in terms of y.

53. Business and Finance The price of an item is given by $p = 2x - 10$. If the revenue generated is found by multiplying the number of items (x) sold by the price of an item, find the polynomial that represents the revenue.

54. Business and Finance The price of an item is given by $p = 2x^2 - 100$. Find the polynomial that represents the revenue generated from the sale of x items.

Answers

1. $-28b^{10}$ **3.** $40p^{13}$ **5.** $-6m^9n^3$ **7.** $10x + 30$ **9.** $12a^2 + 15a$
11. $12s^4 - 21s^3$ **13.** $8x^3 - 4x^2 + 2x$ **15.** $6x^3y^2 + 3x^2y^3 + 15x^2y^2$
17. $x^2 + 5x + 6$ **19.** $m^2 - 14m + 45$ **21.** $p^2 - p - 56$
23. $3x^2 - 29x + 40$ **25.** $6x^2 - x - 12$ **27.** $21p^2 - 13pq - 20q^2$
29. $x^2 + 10x + 25$ **31.** $y^2 - 18y + 81$ **33.** $36m^2 + 12mn + n^2$
35. $a^2 - 25$ **37.** $x^2 - 4y^2$ **39.** $3x^2 + 11x + 10$ **41.** $6m^2 - m - 35$
43. $a^4 - 2a^3b - 15a^2b^2 + 8ab^3 - b^4$ **45.** $x^3 - 8y^3$ **47.** False   **49.** True
51. $6x^2 - 11x - 35$ cm^2 **53.** $2x^2 - 10x$

10.5 Introduction to Factoring Polynomials

10.5 OBJECTIVES

1. Determine the greatest common factor (GCF) of a list of terms
2. Factor the GCF from the terms of a polynomial

In Section 10.4, you were given factors and asked to find a product. We are now going to reverse the process. You will be given a polynomial and asked to find its factors. This is called **factoring.**

We start with an example from arithmetic. To *multiply* $5 \cdot 7$, you write

$$5 \cdot 7 = 35$$

To *factor* 35, you would write

$$35 = 5 \cdot 7$$

Factoring is the *reverse* of multiplication.

Now look at factoring in algebra: You have used the distributive property as

$$a(b + c) = ab + ac$$

For instance,

NOTE 3 and $x + 5$ are the factors of $3x + 15$.

$$3(x + 5) = 3x + 15$$

To use the distributive property in factoring, we apply that property in the opposite fashion, as

$$ab + ac = a(b + c)$$

This property lets us remove the common monomial factor a from the terms of $ab + ac$. To use this in factoring, the first step is to see whether each term of the polynomial has a common monomial factor. We can apply this to the polynomial $3x + 15$.

$$3x + 15 = 3 \cdot x + 3 \cdot 5$$

Common factor

So, by the distributive property,

$$3x + 15 = 3(x + 5)$$ The original terms are each divided by the greatest common factor to determine the terms in parentheses.

NOTE Factoring is the reverse of multiplication.

To check this, multiply $3(x + 5)$.

Multiplying

NOTE This diagram relates the ideas of multiplication and factoring.

$$3(x + 5) = 3x + 15$$

Factoring

The first step in factoring is to identify the *greatest common factor* (GCF) of a set of terms. This is the monomial with the largest common numerical coefficient and the largest power common to any variables.

> **Definition:** Greatest Common Factor
>
> The **greatest common factor (GCF)** of a polynomial is the monomial with the largest powers and the largest numerical coefficient that is a factor of each term of the polynomial.

OBJECTIVE 1

> **Example 1   Finding the GCF**

Find the GCF for each set of terms.

(a) 9 and 12

RECALL We studied this technique for finding the GCF in Chapter 3.

The prime factorizations of 9 and 12 are

$$9 = 3 \cdot ③$$
$$12 = 2 \cdot 2 \cdot ③$$

We circled the common prime factors of 9 and 12, and we see that the GCF of 9 and 12 is 3.

(b) 10, 25, and 150

The prime factorizations of 10, 25, and 150 are

$$10 = 2 \cdot ⑤$$
$$25 = ⑤ \cdot 5$$
$$150 = 2 \cdot 3 \cdot ⑤ \cdot 5$$

We see from this that the GCF is 5.

(c) x^4 and x^7

The prime factorizations of x^4 and x^7 are

$$x^4 = ⓧ \cdot ⓧ \cdot ⓧ \cdot ⓧ$$
$$x^7 = ⓧ \cdot ⓧ \cdot ⓧ \cdot ⓧ \cdot x \cdot x \cdot x$$

We see from this that the GCF is $x \cdot x \cdot x \cdot x$, or x^4.

(d) $12a^3$ and $18a^2$

NOTE With practice, you will be able to find the GCF entirely in your head.

The prime factorizations of $12a^3$ and $18a^2$ are

$$12a^3 = ② \cdot 2 \cdot ③ \cdot ⓐ \cdot ⓐ \cdot a$$
$$18a^2 = ② \cdot ③ \cdot 3 \cdot ⓐ \cdot ⓐ$$

We see from this that the GCF is $2 \cdot 3 \cdot a \cdot a$, or $6a^2$.

CHECK YOURSELF 1 _____

Find the GCF for each set of terms.

(a) 14, 24   **(b)** 9, 27, 81
(c) a^9, a^5 **(d)** $10x^5, 35x^4$

Step by Step: To Factor a Monomial from a Polynomial

NOTE Checking your answer is always important and perhaps is never easier than after you have factored.

Step 1   Find the GCF for all the terms.
Step 2   Factor the GCF from each term and apply the distributive property.
Step 3   Mentally check your factoring by multiplication.

OBJECTIVE 2

Example 2   Finding the GCF of a Binomial

(a) Factor $8x^2 + 12x$.

The largest common numerical factor of 8 and 12 is 4, and the greatest common variable factor is x. So $4x$ is the GCF. Write

NOTE It is always a good idea to check your answer by multiplying to make sure that you get the original polynomial. Try it here. Multiply $4x$ by $2x + 3$.

$$8x^2 + 12x = 4x \cdot 2x + 4x \cdot 3$$

GCF

Now, by the distributive property, we have

$$8x^2 + 12x = 4x(2x + 3)$$

(b) Factor $6a^4 - 18a^2$.

The GCF in this case is $6a^2$. Write

NOTE It is also true that $6a^4 + (-18a^2) = 3a(2a^3 + (-6a))$. However, this is *not completely factored*. Do you see why? You want to find the common monomial factor with the *largest possible* coefficient and the *largest* exponent, in this case $6a^2$.

$$6a^4 + (-18a^2) = 6a^2 \cdot a^2 + 6a^2 \cdot (-3)$$

GCF

Again, using the distributive property yields

$$6a^4 - 18a^2 = 6a^2(a^2 - 3)$$

You should check this by multiplying.

CHECK YOURSELF 2

Factor each of the polynomials.

(a) $5x + 20$ **(b)** $6x^2 - 24x$ **(c)** $10a^3 - 15a^2$

The process is exactly the same for polynomials with more than two terms. Consider Example 3.

Example 3   Finding the GCF of a Polynomial

(a) Factor $5x^2 - 10x + 15$.

NOTE The GCF is 5.

$$5x^2 - 10x + 15 = 5 \cdot x^2 - 5 \cdot 2x + 5 \cdot 3$$

GCF

$$= 5(x^2 - 2x + 3)$$

NOTE The GCF is 3*a*.

(b) Factor $6ab + 9ab^2 - 15a^2$.

$$6ab + 9ab^2 - 15a^2 = 3a \cdot 2b + 3a \cdot 3b^2 - 3a \cdot 5a$$

GCF

$$= 3a(2b + 3b^2 - 5a)$$

(c) Factor $4a^4 + 12a^3 - 20a^2$.

NOTE The GCF is $4a^2$.

$$4a^4 + 12a^3 - 20a^2 = 4a^2 \cdot a^2 + 4a^2 \cdot 3a - 4a^2 \cdot 5$$

GCF

$$= 4a^2(a^2 + 3a - 5)$$

NOTE In each of these examples, you should check the result by multiplying the factors.

(d) Factor $6a^2b + 9ab^2 + 3ab$.

Mentally note that 3, *a*, and *b* are factors of each term, so

$$6a^2b + 9ab^2 + 3ab = 3ab(2a + 3b + 1)$$

CHECK YOURSELF 3

Factor each of the polynomials.

(a) $8b^2 + 16b - 32$

(b) $4xy - 8x^2y + 12x^3$

(c) $7x^4 - 14x^3 + 21x^2$

(d) $5x^2y^2 - 10xy^2 + 15x^2y$

Each case of factoring a polynomial that we have studied in this section involves finding the greatest common factor (GCF) and factoring it out using the distributive property. In an algebra class, you will use this as well as more advanced methods of factoring.

READING YOUR TEXT

The following fill-in-the-blank exercises are designed to assure that you understand the key vocabulary used in this section. Each sentence comes directly from the section. You will find the correct answers in Appendix C.

Section 10.5

(a) Factoring is the reverse of _____.

(b) The greatest common factor of a polynomial is a _____ with the largest common numerical coefficient and the largest power common to all variables.

(c) After factoring the GCF from each term, apply the _____ property.

(d) You can check your answer to a factoring problem by _____ the factors.

CHECK YOURSELF ANSWERS

1. (a) 2; **(b)** 9; **(c)** a^5; **(d)** $5x^4$ **2. (a)** $5(x + 4)$; **(b)** $6x(x - 4)$; **(c)** $5a^2(2a - 3)$

3. (a) $8(b^2 + 2b - 4)$; **(b)** $4x(y - 2xy + 3x^2)$; **(c)** $7x^2(x^2 - 2x + 3)$;

    (d) $5xy(xy - 2y + 3x)$

10.5 Exercises

Find the greatest common factor for each of the sets of terms.

1. 10, 12

2. 15, 35

3. 16, 32, 88

4. 55, 33, 132

5. x^2, x^5

6. y^7, y^9

7. a^3, a^6, a^9

8. b^4, b^6, b^8

9. $5x^4$, $10x^5$

10. $8y^9$, $24y^3$

11. $8a^4$, $6a^6$, $10a^{10}$

12. $9b^3$, $6b^5$, $12b^4$

13. $9x^2y$, $12xy^2$, $15x^2y^2$

14. $12a^3b^2$, $18a^2b^3$, $6a^4b^4$

15. $15ab^3$, $10a^2bc$, $25b^2c^3$

16. $9x^2$, $3xy^3$, $6y^3$

17. $15a^2bc^2$, $9ab^2c^2$, $6a^2b^2c^2$

18. $18x^3y^2z^3$, $27x^4y^2z^3$, $81xy^2z$

19. xy^2z^3, x^3y^2z

20. $12ab^9$, $9ab^{12}$

Factor each of the polynomials.

21. $8a + 4$

22. $5x - 15$

23. $24m - 32n$

24. $7p - 21q$

25. $12m^2 + 8m$

26. $24n^2 - 32n$

27. $10s^2 + 5s$

28. $12y^2 - 6y$

29. $12x^2 + 24x$

30. $14b^2 - 28b$

31. $15a^3 - 25a^2$

32. $36b^4 + 24b^2$

33. $6pq + 18p^2q$

34. $8ab - 24ab^2$

Name _____

Section _____ Date _____

ANSWERS

1.	2.
3.	4.
5.	6.
7.	8.
9.	10.
11.	12.
13.	14.
15.	16.
17.	18.
19.	20.
21.	
22.	
23.	
24.	
25.	
26.	
27.	
28.	
29.	
30.	
31.	
32.	
33.	
34.	

35. _____

36. _____

37. _____

38. _____

39. _____

40. _____

41. _____

42. _____

43. _____

44. _____

45. _____

46. _____

47. _____

48. _____

49. _____

50. _____

51. _____

52. _____

53. _____

54. _____

35. $7m^3n - 21mn^3$ **36.** $36p^2q^2 - 9pq$

37. $6x^2 - 18x + 30$ **38.** $7a^2 + 21a - 42$

39. $3a^3 + 6a^2 - 12a$ **40.** $5x^3 - 15x^2 + 25x$

41. $6m + 9mn - 15mn^2$ **42.** $4s + 6st - 14st^2$

43. $10x^2y + 15xy - 5xy^2$ **44.** $3ab^2 + 6ab - 15a^2b$

45. $10r^3s^2 + 25r^2s^2 - 15r^2s^3$ **46.** $28x^2y^3 - 35x^2y^2 + 42x^3y$

47. $9a^5 - 15a^4 + 21a^3 - 27a$ **48.** $8p^6 - 40p^4 + 24p^3 - 16p^2$

49. $15m^3n^2 - 20m^2n + 35mn^3 - 10mn$ **50.** $14ab^4 + 21a^2b^3 - 35a^3b^2 + 28ab^2$

51. Geometry The area of a rectangle with width t is given by $33t - t^2$. Factor the expression and determine the length of the rectangle in terms of t.

52. Geometry The area of a rectangle of length x is given by $3x^2 + 5x$. Find the width of the rectangle.

53. The height of an object dropped from 1,000 feet is given by $h = -16t^2 + 1,000$. It is often easier to work with polynomials when they are in factored form. Factor the given polynomial.

54. The height of an object thrown downward with an initial velocity of 30 ft/s from a height of 1,000 feet is given by $h = -16t^2 - 30t + 1,000$. Factor this polynomial.

Answers

1. 2   **3.** 8   **5.** x^2 **7.** a^3 **9.** $5x^4$ **11.** $2a^4$ **13.** $3xy$
15. $5b$ **17.** $3abc^2$ **19.** xy^2z **21.** $4(2a + 1)$ **23.** $8(3m - 4n)$
25. $4m(3m + 2)$ **27.** $5s(2s + 1)$ **29.** $12x(x + 2)$ **31.** $5a^2(3a - 5)$
33. $6pq(1 + 3p)$ **35.** $7mn(m^2 - 3n^2)$ **37.** $6(x^2 - 3x + 5)$
39. $3a(a^2 + 2a - 4)$ **41.** $3m(2 + 3n - 5n^2) = 3m(1 - n)(2 + 5n)$
43. $5xy(2x + 3 - y)$ **45.** $5r^2s^2(2r + 5 - 3s)$ **47.** $3a(3a^4 - 5a^3 + 7a^2 - 9)$
49. $5mn(3m^2n - 4m + 7n^2 - 2)$ **51.** $t(33 - t)$; $33 - t$
53. $h = -8(2t^2 - 125)$

ACTIVITY 10: THE GRAVITY MODEL

Each chapter in this text concludes with a chapter activity such as this one. These activities provide you with the opportunity to apply the math you studied in that chapter to a relevant topic.

Your instructor will determine how best to use these activities with your class. You may find yourself working in class or outside of class; you may find yourself working alone or in small groups; or you may even be asked to perform research in a library or on the Internet.

This activity requires two to three people and an open area outside. You will also need a ball (preferably a baseball or some other small, heavy ball), a stopwatch, and a tape measure.

1. Designate one person as the "ball thrower."
2. The ball thrower should throw the ball straight up into the air, as high as possible. While this is happening, another group member should take note of exactly where the ball thrower releases the ball. Be careful that no one gets hit by the ball as it comes back down.
3. Use the tape measure to determine the height above the ground that the thrower released the ball. Convert your measurement to feet, using decimals to indicate parts of a foot. For example, 3 inches is 0.25 ft. Record this as the "initial height."

4. Repeat steps 2 and 3 to ensure that you have the correct initial height.
5. The thrower should now throw the ball straight up, as high as possible. The person with the stopwatch should time the ball until it lands.
6. Repeat step 5 twice more, recording the time.
7. Take the average (mean) of your three recorded times. We will use this number as the "hang time" of the ball for the remainder of this activity.

According to Isaac Newton (1642–1727), the height of an object with initial velocity v_0, and initial height s_0, is given by

$$s = -16t^2 + v_0t + s_0$$

in which the height s is measured in feet and t represents the time, in seconds.

The initial velocity v_0 is positive when the object is thrown upward, and it is negative when the object is thrown downward.

For example, the height of a ball thrown upward from a height of 4 ft with an initial velocity of 50 ft/s is given by

$$s = -16t^2 + 50t + 4$$

To determine the height of the ball 2 seconds after release, we evaluate the polynomial for $t = 2$.

$$= -16(2)^2 + 50(2) + 4$$
$$= 40$$

The ball is 40 feet high after 2 seconds.

Take note that the height of the ball when it lands is zero.

8. Substitute the time found in step 7 for t, your initial height for s_0, and 0 for s in the height equation. Solve the resulting equation for the initial velocity, v_0.

9. Now write your height equation using the initial velocity found in step 8 along with the initial height. Your equation should have two variables, s and t.

10. What is the height of the ball after 1 second?

11. The maximum height of the ball will be attained when $t = \dfrac{v_0}{32}$. Determine this time and find the maximum height of the thrown ball.

12. If a ball is dropped from a height of 256 feet, how long will it take to hit the ground? (*Hint:* If a ball is dropped, its initial velocity is 0 ft/s.)

10 Summary

DEFINITION/PROCEDURE	EXAMPLE	REFERENCE
Properties of Exponents		**Section 10.1**
1. $a^m a^n = a^{m+n}$ **2.** $(a^m)^n = a^{mn}$ **3.** $(ab)^m = a^m b^m$ **4.** $\dfrac{a^m}{a^n} = a^{m-n} \ (a \neq 0)$ **5.** $\left(\dfrac{a}{b}\right)^m = \dfrac{a^m}{b^m} \ (b \neq 0)$	$x^2 \cdot x^3 = x^5$ $(x^2)^3 = x^6$ $(4x)^3 = 4^3 x^3 = 64x^3$ $\dfrac{x^7}{x^2} = x^5$ $\left(\dfrac{x}{y}\right)^3 = \dfrac{x^3}{y^3}$	*p. 786*
Introduction to Polynomials		**Section 10.2**
Term A term is a number or the product of a number and variables.		*p. 790*
Polynomial A polynomial is an algebraic expression made up of terms in which the exponents are whole numbers. These terms are connected by plus or minus signs. Each sign ($+$ or $-$) is attached to the term following that sign.	$4x^3 - 3x^2 + 5x$ is a polynomial. The terms of $4x^3 - 3x^2 + 5x$ are $4x^3$, $-3x^2$, and $5x$.	
Coefficient In each term of a polynomial, the number in front of the variable is called the **numerical coefficient** or, more simply, the **coefficient** of that term.	The coefficients of $4x^3 - 3x^2$ are 4 and -3.	*p. 790*
Types of Polynomials A polynomial can be classified according to the number of terms it has. A **monomial** has one term. A **binomial** has two terms. A **trinomial** has three terms.	$2x^3$ is a monomial. $3x^2 - 7x$ is a binomial. $5x^5 - 5x^3 + 2$ is a trinomial.	*p. 791*
Addition and Subtraction of Polynomials		**Section 10.3**
Removing Signs of Grouping **1.** If a plus sign ($+$) or no sign at all appears in front of parentheses, just remove the parentheses. No other changes are necessary.	$3x + (2x - 3)$ $= 3x + 2x - 3$	*p. 797*
2. If a minus sign ($-$) appears in front of parentheses, the parentheses can be removed by changing the sign of each term inside the parentheses.	$2x - (x - 4)$ $= 2x - x + 4$	*p. 799*
Adding Polynomials Remove the signs of grouping. Then collect and combine any like terms.	$(2x + 3) + (3x - 5)$ $= 2x + 3 + 3x - 5 = 5x - 2$	*p. 797*
Subtracting Polynomials Remove the signs of grouping by changing the sign of each term in the polynomial being subtracted. Then combine any like terms.	$(3x^2 + 2x) - (2x^2 + 3x - 1)$ $= 3x^2 + 2x - 2x^2 - 3x + 1$ Signs change $= 3x^2 - 2x^2 + 2x - 3x + 1$ $= x^2 - x + 1$	*p. 799*

Continued

DEFINITION/PROCEDURE	EXAMPLE	REFERENCE
Multiplying Polynomials		**Section 10.4**
To Multiply a Polynomial by a Monomial Multiply each term of the polynomial by the monomial and simplify the results.	$3x(2x + 3)$ $= 3x \cdot 2x + 3x \cdot 3$ $= 6x^2 + 9x$	*p. 807*
To Multiply a Binomial by a Binomial Use the FOIL method: $\qquad\quad$ F $\quad$ O $\quad$ I $\quad$ L $(a + b)(c + d) = a \cdot c + a \cdot d + b \cdot c + b \cdot d$	$(2x - 3)(3x + 5)$ $= 6x^2 + 10x - 9x - 15$ $\quad$ F $\quad\;$ O $\quad\;$ I $\quad\;$ L $= 6x^2 + x - 15$	*p. 811*
To Multiply a Polynomial by a Polynomial Arrange the polynomials vertically. Multiply each term of the upper polynomial by each term of the lower polynomial and add the results.	$\begin{array}{r} x^2 - \;\; 3x + \;\; 5 \\ 2x - \;\; 3 \\ \hline -3x^2 + \;\; 9x - 15 \\ 2x^3 - 6x^2 + 10x \qquad\;\; \\ \hline 2x^3 - 9x^2 + 19x - 15 \end{array}$	*p. 812*
Introduction to Factoring Polynomials		**Section 10.5**
Common Monomial Factor A common monomial factor is a single term that is a factor of every term of the polynomial. The **greatest common factor (GCF)** of a polynomial is the common monomial factor that has the largest possible numerical coefficient and the largest possible exponents.	$4x^2$ is the greatest common factor of $8x^4 - 12x^3 + 16x^2$.	*p. 818*
Factoring a Monomial from a Polynomial **1.** Determine the GCF for all terms. **2.** Factor the GCF from each term and then apply the distributive property in the form $\qquad ab + ac = a(b + c)$ $\qquad\qquad\qquad\quad\uparrow$ $\qquad$ The greatest common factor **3.** Mentally check by multiplication.	$8x^4 - 12x^3 + 16x^2$ $= 4x^2(2x^2 - 3x + 4)$	*p. 819*

Summary Exercises

This summary exercise set is provided to give you practice with each of the objectives of this chapter. Each exercise is keyed to the appropriate chapter section. When you are finished, you can check your answers to the odd-numbered exercises against those presented in the back of the text. If you have difficulty with any of these questions, go back and reread the examples from that section. The answers to the even-numbered exercises appear in the *Instructor's Solutions Manual*. Your instructor will give you guidelines on how to best use these exercises in your class.

[10.1] Simplify each of the expressions.

1. $(2ab)^2$

2. $(p^2q^3)^3$

3. $(2x^2y^2)^3(3x^3y)^2$

4. $(4w^2t)^2\,(3wt^2)^3$

5. $(y^3)^2(3y^2)^3$

[10.2] Classify each of the polynomials as monomial, binomial, or trinomial, where possible.

6. $5x^3 - 2x^2$

7. $7x^5$

8. $4x^5 - 8x^3 + 5$

9. $x^3 + 2x^2 - 5x$

10. $9a^2 - 18a^2$

Find the value of each of the polynomials for the given value of the variable.

11. $5x + 1; x = -1$

12. $2x^2 + 7x - 5; x = 2$

13. $-x^2 + 3x - 1; x = 6$

14. $4x^2 + 5x + 7; x = -4$

[10.3] Add.

15. $9a^2 - 5a$ and $12a^2 + 3a$

16. $5x^2 + 3x - 5$ and $4x^2 - 6x - 2$

17. $5y^3 - 3y^2$ and $4y + 3y^2$

Subtract.

18. $4x^2 - 3x$ from $8x^2 + 5x$

19. $2x^2 - 5x - 7$ from $7x^2 - 2x + 3$

20. $5x^2 + 3$ from $9x^2 - 4x$

Perform the indicated operations.

21. Subtract $5x - 3$ from the sum of $9x + 2$ and $-3x - 7$.

22. Subtract $5a^2 - 3a$ from the sum of $5a^2 + 2$ and $7a - 7$.

23. Subtract the sum of $16w^2 - 3w$ and $8w + 2$ from $7w^2 - 5w + 2$.

Add, using the vertical method.

24. $x^2 + 5x - 3$ and $2x^2 + 4x - 3$

25. $9b^2 - 7$ and $8b + 5$

26. $x^2 + 7$, $3x - 2$, and $4x^2 - 8x$

Subtract, using the vertical method.

27. $5x^2 - 3x + 2$ from $7x^2 - 5x - 7$

28. $8m - 7$ from $9m^2 - 7$

[10.4]   Multiply.

29. $(5a^3)(a^2)$

30. $(2x^2)(3x^5)$

31. $(-9p^3)(-6p^2)$

32. $(3a^2b^3)(-7a^3b^4)$

33. $5(3x - 8)$

34. $4a(3a + 7)$

35. $(-5rs)(2r^2s - 5rs)$

36. $7mn(3m^2n - 2mn^2 + 5mn)$

37. $(x + 5)(x + 4)$

38. $(w - 9)(w - 10)$

39. $(a - 7b)(a + 7b)$

40. $(p - 3q)(p - 3q)$

41. $(a + 4b)(a + 3b)$

42. $(b - 8)(2b + 3)$

43. $(3x - 5y)(2x - 3y)$

44. $(5r + 7s)(3r - 9s)$

45. $(y + 2)(y^2 - 2y + 3)$

46. $(b + 3)(b^2 - 5b - 7)$

47. $(x - 2)(x^2 + 2x + 4)$

48. $(m^2 - 3)(m^2 + 7)$

49. $2x(x + 5)(x - 6)$

50. $a(2a - 5b)(2a - 7b)$

51. $(x + 7)(x + 7)$

52. $(a - 8)(a - 8)$

53. $(2w - 5)(2w - 5)$

54. $(3p + 4)(3p + 4)$

55. $(a + 7b)(a + 7b)$

56. $(8x - 3y)(8x - 3y)$

57. $(x - 5)(x + 5)$

58. $(y + 9)(y - 9)$

59. $(2m + 3)(2m - 3)$

60. $(3r - 7)(3r + 7)$

61. $(5r - 2s)(5r + 2s)$

62. $(7a + 3b)(7a - 3b)$

[10.5]   Factor each of the polynomials.

63. $18a + 24$

64. $9m^2 - 21m$

65. $24s^2t - 16s^2$

66. $18a^2b + 36ab^2$

67. $35s^3 - 28s^2$

68. $3x^3 - 6x^2 + 15x$

69. $18m^2n^2 - 27m^2n + 45m^2n^3$

70. $121x^8y^3 + 77x^6y^3$

71. $8a^2b + 24ab - 16ab^2$

72. $3x^2y - 6xy^3 + 9x^3y - 12xy^2$

Self-Test for Chapter 10

The purpose of this self-test is to help you check your progress so that you can find sections and concepts that you need to review before the next in-class exam. Allow yourself about an hour to take this test. At the end of that hour, check your answers against those given in the back of this text. If you missed an answer, notice the section reference that accompanies the question. Go back to that section and reread the examples until you have mastered that particular concept.

Classify each of the polynomials as monomial, binomial, or trinomial.

1. $6x^2 + 7x$

2. $5x^2 + 8x - 8$

3. Find the value of the polynomial $y = -3x^2 - 5x + 8$ if $x = -2$.

Add.

4. $3x^2 - 7x + 2$ and $7x^2 - 5x - 9$

5. $7a^2 - 3a$ and $7a^3 + 4a^2$

Subtract.

6. $5x^2 - 2x + 5$ from $8x^2 + 9x - 7$

7. $2b^2 + 5$ from $3b^2 - 7b$

8. $5a^2 + a$ from the sum of $3a^2 - 5a$ and $9a^2 - 4a$

Add, using the vertical method.

9. $x^2 + 3$, $5x - 7$, and $3x^2 - 2$

Subtract, using the vertical method.

10. $3x^2 - 5$ from $5x^2 - 7x$

Simplify each of the expressions.

11. $a^5 \cdot a^9$

12. $3x^2y^3 \cdot 5xy^4$

13. $(3x^2y)^3$

14. $(2x^3y^2)^4(x^2y^3)^3$

Name _____

Section _____ **Date** _____

ANSWERS

1. _____
2. _____
3. _____
4. _____
5. _____
6. _____
7. _____
8. _____
9. _____
10. _____
11. _____
12. _____
13. _____
14. _____

15. _____

16. _____

17. _____

18. _____

19. _____

20. _____

21. _____

22. _____

23. _____

24. _____

25. _____

Multiply.

15. $5ab(3a^2b - 2ab + 4ab^2)$

16. $(x - 2)(3x + 7)$

17. $(a - 7b)(a + 7b)$

18. $(4x + 3y)(2x - 5y)$

19. $x(3x - y)(4x + 5y)$

20. $(3m + 2n)(3m + 2n)$

21. $(2x + y)(x^2 + 3xy - 2y^2)$

Factor each of the polynomials.

22. $12b + 18$

23. $9p^3 - 12p^2$

24. $5x^2 - 10x + 20$

25. $6a^2b - 18ab + 12ab^2$

Practice Final Exam

Name _____

Section _____ Date _____

This test is provided to help you in the process of reviewing Chapters 1 through 10. It should help you prepare for a final exam in your course. Answers are provided in the back of the book. If you missed any answers, be sure to go back and review the appropriate chapter sections.

Write the word name for each number.

1. 806,015

2. 17,608

In exercises 3 to 5, perform the indicated operations.

3. $425 - 67$

4. $48 \cdot 65$

5. $247 \div 13$

Solve the exercises.

6. Carlotta made car payments of $355 each month for 48 months. What is the total amount of money paid?

7. Minh's annual salary of $30,240 is paid in 12 equal installments. How much does he receive each month?

Evaluate.

8. $3 \cdot 5^2$

9. $9 - 2^3$

10. $18 \div 2 \cdot 3 - 1 + 2$

In exercises 11 to 21, evaluate each expression.

11. $-7 - (-4)$

12. $8 - 15 + 3$

13. $(-6) \cdot (-9)$

14. $\dfrac{4}{5} \div \left(\dfrac{-7}{15}\right)$

15. $-24 \div (8 - 12)$

16. $-18 - 11$

17. $23 - (-8)$

18. $|-16| - |-7|$

ANSWERS

1. _____

2. _____

3. _____

4. _____

5. _____

6. _____

7. _____

8. _____

9. _____

10. _____

11. _____

12. _____

13. _____

14. _____

15. _____

16. _____

17. _____

18. _____

19. _____

20. _____

21. _____

22. _____

23. _____

24. _____

25. _____

26. _____

27. _____

28. _____

29. _____

30. _____

31. _____

32. _____

33. _____

34. _____

35. _____

36. _____

37. _____

19. $(-2)(-4)(-7)$

20. $\dfrac{8 - (4)(-5)}{(5)(-2) + 6}$

21. $9 - 5 \cdot 3^2$

In exercises 22 to 24, evaluate the expressions for the given values of the variables.

22. $b^2 - 4ac$, for $a = 2$, $b = -5$, and $c = -3$

23. $-2x^2 - 3x + 5$, for $x = -3$

24. $x^2 + 2z^2$, for $x = -5$ and $z = 3$

25. Write the prime factorization of 1,260.

26. Find the least common multiple (LCM) of 4, 18, and 45.

27. Reduce this fraction to lowest terms: $\dfrac{60}{132}$.

In exercises 28 to 36, perform the indicated operations.

28. $\dfrac{4}{15} + \dfrac{8}{15}$

29. $\dfrac{3}{7} \cdot \dfrac{5}{12}$

30. $4\dfrac{1}{3} \cdot 2\dfrac{3}{5}$

31. $\dfrac{2}{3} \div \dfrac{4}{9}$

32. $\dfrac{1}{5} \div 5$

33. $\dfrac{3}{4} + \dfrac{5}{8}$

34. $\dfrac{11}{12} - \dfrac{2}{3}$

35. $8\dfrac{1}{4} - 2\dfrac{7}{8}$

36. $\dfrac{5}{9} \cdot \dfrac{36}{15}$

37. Marissa has $6\dfrac{1}{2}$ yd of material. Her new skirt will take $1\dfrac{4}{5}$ yd. How much material will she have left after the skirt is made?

In exercises 38 to 43, solve each equation for x.

38. $x + 3 = -5$

39. $-6x = 42$

40. $5 - 4x = -7$

41. $\dfrac{x}{4} - 9 = -3$

42. $4x - 5 = -6x + 3$

43. $4x - 2(x - 5) = 16$

44. Write 0.125 as a reduced fraction.

In exercises 45 to 50, perform the indicated operations.

45. $17.289 + 4.93$

46. $0.62 \cdot 0.095$

47. $2.4 \div 0.08$

48. $239.95 - 35.99$

49. $2{,}537.74 \div 113.8$

50. $9.35 \cdot 34.06$

Solve the exercises.

51. Toni had $189.75 in her checking account. She then wrote a check for $39.95. How much money did she have left in her account?

52. A car rental agency charges $45 per day, plus $0.29 per mile for a certain type of car. How much is the rental charge for a 6-day trip of 462 mi?

53. Round 0.426839 to the nearest ten-thousandth.

54. Write $\dfrac{3}{8}$ in decimal form.

55. Simplify: $\sqrt{81}$.

56. The length of one of the legs in a right triangle is 8 in. If the hypotenuse is 17 in. long, find the length of the other leg.

57. Solve the proportion for n: $\dfrac{6}{n} = \dfrac{9}{57}$.

ANSWERS

38. _____

39. _____

40. _____

41. _____

42. _____

43. _____

44. _____

45. _____

46. _____

47. _____

48. _____

49. _____

50. _____

51. _____

52. _____

53. _____

54. _____

55. _____

56. _____

57. _____

58. _____

59. _____

60. _____

61. _____

62. _____

63. _____

64. _____

65. _____

66. _____

67. _____

68. _____

69. _____

70. _____

71. _____

58. Express $\dfrac{17}{20}$ as a percent.

59. Write 185% as a decimal.

60. Write 0.08 as a percent.

61. Write $1\dfrac{1}{5}$ as a percent.

62. Marcia worked on her homework from 7 P.M. until 11 P.M. What percent of the total day did she spend working on her homework?

63. What is 16% of 80?

64. 64 is 4% of what number?

65. Johanna bought a pair of slacks priced at $42. If the sales tax rate in Johanna's state is 7%, what is the sales tax on the purchase?

66. What is a salesperson's commission on a $1,200 sale if the commission rate is 15%?

67. An article regularly selling for $90 is advertised at 25% off. Find the sales price.

68. A store marks up items to make a 20% profit. If an item sells for $22.20, what was the cost before the markup?

69. Convert 630 cm to m.

70. Find the perimeter of a rectangle 84.5 cm long and 52.5 cm wide.

71. Find the area of the triangle shown.

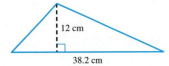

ANSWERS

72.

73.

74.

75.

76.

72. The table below compares wheat production in Kansas and the United States. What percentage of acreage harvested in the U.S. is Kansas responsible for?

Wheat Production: U.S. and Kansas		
	U.S.	**Kansas**
Acreage planted	62.6	9.8
Acreage harvested	53.1	9.4
Yield	42.0	37.0

Sources: Farm Service Agency; U.S. Department of Agriculture (8/03)
Kansas Agricultural Statistics Service; Kansas Department of
Agriculture (9/02)

73. The bar graph shown indicates the number of words Joseph learned to spell in a 5-week period. How many words did Joseph learn after the second week?

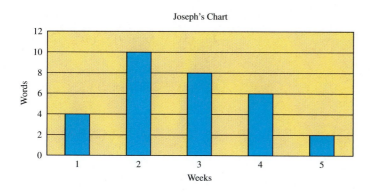

Joseph's Chart

74. The line graph shows sales (in millions of dollars) for XYZ Corporation. How much did sales increase between 2000 and 2002?

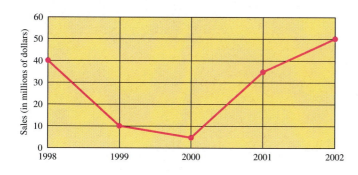

75. Hiro's scores on his first four math tests were 82, 95, 77, and 86. What is the mean of his scores?

76. Hiro's scores on his first four math tests were 82, 95, 77, and 86. What is the median of his scores?

ANSWERS

77. _____

78. _____

79. _____

80. _____

81. _____

82. _____

83. _____

77. Graph the line whose equation is $2x - 3y = 6$.

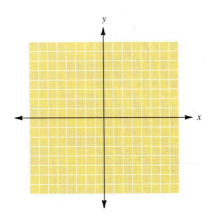

In exercises 78 to 83, perform the indicated operations. Write each answer in simplified form.

78. $3x(2x^2 - x + 4)$

79. $(3x^2 - 5x - 9) - (4x^2 + 2x - 13)$

80. $(5x - 2)(6x + 5)$

81. $(7x - 2y)(7x + 2y)$

82. $(3x - 4y)(3x - 4y)$

83. $2x(x + 3y) - 4y(2x - y)$

1. Eight hundred six thousand, fifteen
2. Seventeen thousand, six hundred eight   3. 358   4. 3,120
5. 19   6. $17,040   7. $2,520   8. 75   9. 1   10. 28
11. -3 12. -4 13. 54   14. $-\dfrac{12}{7}$ 15. 6
16. -29 17. 31   18. 9   19. -56 20. -7
21. -36 22. 49   23. -4 24. 43
25. $2 \cdot 2 \cdot 3 \cdot 3 \cdot 5 \cdot 7$ 26. 180   27. $\dfrac{5}{11}$ 28. $\dfrac{4}{5}$
29. $\dfrac{5}{28}$ 30. $11\dfrac{4}{15}$ 31. $1\dfrac{1}{2}$ 32. $\dfrac{1}{25}$ 33. $1\dfrac{3}{8}$
34. $\dfrac{1}{4}$ 35. $5\dfrac{3}{8}$ 36. $1\dfrac{1}{3}$ 37. $4\dfrac{7}{10}$ yd   38. $x = -8$
39. $x = -7$ 40. $x = 3$ 41. $x = 24$ 42. $x = \dfrac{4}{5}$
43. $x = 3$ 44. $\dfrac{1}{8}$ 45. 22.219   46. 0.0589   47. 30
48. 203.96   49. 22.3   50. 318.461   51. $149.80
52. $403.98   53. 0.4268   54. 0.375   55. 9   56. 15 in.
57. $n = 38$ 58. 85%   59. 1.85   60. 8%   61. 120%
62. $16\dfrac{2}{3}\%$ 63. 12.8   64. 1,600   65. $2.94

66. $180   67. $67.50   68. $18.50   69. 6.3 m
70. 274 cm   71. 229.2 cm^2 72. 17.7%   73. 16 words
74. 45 million dollars   75. 85   76. 84
77.

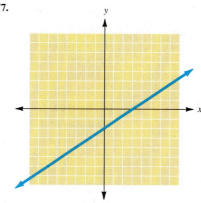

78. $6x^3 - 3x^2 + 12x$ 79. $-x^2 - 7x + 4$
80. $30x^2 + 13x - 10$ 81. $49x^2 - 4y^2$
82. $9x^2 - 24xy + 16y^2$ 83. $2x^2 - 2xy + 4y^2$

The Internet

There are many resources available to help you when you have difficulty with your math. Your instructor can answer many of your questions. Studying with friends and classmates is another great way to learn math. Your school may have a "math lab" where instructors or peers provide tutoring services. In addition, this text provides examples and exercises to help you learn and understand new concepts.

The Internet can also provide you with math help. There are numerous math tutorials on the Web. This activity is designed to introduce you to searching the Web and evaluating what you find there.

If you are new to computers or the Internet, your instructor or a classmate can help you get started. You will need to access the Internet through one of the many Web browsers, such as Microsoft's Internet Explorer, Mozilla Firefox, Netscape Navigator, AOL's browser, or Opera.

First, you need to connect to the Internet. Then, you need to access a page containing a **search engine** that searches the World Wide Web. Many *default* home pages (such as www.aol.com and www.msn.com) contain a *search* field. If yours does not, several of the more popular sites are:

http://www.altavista.com

http://www.google.com

http://www.lycos.com

http://www.yahoo.com

Access one of these websites, or use some other search engine for this activity.

1. Type the word *integers* in the search field. After a brief delay, while the engine searches Web pages, you should get a (long) list of websites related to your search.
2. Look at the page titles and any descriptions given. Find a page that offers an introduction to integers and click on that page.

3. Write two or three sentences describing the layout of the web page. Is it "user friendly"? Are the topics presented in an easy-to-find and useful way? Are the colors and images helpful?

4. Choose a topic such as integer multiplication or some math game offered. Describe the instruction that the website offers for your topic. In what format is the information provided? Is there an interactive component to the instruction?

5. Does the website offer free tutoring services? If so, try to get some help with a homework problem that you did not answer correctly. Briefly evaluate the tutoring services.

6. Chapter 3 in our text introduces you to fractions. Are there activities and Web pages on the website that relate to fractions? Do they seem like they might be helpful if you find yourself having difficulty?

7. Return to your search engine. Find a second math Web page by typing *fractions* into the search field. Choose a page that offers instruction, tutoring, and activities on fractions. "Bookmark" (or *add to favorites* or *create preferences*) this page. If you have trouble with Chapter 3, try using this page to get some additional help.

Circle Graphs

APPENDIX B OBJECTIVES

1. Interpret a circle graph
2. Create a circle graph

When we need a graph to represent how some unit is divided, we frequently use a **circle graph.** As you might expect, a circle graph is created using a circle. Wedges, called **sectors,** are drawn in the circle to show how much of the whole each part makes up.

OBJECTIVE 1

> **Example 1   Reading a Circle Graph**

This circle graph represents the results of a survey that asked students how they get to school most often.

NOTE The percentages in a circle graph should add up to 100%.

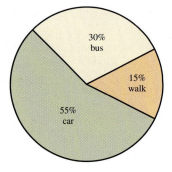

(a) What percent of the students walk to school?

We see that 15% walk to school.

(b) What percent of the students do not arrive by car?

Because 55% arrive by car, there are 100% − 55% = 45% who do not.

 CHECK YOURSELF 1 _____

This circle graph represents the results of a survey that asked students whether they bought lunch, brought it, or skipped lunch altogether.

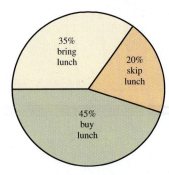

(a) What percent of the students skipped lunch?
(b) What percent of the students did not buy lunch?

If we know what the whole circle represents, we can also find out more about what each sector represents. Example 2 illustrates this point.

Example 2   Interpreting a Circle Graph

This circle graph shows how Sarah spent her $12,000 college scholarship.

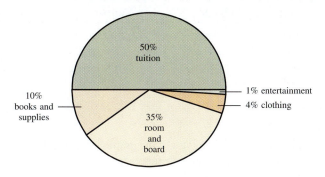

(a) How much money did she spend on tuition?

50% of her $12,000 scholarship, or $6,000.

(b) How much money did she spend on clothing and entertainment?

Together, 5% of the money was spent on clothing and entertainment, and $0.05 \cdot 12,000 = 600$. Therefore, $600 was spent on clothing and entertainment.

 CHECK YOURSELF 2

This circle graph shows how Rebecca spends an average 24-h school day.

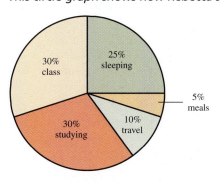

(a) How many hours does she spend sleeping each day?

(b) How many hours does she spend altogether studying and in class?

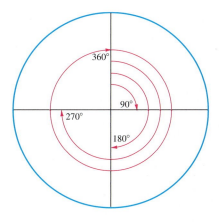

When we are creating a circle graph, how do we know how much of the circle to use for each piece? To make this decision, the circle must be scaled. A standard scale has been established for all circles. As we saw in Chapter 8, each circle has 360°. That means that $\frac{1}{4}$ of the circle has $\frac{1}{4}$ of 360°, which is 90°.

With a *protractor,* we can now create our own circle graph.

OBJECTIVE 2   **Example 3   Creating a Circle Graph**

The table represents the source of automobiles purchased in the United States in one year. Create a circle graph that represents the same data.

Source of Automobiles Purchased

Country of Origin	Number	% of Total
U.S.	6,500,000	80
Japan	800,000	10
Germany	400,000	5
All others	400,000	5

(*Source:* Amer. Auto. Manuf. Assn.)

To find the size of the sector of the graph for each country, we take the given percent of 360°. We create another column to represent the degrees needed.

Source of Automobiles Purchased

Country of Origin	Number	% of Total	Degrees
U.S.	6,500,000	80	288
Japan	800,000	10	36
Germany	400,000	5	18
All others	400,000	5	18

(*Source:* Amer. Auto. Manuf. Assn.)

NOTE 80% of 360° is
0.80 · 360° = 288°

Using a protractor, we start with Japan and mark a sector that is 36°.

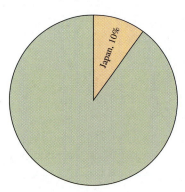

Again, using the protractor, we mark the 18° sector for Germany and the 18° sector for the other countries.

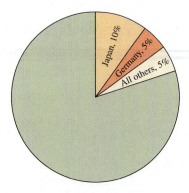

There is no need to measure the remainder of the circle. What is left is the 288° sector for U.S.-made cars. Notice that we saved the largest sector for last. It is much easier to mark the smaller sectors and leave the largest for last.

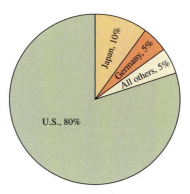

 CHECK YOURSELF 3

Create a circle graph for the table, which shows TV ownership for all United States homes.

TV Ownership	
Number of TVs	**% of U.S. Homes**
0	2
1	22
2	34
3 or more	42
(*Source:* Nielsen Media Research.)	

READING YOUR TEXT

The following fill-in-the-blank exercises are designed to assure that you understand the key vocabulary used in this section. Each sentence comes directly from the section. You will find the correct answers in Appendix C.

Appendix B

(a) A _____ is frequently used to show how some unit is divided.

(b) In a circle graph, wedges, also called _____, are drawn in the circle to show how much of the whole each part makes up.

(c) The percentages in a circle graph should add up to _____.

(d) In the standard scale, one full revolution around a circle is _____.

CHECK YOURSELF ANSWERS _____

1. **(a)** 20%; **(b)** 55%   **2.** **(a)** 6 h; **(b)** 14.4 h   **3.**

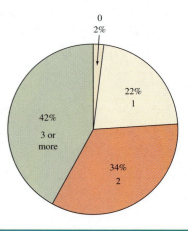

Appendix B Exercises

Business and Finance   The circle graph shows the budget for a local company. The total budget is $600,000. Find the dollar amount budgeted in each of the categories.

1. Production

2. Taxes

3. Research

4. Operating expenses

5. Miscellaneous

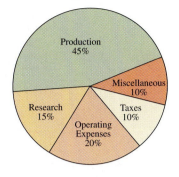

Business and Finance   The circle graph shows the distribution of a person's total yearly income of $24,000. Find the dollar amount budgeted for each category.

6. Food

7. Rent

8. Utilities

9. Transportation

10. Clothing

11. Entertainment

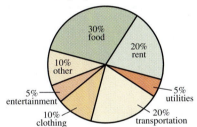

12. **Social Science**   The table represents women on active duty in the U.S. military in one year.

Service	Number of Women
Army	56,800
Navy	45,000
Marines	18,600
Air Force	65,700
Coast Guard	36,000

Create a circle graph for the information.

13. **Statistics**   The table represents the number of Nobel Prize laureates during some period.

Country	Number
United States	170
United Kingdom	69
Germany	59
France	24
USSR	10
Others	88

Create a circle graph for the information.

Answers

1. $270,000 **3.** $90,000 **5.** $60,000 **7.** $4,800 **9.** $4,800

11. $1,200 **13.**

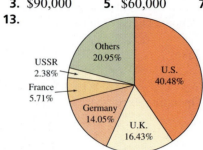

Reading Your Text Answers

CHAPTER 1

Section 1.1	a. Commas b. Arabic c. expanded d. ones
Section 1.2	a. natural b. addends c. zero d. sum
Section 1.3	a. opposite b. minuend c. denominate d. reasonable
Section 1.4	a. place b. rounding c. less d. left
Section 1.5	a. result b. units c. zeros d. reasonable
Section 1.6	a. vocabulary b. subtraction c. multiplication d. zero
Section 1.7	a. area b. zero c. scientific d. miles
Section 1.8	a. operations b. operation c. absolute d. keys
Section 1.9	a. questions b. equation c. true d. sentence

CHAPTER 2

Section 2.1	a. negative b. left c. ascending d. distance
Section 2.2	a. negative b. subtract c. inverse d. zero
Section 2.3	a. addition b. opposite c. two d. subtraction
Section 2.4	a. negative b. positive c. negative d. 1.8
Section 2.5	a. negative b. positive c. zero d. grouping
Section 2.6	a. variables b. expression c. product d. equal
Section 2.7	a. variable b. fraction c. negative d. Greek
Section 2.8	a. term b. coefficient c. like d. single
Section 2.9	a. expressions b. solution c. Linear d. first
Section 2.10	a. equivalent b. weight c. simplify d. check

CHAPTER 3

Section 3.1	a. denominator b. numerator c. proper d. improper
Section 3.2	a. factor b. prime c. composite d. product
Section 3.3	a. cross-products b. equivalent c. Fundamental d. common
Section 3.4	a. multiply b. reciprocal c. product d. reciprocal
Section 3.5	a. reciprocal b. divide c. multiply d. distributive
Section 3.6	a. addition b. multiplication c. positive d. identity

CHAPTER 4

Section 4.1	a. denominator b. magnitude c. multiples d. common
Section 4.2	a. mixed b. whole c. improper d. denominator
Section 4.3	a. denominate b. reasonable c. area d. speed
Section 4.4	a. LCD b. reciprocal c. true d. equivalent
Section 4.5	a. word b. substitution c. consecutive d. perimeter
Section 4.6	a. complex b. reciprocal c. grouping d. LCD

CHAPTER 5

Section 5.1   a. decimal   b. decimal   c. places   d. approximate
Section 5.2   a. points   b. places   c. borrowing   d. tool
Section 5.3   a. add   b. zeros   c. negative   d. right
Section 5.4   a. quotient   b. remainder   c. left   d. expression
Section 5.5   a. equivalent   b. repeating   c. places   d. lowest
Section 5.6   a. addition   b. multiplication   c. parentheses   d. opposite
Section 5.7   a. radical   b. squared   c. hypotenuse   d. perfect
Section 5.8   a. decimal   b. perimeter   c. length   d. kilometer

CHAPTER 6

Section 6.1   a. ratio   b. improper   c. one   d. units
Section 6.2   a. rate   b. unit   c. mixed   d. unit price
Section 6.3   a. proportion   b. true   c. proportional   d. coefficient
Section 6.4   a. right   b. similar   c. similar   d. hypotenuse
Section 6.5   a. unknown   b. units   c. cost   d. units

CHAPTER 7

Section 7.1   a. hundred   b. right   c. 100   d. denominator
Section 7.2   a. base   b. amount   c. rate   d. proportion
Section 7.3   a. translation   b. commission   c. amount   d. markup
Section 7.4   a. interest   b. principal   c. interest rate   d. compound
Section 7.5   a. percents   b. amount   c. base   d. amount

CHAPTER 8

Section 8.1   a. two   b. parallel   c. complementary   d. equal
Section 8.2   a. perimeter   b. circle   c. diameter   d. diameter
Section 8.3   a. square   b. parallelogram   c. triangles   d. Volume

CHAPTER 9

Section 9.1   a. table   b. two   c. legend   d. line graph
Section 9.2   a. origin   b. greater   c. vertical   d. zero
Section 9.3   a. second   b. ordered pair   c. straight   d. horizontal
Section 9.4   a. average   b. median   c. bimodal   d. mode

CHAPTER 10

Section 10.1   a. base   b. exponent   c. multiply   d. subtract
Section 10.2   a. variables   b. number   c. terms   d. trinomial
Section 10.3   a. plus   b. descending-exponent   c. distributive   d. like
Section 10.4   a. multiply   b. distributive   c. binomials   d. vertical
Section 10.5   a. multiplication   b. monomial   c. distributive   d. multiplying

Appendix B   a. circle graph   b. sectors   c. 100%   d. 360°

Answers to Pretests, Summary and Review Exercises, Self-Tests, and Cumulative Reviews

Pretest Chapter 1

[1.1] 1. One hundred seven thousand, nine hundred forty-five
[1.2] 2. Associative property of addition
[1.3] 3. Commutative property of multiplication
4. 27,614 **[1.5] 5.** 19,992 **6.** 56,100
7. (a) 26; **(b)** 56 **[1.6] 8.** 462 **9.** 1,204 r3
10. 81 r131 **11.** 93 **12.** \$151
[1.8] 13. 6 **14.** $P = 14$ yd; $A = 10$ yd^2
[1.4] 15. Estimate: 2,800; Actual: 2,802

Summary and Review Exercises

1. Hundreds
3. Twenty-seven thousand, four hundred twenty-eight
5. 37,583 **7.** Commutative property of addition
9. 1,416 **11.** 4,801 **13.** 11 **15.** 22 in.
17. \$1,620 **19.** 18,800 **21.** 2,574 **23.** 27
25. \$15,798 **27.** 16,000 **29.** $<$
31. Commutative property of multiplication
33. Associative property of multiplication **35.** 1,856
37. 154,602 **39.** 30,960 **41.** 18 in.^{2} **43.** 0
45. 308 r5 **47.** 497 r1 **49.** 28 mi/gal **51.** 70
53. 81 **55.** 28 **57.** 36 **59.** 3
61. 24 **63.** Expression **65.** Expression
67. True **69.** True **71.** $2^2 + 5 = 9$

Self-Test for Chapter 1

[1.1] 1. Hundred thousands
2. Three hundred two thousand, five hundred twenty-five
3. 2,430,000 **[1.2] 4.** Commutative property of addition
5. Associative property of addition **6.** 1,918
7. 13,103 **8.** 21,696 **9.** 55,978 **[1.3] 10.** 235
11. 12,220 **12.** 30,770 **13.** 40,555 **14.** 72 lb
[1.4] 15. 7,700 **16.** $>$ **17.** $<$
[1.5] 18. Associative property of multiplication
19. Distributive property of multiplication over addition
20. Multiplicative identity **21.** 4,984 **22.** 55,414
23. \$308,750 **[1.6] 24.** 76 r7 **25.** \$223
[1.8] 26. 15 **27.** 12 in. **28.** 12 in.^{2}
[1.9] 29. True **30.** $3^2 - 2 = 7$

Pretest Chapter 2

[2.1] 1.

2. $-4, -2, -1, 0, 1, 5$ **3.** Max: 7; Min: -5 **4.** 5
5. 6 **6.** 6 **7.** 6 **8.** 6 **9.** 16 **10.** -23
[2.6] 11. $x - 8$ **12.** $\dfrac{w}{17x}$ **13.** No **14.** Yes
[2.2 to 2.5] 15. -10 **16.** -1 **17.** -5 **18.** -3
19. -19 **20.** 12 **21.** 0 **22.** 21 **23.** 7
[2.7] 24. -3 **25.** 55 **26.** -7 **[2.8] 27.** $8w^2t$
28. $-a^2 + 4a + 3$

Summary and Review Exercises

1.

3. $-8, -7, -3, 0, 1, 2, 3, 7, 8$ **5.** 63 **7.** 9 **9.** -9
11. 20 **13.** 6 **15.** 14 **17.** 6 **19.** -32
21. -5 **23.** 17 **25.** 0 **27.** -70 **29.** 45
31. 0 **33.** 75 **35.** 400 **37.** 25 **39.** 16
41. 27 **43.** 169 **45.** -9 **47.** 0 **49.** Undefined
51. -1 **53.** $y + 5$ **55.** $8a$ **57.** $5mn$
59. $17x + 3$ **61.** Yes **63.** No **65.** -7
67. 15 **69.** 19 **71.** 25 **73.** 1 **75.** -3
77. $4a^3, -3a^2$ **79.** $5m^2, -4m^2, m^2$ **81.** $12c$
83. $2a$ **85.** $3xy$ **87.** $19a + b$ **89.** $3x^3 + 9x^2$
91. Yes **93.** Yes **95.** No **97.** 2 **99.** -5
101. 5 **103.** 1 **105.** -7

Self-Test for Chapter 2

[2.1] 1.

2. $-6, -3, -2, 0, 2, 4, 5$ **3.** 7 **4.** 7 **5.** 11
6. 11 **7.** -19 **8.** -40 **9.** 19 **[2.2] 10.** -13
11. -3 **12.** -21 **[2.3] 13.** -6 **14.** -24
15. 9 **16.** 0 **[2.4] 17.** -40 **18.** 63 **19.** -24
[2.5] 20. -25 **21.** 3 **22.** -5 **23.** Undefined
24. 3 **25.** 65 **26.** 144 **27.** -9 **[2.6] 28.** -4
29. $a - 5$ **30.** $6m$ **31.** $4(m + n)$ **32.** $\dfrac{a + b}{3}$
[2.7] 33. Not an expression **34.** Expression
[2.8] 35. $15a$ **36.** $19x + 5y$ **37.** $8a^2$ **38.** $2x - 8$
39. $2w + 4$ **[2.9] 40.** No **41.** Yes **[2.10] 42.** 11
43. 12 **44.** 7

Cumulative Review for Chapters 1 and 2

[1.1] 1. Hundred thousands
2. Three hundred two thousand, five hundred twenty-five
3. 2,430,000 **[1.2] 4.** Commutative property of addition
5. Additive identity **6.** Associative property of addition
7. 966 **8.** 23,351 **[1.4] 9.** 5,900 **10.** 950,000
11. 7,700 **[2.1] 12.** 9 **13.** -9 **14.** 3
[2.2] to [2.5] 15. -18 **16.** 0 **17.** -2 **18.** -2
19. 0 **20.** 10 **21.** 96 **22.** -90 **23.** -2
24. Undefined **[2.7] 25.** -60 **26.** 27
27. 24 **28.** 3 **[2.8] 29.** $3x + 19$ **30.** $2x + y$
[2.10] 31. 22 **32.** -8

Pretest Chapter 3

[3.2] 1. $1, 2, 3, 6, 7, 14, 21, 42$ **2.** $2 \cdot 5 \cdot 5 \cdot 7$ **3.** 4
[3.3] 4. Yes **5.** $\dfrac{-3}{5}$ **[3.4] 6.** $\dfrac{-4}{7}$ **7.** $\dfrac{4}{5}$
8. $\dfrac{2}{21}$ **[3.5] and [3.6] 9.** -1 **10.** 7

Summary and Review Exercises

1. Numerator: 5; Denominator: 9   **3.** $\dfrac{3}{8}$; 3; 8

5. $\dfrac{2}{3}, \dfrac{7}{10}, \dfrac{5}{4}, \dfrac{45}{8}, \dfrac{7}{1}, \dfrac{9}{1}, \dfrac{12}{5}$ **7.** 1, 41   **9.** 2 and 5

11. $2 \cdot 2 \cdot 2 \cdot 2 \cdot 3$ **13.** $2 \cdot 2 \cdot 2 \cdot 2 \cdot 3 \cdot 5 \cdot 11$ **15.** 5

17. 8   **19.** 7   **21.** No   **23.** $\dfrac{2}{3}$ **25.** $\dfrac{7}{9}$

27. 3   **29.** 8   **31.** $\dfrac{-1}{9}$ **33.** $\dfrac{3}{2}$ **35.** $\dfrac{-5}{6}$

37. 7   **39.** -4 **41.** 32   **43.** 3   **45.** -2
47. 18   **49.** 6   **51.** 6   **53.** -4

Self-Test for Chapter 3

[3.1] 1. $\dfrac{5}{6}$ 5 is the numerator; 6 is the denominator.

2. $\dfrac{5}{8}$ 5 is the numerator; 8 is the denominator.   **3.** $\dfrac{3}{5}$ 3 is the numerator; 5 is the denominator.

4. Proper $\dfrac{10}{11}, \dfrac{1}{8}$;   Improper $\dfrac{9}{5}, \dfrac{7}{7}, \dfrac{8}{1}$

[3.2] 5. Prime: 5, 13, 17, 31; composite: 9, 22, 27, 45
6. 2 and 3   **7.** $2 \cdot 2 \cdot 2 \cdot 3 \cdot 11$ **8.** 12   **9.** 8

[3.3] 10. Yes   **11.** Yes   **12.** No   **13.** $\dfrac{7}{9}$ **14.** $\dfrac{3}{7}$

15. $\dfrac{8}{23}$ **16.** 4   **17.** 3   **18.** 7   **[3.4] 19.** $\dfrac{9}{16}$

20. $\dfrac{-4}{15}$ **21.** $\dfrac{5}{8}$ **22.** $\dfrac{-3}{20}$ **23.** $\dfrac{2}{5}$ **24.** 1

25. $\dfrac{49}{9}$ **[3.5] and [3.6] 26.** 7   **27.** -12 **28.** 25

29. 3   **30.** 4   **31.** -5 **32.** -5 **33.** 21

34. $\dfrac{1}{2}$ **35.** $\dfrac{-57}{8}$ **[3.3] 36.** $\dfrac{7}{12}$ **37.** $\dfrac{6}{25}$

Cumulative Review for Chapters 1 to 3

[1.2] 1. 7,173   **2.** 1,918   **3.** 2,731   **4.** 13,103
[2.2] 5. -235 **[1.3] 6.** 12,220   **7.** 429   **8.** 3,239
[2.4] 9. -174 **10.** 1,911   **[1.5] 11.** 4,984
12. 55,414   **[1.6] 13.** 24 r191   **14.** 22 r21
15. 209 r145   **[1.7] 16.** 5   **17.** 7   **18.** 3
19. 16   **20.** 20   **21.** 3

[3.1] 22. Proper: $\dfrac{5}{7}, \dfrac{2}{5}$;   Improper: $\dfrac{15}{9}, \dfrac{8}{8}, \dfrac{11}{1}$

[3.3] 23. Yes   **24.** No   **[3.4] 25.** $\dfrac{1}{9}$ **26.** $\dfrac{1}{6}$

27. $\dfrac{-3}{2}$ **28.** $\dfrac{-15}{8}$ **29.** $\dfrac{5}{2}$ **30.** $\dfrac{-3}{13}$ **[2.8] 31.** $5x$

32. $-x + 23$ **33.** $-2x + 1$ **34.** $-2x + 3y$
35. $13x + 12y$ **[3.6] 36.** 5   **[3.5] 37.** -24

[3.6] 38. $\dfrac{5}{4}$ **39.** $\dfrac{-2}{5}$ **40.** 5

Pretest Chapter 4

[4.1] 1. (a) $\dfrac{5}{7}$; **(b)** $\dfrac{5}{9}$ **2.** 48   **3. (a)** 200; **(b)** 120

4. $\dfrac{17}{8}$ **5.** $\dfrac{5}{72}$ **[4.2] 6.** $9\dfrac{1}{4}$ **7.** $\dfrac{46}{7}$

8. $\dfrac{21}{5} = 4\dfrac{1}{5}$ **9.** $\dfrac{2}{21}$ **10.** $6\dfrac{17}{24}$ **11.** $1\dfrac{1}{36}$

[4.3] 12. 14 blocks   **13.** $18\dfrac{1}{4}$ yd^2 **[4.4] 14.** 40

[4.5] 15. 30   **[4.6] 16.** $\dfrac{9}{20}$

Summary and Review Exercises

1. $\dfrac{2}{3}$ **3.** $\dfrac{11}{9}$ **5.** 72   **7.** $\dfrac{7}{12}, \dfrac{5}{8}$

9. $>$ **11.** $<$ **13.** $\dfrac{36}{120}, \dfrac{75}{120}, \dfrac{70}{120}$ **15.** 36

17. 24   **19.** $\dfrac{19}{24}$ **21.** $\dfrac{-7}{12}$ **23.** $\dfrac{109}{60}$ **25.** $\dfrac{3}{10}$

27. $\dfrac{1}{3}$ **29.** $\dfrac{7}{30}$ **31.** $\dfrac{7}{12}$ **33.** $\dfrac{14}{15}$ **35.** -4

37. $11\dfrac{3}{4}$ **39.** $\dfrac{-43}{10}$ **41.** $\dfrac{-164}{13}$ **43.** $11\dfrac{1}{3}$

45. $1\dfrac{1}{2}$ **47.** $10\dfrac{2}{7}$ **49.** $-4\dfrac{1}{3}$ **51.** 220 mi

53. $540   **55.** 56 mi/h   **57.** 48 lots

59. $69\dfrac{1}{16}$ in.   **61.** $\dfrac{3}{4}$ in.   **63.** 15   **65.** 32

67. 12   **69.** 3   **71.** 10   **73.** 8   **75.** 17, 19, 21
77. Susan: 7 years, Larry: 9 years, Nathan: 14 years

79. $\dfrac{5}{6}$ **81.** $\dfrac{52}{163}$

Self-Test for Chapter 4

[4.1] 1. $\dfrac{9}{10}$ **2.** $\dfrac{2}{3}$ **3.** 72   **4.** 60   **5.** 36

6. $\dfrac{4}{5}$ **7.** $\dfrac{-25}{42}$ **8.** $\dfrac{63}{40}$ **9.** $\dfrac{11}{9}$ **10.** $\dfrac{-2}{3}$

11. $\dfrac{23}{30}$ **[4.2] 12.** $4\dfrac{1}{4}$ **13.** $\dfrac{74}{9}$ **14.** $3\dfrac{3}{7}$

15. 4   **16.** $2\dfrac{2}{3}$ **17.** $7\dfrac{11}{12}$ **18.** $1\dfrac{3}{4}$

19. $1\dfrac{8}{15}$ **[4.3] 20.** 47 homes   **21.** 48 books

22. Around 39,000 cups   **[4.4] 23.** -32 **24.** $\dfrac{-4}{3}$

25. 50   **26.** $\dfrac{28}{3}$ **[4.5] 27.** 18 and 19   **28.** $35

[4.6] 29. $\dfrac{4}{5}$ **30.** $\dfrac{18}{43}$

Cumulative Review for Chapters 1 to 4

[2.2] 1. 4   **2.** -12 **[4.1] 3.** 6   **4.** $\dfrac{2}{15}$

[2.4] 5. -18 **[3.4] 6.** $\dfrac{1}{5}$ **7.** $\dfrac{-2}{7}$ **[2.5] 8.** 10

[3.4] 9. $\dfrac{-4}{3}$ **[1.6] 10.** Undefined   **[3.4] 11.** 0

[4.1] 12. $\dfrac{37}{60}$ **[2.4] 13.** 27   **[3.4] 14.** $\dfrac{1}{7}$

[4.2] 15. $1\dfrac{7}{9}$ **16.** $7\dfrac{1}{5}$ **17.** $\dfrac{23}{4}$ **18.** $\dfrac{55}{9}$

[4.4] 19. Expression   **20.** Equation   **21.** Expression

22. Equation   **[4.2] 23.** $-2\frac{1}{8}$ **24.** $11\frac{1}{3}$ **25.** $6\frac{2}{7}$

26. $-4\frac{5}{9}$ **27.** $3\frac{5}{8}$ **28.** $2\frac{13}{24}$ **29.** $\frac{-3}{4}$ **30.** $-1\frac{1}{2}$

[4.3] 31. $\frac{5}{8}$ in.   **32.** $1\frac{11}{12}$ h   **[3.6] 33.** -3

[4.4] 34. 8   **[4.5] 35.** 13   **36.** 46, 47   **37.** 7
38. $420   **39.** 5 cm, 17 cm   **40.** 8 in., 13 in., 16 in.

Pretest Chapter 5

[5.1] 1. Thousandths
2. 2.371; two and three hundred seventy-one thousandths
[5.2] 3. (a) 52.31; **(b)** 2.375   **[5.8] 4.** $2.36
[5.3] 5. (a) -0.86037; **(b)** 536.2   **[5.1] 6.** 2.36
[5.8] 7. $39.55   **[5.6] 8.** $x = -5$ **9.** $x = -3.536$
[5.4] 10. 4.25   **11.** 2.435   **[5.8] 12.** $11.60
[5.4] 13. 0.0534   **[5.5] 14. (a)** 0.375; **(b)** 0.29
[5.7] 15. 10

Summary and Review Exercises

1. Hundredths   **3.** 0.37   **5.** Seventy-one thousandths
7. 4.5   **9.** $>$ **11.** $<$ **13.** 5.84   **15.** 4.876
17. 18.852   **19.** 20.533   **21.** 4.245   **23.** 31.38
25. 0.000261   **27.** 0.0012275   **29.** 450   **31.** 4.65
33. 2.664   **35.** 1.273   **37.** 0.76   **39.** 0.0457
41. 0.429   **43.** 3.75   **45.** $\frac{21}{250}$ **47.** $\frac{67}{10,000}$
49. 4   **51.** 1.65   **53.** -4.8 **55.** 18   **57.** 55
59. 22.2 mi   **61.** $61.75   **63.** $345.95   **65.** $152.10
67. $23.45   **69.** 54 lots   **71.** $7.09   **73.** 8.7 cm

Self-Test for Chapter 5

[5.1] 1. Ten-thousandths   **2.** 0.049
3. Two and fifty-three hundredths   **4.** 12.017   **5.** $<$
6. $>$ **[5.2] 7.** 16.64   **8.** 27.547   **9.** 12.803
[5.1] 10. 0.598   **11.** 23.57   **[5.2] 12.** 10.54
13. 24.375   **14.** -3.888 **[5.3] 15.** 17.437
16. 0.02793   **17.** -1.4575 **18.** 735   **19.** 12,570
[5.4] 20. 0.465   **21.** 2.35   **22.** 0.051   **23.** 2.55
24. 2.385   **25.** -7.35 **26.** 0.067   **27.** 0.004983
28. 0.00523   **[5.5] 29.** 0.5625   **30.** 0.571   **31.** $0.6\overline{3}$
32. $\frac{9}{125}$ **33.** $4\frac{11}{25}$ **34.** $>$ **35.** $\frac{229}{500}$
[5.6] 36. 1.45   **37.** -3.6 **38.** -1.07 **39.** 2.15
40. -5.3 **[5.7] 41.** 21   **42.** 65 m   **43.** 112 mm
[5.8] 44. 50.2 gal   **45.** $3.06   **46.** 7.525 in.^{2}
47. $543   **48.** 32 lots   **49.** $573.40

Cumulative Review for Chapters 1 to 5

[2.3] 1. 12   **[2.4] 2.** 20   **[2.5] 3.** 0   **[1.7] 4.** 15
5. -16 **[3.4] 6.** $\frac{3}{4}$ **[4.1] 7.** $\frac{-13}{12}$ **[3.4] 8.** $\frac{-4}{3}$
[5.2] 9. 20.82   **10.** 10.83   **[5.3] 11.** 12.75   **12.** 4.2
[5.4] 13. 0.5326   **[5.1] 14.** 6.821   **15.** 6.0   **16.** $<$
17. $=$ **[5.5] 18.** $\frac{3}{20}$ **19.** $\frac{2}{25}$ **20.** 0.625

21. 0.86   **[3.2] 22.** $2 \cdot 3 \cdot 5 \cdot 7$ **23.** 6   **[4.1] 24.** 90
[4.5] 25. Equation   **26.** Expression   **[2.7] 27.** 27
28. -3 **[2.8] 29.** $3x - y$ **30.** $2x + 7$
[3.6] 31. $x = -4$ **32.** $x = -5$ **[4.5] 33.** $x = 48$
34. $x = -25$ **[5.6] 35.** $x = -2.5$ **36.** $x = -8$
[5.8] 37. 26.4 ft   **38.** 41 ft^2 **[4.6] 39.** 15   **40.** $550
[5.8] 41. $18.2 \frac{\text{mi}}{\text{gal}}$ **[4.6] 42.** $W = 12$ cm; $L = 32$ cm
[5.7] 43. 17   **44.** 22.2 in.

Pretest Chapter 6

[6.1] 1. $\frac{7}{10}$ **2.** $\frac{4}{3}$ **[6.2] 3.** $29 \frac{\text{mi}}{\text{gal}}$ **4.** $\frac{\$0.79}{\text{can}}$
[6.3] 5. Yes   **6.** No   **7.** 10   **8.** 15   **9.** 12
[6.3] 10. 132 ft   **11.** $6.30   **12.** 32   **13.** 11 gal
[6.4] 14. 12

Summary and Review Exercises

1. $\frac{4}{17}$ **3.** $\frac{5}{8}$ **5.** $\frac{4}{9}$ **7.** $\frac{7}{36}$ **9.** $100 \frac{\text{mi}}{\text{h}}$
11. $50 \frac{\text{calories}}{\text{oz}}$ **13.** $200 \frac{\text{ft}}{\text{s}}$ **15.** $6\frac{1}{2} \frac{\text{hits}}{\text{game}}$
17. Marisa   **19.** $\frac{9\cancel{c}}{\text{oz}}$ **21.** $\frac{9.5\cancel{c}}{\text{oz}}$ **23.** $\frac{\$14.95}{\text{CD}}$
25. $\frac{4}{9} = \frac{20}{45}$ **27.** $\frac{110 \text{ mi}}{2 \text{ h}} = \frac{385 \text{ mi}}{7 \text{ h}}$ **29.** no
31. yes   **33.** yes   **35.** yes   **37.** $m = 2$
39. $t = 4$ **41.** $p = 16$ **43.** $a = 30$ **45.** $s = 0.5$
47. 256 first-year students   **49.** 220 drives   **51.** 120 mi
53. 8   **55.** 15   **57.** 15   **59.** 132   **61.** 4 ft 11 in.
63. 9 ft 7 in.   **65.** 5 ft 7 in.   **67.** 4 ft 3 in.   **69.** Yes
71. (b)   **73.** (b)   **75.** km   **77.** 2,000   **79.** 3
81. 0.06

Self-Test for Chapter 6

[6.1] 1. $\frac{7}{19}$ **2.** $\frac{5}{3}$ **3.** $\frac{2}{3}$ **4.** $\frac{1}{12}$ **5.** $\frac{26}{33}, \frac{26}{7}$
[6.2] 6. $4.8 \frac{\text{mi}}{\text{gal}}$ **7.** $8.25 \frac{\text{dollars}}{\text{h}}$ **8.** $\frac{\$2.56}{\text{gal}}$
[6.3] 9. yes   **10.** no   **11.** yes   **12.** no
13. $x = 20$ **14.** $a = 18$ **15.** $p = 3$ **16.** $a = 48$
17. $x = 6$ **18.** $m = 16$ **19.** $2.28   **20.** 644 points
21. 576 mi   **22.** 3,000 no votes   **23.** 600 mufflers
24. 24 tsp   **[6.4] 25.** 55 ft   **26.** 20 m
[6.5] 27. 11 ft 5 in.   **[6.5] 28.** 2 lb 9 oz   **29.** $1,050
30. (b)   **31.** (b)   **32.** (b)

Cumulative Review for Chapters 1 to 6

[1.2] 1. Associative prop. of addition
[1.5] 2. Commutative prop. of multiplication
3. Distributive prop.   **[1.4] 4.** 5,900   **5.** 950,000
[1.8] 6. 8   **[3.2] 7.** $2 \cdot 2 \cdot 2 \cdot 3 \cdot 11$ **[4.1] 8.** 90
[4.2] 9. $3\frac{1}{7}$ **10.** $\frac{53}{8}$ **11.** $\frac{3}{4}$ **12.** $1\frac{1}{2}$ **13.** $8\frac{1}{24}$
14. $3\frac{5}{8}$ **[4.4] 15.** $\frac{5}{8}$ in.   **[5.2] 16.** $3.06

[5.3] 17. 8.04 ft² [5.4] 18. 32 [5.5] 19. $0.\overline{72}$
[6.3] 20. $m = 112.5$ [6.4] 21. 12 in. [2.1] 22. −8
23. 20 24. 12 25. 5 [2.2] 26. −18
[2.3] 27. −4 [2.4] 28. −90 [2.5] 29. −4
[2.6] 30. $3(x + y)$ 31. $\dfrac{n-5}{3}$ [2.10] 32. 5
[4.5] 33. −24 [3.6] 34. $-\dfrac{2}{5}$ 35. 5

Pretest Chapter 7

[7.1] 1. $\dfrac{7}{100}$ 2. 0.23 3. 3.5% or $3\frac{1}{2}$% 4. 80%
[7.2] 5. 63 6. 9% 7. 600 [7.4] 8. $560
[7.3] 9. 5% [7.5] 10. $1,200 11. $11.31
[7.4] 12. $5,955.08 [7.3] 13. $416.50

Summary and Review Exercises

1. 75% 3. $\dfrac{1}{5}$ 5. $1\frac{1}{2}$ 7. 3 9. 0.04
11. 0.135 13. 2.25 15. 37.5% 17. 700%
19. 0.5% 21. 70% 23. 125% 25. 27.3%
27. 140% 29. 12.5% 31. 75 33. 75
35. 175% 37. 3% 39. 660 41. 22%
43. $130 45. $3,000 47. 7.5% 49. $2,800
51. 6.5% 53. 500 s (8 min 20 s) 55. $114.50

Self-Test for Chapter 7

[7.1] 1. 80% 2. $\dfrac{7}{100}$ 3. $\dfrac{72}{100}$ or $\dfrac{18}{25}$ 4. 0.42
5. 0.06 6. 1.6 7. 3% 8. 4.2% 9. 40%
10. 62.5% [7.2] 11. Rate: 25%; base: 200; amount: 50
12. Rate: 8%; base: 500; amount: unknown
13. Rate: 6%; base: amount of purchase; amount: $30
14. 11.25 15. 500 16. 750 17. 20%
18. 7.5% 19. 175% 20. 800 21. 300
[7.3] 22. $4.96 23. 60 questions 24. $70.20
[7.4] 25. 12% [7.3] 26. 24% 27. 8%
28. $18,000 [7.5] 29. 6,400 students
[7.4] 30. $18,500 [7.3] 31. $17,500
[7.4] 32. $3,246.37

Cumulative Review for Chapters 1 to 7

[1.1] 1. Thousands [1.5] 2. 11,368 [1.6] 3. 89
[1.7] 4. 6 5. −4 6. −1 7. 17
[3.2] 8. $2 \cdot 2 \cdot 5 \cdot 13$ 9. 28 [4.1] 10. 180
[4.2] 11. $8\frac{1}{2}$ 12. $1\frac{1}{3}$ 13. $8\frac{7}{12}$ 14. $4\frac{19}{24}$
[4.4] 15. $286 16. $54\dfrac{\text{mi}}{\text{h}}$ 17. $41\frac{1}{4}$ in.
[5.1] 18. Hundredths 19. Ten thousandths 20. <
21. = [5.5] 22. $\dfrac{9}{25}$ 23. $5\frac{1}{8}$ [5.3] 24. 11.284
[5.4] 25. 17.04 [5.8] 26. $108.05 [2.10] 27. 5
[3.6] 28. −36 29. $-\dfrac{2}{5}$ 30. 5 [6.3] 31. $18\frac{2}{3}$
32. 0.4 [6.5] 33. 350 mi 34. $140

[7.1] 35. 0.34; $\dfrac{17}{50}$ 36. 0.55; 55% [7.2] 37. 45
38. 500 [7.3] 39. 125 40. 8.5%

Pretest Chapter 8

[8.1] 1.

$\overset{\bullet}{A} \longrightarrow \overset{\bullet}{B}$

2. 100° 3. 40° 4. 140° [8.2] 5. 15 m
6. 25.1 cm [8.3] 7. 12 ft²

Summary and Review Exercises

1. ∠AOB; acute; 70° 3. ∠AOC; obtuse; 100°
5. ∠XYZ; straight; 180° 7. 135° 9. 13.6 in.
11. 8π m ≈ 25.12 13. 11.56 in.² 15. 16π m² ≈ 50.2 m
17. 37.7 ft 19. 314 ft² 21. 400 in.² 23. $787.50
25. $36,000\pi$ m³ ≈ 113,040 m³

Self-Test for Chapter 8

[8.1] 1. (a) Parallel; (b) neither; (c) perpendicular; (d) neither
2. (a) Straight; (b) obtuse; (c) right 3. 50° 4. 135°
5. 300° [8.2] 6. 16.8 cm 7. 21 in. 8. 8.2 m
9. $21\frac{1}{2}$ in. 10. 18 cm 11. 10π ≈ 31.4 ft
[8.3] 12. 17.64 cm² 13. $27\frac{9}{16}$ in.² 14. 4.14 m²
15. $24\frac{3}{8}$ in.² 16. 13.5 cm² 17. 25π ≈ 78.5 ft²
18. 15.625 m³ 19. 30 ft³ 20. ≈14.1 ft³
21. ≈4,521.6 cm³

Cumulative Review for Chapters 1 to 8

[1.5] 1. $896 [1.6] 2. $62 [2.2] 3. 24 4. −21
[2.5] 5. −16 6. $\dfrac{1}{5}$ [3.2] 7. $2 \cdot 2 \cdot 2 \cdot 3 \cdot 7$
8. 4 [4.1] 9. $\dfrac{3}{5}, \dfrac{5}{8}, \dfrac{2}{3}$ [4.2] 10. $\dfrac{3}{4}$ 11. $\dfrac{5}{12}$
[4.1] 12. 60 13. $\dfrac{31}{30}$ [4.2] 14. $\dfrac{85}{24}$ [2.10] 15. −5
16. −8 [3.6] 17. 14 18. $\dfrac{2}{3}$ [5.8] 19. $18.30
[8.3] 20. 27.84 cm² [8.2] 21. 21.5 cm 22. 20.1 ft
[5.5] 23. 0.5625 24. 0.538 [5.7] 25. 17 ft
[6.3] 26. $x = 24$ [6.5] 27. 400 mi 28. 450 mi
[7.1] 29. $\dfrac{1}{8}$ [7.2] 30. 3,526 31. 225% 32. 150
[7.3] 33. $142,500 [6.6] 34. 21 lb 11 oz 35. 2 ft 9 in.
36. 62,000 37. 74 [8.1] 38. 43° 39. 73°
[6.4] 40. $x = 40$

Pretest Chapter 9

[9.1] 1. $12,000 2. $19,400 3. Approximately 25,000
4. −20,000 5. 1980–1985; 20,000 6. 1,500,000
7. 750% [9.3] 8. (15, 3); (18, 6) 9. (0, 3); (2, 0)

10. $(1, 3); (0, 5); (-3, 11)$ **11.** Answers vary

[9.2] 12. $A(2, 3), B(0, -5), C(-3, 5)$

13.

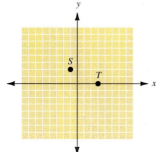

[9.3] 14.

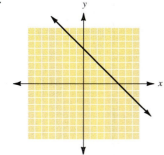

15.

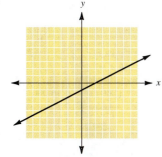

16.

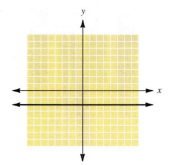

[9.4] 17. mean 27; median 24; mode 42

Summary and Review Exercises

1. General Motors **3.** 53.8% **5.** La Jolla, CA

7. 21.4% **9.** 5,000

11.

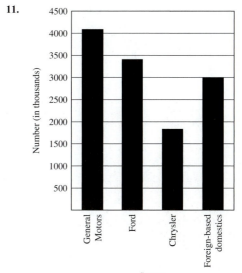

13. 500%

15.

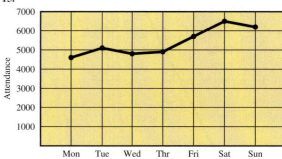

17. $(0, 3)$

19.

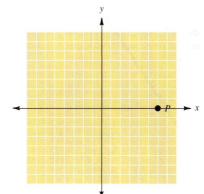

21.

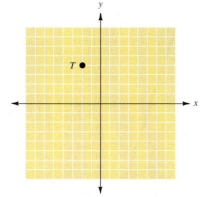

23. No   **25.** No   **27.** $(3, 0), (-3, 4), (0, 2)$

29. $(0, -5), (12, 1), (6, -2), (8, -1)$

31. $(2, 10), (1, 7), \left(\frac{1}{3}, 5\right), \left(\frac{4}{3}, 8\right)$

33. $(0, -2), (3, 0), (6, 2), (9, 4)$

35.

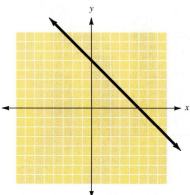

37.

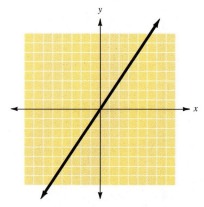

39.

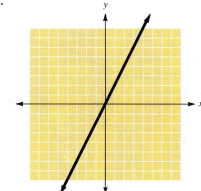

41.

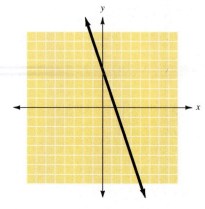

43.

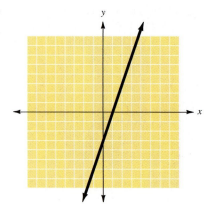

45.

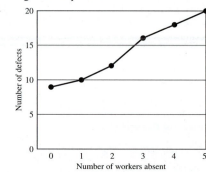

47. 6   **49.** 120   **51.** 85   **53.** 19; 20

55. 29; 28   **57.** 92

Self-Test for Chapter 9

[9.1] 1. $3,955   **2.** 2.9% more affordable   **3.** $5,775

4. 4.5% more affordable   **5.** 16,000   **6.** 16,000

7. 2000 to 2001   **8.**

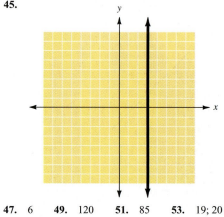

9. **(a)** 700,000; **(b)** 20.6%   **10.** December

11. August and September

12.

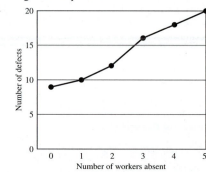

[9.2] 13. $(4, 2)$ **14.** $(-4, 6)$ **15.** $(0, -7)$

16.

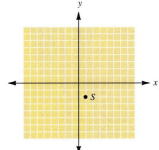

17.

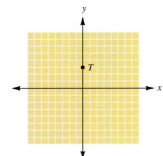

18.

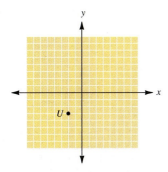

[9.3] 19. $(4, 0), (5, 4)$ **20.** $(3, 3), (6, 2), (9, 1)$
21. Different answers are possible
22.

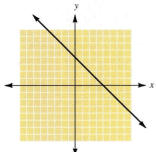

23.

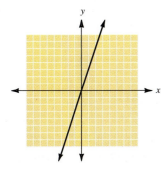

24.

25.

[9.4] 26. 15   **27.** 11   **28.** 6   **29.** 204
30. Brown

Cumulative Review for Chapters 1 to 9

[1.1] 1. Thousands   **[2.5] 2.** -5 **[1.6] 3.** Undefined

[5.3] 4. 75.215   **[3.4] 5.** $\dfrac{2}{3}$ **6.** $\dfrac{6}{11}$ **[4.2] 7.** $7\dfrac{1}{12}$

[2.7] 8. -40 **9.** -28 **[2.8] 10.** $6x$ **11.** $4x - 3y$

[3.5] 12. -24 **[6.3] 13.** $\dfrac{5}{4}$ **14.** 14   **15.** 9

[7.1] 16. $0.18; \dfrac{9}{50}$ **17.** $0.425; 42.5\%$

[6.6] 18. 11 lb 5 oz   **19.** 8,000   **20.** 3   **21.** 5
22. 250   **[9.1] 23.** 1998 and 1999
24.

[9.4] 25. Mean: 9; Median: 9; Mode: 11   **[6.5] 26.** 70

[8.3] 27. \$408   **[6.2] 28.** $48\dfrac{\text{mi}}{\text{h}}$ **[8.2] 29.** $28\dfrac{3}{5}$ cm

[7.3] 30. $55\dfrac{1}{3}$ ft   **31.** 133   **32.** 0.2%   **33.** 185

[4.2] 34. $4\dfrac{1}{3}$ **[7.5] 35.** 2,800 students

Pretest Chapter 10

[10.1] 1. **(a)** x^7; **(b)** $m^{21}n^{12}$
[10.2] 2. **(a)** binomial; **(b)** trinomial **3.** -6
[10.3] 4. $2x^2 - 2x - 2$ **5.** $9x^2 - 11x + 6$ **[10.1] 6.** $8x^5y^7$
[10.4] 7. $12x^3y^3 - 6x^2y^2 + 21x^2y^4$ **8.** $6x^2 - 11x - 10$
9. $x^2 - 4y^2$ **10.** $16m^2 + 40m + 25$
11. $3x^3 - 14x^2y + 17xy^2 - 6y^3$ **[10.5] 12.** $5(3c + 7)$
13. $4q^3(2q - 5)$ **14.** $6(x^2 - 2x + 4)$
15. $7cd(c^2d - 3 + 2d^2)$

Summary and Review Exercises

1. $4a^2b^2$ **3.** $72x^{12}y^8$ **5.** $27y^{12}$ **7.** Monomial
9. Trinomial **11.** -4 **13.** -19 **15.** $21a^2 - 2a$
17. $5y^3 + 4y$ **19.** $5x^2 + 3x + 10$ **21.** $x - 2$
23. $-9w^2 - 10w$ **25.** $9b^2 + 8b - 2$ **27.** $2x^2 - 2x - 9$
29. $5a^5$ **31.** $54p^5$ **33.** $15x - 40$
35. $-10r^3s^2 + 25r^2s^2$ **37.** $x^2 + 9x + 20$ **39.** $a^2 - 49b^2$
41. $a^2 + 7ab + 12b^2$ **43.** $6x^2 - 19xy + 15y^2$

45. $y^3 - y + 6$ **47.** $x^3 - 8$ **49.** $2x^3 - 2x^2 - 60x$
51. $x^2 + 14x + 49$ **53.** $4w^2 - 20w + 25$
55. $a^2 + 14ab + 49b^2$ **57.** $x^2 - 25$ **59.** $4m^2 - 9$
61. $25r^2 - 4s^2$ **63.** $6(3a + 4)$ **65.** $8s^2(3t - 2)$
67. $7s^2(5s - 4)$ **69.** $9m^2n(5n - 3)(n + 1)$
71. $8ab(a + 3 - 2b)$

Self-Test for Chapter 10

[10.2] 1. Binomial **2.** Trinomial **3.** 6
[10.3] 4. $10x^2 - 12x - 7$ **5.** $7a^3 + 11a^2 - 3a$
6. $3x^2 + 11x - 12$ **7.** $b^2 - 7b - 5$ **8.** $7a^2 - 10a$
9. $4x^2 + 5x - 6$ **10.** $2x^2 - 7x + 5$ **[10.1] 11.** a^{14}
12. $15x^3y^7$ **13.** $27x^6y^3$ **14.** $16x^{18}y^{17}$
[10.4] 15. $15a^3b^2 - 10a^2b^2 + 20a^2b^3$ **16.** $3x^2 + x - 14$
17. $a^2 - 49b^2$ **18.** $8x^2 - 14xy - 15y^2$
19. $12x^3 + 11x^2y - 5xy^2$ **20.** $9m^2 + 12mn + 4n^2$
21. $2x^3 + 7x^2y - xy^2 - 2y^3$ **[10.5] 22.** $6(2b + 3)$
23. $3p^2(3p - 4)$ **24.** $5(x^2 - 2x + 4)$
25. $6ab(a - 3 + 2b)$

Index

A

Abscissa, 731
Absolute value, 124–125, 203
Absolute value bars, 95, 125
Absolute value expressions, 125
Acre, 665
Acute angles, 638, 679
Addends, definition of, 12, 109
Addition
  of absolute values, 125
  associative property of, 14, 109
  basic facts of, 13
  calculator for, 21, 141
    of decimals, 408–409
    of fractions, 305–306
  carrying in, 16–17
    of decimals, 404
  commutative property of, 13, 109
  of decimals, 403–405, 485
    calculator for, 408–409
    carrying in, 404
    signed, 404
    in word problems, 463–466
  definition of, 11
  of denominate numbers, 31
  distributive property of
    multiplication over, 52, 110
    combining like terms and, 181
    for solving equations,
      196–198, 284
  estimating, 43–44
  in expanded form, 16
  expressions indicating, 158
  of fractions
    calculator for, 305–306
    like, 293–295, 379
    negative, 296
    unlike, 303–304, 379
    in word problems, 334–335
  grouping symbols and, 14
  identity property in, 14, 109
  of integers, 132–136, 171, 203
  of measurement units in English
    system, 547, 562
  of mixed numbers, 320–321, 380
  multiplication as repeated, 50
  on number line, 12–13
  in order of operations, 95, 111
  for perimeter, 18–20, 109
  place value and, 15
  of polynomials, 797–802, 825
    horizontal method for, 797–799
    simplification in, 802
    vertical method for, 801
  in short form, 17
  solving equations with, 193–195,
    272–276
  subtraction as opposite of, 28
  symbol for, 12
  of whole numbers, 11–21, 109
  in word problems, 20, 135, 334–335,
    463–466
  words for, 16
  of zero, 14
Addition property, of equality,
  192–198, 204
Additive inverse, 134
Adjacent angles, 642, 643
Al-Kashi, 655
Amount, in percent problems,
  588, 623
Angle(s)
  acute, 638, 679
  adjacent, 642, 643
  complementary, 640–641
  definition of, 636, 679
  exterior, 650
  measurement of, 638, 639–640
  naming, 637–638
  obtuse, 638, 679
  right, 637, 638, 679
  in right triangle, 451
  straight, 638
  supplementary, 641
  in triangles, 643–644
  vertex of, 636
  vertical, 642–643
Approximate numbers, 394
Archimedes, 90, 655, 668
Architecture, 633
Area
  of circle, 667, 680
  definition of, 59, 662, 680
  finding with multiplication, 59, 111
  of oddly shaped figure, 61
  of parallelogram, 664, 680
  of rectangle, 60, 111, 662, 663, 680
  of square, 59, 60–61, 88,
    662–663, 680
  total surface, 70
  of triangle, 665, 680
  units of, 60, 665–666
  in word problems, 666
Arithmetic, fundamental theorem
  of, 227
Associative property
  of addition, 14, 109
  of multiplication, 58–59, 111
Average. *See also* Mean; Median;
  Mode
  calculator for, 758–759
  choosing appropriate measure
    of, 757
  definition of, 753
Axes
  of bar graph, 694
  of coordinate system, 715, 767

B

Babylonians, 3, 452
Bar graph, 694–697
  axes of, 694
  creating, 695–697
  definition of, 694, 767
  legend of, 695
  reading, 694
Base
  of exponent, 88, 111
  in percent problems, 588, 623
Bimodal set, 756
Binomial. *See also* Polynomials
  definition of, 791, 825
  greatest common factor of, 819
  multiplication of, 809–812
    by monomial, 808
Borrowing, in subtraction, 30–31
  of decimals, 406
Braces, 122
Brackets, 95

C

Calculator
  for addition, 21, 141
    of decimals, 408–409
    of fractions, 305–306
    of mixed numbers, 323–324

Calculator—*Cont.*
  for division, 80–82, 153–154
    of decimals, 429–430, 471
    of fractions, 258–259
  for evaluating expressions, 97–98, 169
  for exponents, 91
  for fraction simplification, 241–242
  for integers, 141–142, 153–154
  for mean, 758–759
  for multiplication
    of decimals, 417–419
    of fractions, 257–258
  for square root, 455–456
  for subtraction, 32–33, 141–142
    of decimals, 408–409
    of fractions, 305–306
    of mixed numbers, 323–324
Capacity, units of measurement for, 545, 562
Carrying, in addition, 16–17
  of decimals, 404
Cartesian coordinate system. *See* Coordinate system
Cell, definition of, 693
Celsius, conversion of Fahrenheit to, 176
Centi, 550, 553, 562
Centimeter (cm), 550, 562
Chain, 555
Circle
  area of, 667, 680
  circumference of, 655, 656, 680
  diameter of, 655
  origin of, 655
  radius of, 655
Circle graph, 840–844
  creating, 842–843
  definition of, 840
  interpreting, 841
  reading, 840
  sectors of, 840
Circular cylinder, right, volume of, 669
Circumference, of circle, 655, 656, 680
Coefficient
  definition of, 178, 790, 825
  identifying, 178, 790–791
Comma, four-digit numbers written with, 4
Common denominators, 300–301
Common factors
  definition of, 228
  finding, 228
  in fraction simplification, 238
  greatest, 228–230

definition of, 228, 283, 818
  finding, 229–230, 283
  of polynomials, 817–820, 826
Common fraction, 215
Common multiples, 298
Commutative property
  of addition, 13, 109
  of multiplication, 50, 110
Complementary angles, 640–641
Complex fractions
  definition of, 373, 380
  simplifying, 373–375, 380
  writing quotient as, 255
Composite numbers
  definition of, 225, 283
  factoring, 225–226
  identifying, 225
Compound interest, 608, 624
Conditional equation, 186
Consecutive integers, 362–363, 380
Continued division, 227–228
Coordinate system, 715–720, 767
  axes of, 715, 767
  graphing points in, 718–719, 767
  identifying coordinates in, 716–717
  origin of, 715, 767
  quadrants of, 715
Counting, using multiplication to simplify, 51
Counting numbers. *See* Natural numbers
Cross products, 235–236
Cube, volume of, 680
Cubic units, 667
Cubit, 497, 560
Cylinder, volume of, 669, 680

D
Decagon, 653
Deci, 553, 562
Decimal(s), 389–494
  addition of, 403–405, 485
    calculator for, 408–409
    carrying in, 404
    signed, 404
    in word problems, 463–466
  calculator for, 407–408
    addition of, 408–409
    division of, 429–430, 471
    multiplication of, 417–419
    subtraction of, 408–409
    in word problems, 471

comparing, 394
  to fractions, 438–439
converting
  to fractions, 437–438, 486
  fractions to, 434–437, 439, 440, 486
  to mixed numbers, 393, 438
  mixed numbers to, 437, 440
  percent to, 575–576, 623
division of, 423–430, 486
  calculator for, 429–430, 471
  by decimal, 425–426
  by powers of ten, 426–427, 486
  signed, 428
  by whole number, 423–425
  in word problems, 469–470
in equations, 446–448
multiplication of, 414–419, 485
  calculator for, 417–419
  estimating, 415
  by powers of ten, 416–419, 468, 485
  signed, 416
  in word problems, 466–468
on number line, 394
order of operations with, 428–429
power of, 417
in ratios, 501
repeating
  converting fractions to, 435–437
  definition of, 435
rounding, 394–396, 485
  to nearest hundredth, 395–396
  to nearest tenth, 395
  to specified decimal place, 396
signed
  addition of, 404
  division of, 428
  multiplication of, 416
  subtraction of, 407
subtraction of, 404–407, 485
  borrowing in, 406
  calculator for, 408–409
  signed, 407
  in word problems, 464–466
terminating
  converting to fractions, 437, 486
  definition of, 434
in word problems, 463–473
in words, 392–393, 485
writing numbers in, 391–392, 485
Decimal equivalent, 434
Decimal fractions, 391, 393, 485
Decimal places, 393, 485

Decimal place-value system, 3.
    See also Place value
Decimal point, 391, 393
"Decreased by," as term for
    subtraction, 29
Degree, as angle measure, 638
Deka, 553, 562
Denominate numbers. *See also* Unit(s)
    addition of, 31
    division of, 72, 330, 332
    multiplication of, 53, 59, 330
    ratio of, 501–502
    subtraction of, 31
Denominator(s). *See also* Fraction(s)
    common, 300–301
    definition of, 215, 283
    in improper fraction, 218
    least common. *See* Least common
        denominator
    one as, 218
    power of ten as, 391
    in proper fraction, 217
Descartes, René, 88, 715
Diagonal of rectangle, 471–472
Diameter, of circle, 655
Difference, 28, 110. *See also*
    Subtraction
Digits
    definition of, 109
    in number system, 3
    place value of, 4–5
    significant, 90
Distributive property of multiplication
        over addition, 52, 110
    combining like terms and, 181
    for solving equations, 196–198, 284
Dividend
    definition of, 72, 111
    large, 78–79
Division
    calculator for, 80–82, 153–154
        of decimals, 429–430
        of fractions, 258–259
    checking, by multiplication, 73–74
    continued, 227–228
    of decimals, 423–430, 486
        calculator for, 429–430
        by decimal, 425–426
        by powers of ten, 426–427, 486
        signed, 428
        by whole number, 423–425
        in word problems, 469–470
    of denominate numbers, 72, 330, 332
    estimating, 79–80
    expressions indicating, 161

factoring by, 227–228
of fractions, 255–257, 284
    calculator for, 258–259
fractions as, 217, 255
of integers, 151–155, 204
with large dividends, 78–79
long, 75–79
of measurement units in English
        system, 548, 562
of mixed numbers, 318–319, 380
in order of operations, 95, 111, 153
remainders in, 74, 111
as repeated subtraction, 72
by single-digit number, 74, 75–77
statements, writing, 73
by two-digit number, 77–78
using subtraction, 72, 74
of whole numbers, 71–83, 111
in word problems, 79–80, 333,
    469–470
zero in, 75, 111, 152
Divisor
    common. *See* Common factors
    definition of, 72, 111

E

Egypt, ancient, numbers in, 3
Ellipsis, 11, 122
English system of measurement
    addition of units in, 547, 562
    conversions within, 545–546
    division of units in, 548, 562
    multiplication of units in,
        548–549, 562
    simplifying units in, 546–547
    subtraction of units in,
        547–548, 562
    units in, 545, 562
Equality
    addition property of, 192–198, 204
    multiplication property of,
        264–269, 284
Equals sign, 102, 186
Equations. *See also* Linear equations in
        one variable; Linear equations
        in two variables
    conditional, 186
    coordinate system and, 715–719
    decimals in, 446–448, 486
    definition of, 102, 112, 186, 204
    equivalent, 192
    fractional, 351–356
        definition of, 351

vs. simplifying fractional
        expressions, 354–356
    solving, 351–354, 380
identifying, 102–103, 188, 354
identity, 277
left side of, 186
with no solutions, 278
percent in, 597–603, 624
proportions as, 517–527
right side of, 186
solution of, 186–188
    definition of, 187, 204
    verifying, 187–188
solving, 272–278, 284
    addition property for, 193–195,
        272–276
    combining like terms for, 196,
        268, 284
    decimals in, 446–448, 486
    distributive property for,
        196–198, 284
    multiplication property for,
        266–268, 272–276
    order of operations in, 277
translating sentences into, 103
true or false, 103, 112
in word problems, 360–366, 380
Equivalent equations, 192
Equivalent fractions, 235–242
    definition of, 235, 284
    finding, 240
    identifying, 235–236, 518
    negative, 240
Eratosthenes, sieve of, 224
Estimating
    addition, 43–44
    definition of, 43
    division, 79–80
    metric length, 549, 550, 551–552
    multiplication, 56, 318, 415
    percent, 592
    by rounding, 43–44, 56
    in word problem, 44
Euler, Leonhard, 655
Evaluation
    of exponents, 88–89
    of expressions, 95–98, 167–172, 204
    of polynomials, 791
Expanded form
    addition in, 16
    for numbers, 4
Exponents. *See also* Square root
    calculator for, 91
    definition of, 88, 111
    evaluation of, 88–89

Exponents—*Cont.*
  integers with, 147
  notation of, 88
  one as, 89
  in order of operations, 95, 111, 147
  in polynomials, 790
  properties of, 783–787, 825
    power, 784, 825
    power of a product, 784–785, 825
    product, 783, 825
    quotient, 786, 825
    using, 785, 786
  ten as. *See* Powers of ten
  whole numbers and, 88–91, 111
  zero as, 89
Expressions
  absolute value, 125
  for addition, 158
  calculator for, 97–98, 169
  definition of, 102, 112, 159
  for division, 161
  evaluating, 95–98, 167–172, 204
  geometric, 161
  identifying, 102–103, 160,
    188, 354
  with multiple operations, 160
  for multiplication, 159
  simplifying, 177–182, 204
  for subtraction, 158–159
  terms in. *See* Term(s)
Exterior angle, 650
Extrapolation, 700
Extreme values, 123–124

F
Factor(s). *See also* Multiplication
  common. *See* Common factors
  definition of, 50, 110, 223
  finding, 223–224
  prime, 226, 227–228, 283
Factoring
  composite numbers, 225–226
  definition of, 225, 817
  by division, 227–228
  in fraction simplification, 237–238
  polynomials, 817–820, 826
Factorization, prime, 226, 227–228
Fahrenheit, conversion to Celsius, 176
False proportion, 518
First-degree equations. *See* Linear
    equations in one variable
FOIL method, 809–811, 826
Foot, 545, 562

Formula
  for area
    of circle, 667, 680
    of parallelogram, 664, 680
    of rectangle, 60, 111, 662, 680
    of square, 60, 88, 662, 680
    of triangle, 665, 680
  for circumference of circle, 655, 680
  definition of, 19
  for perimeter
    of rectangle, 19, 109, 177, 652, 680
    of square, 652, 680
    of triangle, 652, 680
  for volume
    of cube, 680
    of cylinder, 669, 680
    of rectangular solid, 668, 680
    of sphere, 669, 680
Fraction(s). *See also* Mixed numbers
  addition of
    calculator for, 305–306
    like, 293–295, 379
    negative, 296
    unlike, 303–304, 379
    in word problems, 334–335
  calculator for
    addition of, 305–306
    division of, 258–259
    multiplication of, 257–258
    simplification of, 241–242
    subtraction of, 305–306
  categorizing, 218
  common, 215
  comparing decimals to, 438–439
  complex
    definition of, 373, 380
    simplifying, 373–375, 380
    writing quotient as, 255
  converting
    to decimals, 434–437, 439,
      440, 486
    decimals to, 437–438, 486
    to percent, 576–577, 623
    percent to, 574, 623
  decimal, 391, 393, 485
  definition of, 215, 283
  with denominator of one, 218
  as division, 217, 255
  division of, 255–257, 284
    calculator for, 258–259
  equivalent, 235–242
    definition of, 235, 284
    finding, 240
    identifying, 235–236, 518
    negative, 240

  fundamental principle of, 236,
    240, 284
  improper, 218, 283
    converting mixed number to,
      315, 379
    converting to mixed number,
      313–314, 379
  inequality symbols with, 301–302
  like
    addition of, 293–295, 379
    definition of, 293
    subtraction of, 295–296, 379
    in word problems, 334
  multiplication of, 250–254, 284
    calculator for, 257–258
    by mixed numbers, 316
    simplification before, 252–253
    by whole number, 252
    in word problems, 331
  negative
    addition of, 296
    equivalent, 240
    simplifying, 239
    in standard form, 239
    subtraction of, 296–297
  ordering, 300–301
  proper, 217, 283
  proportional, 518
  ratios as, 499
  reciprocal of, 254–255, 264–265
  simplifying, 236–239
    in addition, 294, 304
    calculator for, 241–242
    common factors in, 238
    factoring in, 237–238
    before multiplication, 252–253
    negative fractions in, 239
  subtraction of
    calculator for, 305–306
    like, 295–296, 379
    negative, 296–297
    unlike, 305, 379
    in word problems, 335–336
  units, 546, 562
  unlike
    addition of, 303–304, 379
    subtraction of, 305, 379
    in word problems, 334–336
  in word problems, 330–331,
    334–336
Fractional equations, 351–356
  definition of, 351
  vs. simplifying fractional
    expressions, 354–356
  solving, 351–354, 380

Fundamental principle of fractions, 236, 240, 284
Fundamental theorem of arithmetic, 227

G

Gallon, 545, 562
Geometric expressions, 161
Geometry, definition of, 635
Girth, 108
Googol, 10
Gothic architecture, 633
Graph(s)/graphing
  bar. *See* Bar graph
  circle. *See* Circle graph
  definition of, 767
  line. *See* Line graph
  linear equations in two variables, 733–741, 768
    horizontal line result of, 740–741
    vertical line result of, 739–740, 741
  in media, 766
  pictograph, 697–698
  points in coordinate system, 718–719, 767
  uses of, 694
Graphing calculator
  for addition
    of fractions, 305–306
    of mixed numbers, 323–324
  for division of fractions, 259
  for fraction simplification, 241–242
  for multiplication of fractions, 258
  for subtraction
    of fractions, 305–306
    of mixed numbers, 323–324
Gravity model, 823–824
Greater than symbol (>)
  with fractions, 301–302
  with whole numbers, 44–45, 110
Greatest common factor (GCF), 228–230
  definition of, 228, 283, 818
  finding, 229–230, 283
  of polynomials, 817–820, 826
Greece, ancient, numbers in, 3
Grouping symbols
  addition and, 14
  examples of, 95
  order of operations and, 94–98, 111
  removing, 797, 799, 825
  using, 95

H

Hecto, 553, 562
Hexagon, 651, 653
Hieroglyphics, 3
Hindu-Arabic numeration system, 4
Horizontal line, as result of equation graphing, 740–741
Horizontal method
  for addition of polynomials, 797–799
  for subtraction of polynomials, 799–800
Hundreds digit, 4
Hypotenuse of right triangle, 451–452, 538–540

I

Identity, definition of, 277
Identity property
  in addition, 14, 109
  in multiplication, 53, 110
Improper fraction, 218, 283
  converting mixed number to, 315, 379
  converting to mixed number, 313–314, 379
"Increased by," as term for addition, 16
Indeterminate form, 152
Inequalities, 45
Inequality symbols
  with fractions, 301–302
  with whole numbers, 44–45, 110
Integers
  addition of, 132–136, 171, 203
  calculator for, 141–142, 153–154
  consecutive, 362–363, 380
  definition of, 122, 203
  division of, 151–155, 204
  with exponents, 147
  in fractional equations, 355–356
  identifying, 124
  multiplication of, 145–148, 203
  on number line, 122–123, 203
  order for, 123
  order of operations with, 147–148, 153
  set of, 122, 203
  subtraction of, 139–142, 203
  in word problems, 135, 140
Interest
  compound, 608, 624
  definition of, 607
  simple, 608, 624
Interest rate, definition of, 607

International System of Units (SI), 549
Internet, 838–839
Inverse, additive, 134
Irrational numbers, 655

J

Jones, William, 655

K

Kashi, Al-, 655
Kilo, 551, 553, 562
Kilometer (km), 551, 562

L

Least common denominator (LCD)
  in addition, 303–304, 320, 379
  definition of, 302
  finding, 302
  solving fractional equations with, 351–354
  in subtraction, 305, 322, 379
Least common multiple (LCM), 298–300, 379
Left side, of equations, 186
Legend, of graph, 695
Legs of right triangle, 451
Length
  estimating metric, 549, 550, 551–552
  of rectangle, 158
  units of measurement for, 545, 549, 562
"Less than," as term for subtraction, 29
Less than symbol (<)
  with fractions, 301–302
  with whole numbers, 44–45, 110
Like fractions
  addition of, 293–295, 379
  definition of, 293
  subtraction of, 295–296, 379
  in word problems, 334
Like terms, 178–179, 196, 797–798
Line(s)
  definition of, 635, 679
  intersecting, 642
  parallel, 636–637, 679
  perpendicular, 636–637, 679
Linear equations in one variable, 186–189, 204
  decimals in, 446–448, 486

Linear equations in one variable—*Cont.*
  definition of, 188
  identifying, 188
  identity, 277
  with no solutions, 278
  percent in, 597–603, 624
  solving, 272–278, 284
    addition property for, 193–195,
      272–276
    combining like terms for, 196,
      268, 284
    decimals in, 446–448, 486
    distributive property for,
      196–198, 284
    multiplication property for,
      266–268, 272–276
    order of operations in, 277
  in word problems, 360–366, 380
Linear equations in two variables,
    729–742, 767–768
  form of, 734, 768
  horizontal line result of, 740–741
  solving
    for corresponding values,
      729–730
    by graphing, 733–741, 768
    by ordered-pair notation,
      730–733, 767
  vertical line result of, 739–740, 741
Line graph
  advantages of, 202
  creating, 703–704
  definition of, 698, 767
  making predictions using, 700–703
  reading, 699
Line segment, definition of, 635, 679
Long division, 75–79
Lowest terms, 237

M
Magic square, 25, 175–176
Maximum values, 123–124
Mayans, 3
Mean
  calculator for, 758–759
  finding, 753–754, 768
  vs. median, 755
Measurement, 545–554, 562
  of angles, 638, 639–640
  English system of. *See* English
      system of measurement
  history of, 497

metric system of. *See* Metric system
    of measurement
Median
  finding, 754–755, 768
  vs. mean, 755
Meter (m), 549, 562
Method of translation, 597
Metric system of measurement, 549
  abbreviation in, 549
  advantages of, 549
  conversions within, 552–553
  estimation in, 549, 550, 551–552
  prefixes in, 553, 562
  units in, 549, 550, 551,
      553, 562
Mile, 545, 562
Milli, 550, 553, 562
Millimeter (mm), 550, 562
Minimum values, 123–124
Minuend, definition of, 28, 110
Minus sign, 28, 799, 825
Mixed numbers, 313–324, 379–380
  addition of, 320–321, 380
  calculator for, 323–324
  converting
    to decimals, 437, 440
    decimals to, 393, 438
    to improper fractions, 315, 379
    improper fractions to,
      313–314, 379
    to percent, 577
    percent to, 574
  definition of, 313
  division of, 318–319, 380
  identifying, 313
  multiplication of, 316–318, 380
  subtraction of, 321–322, 380
  in word problems, 331–333, 337
Mixed operations, 111
Mode
  definition of, 756
  finding, 756, 768
Modeling, 781, 823–824
Monomial. *See also* Polynomials
  definition of, 791, 825
  factoring, 819, 826
  multiplication of, 807–808
    by binomial, 808
"More than," as term for addition, 16
Multiples
  common, 298
  definition of, 297
  least common, 298–300, 379
  listing, 297

Multiplication
  for area, 59, 111
  associative property of, 58–59, 111
  basic facts of, 50–51
  calculator for
    of decimals, 417–419
    of fractions, 257–258
  checking division with, 73–74
  commutative property of, 50, 110
  of decimals, 414–419, 485
    calculator for, 417–419
    estimating, 415
    by powers of ten, 416–419, 468, 485
    in word problems, 466–468
  of denominate numbers, 53, 59, 330
  distributive property of, 52, 110
    combining like terms and, 181
    for solving equations,
      196–198, 284
  estimating, 56, 317, 415
  exponents for repeated, 88
  expressions indicating, 159
  factoring as reverse of, 817
  of fractions, 250–254, 284
    calculator for, 257–258
    by mixed numbers, 316
    simplification before, 252–253
    by whole number, 252
    in word problems, 331
  identity property in, 53, 110
  of integers, 145–148, 203
  of measurement units in English
      system, 548–549, 562
  of mixed numbers, 316–318, 380
  by one, 53, 110, 146
  in order of operations, 95, 111
  of polynomials, 807–813, 826
    by binomial, 809–812, 826
    FOIL method for, 809–811, 826
    by monomial, 807–808, 826
    vertical method for, 812
  by powers of ten, 55
  by reciprocal, 264–265
  as repeated addition, 50
  simplifying counting with, 51
  by single-digit numbers, 50, 52–53
  solving equations with, 266–268,
      272–276
  symbols for, 50, 159
  by ten, 54–55
  of three-digit numbers, 57–58
  by two-digit numbers, 56–57
  of whole numbers, 50–62, 110–111
    by fractions, 252

in word problems, 54, 331–332
by zero, 53, 110, 146
Multiplication property of equality,
    264–269, 284

N

Natural numbers, 11, 223
Negative fractions
    addition of, 296
    equivalent, 240
    simplifying, 239
    in standard form, 239
    subtraction of, 296–297
Negative numbers, 121, 203. *See also*
        Integers
    absolute value of, 124–125
    entering into calculator, 141
    on number line, 122, 203
Negative sign, 122
Newton, Sir Isaac, 781
Norman architecture, 633
Number line
    absolute value on, 125
    addition on, 12–13
    decimals on, 394
    integers on, 122–123, 203
    negative numbers on, 122, 203
    order for whole numbers on, 44, 110
    origin of, 12, 122
    positive numbers on, 122, 203
    rounding on, 41
Numbers
    absolute value of, 124–125, 203
    approximate, 394
    composite. *See* Composite numbers
    decimal. *See* Decimal(s)
    denominate. *See* Denominate
        numbers
    expanded form for, 4
    fractions. *See* Fraction(s)
    history of, 3
    integers. *See* Integers
    irrational, 655
    mixed. *See* Mixed numbers
    natural (counting), 11, 223
    negative. *See* Negative numbers
    opposite of, 124, 134
    positive. *See* Positive numbers
    prime. *See* Prime numbers
    real, 122, 124
    standard form for, 4
    translating words into, 5

value of, 109
whole. *See* Whole numbers
writing, in words, 4–5
Numerator. *See also* Fraction(s)
    definition of, 215, 283
    in improper fraction, 218
    in proper fraction, 217
Numerical coefficient. *See* Coefficient

O

Obtuse angle, 638, 679
Octagon, 654
One (1)
    as denominator, 218
    as exponent, 89
    multiplication by, 110, 146
    multiplicative property of, 53
    as neither prime nor composite, 225
Ones digit, 4
Opposite, of numbers, 124, 134
Order
    for integers, 123
    for whole numbers, 44–45, 110
Ordered pair, 715, 767
Ordered-pair notation, 730–733, 767
Order of operations, 95
    with decimals, 428–429
    grouping symbols and, 94–98, 111
    with integers, 147–148, 153
    in solution of equations, 277
Ordinate, 731
Origin
    of circle, 655
    of coordinate system, 715, 767
    of number line, 12, 122

P

Parallel lines, 636–637
Parallelogram
    area of, 664, 680
    definition of, 653
    perimeter of, 654
Parentheses, 13, 95, 111, 277, 447, 797,
    799, 825
Pentagon, 651, 653
Percent
    amount of, 588, 623
    base of, 588, 623
    converting
        to decimal notation, 575–576, 623

to fractions, 574, 623
fractions to, 576–577, 623
to mixed numbers, 574
mixed numbers to, 577
ratios to, 573
definition of, 573, 623
in equations, 597–603, 624
estimating, 592
notation for, 573
proportions and, 589–592, 623
rate of, 588, 623
symbol for, 573
units and, 578
unknown values in, 589–592,
    600–602, 607, 612
in word problems, 597–615
Perfect square, 451
Perfect triple, 452
Perimeter
    of circle. *See* Circumference,
        of circle
    of curved shapes, 656–657
    definition of, 18, 652, 680
    finding with addition, 18–20, 109
    of octagon, 654
    of parallelogram, 654
    of polygon, 653, 654
    of rectangle, 18–20, 109, 177,
        652, 680
    of square, 652, 680
    of trapezoid, 654
    of triangle, 652, 680
    in word problems, 656–657
Perpendicular lines, 636–637, 679
Pi (p), 655
Pictograph, 697–698
Pint, 545, 562
Placeholder, decimal point as, 393
Place value, 3–6
    addition of digits of same, 15
    definition of, 109
    identifying, 391, 392
Place-value diagram, 4–5
Plus sign, 12, 797, 825
Point, definition of, 635
Point plotting, 718
Polygon
    definition of, 652, 680
    perimeter of, 653, 654
Polynomials, 781–830
    addition of, 797–802, 825
        horizontal method for, 797–799
        simplification in, 802
        vertical method for, 801

Polynomials—*Cont.*
  definition of, 790, 825
  evaluating, 791
  exponents in, 790
  factoring, 817–820, 826
  greatest common factor of, 817–820, 826
  identifying, 790–791
  multiplication of, 807–813, 826
    by binomial, 809–812, 826
    FOIL method for, 809–811, 826
    by monomial, 807–808, 826
    vertical method for, 812
  subtraction of, 797–802, 825
    horizontal method for, 799–800
    simplification in, 802
    vertical method for, 801
  types of, 791, 825
  use of, 781
Positive numbers, 121, 203. *See also* Integers
  absolute value of, 124–125
  on number line, 122, 203
Pound, 545, 562
Power of a product property of exponents, 784–785, 825
Power property of exponents, 784, 825
Powers of ten
  in conversion of metric units, 552
  definition of, 90
  as denominator, 391
  division of decimals by, 426–427, 486
  list of, 89
  multiplying by, 55
    of decimals, 416–419, 468, 485
  in word problems, 468, 470–471
Price, unit, 509–510, 525–527, 561
Prime factorization, 226, 227–228, 283
Prime numbers
  definition of, 224, 283
  identifying, 224–225
  largest known, 224
  relatively, 230
  twin, 233
Principal, definition of, 607
Product(s). *See also* Multiplication
  cross, 235–236
  definition of, 50, 110
  reciprocal, 254
Product property of exponents, 783, 825
Proper fraction, 217, 283
Proportion(s), 517–527
  definition of, 517, 561
  false, 518

percent and, 589–592, 623
  rates in, 518, 519–520
  rule of, 518, 561
  similar triangles and, 537–540, 561
  solving, 520–522, 561
  true, 518
  unknown values in, 521–525
  verifying, 518–520
  in word problems, 519, 522–525
  writing, 517
Proportional fractions, 518
Protractor, 639, 841
Pythagoras, 452
Pythagorean theorem, 452–454, 486
Pythagorean triple, 452

Q

Quadrants, of coordinate system, 715
Quadrilateral, 651
Qualitative data, 756
Quart, 545, 562
Quotient. *See also* Division
  definition of, 72, 111
Quotient property of exponents, 786, 825

R

Radical sign, 451
Radius, of circle, 655
Rates, 507–511
  definition of, 507, 561
  finding, 507–508
  in percent problems, 588, 623
  in proportions, 518, 519–520
  unit, 508–509
  in word problems, 508–510
Ratios, 499–502
  converting to percent, 573
  decimals in, 501
  definition of, 499, 561
  of denominate numbers, 501–502
  simplifying, 500–501
  in word problems, 499, 500–502
  writing as fraction, 499
Real numbers, 122, 124
Reciprocal, of fractions, 254–255, 264–265
Reciprocal product, 254
Rectangle
  area of, 60, 111, 662, 663, 680
  definition of, 18

length of, 158
  length of diagonal of, 471–472
  perimeter of, 18–20, 109, 177, 652, 680
  width of, 158
Rectangular coordinate system. *See* Coordinate system
Rectangular solid
  definition of, 667
  volume of, 668, 680
Regrouping, 16
Relatively prime numbers, 230
Remainder, definition of, 74, 111
Renaming, 16
Repeating decimals
  converting fractions to, 435–437
  definition of, 435
Right angle, 637, 638, 679
Right side, of equations, 186
Right triangle
  angles in, 451
  definition of, 451, 537
  hypotenuse of, 451–452, 538–540
  legs of, 451
  Pythagorean theorem and, 452–454, 486
  similar, 537, 538–540, 561
  in word problems, 539–540
Rod, 497
Roman numerals, 3
Root, of equation. *See* Solution
Rounding
  decimals, 394–396, 485
    to nearest hundredth, 395–396
    to nearest tenth, 395
    to specified decimal place, 396
  definition of, 41
  down, 42
  estimating by, 43–44, 56
  to nearest hundred, 41, 42, 110
  to nearest ten, 42, 43
  to nearest thousand, 41, 42, 110
  on number line, 41
  up, 42
  whole numbers, 41–44, 110

S

Scientific notation, 90, 419
Search engine, 838
Sectors, of circle graph, 840
Set
  bimodal, 756
  definition of, 122

of integers, 122
of whole numbers, 11
Short form, addition in, 17
SI. *See* International System of Units
Sieve of Eratosthenes, 224
Significant digit, 90
Simple interest, 608, 624
Simplest form, 237
Simplification
  in addition and subtraction of
     polynomials, 802
  of complex fractions, 373–375, 380
  of expressions, 177–182, 204
  of fractions, 236–239
    in addition, 294, 304
    calculator for, 241–242
    common factors in, 238
    factoring in, 237–238
    before multiplication, 252–253
    negative fractions in, 239
  of measurement units in English
     system, 546–547
  of mixed numbers, before
     multiplication, 317
  of ratios, 500–501
Solid
  definition of, 667
  rectangular, 667
  volume of. *See* Volume
Solution, 186–188
  definition of, 187, 204
  verifying, 187–188
Sphere, volume of, 669, 680
Square (shape)
  area of, 59, 60–61, 88,
     662–663, 680
  perimeter of, 652, 680
Square, perfect, 451
Square foot, 665
Square inch, 60, 665
Square mile, 665
Square root
  approximating, 455
  calculator for, 455–456
  definition of, 451, 486
  finding, 451
  Pythagorean theorem and,
     452–454, 486
Square units, 60, 665
Square yard, 665
Standard form, for numbers, 4
Statement, 103. *See also* Equations
Straight angle, 638
Substitution, solving word
    problems by, 360

Subtraction
  of absolute values, 125
  borrowing in, 30–31
    of decimals, 406
  calculator for, 32–33, 141–142
    of decimals, 408–409
    of fractions, 305–306
  of decimals, 404–407, 485
    borrowing in, 406
    calculator for, 408–409
    signed, 407
    in word problems, 464–466
  of denominate numbers, 31
  division as repeated, 72
  expressions indicating, 158–159
  of fractions
    calculator for, 305–306
    like, 295–296, 379
    negative, 296–297
    unlike, 305, 379
    in word problems, 335–336
  of integers, 139–142, 203
  of measurement units in English
     system, 547–548, 562
  of mixed numbers, 321–322, 380
  as opposite of addition, 28
  in order of operations, 95, 111
  of polynomials, 797–802, 825
    horizontal method for,
     799–800
    simplification in, 802
    vertical method for, 801
  of single-digit number, 28–29
  symbol for, 28
  of whole numbers, 28–33, 110
  in word problems, 32, 140,
     335–336
  words for, 29–30
Subtrahend, definition of, 28, 110
Sum, 12, 109. *See also* Addition
Supplementary angles, 641
Surface area, total, 70

T
Table
  cells in, 693
  definition of, 693, 767
  reading, 693
Ten (10)
  multiplying by, 54–55
  number system based on, 3, 4
  powers of. *See* Powers of ten
Tens digit, 4

Term(s)
  coefficient of. *See* Coefficient
  combining, 179–181, 204
    for solving equations, 196,
     268, 284
  definition of, 177, 204, 790, 825
  identifying, 177–178, 790–791
  like, 178–179, 196, 204, 797–798
  lowest, 237
Terminating decimals
  converting to fractions,
     437, 486
  definition of, 434
Ton, 545, 562
"Total of," as term for addition, 16
Total surface area, 70
Transit, 636
Translation, method of, 597
Trapezoid
  definition of, 653
  perimeter of, 654
Triangle(s)
  angles in, 643–644
  area of, 665, 680
  perimeter of, 652, 680
  right. *See* Right triangle
  similar, 537–540, 561
Trinomial, 791, 825. *See also*
    Polynomials
Triple, perfect, 452
Triple, Pythagorean, 452
True proportion, 518
Truncating, 69
Twin primes, 233

U
Unit(s). *See also* Denominate numbers;
    Measurement
  in addition, 31
  of area, 60, 665–666
  in division, 72, 330, 332
  history of, 497
  in multiplication, 53, 59, 330
  percent and, 578
  in subtraction, 31
  of volume, 667
Unit price, 509–510, 525–527, 561
Unit rate, 508–509
Units fraction, 546, 562
Unlike fractions
  addition of, 303–304, 379
  subtraction of, 305, 379
  in word problems, 334–336

V

Value(s)
extreme, 123–124
of numbers, 109
Variables
definition of, 158, 186
in evaluation of expressions, 167–172, 204
in linear equations, 186
Vertex, of angle, 636
Vertical angles, 642–643
Vertical line, as result of equation graphing, 739–740, 741
Vertical method
for addition of polynomials, 801
for multiplication of polynomials, 812
for subtraction of polynomials, 801
Volume
of cube, 680
of cylinder, 669, 680
definition of, 667
of rectangular solid, 668, 680
of sphere, 669, 680
units of, 667

W

Websites, 838–839
Weight, units of measurement for, 545, 562
Whole numbers, 1–118
addition of, 11–21, 109
division of, 71–83, 111

division of decimals by, 423–425
exponents and, 88–91, 111
multiplication of, 50–62, 110–111
by fractions, 252
order for, 44–45, 110
place value and, 3–6
rounding, 41–44, 110
set of, 11
subtraction of, 28–33, 110
Width, of rectangle, 158
Word(s)
for addition, 16
decimals in, 392–393, 485
for subtraction, 29–30
translating
into algebraic terms, 361
into equations, 103
into numbers, 5
writing numbers as, 4–5
Word problems
addition in, 20, 135, 334–335, 463–466
area in, 666
decimals in, 463–473
division in, 79–80, 333, 469–470
estimating in, 44
fractions in, 330–331, 334–336
integers in, 135, 140
linear equations in, 360–366, 380
mixed numbers in, 331–333, 337
multiplication in, 54, 331–332
percent in, 597–615
perimeter in, 656–657

powers of ten in, 468, 470–471
proportions in, 519, 522–525
rates in, 508–510
ratios in, 499, 500–502
similar triangles in, 539–540
solving by substitution, 360
subtraction in, 32, 140, 335–336

X

x-axis, 715, 767
x-coordinate, 730, 731, 767

Y

Yard, 497, 545, 562
y-axis, 715, 767
y-coordinate, 730, 731, 767

Z

Zero (0)
absolute value of, 125
adding, 14
in division, 75, 111, 152
as exponent, 89
history of, 10
multiplicative property of, 53, 110, 146
on number line, 12, 122
in set of whole numbers, 11
Zu Chongzhi, 655